115 Notes on Numerical Fluid Mechanics and Multidisciplinary Design (NNFM)

Computational Science and High Performance Computing IV

The 4th Russian-German Advanced Research Workshop, Freiburg, Germany, October 12 to 16, 2009

Egon Krause, Yurii Shokin,
Michael Resch, Dietmar Kröner,
Nina Shokina (Eds.)

Springer

Prof. em. Prof. h.c. Dr. Egon Krause
RWTH Aachen
Aerodynamisches Institut
Wuellnerstr. zw. 5 u. 7
52062 Aachen
Germany

Prof. Dr. Yurii Shokin
Institute of Computational Technologies of SB RAS
Ac. Lavrentiev Ave. 6
630090 Novosibirsk
Russia

Prof. Dr.-Ing. Michael Resch
University of Stuttgart
High Performance Computing Center Stuttgart
Nobelstrasse 19
70569 Stuttgart
Germany
E-mail: mr@ihr.uni-stuttgart.de

Prof. Dr. rer. nat. Dietmar Kröner
University of Freiburg
Section of Applied Mathematics
Hermann-Herder-Str. 10
79104 Freiburg
Germany
E-mail:
dietmar@mathematik.uni-freiburg.de

Dr.-Ing. Nina Shokina
University of Freiburg
Section of Applied Mathematics
Hermann-Herder-Str. 10
79104 Freiburg
Germany
E-mail:
shokina@mathematik.uni-freiburg.de

ISBN 978-3-642-17769-9 e-ISBN 978-3-642-17770-5

DOI 10.1007/978-3-642-17770-5

Notes on Numerical Fluid Mechanics and Multidisciplinary Design ISSN 1612-2909

Typeset & Cover Design: Scientific Publishing Services Pvt. Ltd., Chennai, India.

Printed on acid-free paper

5 4 3 2 1 0

springer.com

NNFM Editor Addresses

Preface

This volume is published as the proceedings of the fourth Russian-German Advanced Research Workshop on Computational Science and High Performance Computing in Freiburg, Germany, in October 2009.

The contributions of these proceedings were provided and edited by the authors, chosen after a careful selection and reviewing.

The workshop was organized by the High Performance Computing Center Stuttgart (Stuttgart, Germany), the Institute of Computational Technologies of the Siberian Branch of the Russian Academy of Sciences (Novosibirsk, Russia) and the Section of Applied Mathematics of the University of Freiburg (Freiburg, Germany) in the framework of activities of the German-Russian Center for Computational Technologies and High Performance Computing.

The current event is the continuation of the good tradition for German and Russian scientists of holding such a workshop biannually, alternating host countries: Russia and Germany. The first workshop took place in September 2003 in Novosibirsk and was hosted by the Institute of Computational Technologies SB RAS. The second workshop was hosted by the High Performance Computing Center Stuttgart in March 2005. The third workshop was held in July 2007, again in Novosibirk, also hosted by the Institute of Computational Technologies SB RAS. The current fourth workshop was hosted by the Section of Applied Mathematics of the University of Freiburg in Freiburg.

All workshops gave the possibility of sharing and discussing the latest results and developing further scientific contacts in the field of computational science and high performance computing.

The topics of the current workshop included software and hardware for high performance computation, numerical modelling in geophysics and computational fluid dynamics, mathematical modelling of tsunami waves, simulation of fibrous composites, simulation of reacting flows, numerical modelling in electromagnetics and medicine and numerical analysis. The highlight of the workshop was the special session on the shallow water modelling.

The participation of representatives of major research organizations engaged in the solution of the most complex problems of mathematical modelling, development of new algorithms, programs and key elements of information technologies, elaboration and implementation of software and hardware for high performance computing systems, provided a high level of competence of the workshop.

The heads and leading specialists of Section of Applied Mathematics of University of Freiburg, High Performance Computing Center Stuttgart (HLRS) (University of Stuttgart), Goethe Center for Scientific Computing (G-CSC) (Goethe University Frankfurt am Main), Alfred Wegener Institute for Polar and Marine Research (Bremerhaven), University of Hamburg, Institute for Technical Thermodynamics (Karlsruhe Institute of Technology), Copf-Bionic GmbH (Ludwigsburg), Institute of Numerical Simulation (Hamburg University of Technology), Institute for Computational Modeling in Civil Engineering (Technical University of Braunschweig), Institute of Aerodynamics (RWTH Aachen), LASSO Ingenieurgesellschaft mbH (Leinfelden-Echterdingen), Department of Mathematics and Computer Science (Free University of Berlin) participated in the workshop from German side.

From Russian side the heads and leading specialists of Institute of Computational Technologies SB RAS (Novosibirsk), Kutateladze Institute of Thermophysics SB RAS (Novosibirsk), Lavrentiev Institute of Hydrodynamics SB RAS (Novosibirsk), Institute of Computational Modelling SB RAS (Krasnoyarsk) and Krasnoyarsk State Agriculture University took part in the workshop.

This workshop continues to broaden the scientific cooperation, involving not only German and Russian colleagues. Kazakh scientists from the Al-Farabi Kazakh National University (Almaty, Kazakhstan), who have been taken part in workshops since 2005 at Prof. Yurii Shokin's suggestion, also participated in the current workshop. And, besides, the scientists from three other countries, namely, France, Israel and USA, presented their works. Therefore, the following institutions were represented: University of Savoy and University of Orléans from France, Department of Fluid Mechanics and Heat Transfer (Faculty of Engineering, Tel Aviv University), Coastal and Marine Engineering Research Institute (Technion City, Haifa). From USA: Department of Mathematics of the North Carolina State University (Raleigh), Department of Mathematics of the University of Louisiana (Lafayette, USA) and Mathematics Department of Tulane University.

We are glad to see the successful continuation and promising perspectives of the highly professional international scientific meetings, which bring together leading scientists from Germany, Russia, Kazakhstan, and now also France, Israel and USA.

The editors would like to express their gratitude to all the participants of the workshop and wish them a further successful and fruitful work.

Freiburg – Novosibirsk
July 2010

Egon Krause
Yurii Shokin
Michael Resch
Dietmar Kröner
Nina Shokina

Contents

List of Contributors

Abugamil U. Abdibekov
Al-Farabi Kazakh National University, Al-Farabi Ave. 71, Almaty, 050038, Kazakhstan
abugamil@mail.ru

Ualikhan S. Abdibekov
Al-Farabi Kazakh National University, Al-Farabi Ave. 71, Almaty, 050038, Kazakhstan
Ualil@mail.ru

Vadym Aizinger
Goethe Center for Scientific Computing (G-CSC), Goethe University Frankfurt am Main, Kettenhofweg 139, Frankfurt am Main, 60325, Germany
vadym.aizinger@csc.uni-frankfurt.de

Alexey Androsov
Alfred Wegener Institute for Polar and Marine Research, Postfach 12-01-61, Bremerhaven, 27515, Germany
alexey.androsov@awi.de

Dmitry G. Arkhipov
Kutateladze Institute of Thermophysics SB RAS, Lavrentiev Ave. 1, Novosibirsk, 630090, Russia
theory@itp.nsc.ru

Jörn Behrens
University of Hamburg, Grindelberg 5, 20144 Hamburg, Germany
joern.behrens@zmaw.de

Sofla A. Belsel
Institute of Computational Technologies SB RAS, Lavrentiev Ave. 6, Novosibirsk, 630090, Russia
beisels@gmail.com

Slavko Brdar
Section of Applied Mathematics, University of Freiburg i. Br., Hermann-Herder-Straße 10, Freiburg i. Br., 79104, Germany
slavko@mathematik.uni-freiburg.de

Didier Bresch
LAMA, équipe EDPs2, Bat. Le Chablais, Campus scientifique, Université de Savoye, Le Bourget du Lac, 73376, France
Didier.Bresch@univ-savoie.fr

Dmitry A. Burakov
Krasnoyarsk State Agriculture University, Mira Av. 90, Krasnoyarsk, 660049, Russia
daburakov@yandex.ru

Viatcheslav Bykov
Institute of Technical Thermodynamics, Karlsruhe Institute of Technology, Engelbert-Arndold-Straße 4, Geb. 10.91, Karlsruhe, 76131, Germany
viatcheslav.bykov@kit.edu

Alina Chertock
Department of Mathematics, SAS Hall, 2311 Stinson Drive, Box 8205, North Carolina State University, Raleigh, NC 27695-8205, USA
chertock@math.ncsu.edu

Alexander A. Chesnokov
M.A. Lavrentiev Institute of Hydrodynamics SB RAS Lavrentiev Ave. 15, Novosibirsk, 630090, Russia; Novosibirsk State University, Pirogova str. 2, Novosibirsk, 630090, Russia
chesnokov@hydro.nsc.ru

Christo Christov
Department of Mathematics, University of Louisiana at Lafayette, Lafayette, LA 70504-1010, USA
christov@louisiana.edu

Dmitry L. Chubarov
Institute of Computational Technologies SB RAS, Lavrentiev Ave. 6, Novosibirsk, 630090, Russia
dchubarov@ict.nsc.ru

Leonid B. Chubarov
Institute of Computational Technologies SB RAS, Lavrentiev Ave. 6, Novosibirsk, 630090, Russia
chubarov@ict.nsc.ru

Peter Copf
Copf-Bionic GmbH, Schloss Monrepos / Gebäude 1, Ludwigsburg, 71634, Germany
p.copf@copf-bionic.de

Nargozy T. Danaev
Al-Farabi Kazakh National University, Al-Farabi Av. 71, Almaty, 050038, Kazakhstan
Nargozy.Danaev@kaznu.kz

Sergey Danilov
Alfred Wegener Institute for Polar and Marine Research, Postfach 12-01-61, 27515 Bremerhaven, Germany
sergey.danilov@awi.de

Andreas Dedner
Section of Applied Mathematics, University of Freiburg i. Br., Hermann-Herder-Straße 10, Freiburg i. Br., 79104, Germany
dedner@mathematik.uni-freiburg.de

Michael Dudzinski
Institute of Numerical Simulation, Hamburg University of Technology, Schwarzenbergstrasse 95, Hamburg, 21073, Germany
michael.dudzinski@tu-harburg.de

Denys Dutykh
LAMA, UMR 5127 CNRS, Université de Savoie, Le Bourget-du-Lac Cedex, 73376, France
Denys.Dutykh@univ-savoie.fr

Mikhail P. Fedoruk
Institute of Computational Technologies SB RAS, Lavrentiev Ave. 6, Novosibirsk, 630090, Russia
mife@ict.nsc.ru

Sören Freudiger
Institute for Computational Modeling in Civil Engineering, Technical University of Braunschweig, Mühlenpfordtstraße 4-5, Braunschweig, 38106, Germany
freud@irmb.tu-bs.de

Martin Geier
Institute for Computational Modeling in Civil Engineering, Technical University of Braunschweig, Mühlenpfordtstraße 4-5, Braunschweig, 38106, Germany
geier@irmb.tu-bs.de

Marguerite Gisclon
LAMA, UMR 5127 CNRS, Université de Savoie, Le Bourget-du-Lac Cedex, 73376, France
Marguerite.Gisclon@univ-savoie.fr

Kseniya S. Golushko
Institute of Computational Technologies SB RAS, Lavrentiev Ave. 6, Novosibirsk, 630090, Russia
golushko@ict.nsc.ru

Sergey K. Golushko
Institute of Computational Technologies SB RAS, Lavrentiev Ave. 6, Novosibirsk, 630090, Russia
golushko@ict.nsc.ru

Daniel Hartmann
Institute of Aerodynamics, RWTH Aachen, Wüllnerstr. 5a, Aachen, 52062, Germany
office@aia.rwth-aachen.de

Ulrich Hindenlang
LASSO Ingenieurgesellschaft mbH, Leinfelderstraße 60, Leinfelden-Echterdingen, 70771, Germany
lasso@lasso.de

Evgeniya D. Karepova
Institute of Computational Modelling SB RAS, Academgorodok, Krasnoyarsk, 660036, Russia
jane@icm.krasn.ru

Georgy A. Khabakhpashev
Kutateladze Institute of Thermophysics SB RAS, Lavrentiev Ave. 1, Novosibirsk, 630090, Russia
theory@itp.nsc.ru

Gayaz S. Khakimzyanov
Institute of Computational Technologies SB RAS, Lavrentiev Ave. 6, Novosibirsk, 630090, Russia
khak@ict.nsc.ru

Eliezer Kit
Department of Fluid Mechanics and Heat Transfer, Faculty of Engineering, Tel Aviv University, Tel Aviv, 69978, Israel
kit@eng.tau.ac.il

Rupert Klein
Department of Mathematics and Computer Science, Free University of Berlin, Arnimallee 6, 14195 Berlin, Germany
Rupert.Klein@math.fu-berlin.de

Robert Klöfkorn
Section of Applied Mathematics, University of Freiburg i. Br., Hermann-Herder-Straße 10, Freiburg i. Br., 79104, Germany
robertk@mathematik.uni-freiburg.de

Manfred Krafczyk
Institute for Computational Modeling in Civil Engineering, Technical University of Braunschweig, Mühlenpfordtstraße 4-5, Braunschweig, 38106, Germany
kraft@irmb.tu-bs.de

Mirko Kränkel
Section of Applied Mathematics, University of Freiburg i. Br., Hermann-Herder-Straße 10, Freiburg i. Br., 79104, Germany
kraenkel@mathematik.uni-freiburg.de

Egon Krause
Institute of Aerodynamics, RWTH Aachen, Wülnnerstr. zw. 5 u.7, Aachen, 52062, Germany
ek.aia@t-online.de

Dietmar Kröner
Section of Applied Mathematics, University of Freiburg i. Br., Hermann-Herder-Straße 10, Freiburg i. Br., 79104, Germany
dietmar@mathematik.uni-freiburg.de

Alexander Kurganov
Mathematics Department, Tulane University, 6823 St. Charles Ave., New Orleans, LA 70118, USA
kurganov@math.tulane.edu

Uwe Küster
High Performance Computing Center Stuttgart (HLRS), University of Stuttgart, Nobelstraße 19, Stuttgart, 70569, Germany
kuester@hlrs.de

Valery Yu. Liapidevskii
M.A. Lavrentiev Institute of Hydrodynamics SB RAS Lavrentiev Ave. 15, Novosibirsk, 630090, Russia
liapid@hydro.nsc.ru

Carine Lucas
MAPMO, Université d'Orléans, UFR Sciences, Bâtiment de mathématiques - Route de Chartres, B.P. 6759 - 45067 Orléans cedex 2, Université de Orléans, France
carine.lucas@univ-orleans.fr

Mária Lukáčová-Medvid'ová
Institute of Numerical Simulation, Hamburg University of Technology, Schwarzenbergstrasse 95, Hamburg, 21073, Germany
lukacova@tu-harburg.de

Alexander S. Lebedev
Institute of Computational Technologies SB RAS, Lavrentiev Ave. 6, Novosibirsk, 630090, Russia
sasa@ict.nsc.ru

Anna Levin
Coastal and Marine Engineering Research Institute, Technion City, Haifa, 32000, Israel
lanna@tx.technion.ac.il

Ulrich Maas
Institute of Technical Thermodynamics, Karlsruhe Institute of Technology, Engelbert-Arndold-Straße 4, Geb. 10.91, Karlsruhe, 76131, Germany
ulrich.maas@kit.edu

Matthias Meinke
Institute of Aerodynamics, RWTH Aachen, Wüllnerstr. 5a, Aachen, 52062, Germany
office@aia.rwth-aachen.de

Yannick Meyapin
LAMA, UMR 5127 CNRS, Université de Savoie, Le Bourget-du-Lac Cedex, 73376, France
Yannick.Meyapin@etu.univ-savoie.fr

Lyudmila Yu. Prokopyeva
Institute of Computational Technologies SB RAS, Lavrentiev Ave. 6, Novosibirsk, 630090, Russia
liudmila.prokopeva@gmail.com

Michael M. Resch
High Performance Computing Center Stuttgart (HLRS), University of Stuttgart, Nobelstraße 19, Stuttgart, 70569, Germany
resch@hlrs.de

Alexander D. Rychkov
Institute of Computational Technologies SB RAS, Lavrentiev Ave. 6, Novosibirsk, 630090, Russia
rych@ict.nsc.ru

Nurziya S. Safarova
Kutateladze Institute of Thermophysics SB RAS, Lavrentiev Ave. 1, Novosibirsk, 630090, Russia
theory@itp.nsc.ru

Ralf Schneider
High Performance Computing Center Stuttgart (HLRS), Nobelstraße 19, Stuttgart, 70569, Germany
schneider@hlrs.de

Wolfgang Schröder
Institute of Aerodynamics, RWTH Aachen, Wüllnerstr. 5a, Aachen, 52062, Germany
office@aia.rwth-aachen.de

Vladimir V. Shaidurov
Institute of Computational Modelling SB RAS, Academgorodok, Krasnoyarsk, 660036, Russia
shidurov@icm.krasn.ru

Yurii I. Shokin
Institute of Computational Technologies SB RAS, Lavrentiev Ave. 6, Novosibirsk, 630090, Russia
shokin@ict.nsc.ru

Nina Shokina
Section of Applied Mathematics, University of Freiburg i. Br., Hermann-Herder-Straße 10, Freiburg i. Br., 79104, Germany
shokina@mathematik.uni-freiburg.de

Michael Sladkevich
Coastal and Marine Engineering Research Institute, Technion City, Haifa, 32000, Israel
michslad@tx.technion.ac.il

Maik Stiebler
Institute for Computational Modeling in Civil Engineering, Technical University of Braunschweig, Mühlenpfordtstraße 4-5, Braunschweig, 38106, Germany
stiebler@irmb.tu-bs.de

Andrey V. Yurchenko
Institute of Computational Technologies SB RAS, Lavrentiev Ave. 6, Novosibirsk, 630090, Russia
yurchenko@ict.nsc.ru

Dauren B. Zhakebaev
Al-Farabi Kazakh National University, Al-Farabi Ave. 71, Almaty, 050038, Kazakhstan
daurjaz@mail.ru

Bakytzhan T. Zhumagulov
Al-Farabi Kazakh National University, Al-Farabi Ave. 71, Almaty, 050038, Kazakhstan
kvt@kaznu.kz

Adaptive Modelling of Two-Dimensional Shallow Water Flows with Wetting and Drying

Andreas Dedner, Dietmar Kröner, and Nina Shokina

Abstract. The current work is done in the framework of the BMBF (Bundesministerium für Bildung und Forschung - the Federal Ministry of Education and Research) project **AdaptHydroMod** – **Adapt**ive **Hydro**logical **Mod**elling with Application in Water Industry [1], which is devoted to the development of generic adaptive approach to modelling of coupled hydrological processes: surface and groundwater flows. The surface water flow is modelled by the two-dimensional shallow water equations and the surface flow – by the Richards equation. The implementation is based within **DUNE** – the **D**istributed and **U**nified **N**umerics **E**nvironment [14]. The surface flow, on which we focus in the presented paper, is numerically solved using the Runge-Kutta discontinuous Galerkin method [10] with modifications to render the scheme well-balanced and for handling correctly possible wetting and drying processes. The newly developed limiter [12] is used for the stabilization of the method. The validation of the code is done using several test problems with known exact solutions. The problem with a mass source term, which is a first step to the coupled simulation of the surface and groundwater flows, is solved numerically.

1 Introduction

Nowadays, the flood forecasting, flood control measures, water protection, water availability and water production are of central importance to the society. It is extremely important to develop new efficient numerical methods allowing to predict the consequences of river development, formation of water reservoirs in the soil, transport of pollutants in the soil, flooding and drying, water pollution and other man-made or natural phenomena.

Andreas Dedner · Dietmar Kröner · Nina Shokina
Section of Applied Mathematics, University of Freiburg,
Freiburg i. Br., Germany
e-mail: andreas.dedner@mathematik.uni-freiburg.de,
dietmar@mathematik.uni-freiburg.de,
shokina@mathematik.uni-freiburg.de

E. Krause et al. (Eds.): Computational Sci., & High Performance Computing IV, NNFM 115, pp. 1–15.
springerlink.com

The BMBF (Bundesministerium für Bildung und Forschung - the Federal Ministry of Education and Research) project **AdaptHydroMod** – **Adapt**ive **Hydro**logical **Mod**elling with Application in Water Industry [1], is devoted to the development of generic adaptive approach to modelling of coupled hydrological processes: surface and groundwater flows.

The current work considers the simulation of surface water flows using the two-dimensional shallow water model together with a new simple well balancing technique, a wetting-drying treatment, which is based on the ideas presented in [8], and the newly developed limiter [12]. The two-dimensional shallow water model can describe a wide range of hydrological phenomena from steady water in small lakes to tsunami wave propagation in oceans.

For the project the coupling of surface flow with equations for the groundwater is a central aspect, and the numerical schemes have to be able to handle problems arising from this coupling, e.g., water rising from the ground through dry regions into wet regions or the drying up in parts of the domain due to infiltration.

The basis of our scheme is the Runge-Kutta discontinuous Galerkin method [10], which proved to be effective for the two-dimensional shallow water equations. See, for example, [17], [20]. The newly developed limiter [12] is used for the stabilization of the method.

Even for a simple 2D simulation of the flow in a natural river bed and especially for the simulation of flooding, the free interface between wet and dry regions has to be handled correctly by a numerical scheme. Our approach for handling wetting and drying processes is based on the algorithm [8].

A numerical scheme has to handle steady states correctly. Those schemes, which treat steady states properly and do not introduce artificial disturbances, are referred to as well-balanced schemes. In the current work we test a new approach to well-balancing.

The implementation is done within **DUNE** – the **D**istributed and **U**nified **N**umerics **E**nvironment [14] – a modular toolbox for solving partial differential equations with grid-based methods [3, 4, 13]. Local adaptivity and parallelization are easy to use features within this framework and the methods presented take advantage of this. The basis for the DG and time discretization methods are part of the DUNE-FEM module [15].

The code is validated using several test problems with known exact solutions. The problem, which includes a mass source term, is numerically solved as a first step to the coupled simulation of the surface and groundwater flows.

2 Runge-Kutta Discontinuous Galerkin Method for Shallow Water Equations

The Runge-Kutta Discontinuous Galerkin method (RK-DG) [10] proved to be effective for the numerical solution of the shallow water equations. Let us provide the formulation of the method for the evolution equations of a very general form (we use the notations from [12]):

$$\partial_t \mathbf{u}(t,\cdot) = \mathscr{L}[\mathbf{u}(t,\cdot)](\cdot) \qquad \text{in } ([0,T)\times\Omega) \subset (\mathbb{R}\times\mathbb{R}^d),\ d \in \{1,2,3\}, \tag{1}$$

with the spatial operator

$$\mathscr{L}[\mathbf{v}] = S(\mathbf{v}) - \nabla\cdot F(\mathbf{v}), \tag{2}$$

where $\mathbf{v} : \Omega \to \Psi \subseteq \mathbb{R}^r$ belongs to some suitable function space V, Ψ is the set of states for a given problem, $S(\mathbf{v})$ is a source term function, and $F(\mathbf{v})$ is the analytical flux function. The appropriate definition of $\mathbf{u}$, S and F gives the 2D shallow water equations:

$$\mathbf{u} = \begin{pmatrix} u_1 \\ u_2 \\ u_3 \end{pmatrix} = \begin{pmatrix} h \\ hu \\ hv \end{pmatrix}, \tag{3}$$

$$F(\mathbf{u}) = \begin{pmatrix} F_1 \\ F_2 \\ F_3 \end{pmatrix} = \begin{pmatrix} hu & hv \\ hu^2 + \frac{1}{2}gh^2 & huv \\ huv & hv^2 + \frac{1}{2}gh^2 \end{pmatrix}, \quad S(\mathbf{u}) = \begin{pmatrix} S_1 \\ S_2 \\ S_3 \end{pmatrix} = \begin{pmatrix} S_h \\ -gh\frac{\partial b}{\partial x} \\ -gh\frac{\partial b}{\partial y} \end{pmatrix}. \tag{4}$$

where $\mathbf{u} = (u,v)$ is the velocity vector, $h = h(t,\mathbf{x})$ is the water depth, measured from the bottom positive upwards, $b = b(\mathbf{x})$ is the bottom height, measured from the reference level positive upwards. The total water height $H = H(t,\mathbf{x}) = h(t,\mathbf{x}) + b(\mathbf{x})$ is also measured from the reference level positive upwards, g is the gravity acceleration.

The source term S_h takes into account sources and sinks due to the coupling with the groundwater flow. Details will be given later. Additional forces, such as bottom friction forces, Coriolis forces, tidal potential forces, wind surface stresses, can be added, but to simplify the presentation, they are neglected here. The proper initial and boundary conditions [26] have to be added to the equations.

For the spatial discretization, a discrete operator $\mathscr{L}_h$ is defined, mapping a discrete function space V_h. We choose:

$$V_h := \{\varphi : \Omega_h \to \mathbb{R} \in L^2(\Omega_h) \mid \varphi|_E \in \mathscr{P}_p(E)\ \ \forall E \in \mathscr{T}_h\},$$

where $\mathscr{P}_p(E$ is the set of all polynomials of an order up p. The set $\Omega_h \subseteq \Omega$ is a polygonal approximation of the domain Ω which is partitioned by a tesselation $\mathscr{T}_h$ in the sense of the grid definition from [3]. The discrete operator is given by

$$\int_{\Omega_h} \mathscr{L}_h[u_h]\varphi\,d\mathbf{x} = \sum_{E\in\mathscr{T}_h}\int_E S(u_h)\varphi\,d\mathbf{x} + \sum_{E\in\mathscr{T}_h}\int_E F(u_h)\cdot\nabla\varphi\,d\mathbf{x} -$$
$$-\sum_{E\in\mathscr{T}_h}\int_{\partial E}\varphi\,G(u_h^+,u_h^-,\ldots)\cdot\mathbf{n}\,d\sigma \qquad \forall\varphi\in V_h. \tag{5}$$

Here, $\mathbf{n}$ is the unit outer normal of the element interface, u_h^+ and u_h^- are the values of the function u_h on both sides of the element interface. $G(u,v,\mathbf{x}) : V \times V \times \mathbb{R}^d \to \mathbb{R}^d$ is a numerical flux function. The discontinuous Galerkin method is completely described by the functions S and F and the numerical flux G and the space V_h. In this work, the Local-Lax-Friedrichs flux function ([19], [21]) is used as the numerical flux.

The RK-DG method is not stable for non-linear problems where strong shocks might appear. In the current work we use the stabilization mechanism [12], which was expanded in the context of our project to a specific criterion such as guaranteeing the conservation of non-negative water level.

A stabilized discrete operator is constructed by concatenation of the DG operator $\mathcal{L}_h$ and a stabilization operator Π_h:

$$\tilde{\mathcal{L}}_h[v_h(t,\cdot)] := (\mathcal{L}_h \circ \Pi_h)[v_h(t,\cdot)].$$

For the time discretization an operator $\bar{\Pi}_h$ is used for the initial data on each time step. The construction of $\Pi_h, \bar{\Pi}_h$ is given in [12].

The space discretization leads to a system of ODEs $\frac{d}{dt}u_h(t) = \tilde{\mathcal{L}}_h[u_h(t)]$ for the coefficients defining $u_h(t)$. An explicit Runge-Kutta method is used with the order $k = p+1$, where p is the polynomial degree of the basis functions used to construct the space V_h.

Let us denote the numerical solution at time step t^n by $u_h^n(\cdot) := u_h(t^n,\cdot)$. The numerical scheme is written as follows, with the time step size $\triangle t^n$ satisfying the stability restrictions of the operator $\mathcal{L}_h$:
$t = 0$:

$$\int_\Omega u_h^0 \varphi \, d\mathbf{x} = \int_\Omega u^0 \varphi \, d\mathbf{x} \; \forall \varphi \in V_h.$$

for all $t^{n+1} := t^n + \triangle t^n < T, \quad n \geq 0$:

$$\begin{aligned} w_0 &= \bar{\Pi}_h(u_h^n), \\ w_1 &= \tilde{\mathcal{L}}_h[w_0(t^n,\cdot)], \\ &\vdots \\ w_k &= \tilde{\mathcal{L}}_h[(w_0 + a_{k,1}w_1 + \ldots + a_{k,k-1}w_{k-1})(t^n + c_k\triangle t^n,\cdot)], \\ u_h^{n+1} &= \mathcal{RK}(u_h^n) := w_0 + b_1 w_1 + \cdots + b_k w_k, \end{aligned} \tag{6}$$

where the appropriate set of coefficients a, b, and c is defined by the order k of the Runge-Kutta method [10].

As we already mentioned before, the numerical algorithm for shallow water flows must successfully overcome two challenges: 1) correct handling of steady states and 2) correct handling of wetting-drying processes.

3 Well-Balancing

There have been many suggestions for the construction of well-balanced schemes in recent years, some applied to the RK-DG scheme (see, for example, [27]). We suggest a new method here, which is easy to implement in the DG framework as implemented in DUNE-FEM [9]. Comparisons with other methods will be presented in a later work.

Our method is based on a simple reformulation of the topography source term in the shallow water model. We focus on the balance law for the discharge $q = (hu, hv)$ and neglect the source term due to the ground water coupling. In its original form this equation reads $\partial_t q + \nabla \cdot F_q = S_q$ where F_q is the flux and $S_q = -gh\nabla b$ (see (1)-(4)). The most important requirement for a well-balanced scheme is to preserve a "lake at rest", i.e., if $u = 0$, $v = 0$, and $h = C - b$ with some constant C, then $F_q = S_q$ should be satisfied also on a discrete level. Inserting $u = 0$ and replacing b with $C - h$ in the equation for the discharge, the balance of flux and source term is given by $\frac{1}{2}\nabla h^2 - h\nabla h = 0$.

We start by rewriting the topology source term: $S_q = -\frac{1}{2}gh\nabla b - \frac{1}{2}\nabla \cdot (ghb) + \frac{1}{2}gb\nabla h$. Moving the divergence term to the left hand side and treating the topography b as an additional unknown we arrive at:

$$\partial_t h + \nabla \cdot q = 0, \tag{7}$$

$$\partial_t b = 0, \tag{8}$$

$$\partial_t q + \nabla \cdot (F_q + \frac{g}{2}hb) = -\frac{g}{2}(h\nabla b - b\nabla h). \tag{9}$$

Now in the DG context, the non-conservative products on the right hand side can be discretized using the approach from [11], treating them as a source term together with a measure on the boundary. We focus on the spatial discretization of the equation for the discharge q, now denoting with h, b the discrete solutions defined by the DG framework, i.e., we assume that h, b are piecewise polynomial functions, where the topography is projected into the same discrete space in which h is defined. The spatial discretization on a single element T of the grid is given by:

$$\begin{aligned} L_T[h,b,q] := &\int_T (F_q + \frac{g}{2}bh) \cdot \nabla\varphi - \int_T \frac{g}{2}(h\nabla b - b\nabla h)\varphi - \\ &\int_{\partial T} (\widehat{F}_q + \frac{g}{2}\widehat{hbn})\,\varphi + \frac{g}{2}(\widetilde{h\varphi}[b] - \widetilde{b\varphi}[h]), \varphi \in V_h. \end{aligned} \tag{10}$$

Here we use the abbreviation $\widehat{F}_q$ and $\widehat{hb}$ to denote numerical flux functions, approximating fluxes in normal direction over the cell boundaries (e.g. a Lax-Friedrichs flux for F_q and an averaging for hb). With $\widetilde{h\varphi}$ and $\widetilde{b\varphi}$ we denote some suitable averages and $[b] = (b^+ - b_T)n$, $[h] = (h^+ - h_T)n$ are the jumps of the (possibly) discontinuous discrete functions b, h in the normal direction n; h_T, b_T denote the discrete functions on the element T and h^+, b^+ the values on neighbouring elements.

The test function φ is assumed to have support only on element T (which means that $\varphi^+ = 0$). It turns out that a very simple averaging process leads to good results, i.e., $\widetilde{h\varphi} = h\widetilde{\varphi} = \frac{1}{4}\varphi(h_T + h^+)$ and $\widetilde{b\varphi} = \frac{1}{4}\varphi(b_T + b^+)$; but note that the method will work with a much more general choice for the averaging procedure.

To achieve well-balancing we focus on the case $q \equiv 0$ and $h + b = C$ on each element, which characterizes a lake at rest. Note that both b and h are assumed to belong to the same discrete space, so that $h + b = C$ holds on the discrete level. In this scenario $[b] = -[h]$ and the only assumption we make on the averaging is that under these circumstances $\widetilde{b\varphi} = \frac{1}{2}C\varphi - \widetilde{h\varphi}$. The only assumption we make on the underlying scheme is that in this situation $\widehat{F_q} = \frac{g}{2}\widehat{h^2}n$. We thus arrive at:

$$\begin{aligned} L_T[h,b,q] &= \int_T \frac{g}{2}(h^2 + (C-h)h)\cdot\nabla\varphi - \int_T \frac{g}{2}(-h\nabla h - (C-h)\nabla h)\varphi - \\ &\quad - \int_{\partial T}\frac{g}{2}(\widehat{h^2} + \widehat{hb})n\varphi + \frac{g}{2}(-\widetilde{h\varphi}[h] - (\frac{1}{2}C\varphi - \widetilde{h\varphi})[h]) = \\ &= \frac{g}{2}C\int_T (h\nabla\varphi + \nabla h\varphi) - \frac{g}{2}\int_{\partial T}(\widehat{h^2} + \widehat{hb} - \frac{1}{2}C(h^+ - h_T))n\varphi = \\ &= -\frac{g}{2}\int_{\partial T}(\widehat{h^2} + \widehat{hb} - \frac{1}{2}C(h^+ - h_T) - Ch_T)n\varphi = \\ &= -\frac{g}{2}\int_{\partial T}(\widehat{h^2} + \widehat{hb} - C\frac{1}{2}(h^+ + h_T))n\varphi\,. \end{aligned} \tag{11}$$

For well-balancing we need $L_T[h,b,q] = 0$, which is satisfied if we choose $\widehat{hb} = -\widehat{h^2} + C\frac{1}{2}(h^+ + h_T)$. Since $b = C - h$ we have $hb = Ch - h^2$ so that this is a reasonable assumption. If, for example, F_q is given by the Lax-Friedrichs scheme, we have (due to $q = 0$) $\widehat{F_q} = \frac{g}{4}(h_T^2 + (h^+)^2)$ so that $\widehat{h^2} = \frac{1}{2}(h_T^2 + (h^+)^2)$. With the simple choice of $\widehat{hb} = \frac{1}{2}(h_T b_T + h^+ b^+) = \frac{1}{2}(Ch_T - h_T^2 + Ch^+ - (h^+)^2) = -\widehat{h^2} + C\frac{1}{2}(h_T + h^+)$ so that the modified scheme is in fact well-balanced.

4 Wetting and Drying Treatment

According to [2] the first example of treatment of moving boundaries seems to be the model of Reid and Bodine (1968) [24].

The work [23] categorize the wetting-drying treatments for finite element hydrodynamic models into three types: mesh adaptation algorithms, mesh reduction algorithms and thin slot algorithms. In [18] wetting-drying treatments for finite element and finite volume formulations are divided into moving mesh methods and fixed mesh methods. The second group of treatments include element (or nodes) deactivation approach, thin slot approach (which is referred to as the most popular method), and the artificial porosity approach. The paper [18] itself describes another, new, approach, where the position of sea bed is allowed to fluctuate in drying areas. See [18] for details and references.

The wetting and drying treatment used in the current work is based on the algorithm [8], belonging to the thin slot algorithms. This approach is also named as "thin layer" technique or "thin water layer" technique (e.g. [8]), or "thin slot" technique (e.g. [23]). Other examples can be found in [5], [7], [17].

Thin water layer approach uses a fixed mesh without eliminating elements or nodes from the domain. A very thin layer of water is kept in dry elements. Therefore, the water depth is kept greater than some given threshold value over the whole domain. When the water depth goes below the threshold, then an element is considered to be "dry". Thus, the shallow water equations can be solved in the whole domain, including possible dry parts.

The thin water layer approach is mass conservative, but, unfortunately, the conservation of momentum is not guaranteed (see [5] for details). Though it is acceptable as long as the error of computation is acceptable.

According to [8] the so-called wet-or-dry status flags are initially set, marking each element E either "wet" (flag = 1) or "dry" (flag = 0), and then updated at each k-th Runge-Kutta intermediate stage. The stability condition for Δt is based on the consideration of the balance between the mass in the element and the outgoing discharge through its boundary. Every intermediate RK-stage must preserve the positivity of a mean water depth in every cell. If this stability condition shows that flows empties an element through a considered boundary, then according to [8] the so-called "reflection numerical flux" is evaluated, which prohibits mass transfer through a boundary. If a considered boundary is an interface between two dry elements, then the reflection numerical flux is used without checking a stability condition in order to do not introduce an artificial mass exchange between dry elements. In our work we don't use this condition for Δt, but apply the reflection flux as soon as empting of an element is detected. This simpler condition nonetheless gives good numerical results.

We implement the requirement as in [8] to cancel the gravity terms in dry elements in order to avoid non-physical oscillations and to prevent dry elements from loosing their mass by using the above described wet-or-dry status flags in the mathematical model, setting $g = g_0$ in an element if it is "wet" and setting $g = 0$ if an element is "dry". Thus, in our implementation the status flags belong to "thin water layer" shallow water mathematical model, which seems to us very logical.

After the positivity of the mean water depth is guaranteed in each element, the so-called Positive Depth operator [8] has to guarantee the positivity of water depth node-wise in each element.

The PD operator Π_h^{PD} is implemented together with the stabilization operator $\tilde{\mathscr{L}}_h[v_h(t,\cdot)] := (\mathscr{L}_h \circ \Pi_h \circ \Pi_h^{PD})[v_h(t,\cdot)]$. Therefore, the scheme is:

$$
\begin{aligned}
w_0 &= \bar{\Pi}_h \Pi_h^{PD}(u_h^n),\\
w_1 &= \tilde{\mathscr{L}}_h[w_0(t^n,\cdot)],\\
&\vdots\\
w_k &= \tilde{\mathscr{L}}_h[(w_0 + a_{k,1}w_1 + \ldots + a_{k,k-1}w_{k-1})(t^n + c_k\Delta t^n,\cdot)],\\
u_h^{n+1} &= \mathscr{R}\mathscr{K}(u_h^n) := w_0 + b_1 w_1 + \cdots + b_k w_k.
\end{aligned}
\tag{12}
$$

See [8] for the details on the PD operator.

Let us note, that currently triangular elements and linear approximations of water depth and discharges are used in this algorithm. But due to the generic DUNE implementation we see it possible to broaden on other elements and higher order approximations in the future.

We can not prove the convergence of the developed algorithm, but the validation of the code shows only a small error between numerical result and known exact solution for the series of test problems. Also the realistic behaviour of the numerical solution for the problem with source/sink terms is shown. See the section 6 for the details.

5 Implementation Notes

The described algorithm is implemented using **DUNE** – the **D**istributed and **U**nified **N**umerics **E**nvironment [14] – a modular toolbox for solving partial differential equations with grid-based methods [14]. The DUNE module DUNE-SWE is currently being developed for numerical simulation of shallow water flows with well-balancing, limiting and taking into account wetting and drying processes. The Runge-Kutta Discontinuous Galerkin method for solving the shallow water equations with wetting and drying is implemented on the basis of DUNE-FEM module [15]. As both stabilization operator and wetting-drying treatment are based on the same concept of keeping some quantity realistic, we implement them within one pass [16].

6 Validation of Code

The developed code is validated by solving numerically different types of problems, where wetting and drying processes occur. The results show good correspondence of numerical results and the exact solution.

6.1 *"Lake at Rest"*

The correct numerical treatment of equilibrium states, more exactly, the correct simulation of still water ("lake at rest") is shown by the developed code. The domain settings and initial data are taken as in [8].

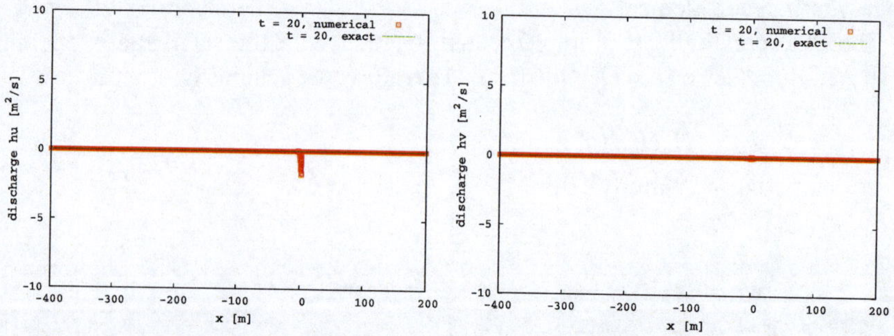

Fig. 1 "Lake at rest" problem: discharges hu and hv for $t = 10$

6.2 Dam Break Problem

The dam break problem with a known exact solution [6] is solved numerically. The domain settings and initial data are taken as in [8]. The figure 2 shows the values of the total water height H and the discharge hu for different time moments. The grid with 240*40 nodes is taken for the simulation. For nearly all time moments the graphs of corresponding exact solution are hidden behind the graphs of numerical solution.

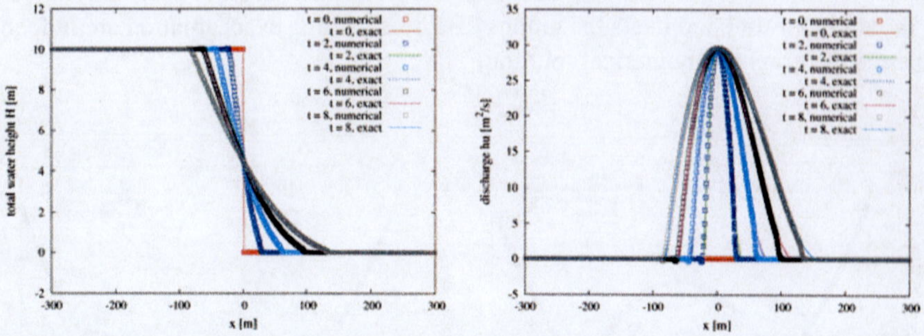

Fig. 2 Dam break problem: total water height H and discharge hu for $t = 0, 2, 4, 6, 8$. For nearly all time moments the graphs of corresponding exact solution are hidden behind the graphs of numerical solution

6.3 Drying Riemann Problem

The drying Riemann problem with a known exact solution [6] is solved numerically. The domain settings and initial data are taken as in [8]. The figure 3 shows the values of the total water height H and the discharge hu for different time moments. The grid with 240*40 nodes is taken for the simulation. For nearly all time moments the graphs of corresponding exact solution are hidden behind the graphs of numerical solution.

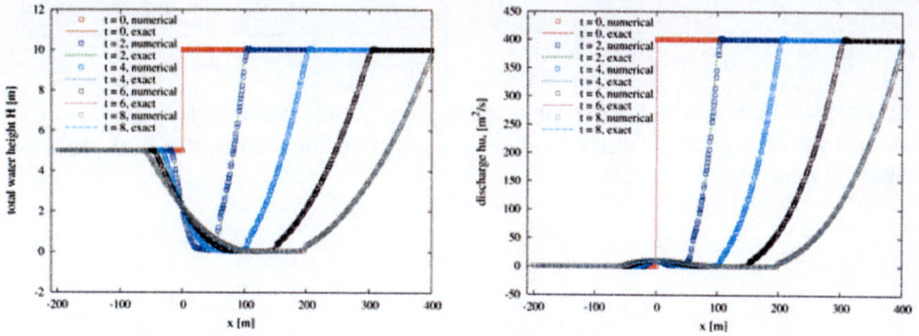

Fig. 3 Drying Riemann problem: total water height H and discharge hu for $t = 0, 2, 4, 6, 8$. For nearly all time moments the graphs of corresponding exact solution are hidden behind the graphs of numerical solution

6.4 Parabolic Bowl Problem

The two-dimensional axisymmetric problem with a known exact solution [25] (a so-called "parabolic bowl" problem) is solved numerically. The domain settings and initial data are taken as in [8] and [17]. The exact solution is periodic in time with a period value τ. The figures 4 and 5 show the values of the total water height H and the discharge hv for different moments of time multiple to a period value. The grids with 80*80, 160*160, 320*320 and 640*640 nodes are taken for the simulation. For nearly all time moments the graphs of corresponding exact solution are hidden behind the graphs of numerical solution.

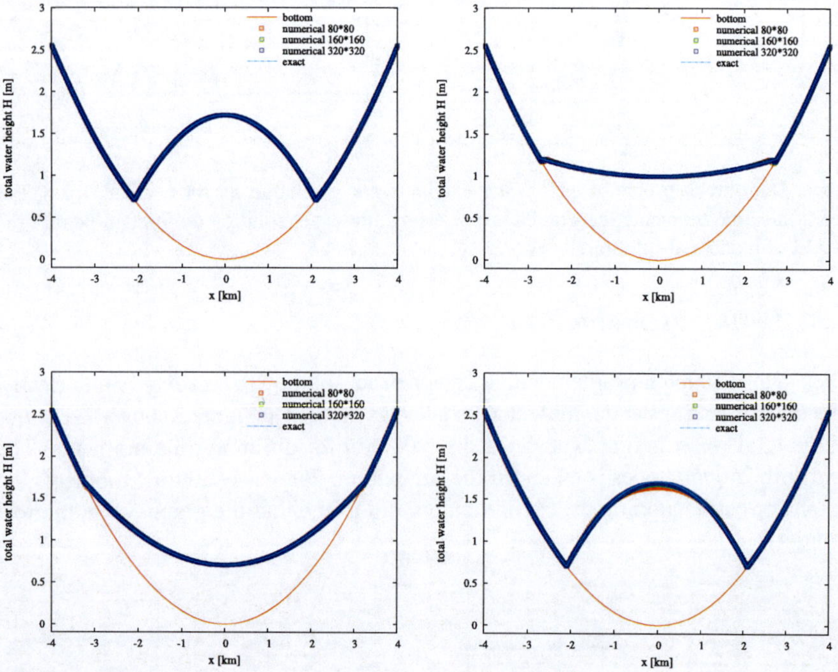

Fig. 4 "Parabolic bowl" problem: total water height H for $t = 0, \tau/4, 3\tau/4, \tau$. For nearly all time moments the graphs of corresponding exact solution are hidden behind the graphs of numerical solution

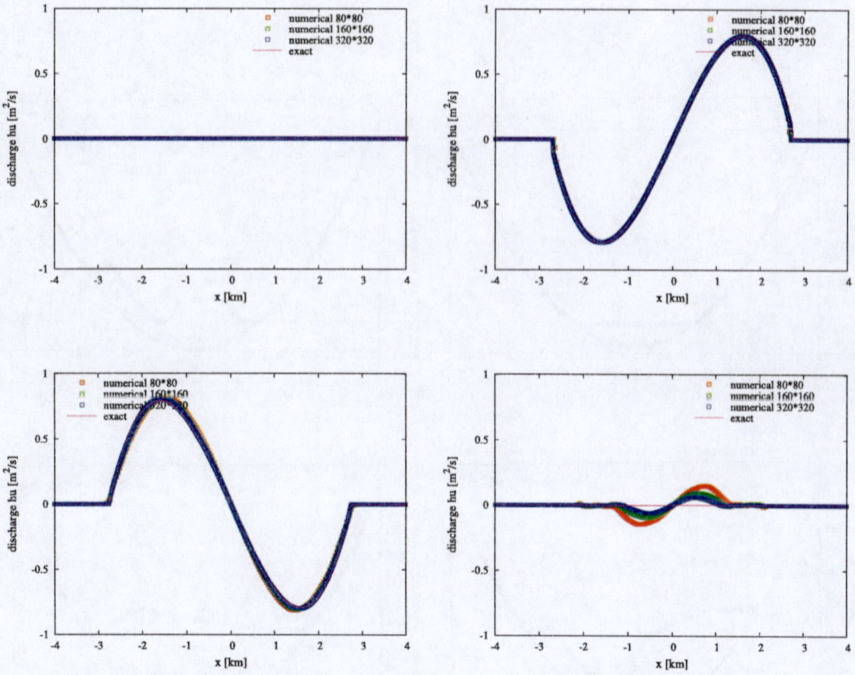

Fig. 5 "Parabolic bowl" problem: discharge hv for $t = 0, \tau/4, 3\tau/4, \tau$. For nearly all time moments the graphs of corresponding exact solution are hidden behind the graphs of numerical solution.

7 "Source-Sink" Problem - First Step towards Coupled Modelling of Surface and Groundwater Flows

The next test problem is the first step towards the coupled modelling of surface and groundwater flows. Initial data is the "lake at rest" problem: the still water with a constant water height. The bottom is a parabola in 1D case and a elliptic paraboloid in 2D case. At $t = 0$ the source of water appears on the right side of the bottom above the steady water, and the water flows down inside the domain till $t = 4$. The water fills the domain, and after $t = 3$ (thus, before closing the source) the sink appears on the left side of the bottom. After $t = 4$ the water outflows only from the domain, thus, the water height decreases until it reaches the position below the sink. The grid with 80*5 nodes was used for the 1D case, and the grid with 100*100 nodes - for the 2D case.

Figure 6 shows the profile of the total water height H for the 1D problem at different time moments. Figure 7 shows the flow evolution for the 2D problem.

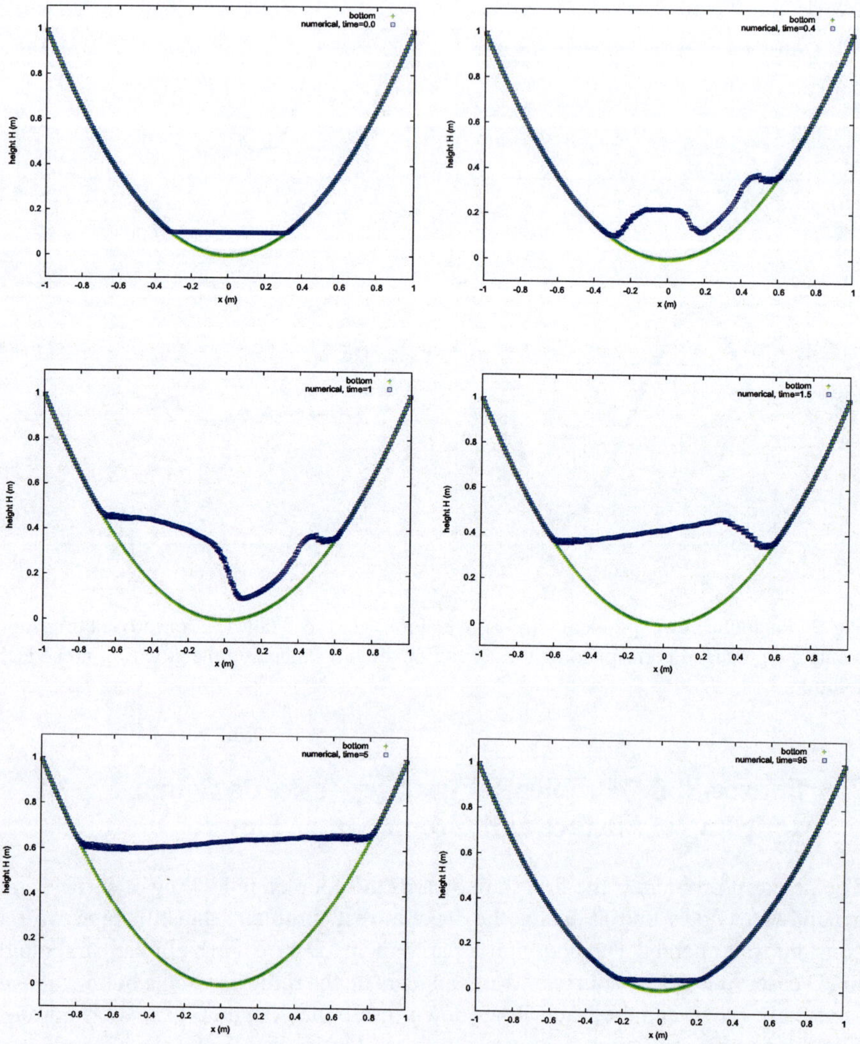

Fig. 6 1D "source-sink" problem: total water height H profile for $t = 0, 0.4, 1, 1.5$

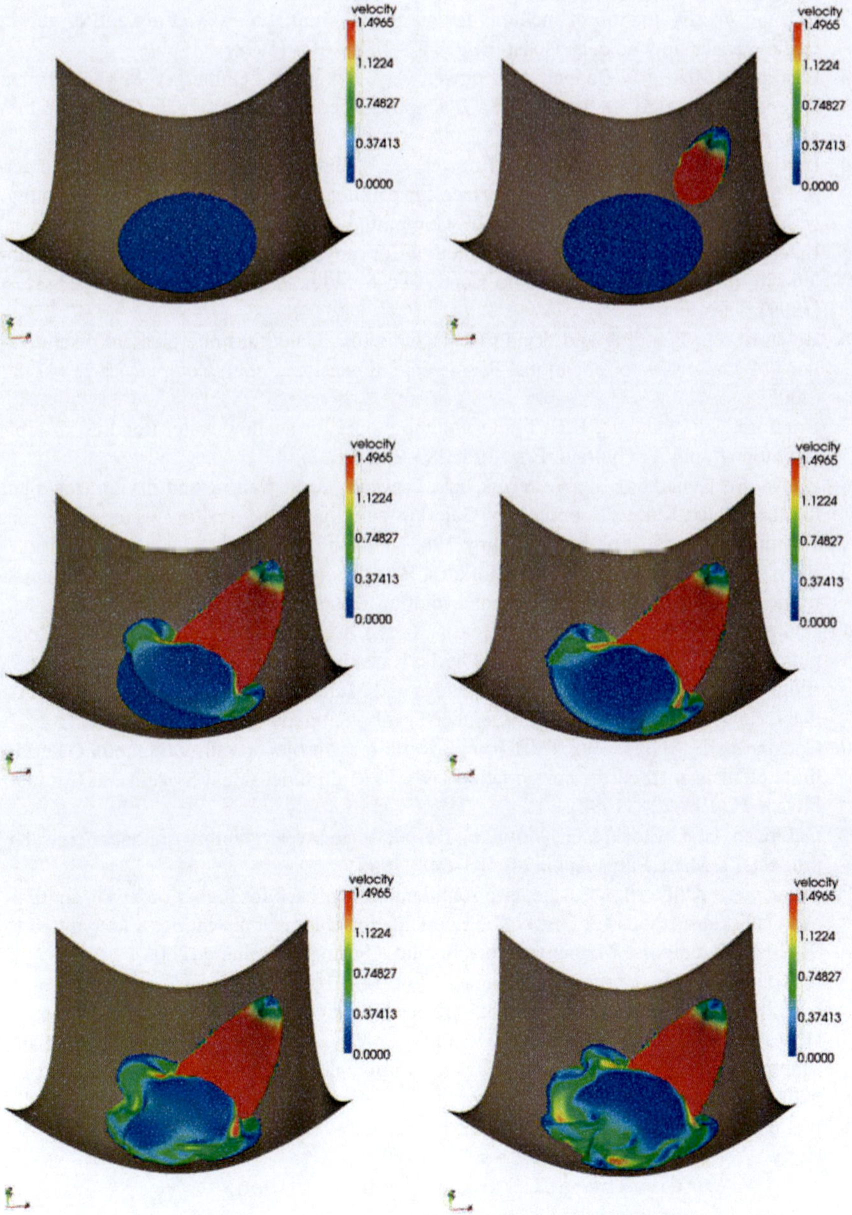

Fig. 7 2D "source-sink" problem: the flow evolution

References

1. http://www.adapthydromod.de
2. Balzano, A.: Evaluation of methods for numerical simulation of wetting and drying in shallow water flow models. Coast. Eng. 34(1-2), 83–107 (1998)
3. Bastian, P., Blatt, M., Dedner, A., Engwer, C., Klöfkorn, R., Ohlberger, M., Sander, O.: A generic grid interface for parallel and adaptive scientific computing. I: Abstract framework. Computing 82(2-3), 103–119 (2008)
4. Bastian, P., Blatt, M., Dedner, A., Engwer, C., Klöfkorn, R., Kornhuber, R., Ohlberger, M., Sander, O.: A generic grid interface for parallel and adaptive scientific computing. II: Implementation and tests in DUNE. Computing 82(2-3), 121–138 (2008)
5. Bates, P.D., Hervouet, J.-M.: A new method for moving-boundary hydrodynamic problems in shallow water. Proc. R. Soc. Lond., Ser. A, Math. Phys. Eng. Sci. 455, 3107–3128 (1999)
6. Bokhove, O.: Flooding and drying in discontinuous Galerkin finite-element discretizations of shallow-water equations. Part 1: One dimension. J. Sci. Comput. 22-23, 47–82 (2005)
7. Bradford, S.F., Sanders, B.F.: Finite-volume model for shallow-water flooding of arbitrary topography. J. Hydraul. Eng. 128, 289–298 (2002)
8. Bunya, S., Kubatko, E.J., Westerink, J.J., Dawson, C.: A wetting and drying treatment for the Runge-Kutta discontinuous Galerkin solution to the shallow water equations. Comput. Methods Appl. Mech. Engrg. 198, 1548–1562 (2009)
9. Burri, A., Dedner, A., Diehl, D., Klöfkorn, R., Ohlberger, M.: A general object oriented framework for discretizing nonlinear evolution equations. In: Shokin, Y.I., Resch, M., Danaev, N., Orunkhanov, M., Shokina, N. (eds.) Advances in High Performance Computing and Computational Sciences. The 1st Kazakh-German Advanced Research Workshop, Notes on Numerical Fluid Mechanics and Multidisciplinary Design (NNFM), Almaty, Kazakhstan, September 25-October 1, vol. 93, pp. 69–87 (2006)
10. Cockburn, B., Shu, C.-W.: TVB Runge-Kutta local projection discontinuous Galerkin finite element method for conservation laws V: Multidimensional Systems. J. Comput. Phys. 141, 199–224 (1998)
11. Dal Maso, G., LeFloch, P.G., Murat, F.: Definition and weak stability of nonconservative products. J. Math. Pures Appl. 74, 483–548 (1995)
12. Dedner, A., Klöfkorn, R.: A generic stabilization approach for higher order Discontinuous Galerkin methods for convection dominated problems. Preprint no. 8 (submitted to SIAM Sci. Comput.), Mathematisches Institut, Unversität Freiburg (2008), http://www.mathematik.uni-freiburg.de/IAM/homepages/robertk/postscript/dedner_kloefkorn_limiter.pdf
13. Dedner, A., Klöfkorn, R., Nolte, M., Ohlberger, M.: A generic interface for parallel and adaptive discretization schemes. Abstraction principles and the dunefem module. Computing (to appear)
14. http://www.dune-project.org
15. http://dune.mathematik.uni-freiburg.de
16. http://dune.mathematik.uni-freiburg.de/doc/html-current/group__FEM.html
17. Ern, A., Piperno, S., Djadel, K.: A well-balanced Runge-Kutta discontinuous Galerkin method for the shallow-water equations with flooding and drying. Int. J. Numer. Meth. Fluids 58, 1–25 (2008)

18. Kärnä, T., de Brye, B., Gourgue, O., Lambrechts, J., Comblen, R., Legat, V., Deleersnijder, E.: A fully implicit wetting-drying method for DG-FEM shallow water models, with an application to the Scheldt Estuary. Comp. Meth. Appl. Mech. Eng. (2010) (accepted manuscript, available online July 11, 2010)
19. Kröner, D.: Numerical Schemes for Conservation Laws. Verlag Wiley & Teubner, Stuttgart (1997)
20. Kubatko, E.J., Bunya, S., Dawson, C., Westerink, J.J.: Dynamic p-adaptive Runge-Kutta discontinuous Galerkin methods for the shallow water equations. Comput. Methods Appl. Mech. Engrg. 198, 1766–1774 (2009)
21. Leveque, R.J.: Finite volume methods for hyperbolic problems. Cambridge Texts in Applied Mathematics. Cambridge University Press, Cambridge (2002)
22. Lynch, D.R., Gray, W.G.: Finite element simulation of flow deforming regions. J. Comp. Phys. 36, 135–153 (1980)
23. Nielsen, C., Apelt, C.: Parameters affecting the performance of wetting and drying in a two-dimensional finite element long wave hydrodynamic model. J. Hydraul. Eng. 129, 628–636 (2003)
24. Reid, R.O., Bodine, B.R.: Numerical model for storm surges in Galveston Bay. J. Waterways Harbors Division, ASCE 94(WW1), 33–57 (1968)
25. Thacker, W.C.: Some exact solutions to the nonlinear shallow-water wave equations. J. Flud Mech. 107, 499–608 (1981)
26. Vreugdenhil, C.B.: Numerical methods for shallow-water flow. Kluwer academic Publishers, Dordrecht (1994)
27. Xing, Y., Shu, C.-W.: High order well-balanced finite volume WENO schemes and discontinuous Galerkin methods for a class of hyperbolic systems with source terms. J. Comput. Phys. 214, 567–598 (2006)

HPC Processor Technologies and Their Impact on Simulation

Michael M. Resch and Uwe Küster

Abstract. Moore's law has come to an end with respect to the clock speed of the single processor. Clock rates are no longer increasing. Parallelism carries the day and accelerators are making the most of this. What is the future of processors for HPC going to look like? This talk will give a short overview and discuss some potential solutions.

1 Introduction

In recent years high performance computing has seen a dramatic change. First and foremost we have seen a change in architecture. Back in 2002 the basic architectural concept was the cluster. The building blocks for high performance computer architectures were standard components. Big clusters were said to dominate the market for a long time to come. The promise for the future was a single, consistent and standardized architecture that would span a range from desktop systems to high end computers. The main problem seen was the software problem.

Looking at the landscape and planning in 2009 the picture has changed completely. The TOP500 list [17] of the fastest systems in the world is led by two high performance computers that are non-standard, the IBM Roadrunner [11] on the one hand and the Cray Jaguar [12] on the other hand. Furthermore the list is dominated by systems like BlueGene [14] with a completely new architectural approach. At the same time we see future planning in large scale projects that aim at creating so called hybrid architectures. The most important of these was the Japanese Next Generation Supercomputer Project. This approach, however, seems to have face a drawback when NEC dropped out of the project.

Michael M. Resch · Uwe Küster
High Performance Computing Center,
Nobelstraße 19, Stuttgart, 70569, Germany
e-mail: resch@hlrs.de,
kuester@hlrs.de

E. Krause et al. (Eds.): Computational Sci., & High Performance Computing IV, NNFM 115, pp. 17–28.
springerlink.com

Rather than providing the community with a single platform hardware vendors offer a basket of solutions that follow entirely different approaches. As a consequence the community has to deal with uncertainty. The problem is further aggravated by the fact that the community is running into an energy problem. Costs for power supply have at least grown by a factor of three over the last five years. This is not to be counted as the least problem for high performance computing simulation. The operational costs of hardware have grown to become as high as the initial investment costs - for some medium sized centres prohibitively high costs.

While the computer science community is struggling with these problems, simulation experts are faced with a different challenge. The level of performance of high performance computers is at least theoretically so high that traditional simulations can be done easily. At least there is no more performance barrier for some of the more common problems. As single phenomena (like computational fluid dynamics, crash, ...) can be simulated on a reasonably large system the trend is towards coupled application. In such applications users have to deal will multiscale problems which make programming further difficult.

2 Hardware Challenges

2.1 Building Blocks

High performance computer hardware is currently often built from standard components. Sometimes they are only integrated into a single system like clusters of PCs. In this case no special hardware effort is made to integrate the components. Other concepts focus on special hardware and software integrating standard components. It therefore makes sense to have a look at these standard components to understand current trends in high performance computing.

2.1.1 Processors

Processors are at the core of each high performance computing system. Over the last 50 years processor speed has substantially increased. We have seen a growth in single processor performance of about 10^9. What we see today, however, is a limit in clock frequency not going beyond 4 GHz which means that we have not made any progress over the last years. Even worse, we have seen 1 GHz already back in 1986 in special purpose processors which means that over the last 20 years we have only seen a factor of four in clock rate for high end processors. The problem of speed has been widely ignored because standard processors had to catch up with high speed processors over these 20 years. Now these standard processors are operating at the same clock frequency as special high end processors.

The reason why we have not seen a further increase in clock rate is the problem of power and cooling. With our current processor technology any increase in clock frequency is going to drive up processor temperatures to a level for which there is currently no cooling method available.

The problem is not in feature size. We still see a continued growth of integrated circuits on a die as first described by Moore [5]. On the contrary, the feature size is still decreasing with a feature size of 90nm in 2003, 45nm in 2010 and an expected feature size of 25nm in 2015 as described by the International Technology Roadmap Committee [2]. The problem is in the leakage current which is growing non-linearly with clock frequency and hence increases the heat.

So, even though we cannot increase the clock rate, the number of integrated circuits on a die is still growing. It is hence an obvious solution to increase the number of cores on one die. As a consequence all standard processors are currently working with so called multi-core architectures. All vendors have figured that by putting two or four cores on a die the overall speed of the processor can be increased. The basic processor building block currently is characterized by the following features.

Table 1 Characteristic numbers for standard processors

Manufacturing Technology	45 nm
Clock Frequency	2.5 - 3.0 GHz
Number of cores	4 - 8
Number of FPUs	4

2.1.2 Accelerators

Given the increasing number of integrated circuits and considering that an increase in number of cores is the best way to increase the overall performance new concepts have emerged. The basic idea is to use many cores on a single chip or board in order to achieve much higher performance.

IBM, Sony, Toshiba Cell Processor

The development of Cell Processor is a collaborative effort of IBM, Sony and Toshiba [6]. The Cell Processor will serve for entertainment electronics. It will empower the future Sony Play station 3 for high performance graphics and will be used for the future computing intensive High Definition Television (HDTV). The processor may be used in medical applications, image recognition and signal processing. Workstation systems for scientific usage can be expected. The Cell Processor will be a cheap device despite its complexity. The machine has an uncommon architecture. The single chip architecture has a PowerPC processor as host and 8 Synergistic Processor Elements (SPE), some specially small and fast processors. The host processor will run the operating system, do the IO and control the different tasks. The PowerPC processor allows for two hardware threads and operates in order. In order suppresses the ability of reordering instructions at run time for better filling the pipelines but simplifies the architecture and allows for higher frequencies. The burden for the compiler is higher as is seen for the IA64 processors.

The Altivec/VMX vector part of the PowerPC allows for relatively high single precision floating point performance. Instruction and data L1-caches have 32 KB. The 512 KB L2-cache is on die and is directly accessible by the SPEs. The system has two XDR Rambus memory interfaces with a high nominal memory bandwidth of 25.6 GB/s. Connected to the system caches is the Element Interconnect Bus (EIB). This bus consists of 2x2 rings for both directions operating at half of the system clock. Its peak bandwidth is 96B/cycle. For a 3.2 GHz system this results in 154 GB/s.

The SPEs are independent vector processors. Each has its own local memory of 256 KB. A register-file with 128 registers of 128 bit width holds the vector operands. Two instructions are issued per clock cycle and executed in order. Four single precision floating point multiply-add instructions may be handled per cycle. The total single precision peak performance of all SPEs of a 3.2 GHz processor is around 200 GFLOP/s. The single precision units deliver results rounded to zero restricting their usability. The SPEs IEEE double precision performance of aggregated 26 GFLOP/s is by far less but still four times more than todayâĂŹs PCs peak performance. The main memory bandwidth peak performance relation of 1 B/FLOP is a reasonable value for double precision results. For single precision results this relation is quite small with 0.1 B/FLOP. A special DMA engine in the SPE maps parts of the main memory to the local memory and initiates and performs block data transfers. Latencies for the access of the main memory are in the same range as for other cache systems. Addresses are automatically mapped. The block data transfers may also be performed between different SPEs of different cells which are connected by the fast FLEXIO interface of aggregated 76 GB/s bandwidth.

The system has no shared memory processors. SPEs and main memory have different addresses. The host processor and the SPEs have even different instructions. The SPEs have no caches. Flexibility is reduced but the performance is increased.

The single chip system has an incredible peak performance. With respect to numerical applications cutting of floating point results has to be considered a deficiency. Iterative schemes in a defect correction approach may tolerate this problem. But even with the actual IEEE double precision performance a very challenging system has to be acknowledged. Raising double precision capabilities can be expected in future systems. Noticing the fast memory system it is important to keep an eye on the decreasing bandwidth peak performance relation. The architects try to smooth the consequences by the SPEs local stores. They enable the programmers to optimize the codes by decoupling heavily used data from main memory. But the programmer is also forced in doing that. Questionable is the ability of porting existing software to this architecture. It cannot be expected that today's packages can be ported without larger effort to the system at least if the high potential of the machine should be realized. Important will be compiler support for auto-vectorization on the SPEs and OpenMP as parallel programming model on the host processor loading the parallel segments to the SPEs. The small SPE memory restricts the memory consumption of data and instructions. Overlay techniques forgotten in the eighties will be reinvented again.

The future of the Cell Processor systems is not yet entirely clear. The architecture is mainly driven by consumer electronics. The processors of the family have to be inexpensive. But the system has also become available for high performance computing simulations. Their high peak performance offers a high potential for the needs of technical computing. If systems will be broadly available at a reasonable price many research groups will test the systems for their purposes. Performance and the architectural restrictions will generate new algorithmic techniques. Vectorization of codes will be important to exhaust the performance potential. Most important seems the ability of the Cell Processor to boost also relatively small numerical problems to high performance.

Clearspeed Accelerator Card CSX 600

The Clearspeed accelerator card CSX 600 consists of two processors with their own memory on a PCI-X card. Cards for other interfaces (Hyper-Transport, PCI Express, and SGI NumaLink) are to be expected. The Clearspeed processor runs at a low frequency of 250 MHz proving that not the frequency is the main limitation for performance. The small frequency requires only 5 Watts power consumption per processor. The chip is composed of 96 SIMD processing elements (PE) which are controlled and directed by a small host reminding of the Connection Machine and Maspar. Each PE has its own fast SRAM memory of (only) 6 KB. The internal memory bandwidth of 1 GB/s per PE is high enough to feed the functional units of 0.25 GFLOP/s (4 Byte per Flop). But the external memory bandwidth of 3.2 GB/s is by far too low to pump the data into the processing elements. For this reason there are only a few exceptions showing the aggregated performance potential of 50 GFLOP/s for the two processor cards. Even DGEMM matrix multiplication reaches the peak performance only for large data sets. Dense matrix operations, Fast Fourier Transforms and some techniques of molecular dynamics may benefit from the card.

Loading the data from the host computer to the accelerators memory may be time consuming at least for solving sets of small problems. This will be better for future interfaces like Hyper-Transport. The parallel programming model is based on a C-dialect with additional syntax elements for variables declaration. Special data movement instructions allow for distributing data to the processing elements. The model emphasizes on running copies of identical procedures with different data (Flynn SPMD). Clearspeed proves high aggregated performance to be reached with small frequency and low power consumption. But we notice an insufficient card memory bandwidth limiting the sustained performance for problems with modest data intensity.

Clearspeed cards have been part of a recent big procurement of the Tokyo Institute of Technology (TITech) consisting on a large number of AMD Opteron based shared memory nodes. So far, however, no results for the usage of Clearspeed cards - other than for Linpack - are yet available for large scale systems.

FPGAs

Freely Programmable Gate Arrays are systems where even the processor hardware architecture has to be implemented by the user. For an appropriate design they will efficiently perform restricted tasks. Compilers for high level languages are on the way. All US vendors offer FPGAs cards for the interfaces of their machines. Usage for numerical algorithms is still questionable. But because of the specialization suitable algorithms may run very fast. Graphic Cards.

Today's graphic cards are freely programmable (e.g. CG programming language) hardware devices offering a very high performance potential. An increasing number of research groups try to use these cards for technical computing. The arithmetic is limited to an accuracy of 4 Bytes. They can be used for acceleration of numerically intensive parts started as "remote procedure calls" form the hosting PC similar to the favourite programming model of the Cell Processor, but they are not as flexible.

2.1.3 Memory Systems

Memory subsystems have been a key problem in high performance computing over the last years. For many applications the speed of the memory is more important than the speed of the processor. For standard micro-processors this problem has hardly ever been an issue. Standard applications of private users - for which most processors and memory subsystems are designed - are not memory bound. Manufacturers are hence not interested in improving the speed of memory for their standard products.

Scientific applications are very often memory bound. A typical number for computational fluid dynamics is that the speed of the memory subsystem should allow a transfer rate of 2 bytes and more for every floating point operation that is performed. For many applications in physics memory speed is not an issue. In engineering and climate and weather research it is an issue, however.

When looking at memory subsystems for these applications one can easily predict the processor performance from the memory speed. Kobayashi [3] has shown how the memory speed decrease from the NEC SX-8 system to the NEC SX-9 system has impacted the performance of a variety of codes. As a rule of thumb it is the memory speed ratio rather than the peak performance ratio of the two systems that gives a good indication for the application performance.

2.1.4 Networks

Networks were once considered to be the most important part of a high performance computer. Following the phrase "The network is the computer" scientists and vendors alike considered the development of fast networks a corner stone of high performance computer development.

This has changed over the last years. There is still agreement that the network is an important part of any system. However, the market has not paid for faster proprietary networking solutions. Instead standardization carries the day. Infiniband

and Gigabit Ethernet have become pervasive. In the most recent TOP500 list [17] we find the following interconnects: 28% Infiniband, 56% Gigabit Ethernet, 2% Myrinet, 1% Quadrics, and only 13% proprietary of some kind.

Neither Infiniband nor Gigabit Ethernet were ever designed with high performance computing in mind. As a consequence we are faced with high latencies and low bandwidths. When comparing parallel systems like the Intel Paragon and the Cray T3E with current systems of 1994/1995 in terms of network performance we find the following. In terms of processor performance recent systems are about 100-1000 times faster. In terms of latency we have only gained a factor of about 20-30. And when we look at bandwidth we only see a factor of 10-20. Interconnects have hence not been able to catch up with the overall speed of high performance computing systems.

2.2 *System Architectures*

As mentioned above high performance computing is still mainly about integrating standard components. The TOP500 list [17] is full of standard clusters using some sort of x86-processor and one of the two big networks described above. The merit of such approaches is in quantity and not in quality. We do, however, see some more innovative and creative approaches which are described here in detail.

2.2.1 Hybrid Concepts

So called hybrid concepts have gained some popularity over the last years. There is no clear definition for a hybrid high performance computer yet. In principal we consider systems to be hybrid if they integrate a combination of different processors into a single high performance computer.

The first such system that became widely known was the Tsubame system at Tokyo Institute of Technology [4]. The system is a combination of AMD Opteron processors and Clearspeed cards. A similar system was implemented in the Road-Runner [11] project where AMD Opteron processors are combined with the Cell Processor. The Japanese Next Generation Supercomputer Project [13] is suggesting a similar approach. The architecture is planned to be a combination of vector processors and microprocessors.

2.2.2 Low Power Systems

Given the problems with memory speed and interconnect speed designers had to rethink their approaches. The most radical solution is to avoid high speed processors and couple a huge number of low speed processors. This was done in the IBM BlueGene [14] project. The benefit of choosing a slow processor is a relatively low power consumption of large scale systems. Typically a factor of 2-4 in power consumption can be saved with a BlueGene system. However, the disadvantage is in the high number of processors for comparable performance. We will see later that this increases the cost for programming for the user.

A very interesting low power approach is currently investigated by the National Energy Research Scientific Computing Center (NERSC) and the Computational Research Division of Lawrence Berkeley National Laboratory (LBNL). In [8] they proposed an innovative way to improve global climate change predictions by using a supercomputer with low-power embedded microprocessors, an approach that would overcome limitations posed by today's conventional supercomputers.

2.3 Operational Issues

As already pointed out above, operation of high performance computers has become fairly expensive. The main problems are power and cooling. As the power consumption of processors increases high performance computer owners are facing power consumption in the range of 2-6 MW. This is not only a problem of power supply but also a cooling problem. Furthermore, the operational costs for a high performance computer have increased over the last years and may become as high as the costs for investment in the years to come.

Air cooling for 2-6 MW is getting difficult. So, most vendors are moving to liquid cooled solutions. Some provide water cooled racks. Others try to cool processors directly - which increases both the complexity of the system and the production costs. Huge investments have been made to build new facilities for most international high performance computing centres. Many centres are planning to upgrade their buildings or cooling facilities.

3 Software Challenges

The biggest challenge in high performance computing is currently in the huge number of processors used in systems. From an operational point of view it is a challenge to control the system and to guarantee availability. Redundancy and fault tolerance has to be built into these architectures.

The much bigger challenge is the software challenge. Over the last years users have become familiar with parallel programming. They have learned to master massively parallel systems with up to 2000 processors. The community has introduced MPI and OpenMP as standards. However, both standards were never designed with 100.000 or more processors in mind. The big challenge is therefore to get performance out of these new systems.

3.1 The Speed Issue

Sustained performance has become a critical number for most high performance computer systems. The gap between peak performance and sustained performance has continuously widened over the last years.

Without very careful and long optimization of applications the level of sustained performance on a state of the art high performance computer has to be expected to be in the range of 5-10% only. Even for very carefully tuned applications a sustained performance of typically 10-20% can be expected. It goes without saying that some applications show much higher sustained performance. Experts are, however, not impressed if the sustained performance is at a level of 97% of peak performance. In such cases it is immediately clear that simplifications to the code had to be made to achieve such results. To adjust the application to the system by simplifying the models in order to get more performance is certainly not the way to go.

The main road blocks to speed are memory speed and interconnect speed. Both cannot easily be improved. On the contrary: we have to expect the situation to become worse over the coming 3-4 years. Sustained performance will hence be a big issue for high performance computing centres and users for years to come.

3.2 The Scalability Issue

Architects have scaled high performance computing systems to hundreds of thousands of cores. We anticipate seeing even 1 million and more cores in the years to come. Programming models have not been designed for this level of massive parallelism. Neither have applications. Even though parallelism is now in use for about 20 years in high performance computing, most applications have not been rewritten substantially over the last years. Most have been parallelized to make use of new hardware. Very few have been completely rewritten with a parallel approach in mind. Even those did not in the first place consider 1 million processors the target architecture of the code. The software base is hence not prepared for such a huge number of processors.

Furthermore architectural concepts have widely ignored the problem of scalability. Most architectures scale with respect to peak performance, Scalability for sustained performance is much more difficult to achieve. The worst problem, however, is I/O. None of the systems that take a lead worldwide seems to be able to guarantee an I/O bandwidth that compares well to the required I/O of standard applications. Scalability at all levels is not an issue as long as the TOP500 list [17] mainly honours peak performance.

4 Solution Approaches

There is currently no silver bullet for high performance computing [7]. Among a variety of approaches there is none that would solve all user problems. The situation is such that for most applications there is an architecture that is well suited for the problem. However, we see a bias towards particle physics applications. On the one hand, this is what the big labs in the US need to simulate on their classified systems. On the other hand, most system architectures are well suited for the low communication requirements of these types of applications.

For the main programming problem a number o solutions have been proposed. The perhaps most interesting one is to rethink MPI. The idea is to come up with a light weight MPI library and with special programming support for very large scale systems. This may help to reduce software latencies on the one hand. On the other hand it may help to ease programming of very large scale systems.

New programming approaches have been developed over the last years. The most interesting one is partitioned global address space (PGAS) languages as a supplement for the complete system but also for the parallelization on multi-core chips. UPC [18] and Co-Array Fortran [10] are of this type. Co-Array Fortran is proposed for the next Fortran standard. The role of new languages like Chapel [9], X-10 [15] and Fortress [17] is not yet fully clear. The portability of these languages to other platforms will be important.

5 Findings

High performance computing is breaking new barriers every 6 months. We have passed through the 1 PFLOP/s barrier easily in June 2008 and have seen more than 1 PFLOP/s for a real application in November 2008. From a political point of view high performance computing is in good shape. A lot of money in spent and a lot of records are broken. National security on the one hand and scientific breakthroughs on the other hand seem to be guaranteed. Nevertheless, something is wrong with the market. If we look more closely at the analysis above we find a number of difficulties on which we will have to work in the future.

5.1 Science and Industry Gap

There is an ever widening gap between scientific simulation and industrial simulation. When clusters were the standard approach technology used in science and industry was the same. Even though this was the case, independent software vendors did not push the limits of big clusters. An investigation by the International Data Corporation (IDC) shows that of 78 ISV codes only five use more than 128 processors for a single job [18]. Most industrial users still prefer small clusters or use larger clusters mainly in throughput mode.

With new very massively parallel systems coming up the situation for industrial simulation is getting even more difficult. New levels of performance can only be achieved if parallelism in the code is driven to at least 10.000 cores. Most commercial codes are not prepared for that. New codes require big efforts in programming and it is unlikely that new codes are able to replace old ones. Most companies - if they do not use in-house codes - have adapted their design and development processes to existing software packages. Some companies have gone as far as to define a small number or even only one software package as a standard tool. It is extremely unlikely that they will switch to new software.

The widening gap between industrial applications and scientific applications is a danger for all high performance computing centres that are not covered by the "national security umbrella". Most of these centres have to justify investments by proofing that the high performance computing systems have created some benefit for the country. Although climate research has become popular over the last years and although particle physics has made some achievements this will not be enough to justify the next wave of investments.

5.2 *Wag the Tail?*

The gap described above is a result of a new phenomenon. For decades the requirements of many applications have driven high performance computing. Today hardware architectures seem to drive applications rather than the other way round. Simulations in fields like computational fluid dynamics or structural mechanics are disappearing from the scene.

This is not because we have solved all the problems in these fields, but because it is getting increasingly difficult for these applications to get solutions from the new systems. Gradually we see a replacement of traditional applications by particle physics and astrophysics on the one hand and by molecular dynamics and bioinformatics on the other hand. There is a lot of good in simulating physical and chemical problems and there is certainly a lot of benefit in bioinformatics. The key issue here is whether a community is driven by hardware or is driving hardware.

Currently we believe that hardware is driving the community. One key problem certainly is the TOP500 list [1] which severely distorts the picture and pushes centres into buying systems which are theoretically fast but practically slow. Even users start thinking mainly about peak performance for Linpack rather than about sustained performance of their own code.

The author was recently challenged by a user complaining about the low rank of his system on the TOP500 list [1]. Not even the hint that the system provided much higher sustained performance than any other system on the list for his special application did console the user. This certainly is an extreme case but it highlights the problem quite well. Unless we are able to put the picture right the high performance computing community will face serious problems. As of today we can speak of a misallocation of funds in the high performance computing market. History tells that misallocation leads to economic failure. For some applications high performance computing will always be important enough to justify investment. If we fail to show the benefits of high performance computing for a wide variety of applications high performance computing might see such an economic failure in the near future.

References

1. IDC, IDC HPC Breakfast Briefing. Supercomputing 2008, Austin, USA, November 17-20 (2008)
2. International Roadmap Committee, International Technology Roadmap for Semiconductors, 2007 Edition, Executive Summary `www.itrs.net` (cited September 11, 2009)
3. Kobayashi, H.: SX-9 first experience. In: 9th Teraflop Workshop, Tohoku University, Sendai, Japan, November 12-13 (2008)
4. Matsuoka, S.: The TSUBAME cluster experience a year later, and onto petascale TSUBAME 2.0. In: Cappello, F., Herault, T., Dongarra, J. (eds.) PVM/MPI 2007. Proceedings LNCS, pp. 8–9. Springer, Heidelberg (2007)

5. Moore, G.E.: Cramming more components onto integrated circuits. Electronics 38(8), 114–117 (1965)
6. Pham, D., Asano, S., Bolliger, M., Day, M.N., Hofstee, H.P., Johns, C., Kahle, J., Kameyama, A., Keaty, J., Masubuchi, Y., Riley, M., Shippy, D., Stasiak, D., Suzuoki, M., Wang, M., Warnock, J., Weitzel, S., Wendel, D., Yamazaki, T., Yazawa, K.: The design and implementation of a first-generation CELL processor. In: Proceedings of the 2005 IEEE International Solid-State Circuits Conference, San Francisco, USA, February 6-10, pp. 182–185 (2005)
7. Resch, M., Küster, U.: Investigating the impact of architectural and programming issues on sustained petaflop performance. In: Bader, D. (ed.) Petascale Computing: Algorithms and Applications. Computational Science series, pp. 485–503. Chapman & Hall / CRC Press, Taylor and Francis Group (2007)
8. Wehner, M., Oliker, L., Shalf, J.: Towards ultra-high resolution models of climate and weather. Int. J. High Perform. Comput. Appl. 22, 149–165 (2008)
9. `http://chapel.cs.washington.edu/` (cited September 11, 2009)
10. `http://www.co-array.org/` (cited September 11, 2009)
11. `http://www.lanl.gov/roadrunner/` (cited September 11, 2009)
12. `http://www.nccs.gov/computing-resources/jaguar/` (cited September 11, 2009)
13. `http://www.nsc.riken.jp/project-eng.html`
14. `www.research.ibm.com/bluegene/` (cited September 11, 2009)
15. `www.research.ibm.com/x10/` (cited September 11, 2009)
16. `http://research.sun.com/projects/plrg/Publications/GSPx-Lecture2006public.pdf` (cited September 11, 2009)
17. `www.top500.org` (cited September 11, 2009)
18. `http://upc.lbl.gov/` (cited September 11, 2009)

A Parallel Implementation of FEM for a Boundary Value Problem for the Shallow Water Equations

Evgeniya D. Karepova and Vladimir V. Shaidurov

Abstract. In the present paper efficiency of several parallel implementations of an algorithm for the numerical solution of a boundary value problem for the shallow water equations with the use of the MPI library for the C language is studied. Theoretical estimates of acceleration for parallel algorithms are given. Numerical results on a special model grid and on a non-structured grid for the Sea of Okhotsk are presented. The calculations ware performed with the MVS1000 cluster of ICM SB RAS and the SKIF Cyberia cluster of the Tomsk state university.

1 Introduction

Shallow water models adequately describe a large class of natural phenomena such as large-scale free surface waves arising in seas and oceans, tsunamis, flood currents, surface and channel run-offs, gravitation oscillation of the ocean surface [7, 1]. In the papers [1, 4, 5] the numerical modelling of free surface waves in large water areas on the basis of the shallow water equations (SWE) is considered taking into account the Earth's sphericity and the Coriolis acceleration. In [1] for the differential formulation of the problem useful a priori estimates which provide stability as well as existence and uniqueness of a solution of the problem are obtained. In [4, 5] for this problem the finite elements method (FEM) is constructed and corresponding a priori estimates are obtained. Besides, numerical results on special model grids and on non-structured grids for water areas of the Sea of Okhotsk and the World Ocean are presented.

In the present paper efficiency of two parallel implementations of the numerical solution of a boundary value problem for SWE with the use of the MPI library for the C language is studied. This approaches differ in a way of decomposition of

Evgeniya D. Karepova · Vladimir V. Shaidurov
Institute of Computational Modeling SB RAS,
Akademgorodok, 660036 Krasnoyarsk, Russia
e-mail: jane@icm.krasn.ru, shidurov@icm.krasn.ru

E. Krause et al. (Eds.): Computational Sci., & High Performance Computing IV, NNFM 115, pp. 29–42.
springerlink.com

a computational domain. The first implementation is based on the decomposition with shadow lines (with overlap). The width of this overlap is defined by a stencil of a discrete problem. In the second implementation, the procedure of calculations in the FEM on individual triangles provides the decomposition without shadow lines.

Numerical experiments were performed on a special model rectangular grid and on non-structured grids for the Sea of Okhotsk. The results of calculations were compared in the number of processes, a way of implementation of communication (blocking and nonblocking send-receive operations), and a way of a decomposition of a computational domain.

2 The Modeling of Surface Waves by FEM

Let (r, λ, θ) be standard spherical coordinates with origin at the center of terrestrial globe. Further we will use the geographic latitude $\varphi = \theta + \pi/2$ instead of angle θ thus $0 \leq \varphi \leq \pi$. We denote geographic longitude by λ and so $0 \leq \lambda \leq 2\pi$. Let us suppose that $r = R_E$ everywhere, where R_E is the radius of the Earth which is assumed to be constant.

We consider the long-waves propagation problem in the following form. Let Ω be a given domain on the sphere with a boundary $\Gamma{=}\Gamma_1 \cup \Gamma_2$, where Γ_1 is a part of boundary passing along the coastline and $\Gamma_2 = \Gamma \setminus \Gamma_1$ is a part of boundary rounded the area of water. We denote the characteristic functions of corresponding parts of the boundary by m_1 and m_2. For simplicity sake we assume that the points $\varphi = 0$ and $\varphi = \pi$ (poles) do not belong to Ω. We write in $\Omega_{R_E} \times (0,T)$ the impulse balance equations and the equation of continuity with respect to the unknown functions $u = u(t,\lambda,\varphi)$, $v = v(t,\lambda,\varphi)$ and $\xi = \xi(t,\lambda,\varphi)$ [1]

$$
\begin{aligned}
\frac{\partial u}{\partial t} &= lv + mg\,\frac{\partial \xi}{\partial \lambda} - R_f u + f_1, \\
\frac{\partial v}{\partial t} &= -lu + ng\,\frac{\partial \xi}{\partial \varphi} - R_f v + f_2, \\
\frac{\partial \xi}{\partial t} &= m\left(\frac{\partial}{\partial \lambda}(Hu) + \frac{\partial}{\partial \varphi}\left(\frac{n}{m}Hv\right)\right) + f_3,
\end{aligned}
\tag{1}
$$

where $u = u(t,\lambda,\varphi)$ and $v = v(t,\lambda,\varphi)$ are longitude and latitude components of the velocity vector $\mathbf{U}$, $\xi = \xi(t,\lambda,\varphi)$ is a fluctuation of a free surface from the non-perturbed level, $H(\lambda,\varphi) > 0$ is the depth of a water body at the point (λ,φ), the function $R_f = r_*|\mathbf{U}|/H$ takes into account the force of friction on the floor, r_* is the friction coefficient, $l = -2\omega\cos\varphi$ is the Coriolis parameter, $m = 1/(R_E \sin\varphi)$, $n = 1/R_E$, g is the acceleration of gravity; $f_1 = f_1(t,\lambda,\varphi)$, $f_2 = f_2(t,\lambda,\varphi)$, and $f_3 = f_3(t,\lambda,\varphi)$ are given functions of the external forces.

We consider the boundary conditions in the following form [1]:

$$
HU_n + \beta m_2 \sqrt{gH}\xi = m_2\sqrt{gH}d \quad .. \quad \Gamma \times (0,T), \tag{2}
$$

where $U_n = \mathbf{U} \cdot \mathbf{n}$, $\mathbf{n} = (n_1, \frac{n}{m} n_2)$ is a vector of an outer normal to the boundary; $0 < \beta < 1$ is a given parameter of the problem, $d = d(t, \lambda, \varphi)$ is a boundary function, defined on the boundary Γ_2.

We also specify the initial conditions

$$u(0,\lambda,\varphi) = u_0(\lambda,\varphi),\ v(0,\lambda,\varphi) = v_0(\lambda,\varphi),\ \xi(0,\lambda,\varphi) = \xi_0(\lambda,\varphi). \tag{3}$$

For the time discretization we subdivide the time interval $[0,T]$ into K subintervals: $0 = t_0 < t_1 < \cdots < t_K = T$ with step $\tau = T/K$. Approximating the derivatives with respect to time with the help of left differences, we consider the system (1)–(2) on the time interval (t_k, t_{k+1})

$$\begin{aligned}
&\left(\frac{1}{\tau} + R_f\right) u - lv - mg\,\frac{\partial \xi}{\partial \lambda} = f_1 + \frac{1}{\tau} u^k \quad \text{in} \quad \Omega,\\
&\left(\frac{1}{\tau} + R_f\right) v + lu - ng\,\frac{\partial \xi}{\partial \varphi} = f_2 + \frac{1}{\tau} v^k \quad \text{in} \quad \Omega,\\
&\frac{1}{\tau}\xi - m\left(\frac{\partial}{\partial \lambda}(Hu) + \frac{\partial}{\partial \varphi}\left(\frac{n}{m} Hv\right)\right) = f_3 + \frac{1}{\tau}\xi^k \quad \text{in} \quad \Omega,
\end{aligned} \tag{4}$$

$$HU_n + \beta m_2 \sqrt{gH}\xi = m_2 \sqrt{gH} d \quad \text{on} \quad \Gamma, \quad k = 0,1,\dots,K-1, \tag{5}$$

where $f(t_k, \lambda, \phi) = f^k$, $f(t_{k+1}, \lambda, \phi) = f^{k+1} = f$. In what follows superscript $k+1$ in difference expressions will be omitted where this could not be ambiguously interpreted. The floor friction is given in the form $R_f = r_* |\mathbf{U}^k| / H$. Notice that the boundary conditions (5) are natural for (4).

To construct the Bubnov - Galerkin scheme, we consider a consistent triangulation $\mathscr{T} = \{\omega_i\}\big|_{i=1}^{N_{el}}$ of the domain Ω, which consists of nondegenerate triangles with rectilinear sides in λ- and φ-coordinates and includes the domain Ω. Consistency means that any side of a triangle either is boundary and belongs to a single triangle or is common for two neighbouring triangles which have no common interior point. In the general case a grid may be nonstructural (Fig. 1).

Let $\bar{\Omega}_h$ be a set of nodes (i.e., of vertices of triangular elements) and their number be equal to N_{nd} and Ω_h be the set of interior nodes.

For any node $z_j \in \bar{\Omega}_h$ we introduce the basis function $\Psi_j(\lambda, \varphi)$ which is equal to one at z_j, vanishes at all other nodes of $\bar{\Omega}_h$, and is linear on each triangle. Denote the span of these functions by $H_h(\Omega_h) = \text{span}\{\Psi_j\}_{j=1}^{N_{nd}}$.

For real vector-functions $\Phi_h = (u^h, v^h, \xi^h)$, $\hat{\Phi}_h = (\hat{u}^h, \hat{v}^h, \hat{\xi}^h) \in \mathbf{H}_h(\Omega_h) \equiv (H_h(\Omega_h))^3$ we consider the discrete analogue of an inner product [1, 4, 5]

$$(\Phi_h, \hat{\Phi}_h)_h = \sum_{i=1}^{N_{el}} \frac{1}{3} S_i \sum_{j=0}^{2} R_E^2 \sin(\varphi_{ij}) \left(H_{ij}(u_{ij}\hat{u}_{ij} + v_{ij}\hat{v}_{ij}) + g\xi_{ij}\hat{\xi}_{ij}\right). \tag{6}$$

Here S_i denotes the area of the i-th triangular element whose vertices are numbered 0, 1, 2, hence, $f_{ij} = f(\lambda_{ij}, \varphi_{ij})$ is the value of the function at the j-th vertex of the i-th element.

In the terms of the inner product (6) we formulate the Bubnov - Galerkin method [4, 5] in the following form: *for a fixed instant find a vector-function* $\Phi_h = (u^h(\lambda,\varphi), v^h(\lambda,\varphi), \xi^h(\lambda,\varphi))$, *where*

$$u^h(\lambda,\varphi) = \sum_{j=1}^{N_{nd}} \alpha_j^u \Psi_j(\lambda,\varphi), \quad v^h(\lambda,\varphi) = \sum_{j=1}^{N_{nd}} \alpha_j^v \Psi_j(\lambda,\varphi), \quad \xi^h(\lambda,\varphi) = \sum_{i=1}^{N_{nd}} \alpha_j^\xi \Psi_j(\lambda,\varphi)$$

such that the identity

$$a^h(\Phi_h, \mathbf{W}_h) = f^h(\mathbf{W}_h) + b^h(d^h, \mathbf{W}_h) \tag{7}$$

holds $\forall\ \mathbf{W}_h = (w_u^h, w_v^h, w_\xi^h) \in \mathbf{H}_h$. For approximation of the integrals in bilinear $a^h(\cdot,\cdot)$ and linear $f^h(\cdot)$, $b^h(\cdot)$ forms the trapezoid formulae and its 2D analogue are used.

Enumerating the notes $\bar{\Omega}_h$ from 1 to N_{nd}, we write the problem (7) in the vector-matrix form: *for a fixed instant* t^{k+1} *find a vector* $\mathbf{V}^{k+1} = (u_1, \ldots, u_{N_{nd}}, v_1, \ldots, v_{N_{nd}}, \xi_1, \ldots, \xi_{N_{nd}})$, *satisfying the system of linear algebraic equations*

$$A^{k+1}\mathbf{V}^{k+1} = \mathbf{F}^{k+1}. \tag{8}$$

In [4] second-order convergence in the norm induced by the inner product (6) on a uniform grid is shown.

To solve the system (8) the Jacobi iterative method is used. This method is effectively paralleled and diagonal predominance for its convergence is easily provided by the choice of a time step τ. In the vector-matrix form the Jacobi method is written as follows:

$$\Phi^{(v+1)} = \Phi^{(v)} - D^{-1}\left(A\Phi^{(v)} - \mathbf{F}\right). \tag{9}$$

Here v is the number of iteration step, superscripts $(k+1)$ for a time step are omitted, in addition, the entries of the global stiffness matrix and the components of the right-hand side vector depend on time and must be recalculated at the beginning of time step.

Notes some features of the implementation caused by FEM. For the Jacobi method on the basis of FEM it is not required to store the global stiffness matrix A. In the program only the entries of local stiffness matrices are calculated, moreover, only the diagonal elements depend on time and are recalculated at each time step. The residual $A\Phi^{(v)} - \mathbf{F}$ in (9) is calculated on triangles with the use of entries of the local matrices.

The iteration process is terminated when the difference $\Phi^{(\nu)}$ for two neighbouring iteration steps becomes small in the discrete analogue of the uniform norm:

$$\max_{1 \le i \le N_{nd}} |\Phi_i^{(\nu+1)} - \Phi_i^{(\nu)}| \le \varepsilon. \tag{10}$$

3 Parallel Algorithm

FEM for the problem (7) yields the seven-point stencil, therefore in the Jacobi method the calculation of the value of $\Phi_i^{(\nu+1)}$ at the $(\nu+1)$-st iteration step requires the values at seven points $\Phi_i^{(\nu)}$ at the ν-th iteration step (Fig.2).

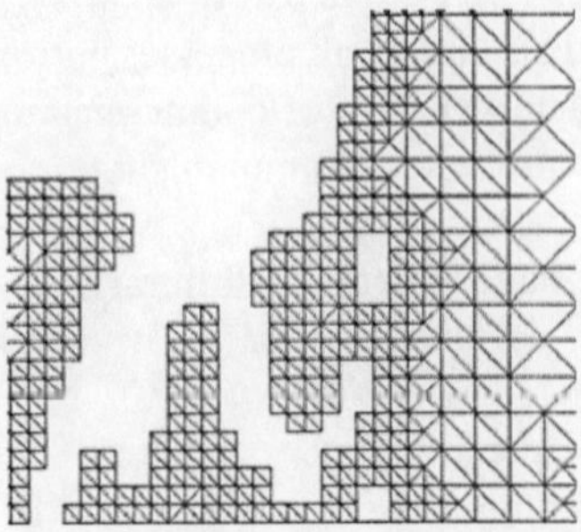

Fig. 1 A fragment of triangulation

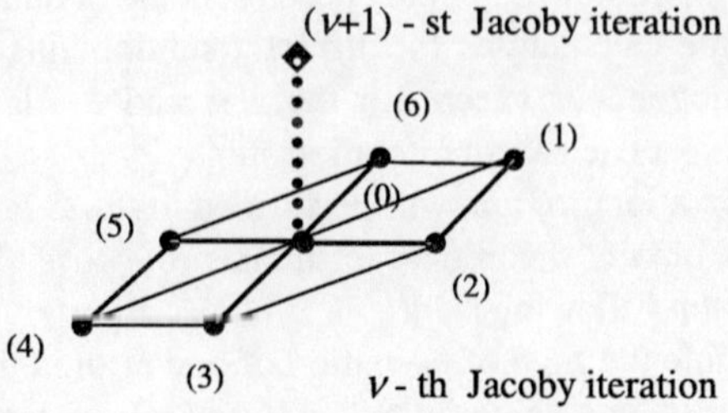

Fig. 2 Dependence in the Jacobi iteration

Using obvious parallelism with respect to data, we can subdivide an original computational domain into several subdomains which intersect each other only in the boundary of the subdivision. Calculations in each subdomain can be performed independently of each other at one iteration step. However, after each iteration step it is required to perform concordance of data at the boundaries of subdomains. At least, two versions of subdivision have place.

1. *Decomposition with shadow lines.* An original domain includes mutually overlapping subdomains, which are defined by a stencil (Fig.2). In addition, residual at boundary points of the subdomain of the i-th process is calculated in the subdomains of the neighbouring processes. Due to the seven-point stencil and consistency of a triangulation, the overlapping of subdomains consists of two layers of grid points. In the exchange of the results of calculations after each Jacobi iteration step, the additional processing is not required, however, for each process it is required to store doubling data.
2. *Decomposition without shadow lines.* An original domain is cut into subdomains which intersect each other only in boundaries of the cut. The contribution to residual at a boundary point of a subdomain is calculated on the triangles which belong to this subdomain and have this point as a vertex. After each

Jacobi iteration step, the exchange of the contributions to residual at boundary points is performed and this contributions are summed up, thus, an additional summation is required.

Notice that the decomposition with shadow lines into p processes requires to store eighteen entries of a stiffness matrix, three values of a vector of a solution for an actual iteration step and for a previous one, and three values of a right-hand side at each boundary point. Therefore this requires $27(p-1)N_{bnd} * \mathsf{SizeOfDouble}$ bytes grater than that for the decomposition without shadow lines. Here N_{bnd} denotes the number of points on a cut, SizeOfDouble denotes the number of bytes for a double precision variable. Thus, the second way of decomposition of a domain is more effective with respect to memory and is simple for coding, however, an additional summation at each Jacobi iteration step is assumed. It has an evident advantage for nonstructured grids, where boundary points are not a successive part of an array.

In the framework of the chosen scheme of data distribution, all processes perform the same calculations for different subdomains. The structure of communications is homogeneous except for the first and the last processes. Communications are required at each Jacobi iteration step.

A parallel program was performed in the C language with the MPI library.

We obtain theoretical estimates of possible acceleration for each parallel algorithm following [10].

Denote the time of execution of one arithmetic operation by t_{op}, and the time of transfer of one value by t_{comm}. Let N_{nd} be a total number of points of a grid of a computational domain, s be a number of operations per a grid point at each Jacobi iteration step, k be a number of time steps, ν be an average number of the Jacobi iteration steps at each time step. Then the total volume V_{calc} of calculations in the algorithm is define by the relation $V_{calc} = k\nu sN_{nd}$, and the execution time for one processor can be estimated as follows: $T_1 \sim k\nu sN_{nd}t_{op}$.

The potential acceleration of an algorithm is estimated as the ratio of the execution time T_1 with one processor to the execution time T_p with p processors: $S_p = \dfrac{T_1}{T_p}$. We obtain theoretical estimates of acceleration for each case of decomposition of a domain taking into account the time required for communications and additional calculations wherever possible.

From our scheme of data distribution it follows that each Jacobi iteration step requires:

1. the global reduction operation for calculation of the uniform norm of the difference of two successive approximations of the Jacobi iterative method;
2. the exchange of boundary elements of a vector of a solution on cuts;
3. additional calculations caused by distributed data.

Assume that in the decomposition we succeed in uniform distribution of the total volume V_{calc} of calculations over processors. Then for acceleration we can write the following formula:

$$S_p = \frac{T_1}{T_1/p + T_{over} + T_{comm}}. \tag{11}$$

Here T_1 is the execution time with one processor, T_{over} is the time for additional calculations related to decomposition, T_{comm} is the time for communications.

Let g be the number of additional calculations related to decomposition which are required for one boundary point of a subdomain. The decomposition with shadow lines does not require direct additional calculations, i.e., $g = 0$. In the decomposition without shadow lines each process calculating residual at a boundary point of a subdomain "processes" its part of triangles. In this case after communication the additional summation of contribution to residual calculated with neighbouring processors is required. At each point residual is calculated for three components of the vector (u, v, ξ), hence, $g = 3$. As a result, the volume of additional calculations is estimated as follows:

$$V_{over} \sim 2k\nu g N_{bnd}(p-1),$$

where

$$g = \begin{cases} 0, & \text{for the decomposition with shadow lines,} \\ 3, & \text{for the decomposition without shadow lines.} \end{cases}$$

Processes perform summation asynchronously and independently therefore the time for additional calculations at N_{bnd} boundary points of a subdomain can be estimated as follows:

$$T_{over} \sim 2k\nu g N_{bnd} t_{op}.$$

For calculation of a termination criterion of an iterative process, at each iteration step it is required to calculate a global maximum over all processes. Assume that implementation of the global reduction operation in MPI is performed according to the optimal dichotomy algorithm that provides the execution time

$$T^1_{comm} = k\nu(t_{op} + t_{comm}) \log_2 p$$

for one reduction operation.

Assume that for each boundary point the passing of m values (in our case $m = 3$) is required. The volume of data passing by each process to a neighbouring one can be estimated in the following way:

$$V_{comm} \sim k\nu m N_{bnd}.$$

When using the MPI library, two radically different procedures for implementation of point-to-point communications (with the use of blocking or nonblocking operations of data exchange) can be applied. We estimate the time required for exchange in each case.

Blocking communications. In Fig. 3 a scheme of data exchange implementation on the chain of processes with the help of MPI_Sendrecv(...) send-receive operations is shown. This scheme shows the influence of blocking operations on performance in the case of eight processes. In Fig. 3 sends to right and left neighbours are denoted by Snd_R and Snd_L, respectively, and receives from right and left neighbours are denoted by Rcv_R and Rcv_L, respectively. Idle time is shown by grey. First all processes except for the last one send data to their right neighbours and wait receiving data from them. Then all processes except for the first one send data to their left neighbours and wait receiving data from them. Since at the first stage the last process does not send data to the right, it successfully performs data exchange with the left (sixth) process, whereas the processes from zeroth to fifth stand idle as long as their right neighbours receive data. At the second stage the sixth process performs data exchange with the fifth one and the processes from zeroth to fourth stand idle. Thus, all exchanges are resolved on chain. All exchanges are accomplished per $p+1$ stages.

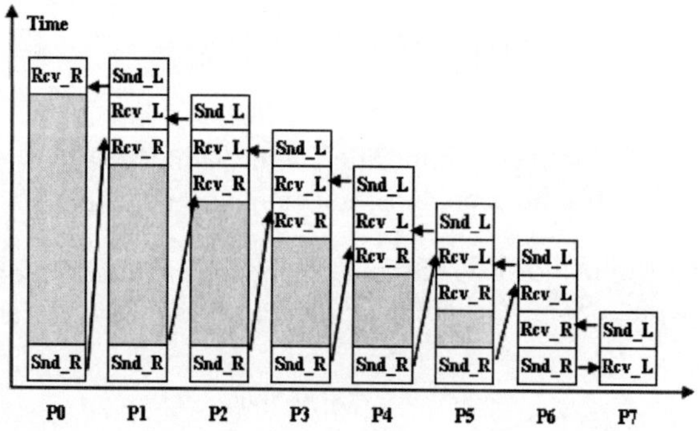

Fig. 3 The influence of blocking communications

In the general case the exchange with blocking communications requires $2(p-1)$ send operations of V_{comm} data units, where p is the number of used processes. Then $T^2_{comm} = 2(p-1)kvmN_{bnd}t_{comm}$ is total time for exchanges in blocking mode.

Taking into account (11) and the above estimates, acceleration with blocking communications is determined by the following relation:

$$S_p^{bl} = \frac{1}{\dfrac{1}{p} + 2\dfrac{g}{s}R + \dfrac{\log_2 p}{sN_{nd}}(1+\varkappa) + 2(p-1)\dfrac{m}{s}R\varkappa}. \tag{12}$$

Here R denotes the ratio of the number of boundary points in a subdomain to the total number of points of a computational domain, i.e., $R = N_{bnd}/N_{nd}$, and $\varkappa$ denotes the ratio of the time required to send one value to the time of execution of one arithmetic operation: $\varkappa = t_{comm}/t_{op}$.

Nonblocking communications. In the general case, the time required for nonblocking communications does not depend on the number of processes taking part in exchanges: $T^2_{comm} = 2V_{comm}t_{comm} = 2kvmN_{bnd}t_{comm}$. The estimate (11) in the case of nonblocking communications takes the form

$$S_p^{unbl} = \frac{1}{\frac{1}{p} + 2\frac{g}{s}R + \frac{\log_2 p}{sN_{nd}}(1+\varkappa) + 2\frac{m}{s}R\varkappa}. \tag{13}$$

Concerning the estimates (12) – (13), we notice the following. First, in decomposition without shadow lines T_{over} is small in comparison with the other terms of the denominator and does not depend on the number of processes. Taking into account saving memory and easy implementation on nonstructured grids, this provides an advantage of this type of decomposition. Second, nonblocking communications even in the absence of simultaneous calculations and communications are preferable than blocking ones. However, the theoretical estimates show that dependence on a type of communications and a way of a decomposition is not significant.

4 Numerical Experiments

In the numerical study of acceleration of the parallel algorithm the following problem is considered. Let Ω be a square on a sphere: $\Omega = [0, \pi/10] \times [\pi/2, \pi/2 + \pi/10]$. The boundaries of Ω are assumed to pass along the coastline. In Ω the problem with the known exact solution [4] is considered. In the computational domain two uniform square grids involving 401×401 and 801×801 points of corresponding consistent triangulation are constructed. In the numerical experiments 1000 time steps were performed.

The computational experiments have been performed with two high performance computer systems.

Firstly, computations were performed with the 99-processor cluster of ICM SB RAS. The MVS-1000/ICM cluster (developed in ICM SB RAS, [3]) involves 27 AMD Athlon64/3500+/1Gb one-processor one-core computational nodes; 12 AMD Athlon64 X2 Dual Core/4800+/2Gb one-processor two-core computational nodes; 12 2X Dual-Core AMD Opteron Processor 2216/4Gb two-processor two-core computational nodes. Athlon64/3500+/1Gb is a manage node, access server, file server with 400 Gb disk space with the Gentoo Linux operating system. A FastEthernet (100 Mbit/sec) is a managing network, a GigaEthernet (1000 Mbit/sec) is an interconnect. The cluster is heterogeneous therefore runtime of a program was determined as an average of several tens of calculations.

Secondly, we used the SKIF Cyberia cluster of the Tomsk state university with 283 IntelXeon 5150 2,66 GHz two-processor two-core computational nodes, with 1136 Gb main memory and 22,56 Tb disk space, a file server is of 10 Tb. A Gigabit Ethernet is a managing network, an InfiniBand is an interconnect.

The computations doubtlessly show advantage of the homogeneous architecture of a high performance computational system.

Fig. 4 shows the dependence of acceleration of computations on the number of processes being used for model grids obtained with the ICM SB RAN cluster. The graphs of accelerations for decomposition with shadow lines and blocking communications essentially coincide with those for decomposition with shadow lines and non-blocking communications, therefore they are not shown.

Analysis of numerical results shows that for a relatively small number of processes ($P \leq 12$) the value of acceleration remains the same for all considered versions, besides, moderate advantage of decomposition without shadow lines can be noticed.

So far, the possibility of combination of computations and communications is not used in our implementation, therefore influence of blocking and non-blocking passings has to be minimal. Besides, on the 401×401 grid blocking communications for decompositions without shadow lines do not provide improving acceleration starting with 12 processes. In the remaining cases on the 401×401 grid, increase of acceleration is observed up to the use of 32 processes. The value of efficiency greater than one with 18-32 processes is likely due to a cache hit. Unstable increase of acceleration on the 801×801 grid with the use of more then 12 processes is related to prevalence of communication time over computation time, the use of more slow one-core nodes, and architecture features of a heterogeneous cluster.

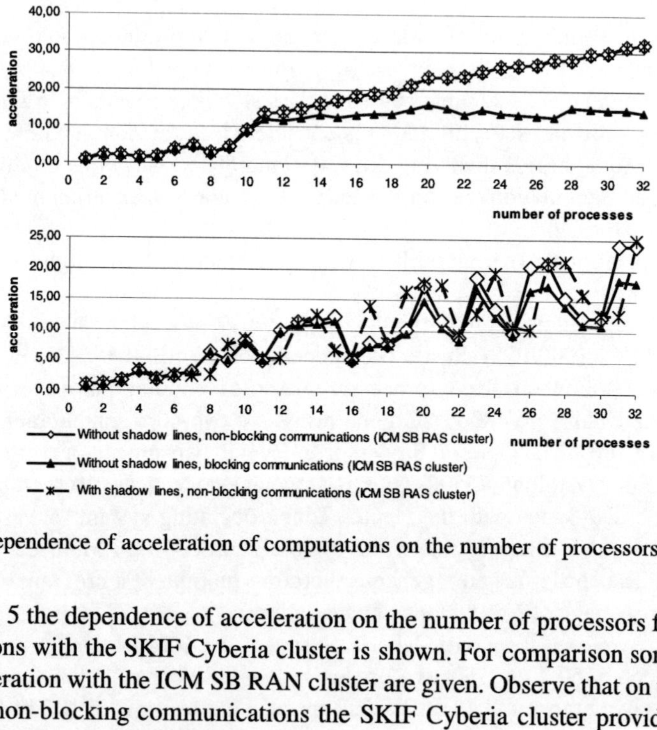

Fig. 4 Dependence of acceleration of computations on the number of processors available

In Fig. 5 the dependence of acceleration on the number of processors for similar calculations with the SKIF Cyberia cluster is shown. For comparison some graphs for acceleration with the ICM SB RAN cluster are given. Observe that on the coarse grid for non-blocking communications the SKIF Cyberia cluster provides higher acceleration than the ICM SB RAN cluster. Besides, in this case blocking communications provide stable acceleration. This fact is related to homogeneous architecture of the SKIF Cyberia cluster and properties of its network. Calculations on the fine grid performed with the SKIF Cyberia cluster show the classic linear acceleration with increasing the number of processes and efficiency about one (efficiency of calculations for 32 processes ≈ 0.85).

As a result, nonstable acceleration in the computations performed with the ICM SB RAN cluster is caused by its heterogeneous architecture.

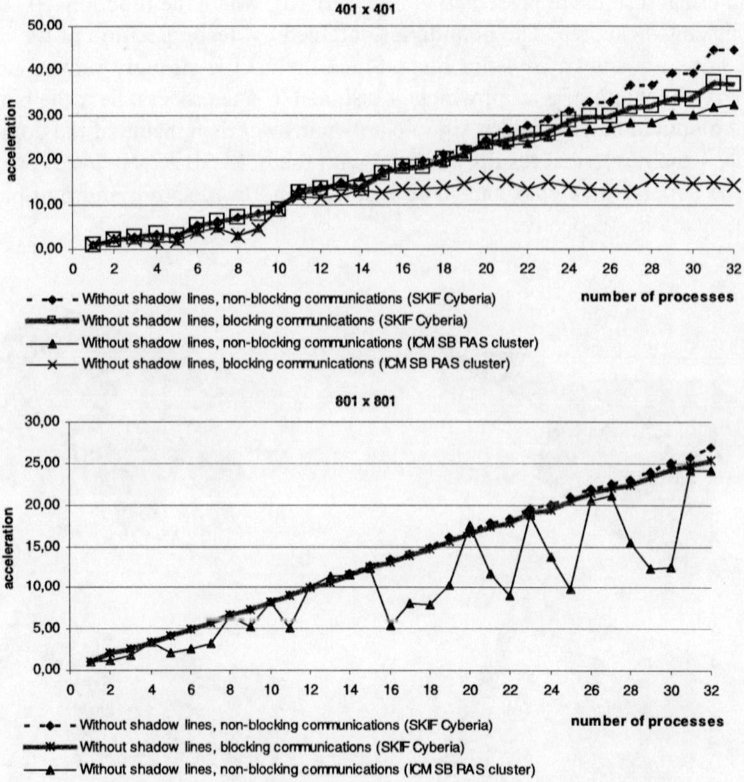

Fig. 5 Dependence of acceleration of computations on the number of processors available and architecture of a cluster

Test calculations for the Sea of Okhotsk were performed on the grids constructed by S.F. Pyataev and I.V. Kireev on the base of the ETOPO2 bathymetric data base [6, 9].

The considered grid involves $N_{nd} = 43768$ nodes and $N_{el} = 78929$ triangles (Fig. 6 a).

The numerical experiment corresponds to initial data with a local peak described by the Gauss function

$$\xi(0,\lambda,\varphi) = A\exp\left(-(\lambda-\lambda_0)^2/(2D)^2 - (\varphi-\varphi_0)^2/(2D)^2\right).$$

In the calculations we take $A = 10$, $\lambda_0 = 149.1°$ of eastern longitude and $\varphi_0 = 53.1°$ of northern latitude, $D = 0.005$, the friction coefficient is $r_* = 0.0026$. For velocities we put $u(0,\lambda,\varphi) = v(0,\lambda,\varphi) = 0$.

In the grid about 6.4 percent of boundary sections (sides of triangular elements) belong to a water area. On this sections $m_2 = 1$ and in the boundary condition (2) we put $\beta = 1$ and $d(t,\lambda,\varphi) \equiv 0$.

Numerical results are presented in Fig. 6 b) - d), where the function $\xi(t,\lambda,\varphi)$ at some instants is shown. The boundaries obtained by decomposition of the domain into 8 parts are shown by solid lines. Since the grid is strongly refined near the boundary, the width of a strip, which is assigned to one process, near the boundary of the computational domain is significantly narrower than inside of the domain. In the Tab. 1 the numerical results obtained with the ICM SB RAN cluster are given. Here the superscribes «*bl*» and «*unbl*» correspond to blocking and non-blocking communications, respectively.

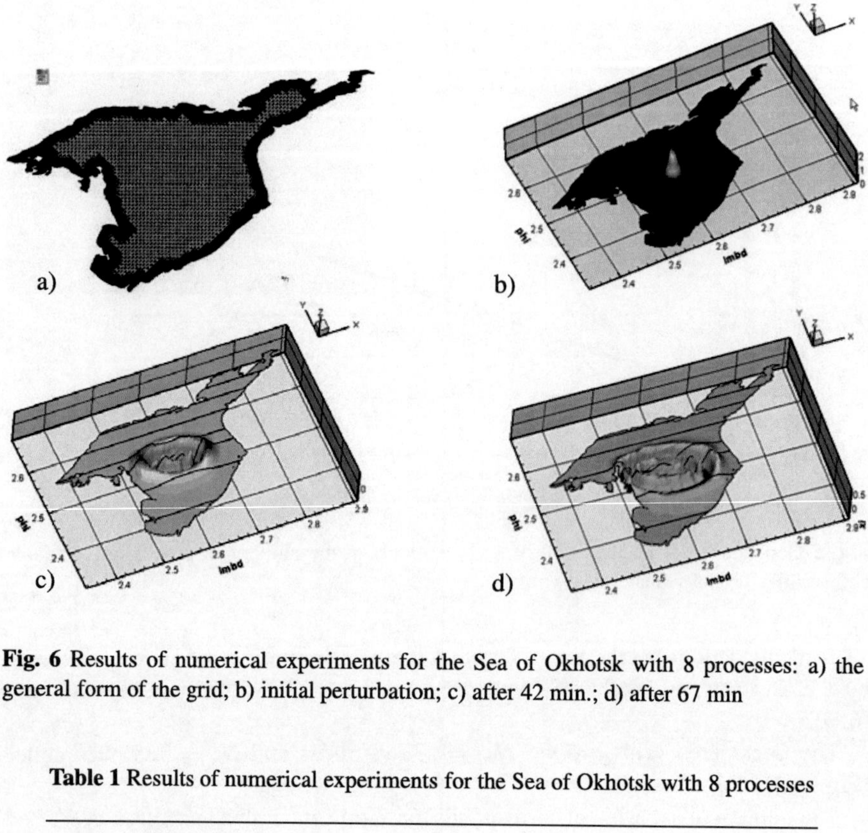

Fig. 6 Results of numerical experiments for the Sea of Okhotsk with 8 processes: a) the general form of the grid; b) initial perturbation; c) after 42 min.; d) after 67 min

Table 1 Results of numerical experiments for the Sea of Okhotsk with 8 processes

number of processes	1	2	4	8
Time t^{bl}	6,8729	4,0690	2,3030	1,6324
Acceleration S^{bl}	1.0000	1,6891	2,9843	4,2103
Efficiency E^{bl}	1.0000	0,8446	0,7461	0,5263
Time t^{unbl}	7,0290	4,1209	2,3389	1,5304
Acceleration S^{unbl}	1.0000	1,7057	3,0053	4,5928
Efficiency E^{unbl}	1.0000	0,8528	0,7513	0,5741

We see from the table that acceleration increases proportionally to the number of processors available. In addition, runtime with the use of nonblocking communications is somewhat less than that with blocking communications.

It should be noticed that the decomposition with shadow lines was not performed because it is a tedious procedure.

5 Conclusion

For the numerical solution of a boundary value problem for the shallow water equations, the technique of development of scalable parallel algorithms for cluster high performance computational systems with the MPI library is considered.

Since for discretization of the problem the FEM with calculations on triangular elements was used, two natural approaches to the decomposition of a computational domain (without shadow lines and with shadow lines) are considered.

The theoretical estimates show that the algorithm has a good structure from the parallelizing point of view and provides acceleration close to linear depending on the number of processors being used.

Numerical experiments show that the nonblocking communication mode, which is allowed in the algorithm, is undoubtedly more effective. Further improvement of efficiency of the algorithm, for example, due to combination of calculations and nonblocking communications, is in project.

It should be noticed that an advantage of the algorithm is in simple organization of decomposition without shadow lines on nonstructured triangulations for actual water basins.

The calculations were performed with two high performance computational systems of different architecture and the advantage of homogeneous structure (of the SKIF Cyberia cluster) over heterogeneous one was demonstrated. In addition, unstable acceleration for heterogeneous structure of a cluster, which is inherent neither in the algorithm nor in the implementation, is shown.

Acknowledgements. The work is supported by Russian Foundation of Fundamental Researches (grant 08-01-00621-a) and by State Contract No. 02.740.11.0621 of Federal Program "Scientific and education personnel of innovative Russia".

The authors are grateful to Prof. A.V. Starchenko and to the staff of the Interregional Computer Center of the Tomsk state university who presented a possibility to perform numerical experiments with the SKIF Cyberia cluster.

References

1. Agoshkov, V.I.: Inverse problems of the mathematical theory of tides: boundary-function problem. Russ. J. Numer. Anal. Math. Modelling 20(1), 1–18 (2005)
2. Balay, S.: Efficient management of parallelism in object-oriented numerical software libraries. In: Balay, S., Gropp, W.D., McInnes, L.C., et al. (eds.) Modern Software Tools in Scientific Computing, pp. 163–202. Birkhauser Press (1997)

3. Isaev, S.V., Malyshev, A.V., Shaidurov, V.V.: Development of the Krasnoyarsk center of parallel computations. Computat. Technologies, Special Issue 11, 28–33 (2006) (in Russian)
4. Kamenshchikov, L.P., Karepova, E.D., Shaidurov, V.V.: Simulation of surface waves in basins by the finite element method. Russian J. Numer. Anal. Math. Modelling 21(4), 305–320 (2006)
5. Kamenshchikov, L.P., Karepova, E.D., Shaidurov, V.V.: Numerical solution of the boundary value problem for shallow water equations for modelling surface waves in world ocean by finite elements methods. In: Proceedings of Fourth International Conference FDM:T&A 2006, Finite Difference Methods: Theory and Applications, Rousse, Bulgaria, pp. 227–233 (2007)
6. Kamenshchikov, L.P., Karepova, E.D., Pyataev, S.F., Shaidurov, V.V.: The modeling of gravitation vawes in the World Ocean with the finite element method with parallelizing. In: Proceeding of the Sixth school 'Distributed and cluster calculations', pp. 52–64. ICM SB RAS, Krasnoyarsk (2006) (in Russian)
7. Marchuk, G.I.: Dynamics of Ocean Tides. Gidrometizdat, Leningrad (1983) (in Russian)
8. McBryan, O.A.: An overwiev of message passing environments. Parallel Computing 20, 417–441 (1994)
9. National Geophysical Data Center, `http://www.ngdc.noaa.gov/ngdc.html`
10. Ortega, J.M.: Introduction to parallel and vector solution of linear systems. Springer, New York (1988)

Parallel FVTD for Solving Maxwell Equations in Dielectric-Metal Composite Media

Yurii I. Shokin, Lyudmila Yu. Prokopyeva, Mikhail P. Fedoruk, Alexander S. Lebedev, and Dmitry L. Chubarov

Abstract. Parallel implementation of finite volume time domain method (FVTD) described in this paper is being developed for numerical simulations in the optics of nano-structured metal-dielectric composites. The problem is complicated by the dispersive behaviour of the dielectric function in metals, which is introduced into the system of equations through the time-convoluted Debye model. Another difficulty is the extreme contrast between the dielectric functions of host media and metallic elements. Thus, unstructured grids are used to fit the boundaries of elementary materials within the composite media, and also a special technique for space gradients calculation is employed. The parallel code is developed using the MPI library for interprocessor communications and was evaluated on analytical models of plane wave scattering from a metal cylinder with uniform or gradient optical properties. In this paper we show another numerical simulation that was performed for imaging with cylindrical hyperlens (imaging device that allows to overcome the diffraction limit using anisotropic metamaterials). Future study will address three-dimensional FVTD parallel code and simulations of light scattering from the core-shell systems, where parallel computing and accurate modelling of metal coat will be of great importance.

1 Introduction

Fundamental theoretical and experimental research in designing new nano-structured optical metal-dielectric composites (plasmonic metamaterials) with nontrivial properties is a very hot topic with many important breakthroughs in the past couple of years [5],[6]. Since expensive fabrication procedures limit a possible volume of direct experimental studies, an accurate numerical simulation is of huge

Yurii I. Shokin · Lyudmila Yu. Prokopyeva · Mikhail P. Fedoruk ·
Alexander S. Lebedev · Dmitry L. Chubarov
Institute of Computational Technologies SB RAS,
Lavrentiev Ave. 6, 630090 Novosibirsk, Russia
e-mail: shokin@ict.nsc.ru

E. Krause et al. (Eds.): Computational Sci., & High Performance Computing IV, NNFM 115, pp. 43–49.
springerlink.com

importance. The complexity of the electromagnetic and multi-physics processes inside even the simplest plasmonic metamaterials is a challenging task for known numerical schemes. The problem is complicated by the dispersive behaviour of the dielectric permittivity in metals, large aspect ratios of nanoscale plasmonic elements, considerable difference between the scale of nano-structured elements and the dimensions of macroscopic objects often combined with metal-dielectric composites. Effective and universal time-domain numerical methods giving accurate description of the light propagation in such heterogeneous and non-linear media are in great demand.

Namely, core advantages of the proposed finite volume approach are (i) ability to handle complex shapes with unstructured non-uniform co-located grids (ii) the conservation of fluxes, which plays a big role at the interface of composite media (iii) explicit algorithm that allow to perform computations in linear time with respect to number of finite volumes, which are normally of triangular (2D) or tetrahedral (3D) shape ([1]-[4], [8], [10]-[12]).

In this extended abstract we give a brief description of the 3D FVTD method with details pertinent to plasmonic nano-structures. Then in the second section we present the speedup of the parallel implementation of FVTD method using MPI library, pictures were obtained on cluster architectures. In the last section we show a two-dimensional demo simulation of imaging with hyperlens.

Future study will address full three-dimensional FVTD parallel code and simulations of core-shell systems, that are essentially dielectric spheres coated with metal.

2 Finite Volume Time Domain Method

In this section we, first, give the basics of FVTD technique for full 3D Maxwell equations in simple dielectric media and then introduce metallic media.

The electromagnetic waves propagating in nonmagnetic dispersive optical media ($\mu = 1$, $div\mathbf{D} = 0$) are described by Maxwell equations:

$$\begin{cases} \dfrac{\partial \mathbf{D}}{\partial t} - \mathrm{rot}\mathbf{H} = 0, \mathbf{D} = \varepsilon_0 \varepsilon_r \mathbf{E}, \\ \dfrac{\partial \mathbf{B}}{\partial t} + \mathrm{rot}\mathbf{E} = 0, \mathbf{B} = \mu_0 \mathbf{H}, \\ \mathrm{div}\mathbf{D} = 0, \ \mathrm{div}\mathbf{B} = 0. \end{cases} \tag{1}$$

For a full three-dimensional model in Cartesian coordinates we rewrite Eq. 1 using matrix notation:

$$\frac{\partial \mathbf{U}}{\partial t} + \frac{\partial}{\partial x} A_1 \mathbf{U} + \frac{\partial}{\partial y} A_2 \mathbf{U} + \frac{\partial}{\partial z} A_3 \mathbf{U} = 0 \tag{2}$$

where

$$
A_1 = \begin{pmatrix} 0 & 0 & 0 & 0 & 0 & 0 \\ 0 & 0 & 0 & 0 & 0 & \frac{1}{\mu_0} \\ 0 & 0 & 0 & 0 & -\frac{1}{\mu_0} & 0 \\ 0 & 0 & 0 & 0 & 0 & 0 \\ 0 & 0 & -\frac{1}{\varepsilon_0\varepsilon_r} & 0 & 0 & 0 \\ 0 & \frac{1}{\varepsilon_0\varepsilon_r} & 0 & 0 & 0 & 0 \end{pmatrix}
A_2 = \begin{pmatrix} 0 & 0 & 0 & 0 & 0 & -\frac{1}{\mu_0} \\ 0 & 0 & 0 & 0 & 0 & 0 \\ 0 & 0 & 0 & \frac{1}{\mu_0} & 0 & 0 \\ 0 & 0 & \frac{1}{\varepsilon_0\varepsilon_r} & 0 & 0 & 0 \\ 0 & 0 & 0 & 0 & 0 & 0 \\ -\frac{1}{\varepsilon_0\varepsilon_r} & 0 & 0 & 0 & 0 & 0 \end{pmatrix}
$$

$$
A_3 = \begin{pmatrix} 0 & 0 & 0 & 0 & \frac{1}{\mu_0} & 0 \\ 0 & 0 & 0 & -\frac{1}{\mu_0} & 0 & 0 \\ 0 & 0 & 0 & 0 & 0 & 0 \\ 0 & -\frac{1}{\varepsilon_0\varepsilon_r} & 0 & 0 & 0 & 0 \\ \frac{1}{\varepsilon_0\varepsilon_r} & 0 & 0 & 0 & 0 & 0 \\ 0 & 0 & 0 & 0 & 0 & 0 \end{pmatrix}
\mathbf{U} = \begin{pmatrix} D_x \\ D_y \\ D_z \\ B_x \\ B_y \\ B_z \end{pmatrix} \qquad (3)
$$

For discretization of the equations, the computational domain is first covered with a tetrahedral mesh (although for some applications prismatic meshes are preferred) whose cell edges are aligned, if necessary, with the lines of jumps in physical properties of the media. After that, integrating Eq.(2) over Δ_i and transforming the integral of the spacial derivatives by the Gauss—Ostrogradsky formula, we obtain the integral equation:

$$
\frac{\partial}{\partial t} \iiint_{\Delta_i} \mathbf{U} dV + \sum_{k=1}^{4} \oiint_{S_i^k} A\mathbf{U} dS = 0, \qquad (4)
$$

where S_i^k are facets of volume Δ_i ($k = \overline{1,4}$), $\mathbf{n} = (n_1, n_2, n_3)$ is the outward normal of S_i^k, $A = n_1 A_1 + n_2 A_2 + n_3 A_3$.

Finally we get numerical scheme that approximates integral Eq. 4 with second order of accuracy for continuous media and that is shown to practically keep this order for composite media in [8]:

$$
v_i \frac{\mathbf{U}_i^{n+1} - \mathbf{U}_i^n}{\tau} + \sum_{k=1}^{4} s_{i,k} \mathbf{F}_{i,k}^{n+1/2} = 0 \qquad (5)
$$

where $\mathbf{U}_i^n$ is EM field at the barycenter $\mathbf{x}_i^b$ of a volume Δ_i for $t_n = n\tau$, τ is time step; $\mathbf{F}_{i,k}$ is field flux AU in the center of the k-th facet of the i-th volume; $s_{i,k}$ is the area of a facet; v_i is volume of the i-th tetrahedron.

We omit here further details of flux computation procedure and needed space gradients computation algorithm, these details could be found in [8],[11]. Briefly, the flux $\mathbf{F}_{i,k}$ is calculated by solving the one-dimensional (with respect to the normal to the kth edge of the ith volume) Riemann problem. Gradient computation on non-structured grid uses specific multiple-stage procedure in case of non-smoothly varying permittivity ε and has been proved to be fairly robust [8].

Metallic elements are introduced to the model through the frequency dependent permittivity using Drude model:

$$\varepsilon(\omega) = \varepsilon_\infty - \frac{\omega_p^2}{\omega(\omega + i\Gamma)}, \tag{6}$$

with ε_∞, Γ, and ω_p being the permittivity at infinite frequencies, the collision frequency, and plasma frequency respectively. Alternatively, a more general Debye model can be employed

$$\varepsilon_r(\omega) = \varepsilon_\infty + \frac{\chi}{1 - i\omega t_0} - \frac{\sigma}{i\omega\varepsilon_0}, \tag{7}$$

where $\chi = -(\omega_p t_0)^2$ is the permittivity step, $\sigma = \omega_p^2 \varepsilon_0 t_0$ is the conductivity and $t_0 = 1/\Gamma$ is the relaxation time.

$$\mathbf{D} = \varepsilon_0\varepsilon_\infty \mathbf{E}(t) + \frac{\varepsilon_0\chi}{t_0}\int_0^t e^{-\frac{t-\tau}{t_0}}\mathbf{E}(\tau)\,d\tau + \sigma\int_0^t \mathbf{E}(\tau)\,d\tau \tag{8}$$

After substituting $D = \varepsilon_0\varepsilon_\infty E$ in Eq. 1, the equations for propagation of electromagnetic waves in metal become as follows

$$\frac{\partial \mathbf{U}}{\partial t} + \frac{\partial}{\partial x}A_1\mathbf{U} + \frac{\partial}{\partial y}A_2\mathbf{U} + \frac{\partial}{\partial z}A_3\mathbf{U} = -\mathbf{I} \tag{9}$$

here

$$\mathbf{I} = \frac{\partial}{\partial t}\left(\frac{\varepsilon_0\chi}{t_0}\int_0^t e^{-\frac{t-\tau}{t_0}}\mathbf{E}(\tau)\,d\tau + \sigma\int_0^t \mathbf{E}(\tau)\,d\tau\right) = \omega_p^2\varepsilon_0\int_0^t e^{-\frac{t-\tau}{t_0}}\mathbf{E}(\tau)\,d\tau. \tag{10}$$

The same FVTD algorithm is employed for metallic elements.

3 Parallel Implementation and Speedup Tests

To parallelize the FVTD method described in previous section we perform a domain decomposition on unstructured grid. Currently the symmetry of the structure is used to guarantee the balancing between sub-domains. To simplify the parallel implementation we use "halo cells" approach. It is employed to replicate the cells participating in inter-processor communications from neighbouring sub-domains to local sub-domain. For communications between the different processors we use MPI.

The parallel speedup that was observed on cluster architectures for one- and two-dimensional FVTD codes is shown below.

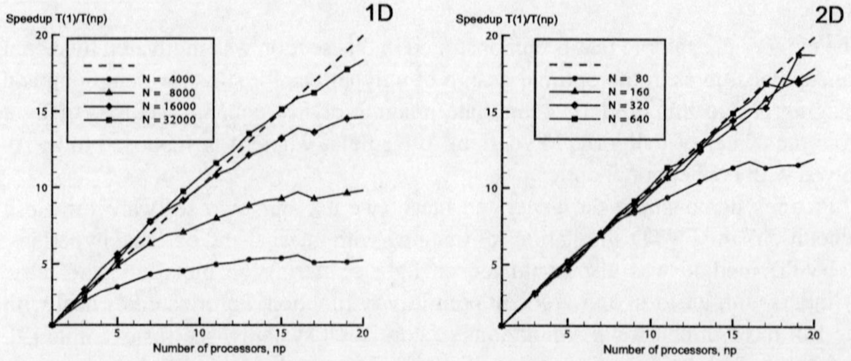

Fig. 1 Parallel speedup of FVTD codes with MPI for one- and two- dimensional models. In the left picture *N* is number of computational nodes. In the right picture computation was performed in a square, decomposition was made with horizontal stripes and *N* is number of edges per side of a square. For both pictures computations were performed on NEC cluster with Intel Xeon nodes and InfiniBand interconnect (http://www.hlrs.de/systems/platforms/nec-linux-cluster-cacau/). One MPI process per dual processor node was run

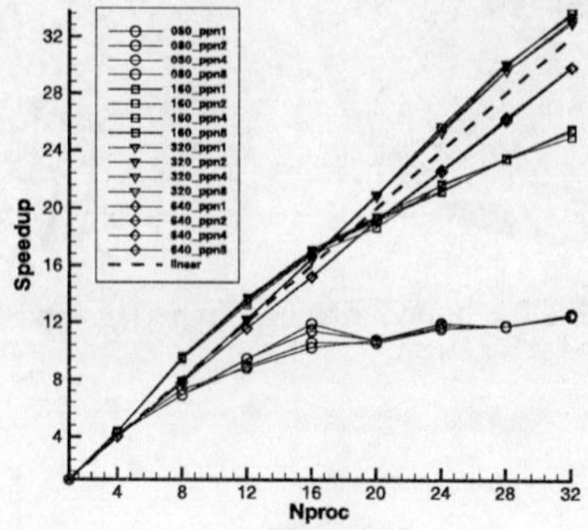

Fig. 2 Parallel speedup of FVTD codes with MPI for two-dimensional model. Square computational domain was decomposed with horizontal stripes. Number of triangles per square side: 80, 160, 320, 640. Labels *ppn*1, *ppn*2, *ppn*4, *ppn*8 correspond to number of MPI processes running on one node. Computations were performed on HP cluster with Intel quad-core processors and InfiniBand interconnect

4 Numerical Results

The numerical example that is demonstrated in this section was motivated by recent conceptual studies in the optimal design of a hyperlens for sub-wavelength optical imaging [7]. In this model we simulate imaging of five coherent sources of light from the center of cylindrical hyperlens to far field, where it is supposed to be resolved with conventional optics.

Pictures demonstrate the generated mesh (we use our own software for mesh generation) and FVTD simulations of imaging with internal and external hyperlens.

FVTD method was also evaluated on light scattering on metal and dielectric cylinders with uniform and gradient permittivity function. Future studies deal with the full three-dimensional simulations of core-shell systems, see for example [9]. For these simulations the power of parallel computations will be extremely important.

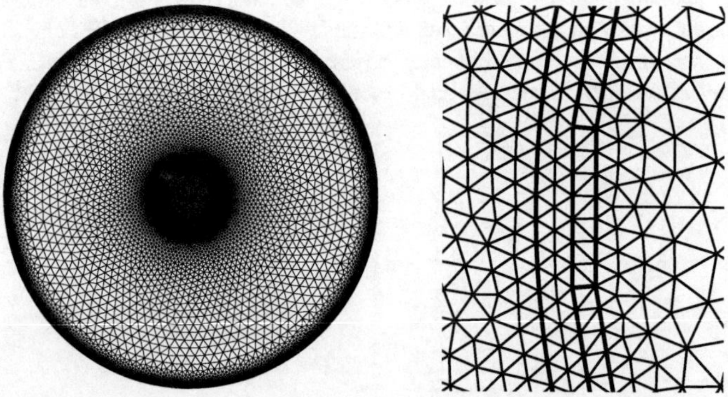

Fig. 3 Generated mesh, that fits the boundaries of a hyperlens and resolves the geometry of light sources

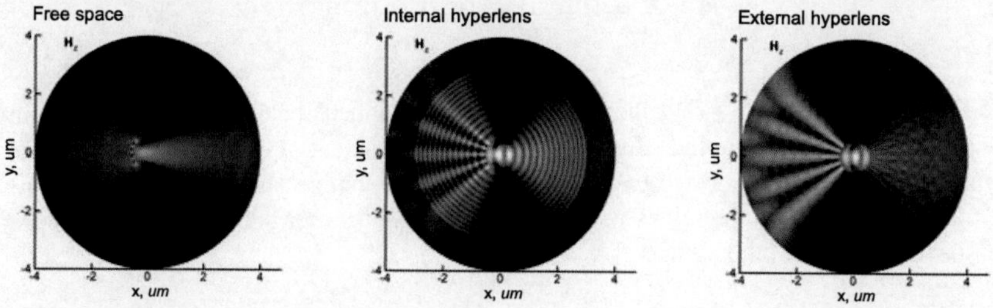

Fig. 4 Simulations of imaging of five sources in free space ($\varepsilon_\rho(\rho) = \varepsilon_\phi(\rho) = 1$), internal hyperlens ($\varepsilon_\rho(\rho) = (r\tau)/\rho$, $\varepsilon_\phi(\rho) = \rho/r$, $a \le \rho \le l$) and external hyperlens ($\varepsilon_\rho(\rho) = (r\tau)/\rho$, $\varepsilon_\phi(\rho) = (\rho/r)/(b/l)$, $a \le \rho \le l$). Where $\varepsilon_\rho(\rho), \varepsilon_\phi(\rho)$ are components of permittivity tensor in cylindrical coordinates, $r(\rho) = \tau^{-1}(\rho - l) + b$, $\tau = (l-a)/(b-a)$, $a = 600nm$, $b = 610nm$, $l = 3\mu m$. Sources are located at the arc of the circle with $\rho = 570nm$

Acknowledgements. This work was supported by the Program of Interdisciplinary Integration Research of the SBRAS No. 113, and the Russian Foundation for Basic Research (No. 09-01-00352).

References

1. Bonnet, P., Ferrieres, X., Michielsen, B., Klotz, P.: In: Rao, S.M. (ed.) Time Domain Electromagnetics. Academic, New York (1997)
2. Bonnet, P., Ferrieres, X., Paladian, F., Grando, J., Alliot, J., Fontaine, J.: Electron Lett. 33, 31–32 (1997)
3. Fedoruk, M.P., Munz, C.D., Omnes, P., Schneider, R.: A Maxwell-Lorentz solver for self-consistent particle-field simulations on unstructured grids. Forschungszentrm Karlsruhe GmbH, Karlsruhe (1998)
4. Fumeaux, C., Baumann, D., Leuchtmann, P., Vahldieck, R.: IEEE Trans. Microwave Theory Tech. 52, 1067–1076 (2004)
5. Kildishev, A.V., Chettiar, U.K., Liu, Z., Shalaev, V.M., Kwon, D.-H., Bayraktar, Z., Werner, D.H.: J. Opt. Soc. Am. B 24, A34–A39 (2007)
6. Kildishev, A.V., Klar, T.A., Drachev, V.P., Shalaev, V.M.: Nanophotonics with surface plasmons. In: Shalaev, V.M., Kawata, S. (eds.) Advances in Nano-Optics and Nano-Photonics, ch. 9. Elsevier, Amsterdam (2007)
7. Kildishev, A.V., Narimanov, E.E.: Opt. Lett. 32, 3432–3434 (2007)
8. Lebedev, A.S., Fedoruk, M.P., Shtyrina, O.V.: Vichisl. Mat. i Mat. Fiz. 46, 1302–1317 (2006) (in Russian)
9. Li, J., Engheta, N.: Antennas and Propagation Society International Symposium. IEEE, Los Alamitos (2007)
10. Madsen, N.K., Ziolkowski, R.W.: Electromagnetics 10, 147–161 (1990)
11. Prokopeva, L.Y., Shokin, Y.I., Lebedev, A.S., Shtyrina, O.V., Fedoruk, M.P.: In: Krause, E., Shokin, Y.I., Resch, M.M., Shokina, N. (eds.) Computational Science and High Performance Computing III. The 3rd Russian-German Advanced Research Workshop, Notes on Numerical Fluid Mechanics and Multidisciplinary Design (NNFM), Novosibirsk, Russia, July 23-27, vol. 101, pp. 122–135 (2008)
12. Shankar, V., Mohammadian, A.H., Hall, W.F.: Electromagnetics 10, 127–145 (1990)

Numerical Modeling of Work of Pulse Aerosol System of Fire Fighting on Computer Clusters

Alexander D. Rychkov and Yurii I. Shokin

Abstract. A model of a Pulse Aerosol System for Fire Fighting (PASFF), that is designed for fighting fires on gas wells, during the tankage of oil and at oil spills. The system represents a separate fillings device, that consists of a charge of solid propellant (the gas generator) and a container with fine dispersed powder of a flame retardant (FR). The combustion products from the gas generator form a low-intensity ballistic wave, that quenches the flame, disperses the flame retardant, and transports it to the combustion zone in the form of an aerosol cloud of finely dispersed particles. Favre-averaged Navier–Stokes equations closed by the $q-\omega$ model of turbulence were used as a model of motion for the carrier gas. The motion of the polydisperse solid phase was described as a stochastic process. The resulting system of equations is solved numerically by the finite volume method with a second order upwind difference scheme with TVD-properties. A software toolkit for the numerical simulation of PASFF on different complexity levels was developed. The toolkit includes parallel implementations of mathematical models of PASFF application on model fires. The dependence of computational speedup on the number of computational nodes was investigated.

1 Introduction

The problem of effective fighting against fires on gas wells and spills of oil and petroleum products that occur as a result of accidents caused by human activity and natural disasters is extremely important. An analysis of the dynamics of initiation and development of such fires showed that fire fighting at the initial stage of ignition has the most chances of success. Therefore much attention is paid to the

Alexander D. Rychkov · Yurii I. Shokin
Institute of Computational Technologies SB RAS,
Lavrentiev Ave. 6, 630090 Novosibirsk, Russia
e-mail: rych@ict.nsc.ru, shokin@ict.nsc.ru

E. Krause et al. (Eds.): Computational Sci., & High Performance Computing IV, NNFM 115, pp. 51–63.
springerlink.com

development and design of effective means of fire fighting that can be deployed at the site of a well and can generate a sufficient amount of flame retardant powder (FR) and transport it to the ignition spot. Such flame retardants can be organophosphorous, organometallic, or halogen-containing compounds. The FRs decompose in the flame zone and form gaseous active compounds, that involve into chemical reactions with radicals (particles responsible for combustion and flame propagation) and convert the latter into inactive particles. Branching chain reactions of combustion are terminated, and the flame is thereby extinguished. The effectiveness of flame extinguishing by means of FRs dispersion is mainly determined by the concentration of the flame retardant powder in the combustion zone [2]. Note that the total surface area of all flame-retardant particles plays an important role in increasing the effectiveness of fire extinguishing in the mechanism described above. Therefore it would be desirable to transport the flame retardant to the combustion zone in the form of an aerosol cloud of fine-dispersed particles. Available powder systems for fighting fires however do not allow for application of aerosols with a particle size of about 10 μm. Commonly used powder systems are gas generators based on unitary aerosol-generating solid-propellant compositions whose combustion products contain fine particles of the flame retardant in high concentrations sufficient for effective suppressing of ignition spots within a comparatively short time [6]. The main drawbacks of such systems are significant technological difficulties associated with the preparation of solid-propellant compositions with necessary properties and a comparatively low gas generation rate. Therefore, they are not sufficiently effective in extinguishing fires in gas wells and fires extended over large areas.

In the present work, we propose to use a pulse aerosol system of fire fighting (PASFF) whose concept was described in [8] -[10]. The system consists of two separate modules where FRs dispersion and its transportation to the combustion zone in the form of a cloud of a finely dispersed aerosol is performed by combustion products of a usual short-duration (pulse) solid-propellant gas generator, that does

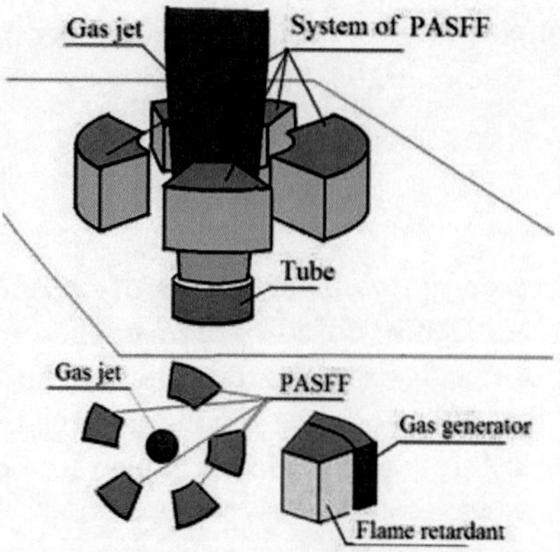

Fig. 1 Scheme of fire fighting of burning gas plume

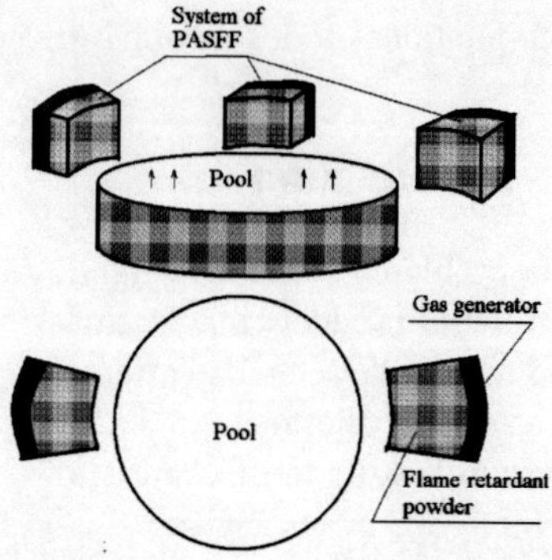

Fig. 2 Scheme of peripheral fire fighting of oil storage

not contain an oxidizer. A system of several devices of this type arranged around the plume can be used for fire extinguishing in gas wells (Fig. 1). Each device is a semi-closed container that contains a charge of a solid propellant and a fine powder of an FR. To extinguish a fire at on an oil spill or oil tankage infrastructure a similar setup can be used with several devices located around the source of fire (Fig. 2). Under the action of the combustion products of the solid-propellant charge of the gas generator the FR particles are compressed and set into motion towards the fire. When the combustion products leave the device, they form a low-intensity ballistic wave, which quenches the flame and assist in spraying and dispersion of the FR transported to the combustion zone in the form of an aerosol cloud of finely dispersed particles. As vaporization of flame-retardant particles is an endothermal process, the temperature in the region of the cloud decreases. Moreover, the oxygen contained in the air is displaced away from the combustion spot since the combustion products of the gas generator do not contain oxygen. A combination of these factors, namely the effect of the ballistic wave, the cooling of the gas in the region of the cloud of particles, the displacement of oxygen from this zone, and the presence of flame-retardant vapours, and an appropriate choice of a necessary mass ratio of the gas generator charge and FR compound can extinguish various types of high-intensity fires.

2 Description of Model and Governing Equations

Gaseous combustion products of the solid-propellant gas generator create a pressure of several tens of atmospheres, and form a turbulent flow directed towards the source of fire. The model of the flow of the carrier gas is based on Favre-averaged Navier-Stokes equations closed by the $q-\omega$ model of turbulence [3]. The motion of the polydisperse solid phase was described within the framework of the stochastic approach [4] with allowance for the influence of the turbulent nature of the carrier gas flow on the motion of particles. The following assumptions were employed to construct the mathematical model.

- The complex structure of PASFF (see Figs. 1, 2) was replaced by a torus with a rectangular cross section, that permits us to consider an axisymmetric flow instead of a three-dimensional flow.
- Natural gas escaping from the gas well and vapours of oil were modelled as gaseous methane. The burning of methane was described by a one-stage global reaction

$$CH_4 + 2O_2 = CO_2 + 2H_2O. \tag{1}$$

- The carrier gas is a chemically reacting mixture consisting of methane, carbon dioxide, water vapours, air, chemically inert combustion products of the gas generator, that do not contain an oxidizer, and products of decomposition of the solid particles of the flame retardant.
- The flow is assumed to be a turbulent, two-phase, axisymmetric flow.

- The FR particles are polydisperse, have a shape of a sphere, and consist of N fractions, each containing particles of identical size. Particle collisions were ignored.
- The temperature distribution over the particle volume was uniform. When a specified value T_{vap} was reached, the FR decomposed with vapour formation, and the linear rate of decomposition was assumed to be constant.
- The influx of the gas from the solid-propellant granules and from the flame-retardant particles is modelled by source terms in the equations of conservation of mass and energy.
- Thermal emission is not accounted for.

The rate of the global reaction of methane combustion was determined by an expression typical for bimolecular reactions [11]

$$W = \rho^2 Y_{O_2} Y_{CH_4} K_0 \exp(-E/(R \cdot T)), \tag{2}$$

where ρ and T are the density of the gas medium and its temperature, Y_{O_2} and Y_{CH_4} are the mass concentrations of oxygen and methane, respectively, and K_0, E and R are the pre-exponent in the Arrhenius law, the activation energy, and the universal gas constant. The effect of the flame-retardant vapours on the combustion process was accounted for by replacing the pre-exponent K_0 in Eq. (2) by with K_0^* calculated by the following formula suggested by O.P. Korobeinichev et al. (private communication):

$$K_0^* = \begin{cases} K_0(1 - 7.071 \cdot Y_{FR}^{1/2})^2, & at\ Y_{FR} \leq 0.02; \\ 0, \quad at\ Y_{FR} > 0.02, \end{cases} \tag{3}$$

where Y_{FR} is the mass concentration of FR vapours.

The system of equations that describe such a flow in a cylindrical coordinate system (x, r) has the form

$$\frac{\partial \rho}{\partial t} + \frac{\partial}{\partial x}(\rho u) + \frac{1}{r}\frac{\partial}{\partial r}(r\rho v) = M_{gen} - \sum_{i=1}^{N} \langle \frac{dm_i}{dt} \rangle n_i; \tag{4}$$

$$\frac{\partial \rho \mathbf{Y}}{\partial t} + \frac{\partial}{\partial x}(\rho u \mathbf{Y} + \mathbf{q}_x) + \frac{1}{r}\frac{\partial}{\partial r} r(\rho v \mathbf{Y} + \mathbf{q}_r) = \mathbf{S}; \tag{5}$$

$$\frac{\partial \rho u}{\partial t} + \frac{\partial}{\partial x}(\rho u^2 + p + \tau_{xx}) + \frac{1}{r}\frac{\partial}{\partial r} r(\rho uv + \tau_{xr}) = -\sum_{i=1}^{N} n_i \langle m_i \frac{du_i}{dt} \rangle; \tag{6}$$

$$\frac{\partial \rho v}{\partial t} + \frac{\partial}{\partial x}(\rho vu + \tau_{rx}) + \frac{1}{r}\frac{\partial}{\partial r} r(\rho v^2 + p + \tau_{rr}) = p - \sum_{i=1}^{N} n_i \langle m_i \frac{dv_i}{dt} \rangle; \tag{7}$$

$$\frac{\partial \rho E}{\partial t}+\frac{\partial}{\partial x}(\rho uH+u\tau_{xx}+v\tau_{xr}+\varphi_x)+\frac{1}{r}\frac{\partial}{\partial r}r(\rho vH+u\tau_{xr}+v\tau_{rr}+\varphi_r)=$$
$$=-\sum_{i=1}^{N} n_i[u\langle m_i\frac{du_i}{dt}\rangle+v\langle m_i\frac{dv_i}{dt}\rangle+\langle\pi d_i\lambda\cdot Nu(T-T_i)\rangle-$$
$$-\langle\frac{dm_i}{dt}C_i(T-T_i)\rangle]+Q_{gen}M_{gen}+WQ_p; \tag{8}$$

$$\frac{\partial \rho q}{\partial t}+\frac{\partial}{\partial x}(\rho uq+\tau_{x,q})+\frac{1}{r}\frac{\partial}{\partial r}r(\rho uq+\tau_{r.q})=S_q; \tag{9}$$

$$\frac{\partial \rho \omega}{\partial t}+\frac{\partial}{\partial x}(\rho u\omega+\tau_{x,\omega})+\frac{1}{r}\frac{\partial}{\partial r}r(\rho u\omega+\tau_{r.\omega})=S_\omega; \tag{10}$$

$$p=\rho TR_0\sum_{i=1}^{7}\frac{Y_i}{M_i};\quad \sum_{i=1}^{7}Y_i=1, \tag{11}$$

$$\mathbf{Y}=\{Y_{CH_4},\,Y_{O_2},\,Y_{H_2O},\,Y_{CO_2},\,Y_{FR},\,Y_M,\,Y_{N_2}\}^T,$$

$$\mathbf{S}=\{-W,\,-2\frac{M_{O_2}}{M_{CH_4}}W,\,2\frac{M_{H_2O}}{M_{CH_4}}W,\,\frac{M_{CO_2}}{M_{CH_4}}W,\,-\sum_{l=1}^{N}\langle\frac{dm_i}{dt}\rangle n_i,\,M_{gen},\,0\}^T,$$

$$H=\int_{T_0}^{T}C_p dT,\quad C_p=\sum_{k=1}^{7}Y_k C_{p,k},\quad S_q=C_{q1}(C_\mu S/\omega^2-2D/(3\omega)-1)\rho\omega q,$$

$$S_\omega=[C_{\omega1}(C_\mu S/\omega^2-C_{\omega3}D/\omega)-C_{\omega2}]\rho\omega^2,$$

Here u and v are the projections of the velocity vector onto the coordinate axes; ρ is the density, M_i are the molar weights of the species in the mixture; T is temperature, Q is the specific heat of FR particles; n_i are the concentration of particles of the ith fraction in the cell at a given time, $\mathbf{Y}$ is the vector of mass concentrations of the species; Y_M is the mass concentration of the inert combustion products of the gas generator; $q=\sqrt{k}$ and $\omega=\varepsilon/k$ are the "turbulent" variables related to the turbulent kinetic energy k and to its dissipation rate ε; $E=C_vT+(u^2+v^2)/2$ is the total specific energy; $H=E+p/\rho$ is the specific enthalpy of the mixture; μ and μ_t are the molecular and turbulent viscosities; λ is the thermal conductivity; p and T are the static pressure and temperature; R_0 is the gas constant; S is the strain tensor; $M_{gen}$and Q_{gen} are the generated mass and energy, that model the influx of high-temperature combustion products from the gas generator; Q_p is the specific heat of methane combustion; W is the rate of the generalized global reaction of methane combustion (2); m_i and d_i are the mass and diameter of particles of the ith fraction.

The terms in broken brackets indicate averaging over the volume of the cell of the difference grid. The following constants were used in the model of turbulence:

$$C_{\omega 1} = 0.55, C_{\omega 2} = 0.833, C_{\omega 3} = 0.666, C_{\mu} = 0.09, \quad C_{q1} = 0.5, \quad \mu_t = C_{\mu}\rho q^2/\omega.$$

$$\phi_x = -\lambda_e \frac{\partial T}{\partial x}, \phi_r = -\lambda_e \frac{\partial T}{\partial r}, \mu_e = \mu + \mu_t, \lambda_e = \lambda + \lambda_t$$

$$\lambda_t = \mu_t C_p \Big/ \Pr_t, \quad D = \frac{\partial u}{\partial x} + \frac{1}{r}\frac{\partial rv}{\partial r}, \quad S = (u_{i,j} + u_{j,i})u_{i,j} - 2/3u_{k,k}^2.$$

$$\mathbf{q}_x = -\left(\frac{\lambda}{C_p} + \frac{\mu_t}{Sc_t}\right)\frac{\partial \mathbf{Y}}{\partial x}, \quad \mathbf{q}_r = -\left(\frac{\lambda}{C_p} + \frac{\mu_t}{Sc_t}\right)\frac{\partial \mathbf{Y}}{\partial r},$$

The equations of motion of the ith FR particle along its trajectory are written in the form

$$\frac{dm_i}{dt} = -\rho_p d_i^2 \pi \cdot r_p; \tag{12}$$

$$\frac{du_i}{dt} = \frac{3}{4}\frac{\mu C_{Di} Re_{p,i}}{\rho_p d_i^2}\left(u + u' - u_i\right); \tag{13}$$

$$\frac{dv_i}{dt} = \frac{3}{4}\frac{\mu C_{Di} Re_{p.i}}{\rho_p d_i^2}\left(v + v' - v_i\right); \tag{14}$$

$$\frac{dT_i}{dt} = \frac{6}{C_i}\frac{\lambda \cdot Nu}{\rho_p d_i^2}(T - T_i); \tag{15}$$

$$\frac{dx_i}{dt} = u_i; \tag{16}$$

$$\frac{dy_i}{dt} = v_i, \tag{17}$$

where x_i and y_i are the particle coordinates; C_{Di} are the drag coefficients of the particles:

$$C_{Di} = \begin{cases} \dfrac{24}{Re_{pi}}\left(1 + \dfrac{Re_{pi}^{2/3}}{6}\right), & for\ Re_{pi} > 10^3, \\ 0.44, & for\ Re_{pi} \le 10^3, \end{cases} \tag{18}$$

$$Re_{pi} = \frac{d_i \rho |\mathbf{V} - \mathbf{V}_i|}{\mu}, \quad \mathbf{V}' = \sqrt{2k/3}\, erfc^{-1}(N_r)\mathbf{V}/|\mathbf{V}|,$$

where $\mathbf{V}$ is the averaged velocity vector; $\mathbf{V}' = \{u', v'\}$ is a random vector of perturbation of the gas velocity; N_r is a random number from the interval $[-1, +1]$, and $erfc^{-1}()$ is an inverse error function.

The carrier-gas flow was calculated by a second-order upstream LU difference TVD scheme, that is similar to the one used in [12]. The particle motion was calculated by an A-stable difference scheme, that also had a second order of accuracy [7]. The domain of integration of the system (1) – (10) was bounded by an impermeable surface from below and by the axis of symmetry of the flow on the left; there were no boundaries on the right and on the top of this domain assuming an infinite spatial extent. The Or axis is aligned in the horizontal direction, and the OX axis is directed vertically. The following boundary conditions are set for the carrier gas: conditions of flow symmetry on the OX axis, no-slip conditions on the PASFF surfaces and on the underlying surface, and non-reflecting boundary conditions on the remaining boundaries. For the system (12) – (17), the underlying surface was subjected to the conditions of inelastic reflection [1]

$$\mathbf{v}_{i,n}^{-} = -(0.993 - 1.76\Theta + 1.56\Theta^2 - 0.49\Theta^3)\mathbf{v}_{i,n}^{+}, \quad \mathbf{v}_{i,\tau}^{-} = \frac{5}{7}\mathbf{v}_{i,\tau}^{+},$$

where $\mathbf{v}_{i,n}^{-}, \mathbf{v}_{i,n}^{+}, \mathbf{v}_{i,\tau}^{-}, \mathbf{v}_{i,\tau}^{+}$ are the normal and the tangential component of the velocity vector of the ith particle; the plus and minus superscripts refer to the incident and reflected particles, respectively; Θ is the angle of incidence.

A granulated charge of the solid propellant was used in the gas generator for obtaining a large area of the burning surface. In this case, the mass income M_{gen} of the combustion products of the gas generator in (4) was determined by solving the equation

$$\frac{dM_{gen}}{dt} = -\rho_{gr} S \cdot r_{bg} N_{gg}^{0}$$

where ρ_{gr}, S,and r_{bg} are the density of the material of the propellant granule, its current surface area, and burning rate; N_{gg}^{0} is the number of granules per unit volume of the solid propellant charge, which remains constant in the course of their combustion and is determined from the initial conditions.

3 Parallel Algorithm

Systems of the equations (4 – 11) and (12 – 17) describing motion of the carrying gas and the particles are coupled through the right parts only. As a matter of convenience they are rewritten in a vector form.

$$\frac{\partial \mathbf{Q}}{\partial t} + \frac{\partial \mathbf{F}}{\partial x} + \frac{1}{r}\frac{\partial (r\mathbf{G})}{\partial r} = \mathbf{R}(\mathbf{Q}, \mathbf{Q}_{pi}) \quad , \tag{19}$$

$$\frac{d\mathbf{Q}_{pi}}{dt} = \mathbf{H}(\mathbf{Q}_{pi}, \mathbf{Q}), \tag{20}$$

were $\mathbf{Q} = (\rho, \rho u, \rho v, \rho E, \rho q, \rho\omega)^T$; $\mathbf{Q}_{pi} = (m_i, u_i, v_i, T_i, x_i, y_i)^T$. The structure of the remaining vectors in (19 – 20) can be inferred from the equations (4 - 17).

As mentioned above the finite volume method using the second order upwind LU difference scheme is applied for solving numerically the system of equations (19 – 20). The system is linearised and written down in the Δ notation (21). The linearisation is applied to the variables on the nth time level, and the dependence of the Jacobi matrices on these variables is not taken into account. Only the terms responsible for the first order approximation of the non-viscous terms, and the terms that approximate the repeated derivatives with respect to the corresponding directions for viscous terms are kept in the left-hand side of the system. Furthermore the derivative with respect to pseudo-time $\dfrac{\partial \mathbf{Q}}{\partial \tau}$ is added to the left-hand side of the system (19). The equations are solved by iteration over this derivative on each time level.

$$\Big[\frac{V_{i,j}}{\Delta\tau}+\frac{3}{2}\frac{V_{i,j}}{\Delta t}+\tilde{A}^{-}_{i+1/2}\Delta_{i+1/2}+\tilde{A}^{+}_{i-1/2}\Delta_{i-1/2}+\tilde{A}^{-}_{j+1/2}\Delta_{j+1/2}+ \\ +\tilde{A}^{+}_{j-1/2}\Delta_{j-1/2}\Big]\Delta\psi^{s+1}=-\frac{3(\mathbf{Q}^{n+1}_{i,j})^{s}-4\mathbf{Q}^{n}_{i,j}+\mathbf{Q}^{n-1}_{i,j}}{2\Delta t}V_{i,j}+(\mathbf{H}^{n+1}_{i,j})^{s}, \tag{21}$$

where $\Delta\psi^{s+1}=(\mathbf{Q}^{n+1})^{s+1}-(\mathbf{Q}^{n+1})^{s}$; $\mathbf{H}_{i,j}=-\big[(\mathbf{F}\cdot S)_{i+1/2}-(\mathbf{F}\cdot S)_{i-1/2}++(\mathbf{G}\cdot S)_{j+1/2}-(\mathbf{G}\cdot S)_{j-1/2}\big]+V_{i,j}\mathbf{R}_{i,j}$; $S_{i\pm1/2,}$ and $S_{j\pm1/2}$ are the areas of the faces of the volume $V_{i,j}$; $(\mathbf{F}\cdot S)_{i\pm1/2,}$ and $(\mathbf{G}\cdot S)_{j\pm1/2}$ are corresponding total (viscous and non-viscous) fluxes through these faces, the s and $s+1$ indices indicate pseudo-time steps. After the iterations with respect to the pseudo-time converge, the left-hand side of the system (21) is reduced to zero since $\Delta\psi^{s+1}=0$. Therefore the system of difference equations (20) approximates the system (19) on the time level $n+1$ with the second order both in the temporal and spatial domains.

The difference equations (21) are solved by a method, based on *LU*-factorization. In order to avoid an inversion of matrices during the solution of (21), the matrices that correspond to the left-hand side of (21) are simplified to diagonal matrices. Naturally, the convergence rate slightly decreases, but the simplicity of matrix structure allows to reduce the solution of (21) to successive recursive scalar operations that can be pipelined.

The *LU* algorithm for solving (21) is realized in two stages:
the *L*–path (sequential increase in the indices i, j):

$$\Delta\psi^{*}_{i,j}=B^{-1}\Big[-\frac{3(\mathbf{Q}^{n+1}_{i,j})^{s}-4\mathbf{Q}^{n}_{i,j}+\mathbf{Q}^{n-1}_{i,j}}{2\Delta t}V_{i,j}+(\mathbf{H}^{n+1}_{i,j})^{s}+\tilde{A}^{+}_{i-1/2}\Delta\psi^{*}_{i-1,j}+ \\ +\tilde{A}^{+}_{j-1/2}\Delta\psi^{*}_{i,j-1}\Big], \tag{22}$$

the *U*–path (sequential decrease in the indices i, j):

$$\Delta\psi^{s+1}_{i,j}=\Delta\psi^{*}_{i,j}-B^{-1}\big[\tilde{A}^{-}_{i+1/2}\Delta\psi^{s+1}_{i+1,j}+\tilde{A}^{-}_{j+1/2}\Delta\psi^{s+1}_{i,j+1}\big]. \tag{23}$$

The matrix B above is also a diagonal matrix and does not require full matrix inversion.

The algorithm (22) – (23) can be implemented on a cluster of symmetric multiprocessors (SMPs) as the following hybrid parallel algorithm. Since the solution domain is a rectangle, the decomposition should be applied along the direction which has the largest number of nodes of computation mesh to provide for a better load balancing of the cluster. Assume that the index i varies along this direction. The mesh is divided into subdomains (in future MPI-domains) according to the number of available processors (CPUs) of the cluster. In this case the L–pass (22) is realized in each of the subdomains as a one-dimensional pipelined calculation in the direction of the increase of the values of the index j starting with $j = 0$.

The scheme of parallelisation is shown in Fig. 3. The vertical double lines are boundaries of the MPI-domains. The first CPU obtains the values $\Delta \psi^*_{-1,j}$ and $\Delta \psi^*_{i,-1}$ from the boundary conditions to start the calculations. When the first CPU achieves the right boundary of its MPI-domain the values $\Delta \psi^*_{m,j}$ are communicated to the next MPI-domain and the second CPU starts processing.

After that the first CPU increments the value of the index j and starts again with the value of index $i = 0$. If each node of the cluster is an SMP there is an opportunity for additional parallelization within each MPI-domain using OpenMP routines. The data dependencies between elements in the domain are show in Fig. 3 with horizontal and vertical arrows. The implementation of the U–pass is completely similar with the only exception that the direction of the calculations is inverted. That is the calculation starts with the last MPI-domain in the direction of the decrease of the index i and j. The set of values $\Delta \psi_{Ni,j}$, $\Delta \psi_{Ni,Nj+1}$ are obtained from the boundary conditions.

The system of equations of motion of particles (20) is integrated along the trajectories. As the interactions between particles of different fractions are not accounted for there is an opportunity to develop an efficient parallelization of the algorithm for their solution.

For the solution of the equations (5) that describe the dynamics of mass concentrations of chemical components in a mixture an approach based on splitting of physical processes can be applied. The values of $\mathbf{Y}^{n+1}$ at the new time step can be calculating using the following sequence of operators with the second order of accuracy

$$\mathbf{Y}^{n+1} = L_1(\Delta t/2) \cdot L_2(\Delta t) \cdot L_1(\Delta t/2)\mathbf{Y}^n,$$

were L_1 and L_2 are two operators. The first operator L_1 corresponds to the convection and diffusion processes. The second operator L_2 takes account of chemical reactions only. In this case the operator L_1 allow can be computed independently on each component of the vector $\mathbf{Y}$. The operator L_2 is a linear transformation that requires the computation of a matrix-vector product and also can be computed in parallel. Note that the interaction between particles of different fractions including processes of their coagulation and breaking up can be accounted for in a similar manner.

Three alternative parallel algorithms have been developed for the solution of the equations (19–20). First algorithm is a pure MPI implementation where one MPI process is placed on each processing element (PE) of an SMP cluster. The second algorithm used pure OpenMP system without the decomposition of the computational domain into several MPI-domains. The solution of the equations (19–20) was performed on a single SMP cluster node (CN) using intra-node communications. The third algorithm was a hybrid one. The calculations were performed on an SMP cluster with HP BL460G5 nodes and InfiniBand DDR interconnect at the scientific computing center of Novosibirsk state university. Each CN has eight CPU cores (PEs). The values of the speed-up K vs. the number of PEs is shown in Fig. 4. The numbers on the curves correspond to the three algorithms outlined above and the dashed curve corresponds to the linear speedup. It can be seen that the best result were obtained with the second algorithm (curve 2) since it uses intra-node communications within the shared memory paradigm. The third (hybrid) algorithm (curve 3) exceeds slightly the speedup of the first algorithm at two or three PEs and gives practically the same result by increasing of number of nodes. The reason for this is the increase in the relative volume of MPI communications due to the increase of the number of MPI-domains and the decrease of length of OpenMP threads in each MPI-domain.

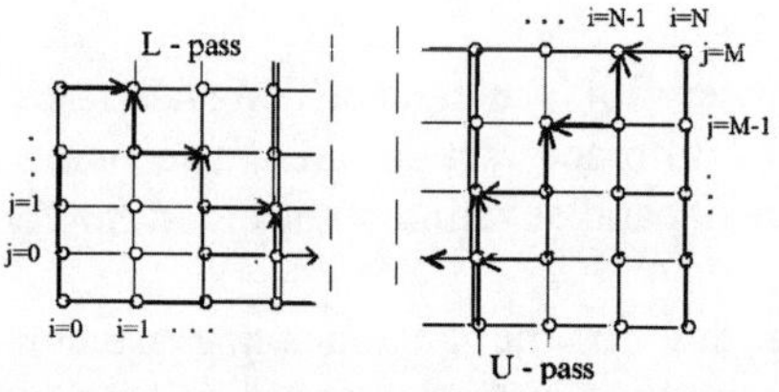

Fig. 3 Scheme of L – pass and U – pass

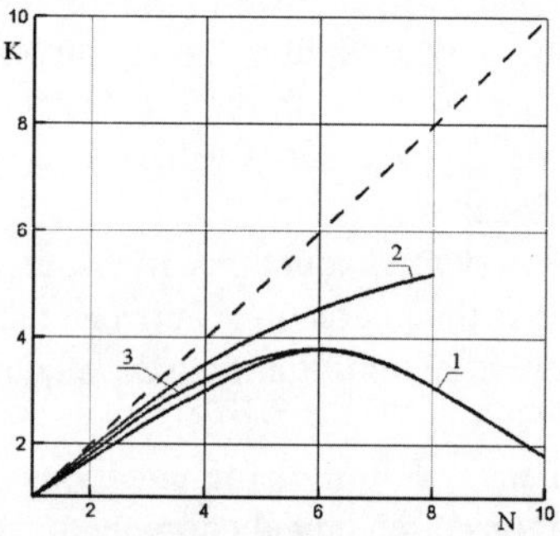

Fig. 4 Coefficient of speedup

4 Discussion of Results

In all calculations described below, we assumed that the rate of particle vaporization was set at the value $r_p = 10\,mm/s$. The number fractions of the polydisperse medium was 10, the temperature of vaporization of the flame-retardant particles was $T_{vap} = 400\,K$, the calorific power of the solid propellant of the gas generator was $Q_{gen} = 800\,kJ/kg$, the diameter of the solid propellant granules was 3 mm, and the burning rate of the granule was $r_{bg} = 20\,mm/sec$. The values of the parameters in Eq. (2) were borrowed from [8]: $K_0 = 10^{10}$, $E/R = 18{,}400\,K$, and the specific heat of methane combustion was set at $Q_p = 50MJ/kg$.

The results of modelling of PASFF on a burning subsonic plume of a mixture of methane (10%) and air exhausting with a velocity of 10 m/sec from a nozzle 0.4 m in diameter are shown on Fig. 5–6 for two values of t. The temperature of the exhausting mixture was equal to 600 K.

As mentioned above for the evaluation of the effect of the PASFF on the plume the PASFF device was modelled by a torus with a rectangular cross section surrounding the plume. The inner diameter of the torus was set at 1.2 m. The device contained 2 kg of the granulated solid propellant and 20 kg of the flame-retardant powder with the mean-mass particle size $d_{43} = 25\mu m$. The time was counted from the moment of the initiation of the gas generator charge. The *OX* axis of the cylindrical coordinate system was aligned in the vertical direction, and the *OY* axis was directed horizontally.

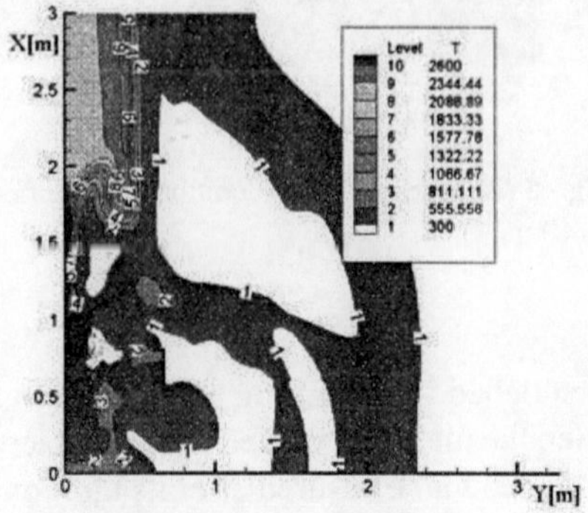

Fig. 5 Isotherms in the plume at the time t = 20 ms, ($d_{43} = 25\mu m$)

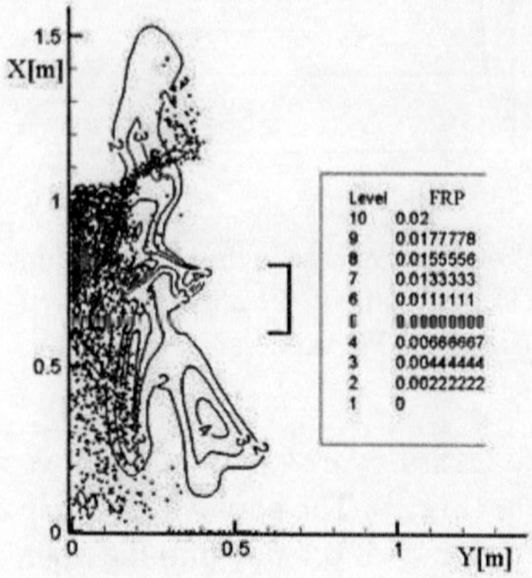

Fig. 6 Isolines of the concentrations of the FR vapours at the time t =20 ms, ($d_{43} = 25\mu m$)

It is seen from Fig. 5 that the plume structure is rapidly destroyed in the zone of the action of the fine FR particles (indicated by black points) and combustion products of the gas generator. The vapours of FRs and the rarefaction wave lead to a rapid decrease in the temperature of the fire and the termination of combustion in this zone. The isolines of concentrations of vapours of FRs for the same value of time plotted in Fig. 6 show that the level of their concentrations sufficient for effective suppression of the combustion reactions is maintained in a rather large volume of the plume. Dark points correspond to the location of FR particles. The total heat release in the case of methane combustion in the plume was determined by the formula

$$Q = \int_V W Q_p dv, \tag{24}$$

where V is the solution domain of (4)-(11). The time evolution of the quantity Q with (curve 1) and without (curve 2) accounting the effect of the FR vapours on the combustion process is shown in Fig. 7. The peaks on the surface are induced by intense influx of oxygen contained in the air to the combustion zone at the initial stage of the PASFF action on the plume.

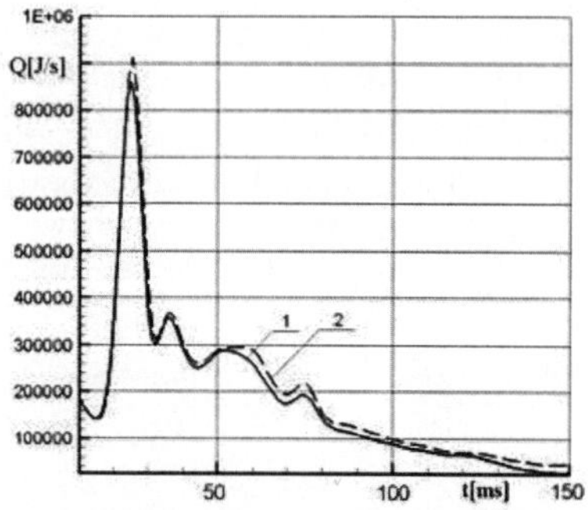

Fig. 7 Total heat release in the burning plume with (1) and without (2) allowance for the influence of the FR vapours

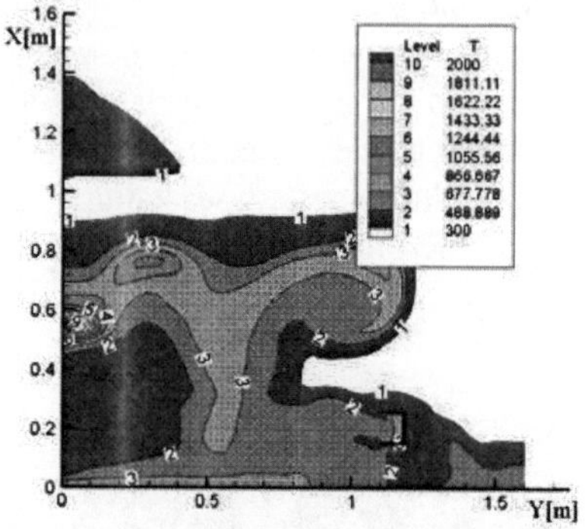

Fig. 8 Isotherms in the combustion zone (t=40ms)

The PASFF work on an oil spill was also modelled for a burning pool of hydrocarbons (Fig. 2). The pool was 1.7 m in diameter, the methane was fed via its surface at the velocity of 0.4 m/s into the open atmosphere, which ensured after its ignition the existence of a quasi-stable combustion regime due to the intermittent regime of turbulent flow. The PASFF in the form of a torus had inner diameter 2.2 m, the weight of the gas generator charge amounted to 2 kg, the FR powder weight was 20 kg. Figures (8 – 9) show the isotherms in the combustion zone and the distribution of concentrations of mass fractions of FR vapours.

It is seen that as a result of the PASFF application the combustion practically stops already at 40 ms, the temperature above the pool rapidly drops below the methane ignition temperature, and a subsequent combustion restart becomes impossible. The integral heat release at the methane combustion was determined by formula (24). The change of the quantity Q in time with regard (curve 1) and without regard (curve 2) for the influence of the vapours of FRs on combustion process is presented in Fig. 10. The peaks in these curves are related to an intense influx of the air oxygen into the combustion zone at the initial stage of the PASFF effect at the expense of an intense mixing caused by the ballistic wave.

The calculations were performed on a rectangular nonuniform difference grid condensed in the domain occupied by the PASFF and near the impermeable boundaries. To estimate the accuracy of the numerical solutions obtained, we calculated one variant on a sequence of three nested difference grids (the main grid had 250 $\times$ 300 nodes in the OXY plane). The calculations showed that the values of temperature in the flow field are determined within 1%. This accuracy is fairly acceptable

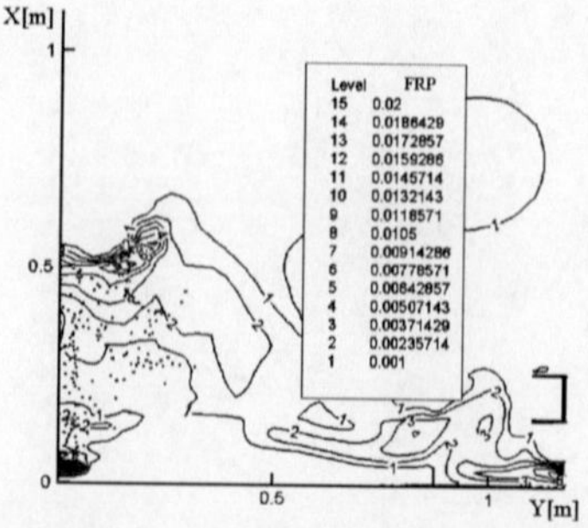

Fig. 9 FR vapours concentration in combustion zone (t=40ms)

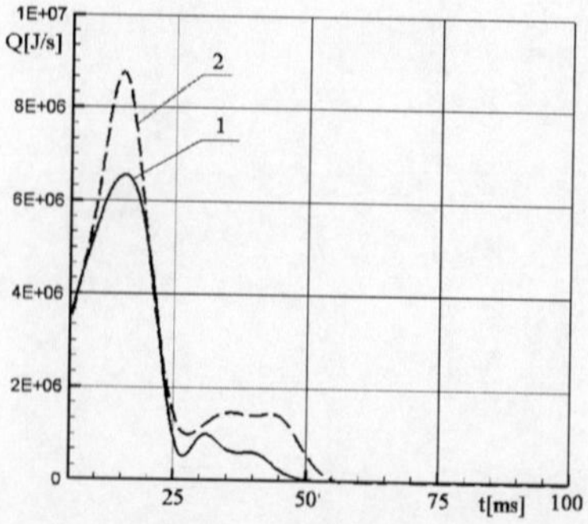

Fig. 10 Variation of the integral heat release

for obtaining qualitative estimates of the main characteristics of interaction between the PASFF and the plume.

Acknowledgements. This work was supported by the Program of Interdisciplinary Integration Research of the SB RAS No. 26, by the Project of the Presidium of the Russian Academy of Science No 1.4 and by Joint Integration Project of the SBRAS with the external scientific organizations No 94.

References

1. Bai, C., Gosman, A.D.: SAE Techn. Paper Series 950283 (1995)
2. Baratov, A.N., Bogman, L.P.: Fire-Retardant Powder Compositions. Stroiizdat, Moscow (1982) (in Russian)
3. Coakley, T.J.: AIAA Paper 92-0436 (1992)
4. Crow, C.T.: Trans. ASME. J. Fluid Eng. 104, 297–303 (1982)
5. Gosman, A.D., Ioannides, S.I.: AIAA J. Energy 7(6), 482–490 (1983)
6. Isavnin, N.V.: Means of Powder Fire Fighting. Stroiizdat, Moscow (1983) (in Russian)
7. Rychkov, A.D.: Mathematical Modeling of Gas-Dynamic Processes in Channels and Nozzles. Nauka, Novosibirsk (1988) (in Russian)
8. Rychkov, A.D.: Combust. Explos. Shock Waves 45(3), 26–34 (2009)
9. Rychkov, A.D.: Thermophys Aeromech 16(2), 34–41 (2009)
10. Rychkov, A.D., Shokin, Y.I.: Comput. Technologies 12(6), 81–89 (2007) (in Russian)
11. Volkov, E.P., Kudryavtsev, N.Y.: Inzh.-Fiz. Zh. 56(6), 885–894 (1989) (in Russian)
12. Yoon, S., Jameson, A.: AIAA Paper 87-600 (1987)

Stagnant Vortex Flow*

Egon Krause

Abstract. With the aid of the vorticity transport equation it is shown that in inviscid, incompressible, axially symmetric vortex flow the axial vorticity component near the axis of the vortex approaches zero if the axial velocity component approaches a stagnation point, and vice versa, the axial vorticity component is increased, if the axial flow is accelerated. This result, obtained in earlier investigations by simplifying the momentum equations for the neighbourhood of the axis of the vortex, is already contained in the vorticity transport equation as formulated by von Helmholtz in 1858. In laminar flow, with viscous forces acting near the stagnation point, the angular velocity does not necessarily vanish with the axial velocity component. These questions are discussed in the following.

1 Introduction

Exactly hundred and fifty years ago Hermann von Helmholtz published a paper entitled "Über Integrale der hydrodynamischen Gleichungen, welche den Wirbelbewegungen entsprechen" [1] in the Journal für die reine und angewandte Mathematik. In conjunction with the date of publication it is of interest to note, that the dreadful medieval witchcraft ordeals had just been stopped a short time before von Helmholtz wrote one of the most fundamental articles in fluid mechanics far reaching into the future. The truly extraordinary pioneering spirit of the paper is best be understood, if one realizes that it was published only 22 years after a woman, believed to be a witch, was put to death in a so-called water ordeal on the Hela Peninsula in Poland. Earlier times still cast deadly shades.

In his paper that later was to become so famous von Helmholtz derived the vorticity transport equation for inviscid flow from the Euler equations and formulated three vortex theorems, later on named after him. They state that a vortex filament

Egon Krause
Aerodynamisches Institut, RWTH Aachen,
Wüllnerstr. 5a, 52062 Aachen, Germany
e-mail: ek.aia@t-online.de

* Originally published in Acta Mechanica 209, pp. 345-351 (2010).

E. Krause et al. (Eds.): Computational Sci., & High Performance Computing IV, NNFM 115, pp. 65–74.
springerlink.com

retains its strength, that it cannot end somewhere in the fluid, and that a fluid cannot acquire vorticity if it does not possess any at its beginning. The theorems, now being taught in every advanced course in fluid mechanics, found numerous applications, perhaps the most famous one in Prandtl's lifting line theory, and many others, which cannot be mentioned here.

In the course of time experimental and analytical studies were carried out to prove the validity of the theorems, for example, N.E. Zhukowski in 1894 proved analytically that a vortex ring could not be cut with a knife, proving the second theorem. In [2] an experiment was described, in which a waterspout was generated with the aid of a fast rotating disk above a free water surface. For sufficiently large speed of rotation a water spout was formed, eventually extending all the way up to the disk, hinting at the strong coupling between azimuthal and axial flow motion, expressed already in von Helmholtz' vorticity transport equation, proving again the second theorem.

Because of the non-linearity of the vorticity transport equation, its application to the solution of actual flow problems had to wait for about hundred years. It was only after computational methods were introduced that the vorticity transport equation, extended to viscous flows, was used in what is now known as "vorticity-stream function formulation". One of the first investigations of this kind was published by J.E. Fromm [3], who in 1963 numerically simulated the von Kármán vortex street downstream from a two-dimensional obstacle with a rectangular cross-section, positioned normal to the oncoming flow at a Reynolds number $Re = 200$.

While von Helmholtz' theorems assured that a vortex could not simply end somewhere in the flow it was embedded in, again almost one hundred years after the theorems were formulated, in 1957, more or less by accident, D.H. Peckham and S.A. Atkinson observed in a wind tunnel experiment, that the core of leading-edge vortices of a delta wing experienced an abrupt change, forming a free stagnation point on the axis of the vortex, and then deforming the core into a spiral or a bubble-like recirculation zone further downstream [4]. Although the experimental results did not imply that the vortices were ending in the flow above the wing, the stagnation points formed in the cores showed that the flow was brought to a halt, the vortex "broke down", as this sudden change of the flow behaviour was called.

The flow structure of the broken-down or burst part of the vortex core can adapt a great variety of shapes and forms, as it is affected by many parameters. For example, the radial profiles of the axial and the azimuthal velocity components of the oncoming flow play a dominant role: If the axial flow exhibits a wake-like velocity defect, then breakdown can occur much easier than in the case of a jet-like velocity overshoot. The formation of a stagnation point as just mentioned can also be influenced by the shape of the radial profile of the azimuthal velocity component. In [5] it was shown that in incompressible flow a positive axial pressure gradient may result from a particular shape of the radial profile of the azimuthal velocity component, forcing the axial flow to be decelerated, eventually leading to a stagnation point. Vice versa, the profile may be such that a negative axial pressure gradient results, and the axial flow is accelerated, not causing any destructive influence on the vortex core, a burst core may even be "healed", as reported in [6].

In compressible flow the core structure is entirely different from that of incompressible flow. In supersonic flow vortex breakdown can be enforced by intersecting the vortex with a normal or an oblique shock. If the shock is strong enough and provides a large enough pressure rise, the vortex may burst, as for example shown in [7]: An oblique shock will deflect the axis of the oncoming vortex in the flow direction downstream from the shock, and the shock itself is bent into an S-shaped form. These few remarks may suffice to point out the many different aspects of the problem of vortex breakdown. It is not intended here to give a comprehensive description of this flow phenomenon of extreme complexity, but rather point to its scientific importance. Detailed elaborates may be found elsewhere.

Since its discovery in 1957, vortex breakdown has been studied in numerous experimental, theoretical, and numerical investigations during the past fifty years. Many details of the local flow behaviour can now accurately be described, but neither fully convincing theoretical models nor comprehensive numerical analyses are yet available to explain all of the experimental data obtained so far. Reviews of the state of art were, for example, provided in 1971 by T. Sarpkaya [8], one year later by M. G. Hall [9], S. Leibovich in 1978 [10], and at the IUTAM Symposium on Dynamics of Slender Vortices, held in Aachen in 1997 [11]. The numerical simulations published were preferably based on solutions of the complete Navier-Stokes equations rather then on solutions of the vorticity transport equation. The vorticity-stream function formulation was mainly used in early studies, when the simulation had to be restricted to axially symmetric flows because of lack of sufficient computational power, as for example, in [12]. Recently the vorticity transport equation was used for interpretation of numerical data, see [13].

The continued interest in attempting to gain a better understanding of the breakdown process to a large extent is motivated by the occurrence of bursting vortical structures in flows in and around technical devices and machines. For example, in hydro turbines at off-design conditions a strong vortex is generated in the draft tube. In the breakdown nomenclature the draft-tube vortex is a longitudinal vortex burst into the spiral type. The vortex core rotates in the draft tube, generating a pressure field that varies locally and in time. It therefore can give rise to excite vibrations. The pressure variations can lead to severe operational problems, and investigations are under way to learn how to predict the characteristic flow behaviour of draft-tube vortices, see for example [14]. The solution of this problem is handicapped by the absence of reliable methods for describing the Reynolds stresses in such flows with adequate accuracy.

Another example for the occurrence of burst vortical flows can be found in combustors of industrial gas turbines. In them air and gaseous fuel are injected into the combustor in the form of a swirling jet. An abrupt enlargement of the cross-sectional area just downstream from the injector into the combustor causes a steep pressure rise, resulting in a stagnation point and a burst region of the swirling flow in form of a bubble-type breakdown. The recirculating flow in the bubble acts as an immaterial flame holder, and it is here, where the gaseous fuel is oxidized [15]. Flow instabilities of the oncoming swirling flow can lead to unsteady vortex breakdown with significant variations of the combustion process. As in the first example, the closure problem to describe the Reynolds stresses is at the present time not satisfactorily controlled.

Vortex breakdown may also occur in turbo engines. It is conjectured in [16], that, when the compressor is operated near the stability limit, the tip leakage vortex may break down, if it is intersected by a shock in the supersonic part of the flow, which may give rise to rotating instabilities and subsequent stall. Perhaps one of the best-known cases of vortex breakdown is that occurring in the supersonic flow over delta wings. If on canard configurations the vortex originated by the front aileron interacts with the shock on the main wing, the vortex may burst. Breakdown may also occur, if on delta wings with leading edge extensions one of the lee-side vortices hits the vertical tail. Severe buffeting may result. This problem has been studied in numerous numerical and experimental investigations, see for example [17].

It is evident that the problem of vortex breakdown is directly related to the contents of von Helmholtz' theorems; but little or no reference to them is made in the breakdown literature (to the author's knowledge): In [10] Leibovich divides the breakdown region into three intervals as observed in experiments, the first of which being characterized by a rapid deceleration of the axial flow and the formation of a stagnation point on the axis of the vortex, the second by flow reversal near the axis for the spiral- as well as for the bubble-type breakdown, and in the third the original direction of the axial flow is restored. Von Helmholtz' theorems are not used in the description of the flow, and on a first glance, the statements concerning the first two intervals seem to disagree with the second theorem, but a close look reveals, that the formation of a stagnation point on the axis of a columnar vortex, with its axis orientated parallel to the oncoming flow, and the halt of the azimuthal motion can directly be concluded from the vorticity transport equation for inviscid flow. The inclusion of viscous forces in the consideration, however, leads to results that differ from those obtained for inviscid flow.

2 Stagnation Point in Inviscid Flow

In order to explain the interdependence between the relations describing the breakdown process and the vorticity transport equation it is necessary to repeat some of the related earlier derivations. As just mentioned the flow conditions leading to the deceleration of the axial flow in the first interval of the breakdown region were described in [10]. A formulation in terms of simplified conservation equations was given in [18]: In that reference an axially symmetric vortex flow was assumed to be parallel to the x-axis with axial, radial, and azimuthal velocity components u, v, and w, in a coordinate frame x and r. The vortex was assumed to be slender with the radial velocity component v of the oncoming flow be small in comparison to u and w. With these assumptions the radial momentum equation for incompressible, inviscid flow was simplified for the neighbourhood of the axis $r \to 0$ to

$$\frac{\rho w^2}{r} = p_r. \tag{1}$$

In (2) the quantity ρ‘ being the fluid density and the subscript r indicating partial differentiation with respect to the radial coordinate. In [18] the simplified radial momentum equation in the form of (1) was differentiated with respect to the axial coordinate x and integrated in the radial direction in the interval $0 \leq r \leq \infty$. For the present derivation it suffices to extend the upper limit only to the radius of the edge of the vortex core R, i. e. the radial position, where the azimuthal velocity attains its maximum value $wmax$. There results

$$p_x(x,R) = p_x(x,0) + 2\rho \int_0^R (ww_x/r)dr. \qquad (2)$$

Again following [18] the x-derivative of the azimuthal velocity component w_x can be substituted by the azimuthal momentum equation, written in the form

$$w_x = -(v/ru)(rw)_r, \qquad (3)$$

and (2) becomes

$$p_x(x,R) = p_x(x,0) - 2\rho \int_0^R (v/u)(w/r^2)(rw)_r dr. \qquad (4)$$

The equation (4), already derived in [18], may now be used to demonstrate the correspondence of the relations given above to von Helmholtz’ vorticity transport equation. Since here only principle arguments are discussed, further approximations can be introduced. In this sense the azimuthal velocity component near the axis is expressed in terms of a Taylor-series expansion in the radial coordinate r. The leading-order term is just the familiar rigid-body rotation, written as $w(x,r) = \Omega(x)r + \ldots$, where Ω is the local angular velocity. With this approximation, recently confirmed in experiments in [19], (4) reduces to

$$p_x(x,R) = p_x(x,0) - 4\rho\Omega^2 \int_0^R (v/u)dr. \qquad (5)$$

In the next step the pressure gradient on the axis $p_x(x,0)$ is related to the flow far away from the axis by assuming the azimuthal velocity component w to be given by a potential vortex, say $w = \Gamma/(2\pi r)$ for $R \leq r$ and extending the integral in (5) to infinity. Other approximations may be used, but they will not basically affect the result. The term $(rw)_r$ then vanishes for $R \leq r$, and the integral in (5) does not add any additional contribution to the axial pressure gradient $p_x(x,0)$:

$$p_x(x,0) = p_x(x,\infty) + 4\rho\Omega^2 \int_0^R (v/u)dr. \qquad (6)$$

The equation (6) shows that $p_x(x,0)$ is affected by the axial flow far away from the axis, i. e. by $p_x(x,\infty)$, further by a contribution that is proportional to the square of the angular velocity Ω, and an integral contribution containing the ratio of the radial and axial velocity components v/u. This ratio, representing the projection of the slope of the stream lines on the meridional plane, is expressed now by the integral of the continuity equation:

$$v/u = -(1/ru)\int_0^r r' u_x dr'. \quad (7)$$

The axial velocity component u and its derivative in (6) are replaced by their mean values $u(x, r_m)$ and $u_x(x, r_m)$ inside the core, i. e. for the interval $0 \leq r \leq R$, and r_m is the radial coordinate, where the mean value is taken. Eqs. (6) and (7) then yield

$$p_x(x, 0) \approx px(x, \infty) - \rho w_{max}^2(x) u_x(x, r_m)/u(x, r_m). \quad (8)$$

The last equation shows, that in comparison to purely axial flow, in swirling flow the pressure gradient on the axis is affected by by the pressure gradient in the far field for $r \to \infty$, and by the product of the square of the azimuthal velocity $w_{max}(x)$, and the ratio $u_x(x, r_m)/u(x, r_m)$.

In the final step the deceleration of the flow to a stagnation point on the axis can be shown by combining the azimuthal momentum equation with the continuity equation, Eq. (3) and Eq. (7), written in differential form, thereby eliminating the term v/r. There results the simple relation

$$u^2(w/u)_x + v^2(w/v)_r = 0. \quad (9)$$

In [20] a similar relation was derived for compressible flow, leading to an analogy to the area-velocity relation of gasdynamics. Here the rigid-body rotation is introduced in (9), which for the vicinity of the axis of the vortex reduces to

$$u\Omega_x - \Omega u_x = 0. \quad (10)$$

Integration of (10) yields

$$\Omega = (\Omega_i/u_i)u, \quad (11)$$

the subscript i indicating some upstream inflow station. Eqs. (10) and (11) were already obtained in [18], derived in a different manner. Eq. (11) relates the local angular velocity Ω near the axis directly to the axial velocity component u, indicating that, if u vanishes, Ω also vanishes: In a columnar slender vortex of inviscid flow the azimuthal motion near the axis of the vortex is stopped, when the axial motion is brought to a halt and a stagnation point is formed.

The result stated here is contained in the vorticity transport equation, as formulated by von Helmholtz in [1]. To show this, the transport equation for the x-component of the vorticity vector ω is introduced for axially symmetric, incompressible, inviscid flow along the x-axis. With the vorticity components

$$\xi = -(1/2r)(rw)_r, \quad \eta = -(1/2)w_x, \quad \zeta = -(1/2)(v_x - u_r) \quad (12)$$

the transport equation for the vorticity component ξ reads

$$\xi_t + u\xi_x + v\xi_r = \xi u_x + \eta u_r. \quad (13)$$

Along the axis for $r \to 0$ the last terms on the left- and on the right-hand side of Eq. (13) vanish because of the symmetry condition $v = u_r = 0$. Then for steady flow with $\partial/\partial t = 0$, (13) reduces to Eq. (10), since only the terms describing the convective acceleration and the vortex stretching are retained, the vorticity component ξ being identical with Ω. Hence the onset of vortex breakdown as formulated by Leibovich in [10] is already contained in the vorticity transport equations, here simplified for the case of a columnar slender vortex with its axis parallel to the oncoming flow in the x-direction. One is tempted to call the contents of Eq. (11) the unformulated fourth von Helmholz's theorem: *In axially symmetric, incompressible, inviscid flow of a columnar vortex with its axis parallel to the main flow the angular velocity near the axis vanishes when the axial velocity component vanishes and a stagnation point is formed.*

If the axial velocity component $u(x,t)$ is known for the case of unsteady flow, the solution of Eq. (13) can be written for $r \to 0$ with the characteristic equation

$$dx/dt = u \tag{14}$$

as

$$\xi = \xi \exp(\int u_x dt). \tag{15}$$

The exponential form of the last equation hints at the rapid motion of the stagnation point along the axis, observed in experiments for certain flow conditions.

3 Stagnant Viscous Flow

The influence of viscous forces on the flow near a free stagnation point on the axis of a slender columnar vortex of steady, incompressible, laminar, axially symmetric flow with the axis parallel to the x-axis of the coordinate system chosen is demonstrated here with the aid of the azimuthal momentum equation:

$$uw_x + vw_r + (w/r)w_\theta + vw/r = \nu\{[(1/r)(rw)_r]_r + w_{xx}\}. \tag{16}$$

In Eq. (16) the quantity ν is the kinematic viscosity. Introduction of the rigid-body rotation for the neighbourhood of the axis$r \to 0$ causes the first term of the Stokes stresses on the right-hand side of Eq. (16) to disappear, so that

$$\nu\Omega_{xx} - u\Omega_x + u_x\Omega = 0. \tag{17}$$

First attempts to show the effect of the Stokes stresses on the flow near the stagnation point for $u = 0$ were discussed in [21]. There it was shown that the x-derivative $(\Omega/u)_x$ does not vanish in viscous flow as it does in inviscid flow, compare Eq. (10). Here an approximate solution of Eq. (17) will be constructed for stagnant axial viscous flow $u(x_s, 0) = 0$ on the axis at x_s and its neighbourhood, where $u(x,0)$ is assumed to also be close to zero. The index s indicates the stagnation point of the axial flow.

An exact solution of Eq. (17) can only be obtained, if $u(x,0)$ is known. Since this information is not available here, an approximate solution is established in the following way: Firstly, it is postulated that the axial velocity component $u(x,0)$ of the inviscid solution is known up to the edge of the viscous region at x_0. The magnitude of the thickness of the viscous region δ, extending upstream of x_s to x_0 can be estimated by non-dimensionalising Eq. (17) and requiring all dimensionless terms to be of order unity. There results $\delta = O(\nu/u_0)$, where u_0 is the inviscid axial velocity component at x_0. Since u_0 has to be of the same order of magnitude as ν for the Stokes stresses to play a role, it follows that the axial extent of the viscous layer is very small. Because of the lack of a solution for $u(x,0)$, it is assumed here that the axial flow in the viscous region is constantly decelerated, given by $u_{x0} = \text{const} < 0$ at x_0, and because of the immediate vicinity of the stagnation point of the axial flow $u(x,0) \approx 0$, so that the term $u\Omega_x$ can be neglected. Then the formal solution of Eq. (17) for the viscous region extending from x_0 to x_s can be written as

$$\Omega = C_1 \exp[\alpha(x - x_s)] + C_1 u \exp[-\alpha(x - x_s)]. \tag{18}$$

The constant α appearing in Eq. (18) is expressed in terms of the constant axial deceleration $u_{x0} < 0$ and the kinematic viscosity ν:

$$\alpha^2 = -u_{x0}/\nu, \quad \alpha = (|u_{x0}|/\nu)^{1/2}. \tag{19}$$

The two boundary conditions, needed for the solution of Eq. (18), are obtained from the inviscid solution for u and Eq. (11). Since a downstream boundary condition for Ω is not known, the x-derivative of Ω at x_0 is used here:

$$x = x_0, \quad \Omega = \Omega_0, \quad \Omega_x = \Omega_{x0}. \tag{20}$$

With Eqs. (19) and (20) Eq. (18) becomes for $x_0 \le x \le x_s$

$$\begin{aligned}\Omega_s(x,0) = \Omega_s = &\{[(|u_{x0}|/\nu)^{1/2}\Omega_0 + \Omega_{x0}]\exp[-(|u_{x0}|/\nu)^{1/2}(u_0/u_{x0}] + \\ &+ [(|u_{x0}|/\nu)^{1/2}\Omega_0 - \Omega_{x0}]\exp[(|u_{x0}|/\nu)^{1/2}(u_0/u_{x0})]\}/[2(|u_{x0}|/\nu)^{1/2}].\end{aligned} \tag{21}$$

To determine the angular velocity at the stagnation point of the axial flow at x_s the axial extent of the viscous region $\delta = x_s - x_0$ must be known. It can be determined consistently with the assumption of the constantly decelerated axial flow. For $u_s = 0$, it is

$$\delta = x_s - x_0 = -u_0/u_{x0}. \tag{22}$$

With Eq. (22) inserted into Eq. (21), the angular velocity Ω at x_s is given by

$$\begin{aligned}\Omega(x_s,0) = \Omega_s = &\{[(|u_{x0}|/\nu)^{1/2}\Omega_0 + \Omega_{x0}]\exp[(|u_{x0}|/\nu)^{1/2}(u_0/u_{x0})] + \\ &+ [(|u_{x0}|/\nu)^{1/2}\Omega_0 - \Omega_{x0}]\exp[(|u_{x0}|/\nu)^{1/2}(u_0/u_{x0})]\}/[2(|u_{x0}|/\nu)^{1/2}].\end{aligned} \tag{23}$$

According to Eq. (23) the angular velocity Ω does not necessarily vanish at the stagnation point of the axial flow. A better insight into the structure of Eq. (23) is obtained by expanding the exponential parts of Eq. (23) up to second order. There is obtained:

$$\Omega_s \approx \Omega_0 u_0^2/(\nu/|u_{x0}|) - |\Omega_{x0}|(u_0/|u_{x0}|). \tag{24}$$

The influence of the viscous forces is manifested by the dimensionless parameter $u_0^2/(\nu/|u_{x0}|)$, and of the inviscid flow by the ratio $u_0/|u_{x0}|$, the axial extent of the viscous region. For the approximations introduced the angular velocity at the stagnation point of the axial flow on the axis vanishes only if the absolute value of the derivative of the angular velocity in the axial direction at the upstream edge of the viscous region at x_0 is equal to the product of the local value of the angular velocity and the axial velocity component, divided by the kinematic viscosity:

$$|\Omega_{x0}| = \Omega_0 u_0/\nu. \tag{25}$$

Eq. (25) can be written in dimensionless form as

$$(|\Omega_{x0}|\nu)/(\Omega_0 u_0) = 1. \tag{26}$$

For the approximations introduced, i. e. the omission of the convective term $u\Omega_x$ and the assumption of a constantly decelerated axial flow along the axis in the viscous region, the left-hand side of Eq. (26) can be considered as a similarity parameter, which, when it equals to unity, assures that in viscous flow the angular velocity near the axis of the vortex vanishes, when the axial velocity component does.

4 Concluding Remarks

It was shown for a slender, columnar vortex with its axis parallel to the oncoming axial flow, that in inviscid flow the azimuthal velocity component near the axis vanishes with the axial velocity component. This close coupling between the axial and the azimuthal flow motion, referred to as the onset of vortex breakdown for vanishing axial flow, is described by the balance of the terms describing the convective acceleration and the vortex stretching in the vorticity transport equation, formulated by von Helmholtz in 1858.

In contrast, if viscous forces are considered, for laminar flow the momentum flux along the axis of slender, columnar vortices is changed by the Stokes stresses. Then the azimuthal flow motion near the axis may not be stopped, when a stagnation point of the axial flow is formed on the axis.

References

1. von Helmholtz, H.: Über Integrale der hydrodynamischen Gleichungen, welche den Wirbelbewegungen entsprechen. J. reine angew. Math. 55, 25–55 (1858)
2. Shukowski, N.J.: Theoretische Grundlagen der Luftfahrt. GTI, Moscow (1925)

3. Fromm, J.E.: A method for computing nonsteady, incompressible viscous fluid flows, LA-2910, UC-32, Mathematics and Computers, TID-4500, 21th edn. (1963)
4. Peckham, D.H., Atkinson, S.A.: Preliminary results of low speed wind tunnel tests on a Gothic wing of aspect ratio 1.0., Aero. Res. Counc. CP 508 (1957)
5. Krause, E.: A contribution to the problem of vortex breakdown. Comput. Fluids 13, 375–381 (1985)
6. Krause, E., Gersten, K. (eds.): Dynamics of Slender Vortices, Proceedings of the IUTAM Symposium held in Aachen, Germany, August 31-September 3. Kluwer Academic Publishers, Dordrecht (1998)
7. Thomer, O., Klaas, M., SchrÃűder, W., Krause, E.: Oblique shock vortex interaction over a wedge. In: Bathe, K.J. (ed.) Proceedings of the MIT Conference on Computational Fluid and Solid Mechanics, Computational Fluid and Solid Mechanics 2003, June 17-20, pp. 1156–1159. Elsevier, Amsterdam (2003)
8. Sarpkaya, T.: Vortex breakdown in swirling conical flows. AIAA J. 9, 1792–1799 (1971)
9. Hall, M.G.: Vortex breakdown. Ann. Rev. Fluid Mech. 4, 195–218 (1972)
10. Leibovich, S.: The structure of vortex breakdown. Ann. Rev. Fluid Mech. 10, 221–245 (1978)
11. Krause, E., Gersten, K. (eds.): Dynamics of Slender Vortices, Proceedings of the IUTAM Symposium held in Aachen, Germany, August 31-September 3. Kluwer Academic Publishers, Dordrecht (1998)
12. Shi, X.: Numerische Simulation des Aufplatzens von Wirbeln, Diss. RWTH Aachen (1983)
13. Kiesewetter, F., Konle, M., Sattelmayer, T.: Analysis of combustion induced vortex breakdown driven flame flashback in a premix burner with cylindrical mixing zone. J. Eng. Gas Turbines Power 129, 929–936 (2007)
14. Göde, E., Ruprecht, A., Lippold, F.: On the part load vortex in draft tubes of hydro electric power plants. In: Krause, E., Shokin, Y.I., Resch, M., Shokina, N. (eds.) Computational Science and High Performance Computing II. The 2nd Russian-German Advanced Research Workshop, Notes on Numerical Fluid Mechanics and Multidisciplinary Design (NNFM), Stuttgart, Germany, March 14-16, vol. 91, pp. 217–231 (2006)
15. Widenhorn, A., Noll, R., Aigner, M.: Numerical characterization of the reacting flow in a swirled gasturbine model. In: Nagel, W.E., Kröner, D.B., Resch, M.M. (eds.) High Performance Computing in Science and Engineering 2008, Transactions of the High Performance Computing Center Stuttgart, pp. 365–380 (2009)
16. Schlechtriem, S., Lötzerich, M.: Breakdown of tip leakage vortices in compressors at flow conditions close to stall. In: Proceedings of IGTI-ASME Conference, Orlando, Florida, USA (June 1997)
17. Kandil, O.A., Massey, S.J., Sheta, E.F.: Structural dynamics/CFD Interaction for computation of vertical tail buffet. In: Proceedings of International Forum on Aeroelasticty and Structural Dynamics, Royal Aeronautical Society, Manchester, U. K., also published in Royal Aeron. J., vol. 100, pp. 297–303 (1996)
18. Krause, E.: Axial flow in slender vortices. J. Eng. Thermophys 11(3), 229–242 (2002)
19. Klass, M.: Experimental investigation of shock vortex interaction, Diss. Aerodyn. Inst., RWTH Aachen (2002)
20. Krause, E.: On the analogy to the area-velocity relation of gasdynamics in slender vortices. Acta Mech. 201, 23–30 (2008)
21. Krause, E.: Breakdown revisited. In: Bathe, K.J. (ed.) Proceedings of the Third MIT Conference on Computational Fluid and Solid Mechanics, Computational Fluid and Solid Mechanics 2005, June 14-17, pp. 706–709. Elsevier, Amsterdam (2005)

Numerical Modelling of Surface Water Waves Arising Due to Movement of Underwater Landslide on Irregular Bottom Slope

Leonid B. Chubarov, Gayaz S. Khakimzyanov, and Nina Shokina

Abstract. The work is devoted to the numerical modelling of surface water waves generated by a moving underwater landslide on irregular bottom. Currently the works of other authors consider flat bottoms only. The modelling is done in the framework of the shallow water model with taking into account bottom mobility. The equations are obtained for an underwater landslide movement under the action of gravity force, buoyancy force, friction force and water resistance force. The predictor-corrector scheme [5], preserving the monotonicity of the numerical solution profiles in a linear case, is used on adaptive grids, which are generated using the equidistribution method [7]. The scheme is tested for the problem with a known analytical solution, describing the wave generation by a nondeformable body, which moves with a constant velocity on a horizontal bottom. The analysis is done for an irregular bottom of the dependencies of wave regime characteristics on bottom slope, initial landslide depth, its length and width.

1 Introduction

An underwater landslide can generate surface waves of rather large amplitudes in natural and artificial reservoirs. These waves can be dangerous for onshore installations. In the recent years many publications have appeared (see the references in [1]), devoted to the results of numerical and laboratory experiments on the

Gayaz S. Khakimzyanov · Leonid B. Chubarov
Institute of Computational Technologies SB RAS,
Lavrentiev Ave. 6, Novosibirsk, 630090, Russia
e-mail: khak@ict.nsc.ru, khak@ict.nsc.ru

Nina Shokina
Section of Applied Mathematics, University of Freiburg i. Br.,
Hermann-Herder-Straße 10, Freiburg i. Br., 79104, Germany
e-mail: shokina@mathematik.uni-freiburg.de

E. Krause et al. (Eds.): Computational Sci., & High Performance Computing IV, NNFM 115, pp. 75–91.
springerlink.com

mechanism of surface wave generation by the movement of non-deformable rigid body on flat underwater slope.

In [2] it has been shown that the substitution of a real landslide by a model non-deformable rigid body leads to the adequate description of wave processes in a wide range of variation of such parameters as inclination of a flat slope, width and length of a landslide, initial landslide depth.

In [2, 3, 4] it has been shown that numerical modelling of generated wave processes on the basis of "full" mathematical models, taking into account fluid vertical movements, allow reproducing amplitude values of generated waves close to experimental values. In these papers (see also the full bibliography from [1]) the possibility of using different approximate mathematical models of hydrodynamics has been investigated. It has shown that the general picture of arising wave regimes is reasonably described even with the shallow water model, where fluid vertical movements are not taken into account.

In the mentioned papers only the movement of an non-deformable landslide has been considered only on a flat slope. The current work is devoted to the modelling of surface wave generation by a landslide, moving on an irregular bottom.

The new equation of a landslide movement is obtained, taking into account bottom irregularity, gravity force, buoyancy force, friction force and water resistance force. For a flat slope the obtained equation coincides with the equation from [2].

The modelling of surface waves, generated by the movement of a landslide on irregular bottom, is done within the framework of the non-linear shallow water model with taking into account bottom mobility. The predictor-corrector scheme [5], preserving the monotonicity of the numerical solution profiles in a linear case, is used on adaptive grids, which are generated using the equidistribution method [7]. The scheme is tested for the problem with a known analytical solution, describing the wave generation by a nondeformable body, which moves with a constant velocity on a horizontal bottom. The analysis is done for an irregular bottom of the dependencies of wave regime characteristics on bottom slope, initial landslide depth, its length and width.

2 Problem Formulation

Let us consider a flat layer of fluid, bounded by free surface from above and by impermeable movable bottom from below. The fluid is assumed to be incompressible, inviscid, and under the action of gravity field. Let the Cartesian coordinate system xOz be chosen such that the equation of free surface of fluid at rest is given by $z = 0$ and all fluid characteristics depend on the variables x and t only.

The non-liner shallow water model is used for modelling of hydrodynamic parameters:

$$\frac{\partial \mathbf{u}}{\partial t} + \frac{\partial \mathbf{f}}{\partial x} = \mathbf{G}, \qquad x \in (0, L), \tag{1}$$

where t is the time, L is the length of domain, $\mathbf{u}$ is the solution vector, $\mathbf{f}$ is the flux vector,

$$\mathbf{u} = \begin{pmatrix} H \\ Hu \end{pmatrix}, \quad \mathbf{f}(\mathbf{u}) = \begin{pmatrix} Hu \\ Hu^2 + gH^2/2 \end{pmatrix}, \quad \mathbf{G} = \begin{pmatrix} 0 \\ gHh_x \end{pmatrix},$$

$u(x,t)$ is the velocity, $H = \eta + h$ is the total depth, $\eta(x,t)$ is the deviation of free surface from the undisturbed level $z = 0$, and g is the gravity acceleration. The equation

$$z = -h(x,t) = h_{\mathrm{bt}}(x) + h_{\mathrm{sl}}(x,t)$$

gives the form of movable bottom, $h_{\mathrm{bt}}(x)$ and $h_{\mathrm{sl}}(x,t)$ are the functions, defining its fixed and movable parts accordingly.

The vertical impermeable wall is assumed to be on the left boundary, thus, the impermeability condition is given there. The right boundary is considered to be open, therefore, the free outflow condition is set on it.

It is assumed that at the initial time moment the fluid is at rest, and the function $z = h_{\mathrm{sl}}^0(x)$ is known, describing the initial form of a landslide:

$$h_{\mathrm{sl}}(x,0) = h_{\mathrm{sl}}^0(x).$$

For $t > 0$ the form and position of a landslide are defined by its motion law, which is obtained in the next section.

The results of the modelling of surface waves generated by a landslide movement on a flat slope, obtained in [3], give grounds to state that the mathematical model, shown above, can help to obtain the general representation of the wave processes under consideration. In the current work this model is used for investigating the surface waves generated by a landslide movement on an irregular slope. The form of the slope is given by the single-valued function:

$$z = h_{\mathrm{bt}}(x). \tag{2}$$

3 Motion Law of Underwater Landslide

One of the main factors, defining the specifics of the wave generation process due to an underwater landslide motion, is the landslide motion law. If a landslide is modelled by a rigid body, then this law is given for one of its characteristic points, for example, for the center of mass. In [2] the motion law is suggested for a landslide on a flat slope. The current work presents the derivation of the motion law for a landslide on an irregular underwater slope.

For the initial time moment it is assumed that a landslide has the finite length b and width T, and its surface is described by the function $z = h_{\mathrm{bt}}(x) + h_{\mathrm{sl}}^0(x)$, where $z = h_{\mathrm{sl}}^0(x)$ is a given non-negative function with a finite support $(x_l^0,\, x_r^0)$ of the length b (along the axis Ox) and maximal value $T > 0$ at some point of the interval $(x_l^0,\, x_r^0)$.

At $t > 0$ a landslide can start to move on an inclined bottom.

The simplified approach to modelling of a landslide process, used in the current work, is based on the following assumptions:

1. for derivation of a landslide motion law, a landslide is identified with some material point $\mathbf{x}_c(t)$ at each time moment. This material point has the abscissa $x_c(t)$ and slides along the irregular base (2) according to the motion law of material point on plane curve. Here $x_c(0) = x_c^0$, where $x_c^0 \in \left(x_l^0,\, x_r^0\right)$;
2. the position of a landslide is defined entirely by the abscissa value $x_c(t)$ of the chosen point $\mathbf{x}_c(t)$. The surface of a landslide for $t > 0$ is described by the function

$$z = h_{\mathrm{bt}}(x) + h_{\mathrm{sl}}(x,t), \tag{3}$$

where $h_{\mathrm{sl}}(x,t) = h^0_{\mathrm{sl}}(x + x_c^0 - x_c(t))$.

Therefore, at the time moment t a landslide is positioned on a bottom slope between the points with abscissas $x_l(t) = x_l^0 - x_c^0 + x_c(t)$ and $x_r(t) = x_r^0 - x_c^0 + x_c(t) = x_l(t) + b$. When moving, the length of a landslide (along the horizontal axis Ox) remains invariant. Also, when moving, the landslide surface is deformed according to the encountered irregularities of the bottom, but the volume V and mass $M = \rho_{sl}V$ of a landslide remain invariant. Here ρ_{sl} is the landslide material density.

Let us describe the motion law for a landslide or, more precisely, let us provide the way for evaluating the function $x = x_c(t)$, because it is this function, which defines the surface (3) of a moving landslide.

Let s be the arc length of the curve (2), counted for its beginning. According to the assumption, the function (2) is the single-valued function of the variable x. Therefore, there is one-to-one correspondence between variables x and s, which is given by the equality:

$$s(x) = \int\limits_0^x \sqrt{1 + \left[\, h'_{\mathrm{bt}}(\xi)\,\right]^2}\, d\xi. \tag{4}$$

Let us assume, that the point $\mathbf{x}_c(0)$, chosen at the initial time moment, corresponds to the parameter $s = S_0$, and the moving point $\mathbf{x}_c(t)$ with the abscissa $x_c(t)$ – to the parameter $s = S(t)$. Then the motion law $s = S(t)$ of the initially motionless material point $\mathbf{x}_c(t)$ along the curve (2) has the following form:

$$m\ddot{S}(t) = F_\tau(t), \qquad S(0) = S_0, \qquad \dot{S}(0) = 0, \tag{5}$$

where $\ddot{S} = d^2S/dt^2$, $\dot{S} = dS/dt$, m is the quantity, which has dimensionality of mass, $F_\tau(t)$ is the tangential force, acting on the moving point $\mathbf{x}_c(t)$ at the time moment t.

As a quantity m the total landslide mass M and added water mass $C_w\rho_wV$ are taken, i.e. $m = M + C_w\rho_wV = (\rho_{sl} + C_w\rho_w)V$, where ρ_w is the water density, taken equal to 1 in our computations, C_w is the added mass coefficient.

Let us consider which quantity is taken in the equation (5) as a force F_τ, acting on the material point $\mathbf{x}_c(t)$. The gravity force and the buoyancy force act in vertical direction on every landslide element with a length dx. At an arbitrary point $x \in (x_l(t),\, x_r(t))$ the component of these two oppositely directed forces, tangential to the curve (2), is evaluated using the formula:

$$(\rho_{sl} - \rho_w)\, h_{\mathrm{sl}}(x,t) W g \sin\theta(x)\, dx,$$

where W is a landslide width in the direction, perpendicular to the plane xOz, $\theta(x)$ is the local bottom angle of slope. Here,

$$\theta(x) = -\arctan h'_{\rm bt}(x), \qquad \sin\theta(x) = -\frac{h'_{\rm bt}(x)}{\sqrt{1+[\,h'_{\rm bt}(x)\,]^2}}\,.$$

Thus, the resultant action of gravity force and buoyancy gravity on a finite-size landslide is defined by the force

$$(\rho_{sl} - \rho_w) W g \int\limits_{x_l(t)}^{x_r(t)} h_{\rm sl}(x,t) \sin\theta(x)\, dx, \tag{6}$$

acting in the tangential direction on the material point $\mathbf{x}_c(t)$, which moves along the curve (2).

The force (6) makes a landslide to accelerate. The deceleration is caused by the water resistance and the bottom friction. The tangential component of the water resistance force is directed oppositely to a landslide movement direction, proportional to the largest cross-sectional area TW of a landslide, and equal to

$$-\frac{1}{2} C_d \rho_w T W (\dot{S})^2, \tag{7}$$

where $\dot{S}$ is the velocity of the point $\mathbf{x}_0(t)$ along the curve (2), C_d is the water resistance coefficient. The resistance force (7) is passive, i.e. it disappears when the motion is absent.

The friction force also decelerates a landslide. The friction force is tangential to the curve (2) to the side, opposite to the landslide motion direction. This force is also passive, because when the motion is absent it can not change a landslide position. The friction force influence on a landslide motion is present only when a landslide moves (at $\dot{S} \neq 0$).

Let us consider a landslide element with the length dx. The friction force is defined using the normal reaction $N(x,t)$, acting on the chosen element from the bottom side:

$$F_f(x,t) = -C_f N(x,t), \tag{8}$$

where C_f is the dynamic friction coefficient (coefficient of sliding friction when moving). The considered simplified model assumes that the numerical value of the coefficient C_f coincides with the value of the coefficient of sliding friction at a standstill. This value is defined by indicating the angle of friction θ_*, i.e. the threshold value of angle of the flat slope, after exceeding of which a landslide starts to slide down on the flat slope. The same angle of friction will be used for irregular slopes. The given assumptions lead to the formula for obtaining the dynamic friction coefficient $C_f = \tan\theta_*$.

The normal reaction $N(x,t)$, acting on the landslide element dx, is defined as the sum of the normal component of gravity force and centrifugal force:

$$N(x,t) = (\rho_{sl} - \rho_w)\, h_{\rm sl}(x,t) W g \cos\theta(x)\, dx + (\rho_{sl} - \rho_w)\, h_{\rm sl}(x,t) W K(x) \left(\dot{S}(t)\right)^2 dx, \tag{9}$$

where $K(x)$ is the fixed sign curvature of the bottom base:

$$K(x) = h''_{\rm bt}(x)\left(\sqrt{1 + \left[\,h'_{\rm bt}(x)\,\right]^2}\right)^{-3}.$$

For a flat bottom $K(x) = 0$ the curvature is equal to zero, therefore, for this case the expression (9) does not contain the second term. The formulas (8) and (9) show that, compared to a flat bottom, the bottom irregularity provides the decrease of the friction force (in absolute magnitude) on the parts of bottom, which are convex upwards ($h''_{\rm bt}(x) < 0$) and, vice versa, the increase of the friction force on the parts of bottom, which are convex downwards ($h''_{\rm bt}(x) > 0$).

Taking into account the expression (9), it is obtained that the following friction force acts on a moving landslide (applied at the point $\mathbf{x}_c(t)$ and tangential to the curve (2) at this point):

$$F_f(t) = -C_f\,(\rho_{sl} - \rho_w)\, W \int\limits_{x_l(t)}^{x_r(t)} h_{\rm sl}(x,t)\left[g\cos\theta(x) + K(x)\left(\dot{S}(t)\right)^2\right] dx. \tag{10}$$

The resultant of forces (6), (7), and (10) is taken as the quantity F_τ from the formula (5).

Thus, taking into account all given assumptions, the problem (5) is written down as follows:

$$(\rho_{sl} + C_w\rho_w)\, V\ddot{S} = (\rho_{sl} - \rho_w)\, W g I_1 - W\left[(\rho_{sl} - \rho_w)\, I_2 + \frac{1}{2} C_d \rho_w T\right](\dot{S})^2, \tag{11}$$

$$S(0) = S_0, \qquad \dot{S}(0) = 0, \tag{12}$$

where

$$I_1(t) = \int\limits_{x_l(t)}^{x_r(t)} h_{\rm sl}(x,t)\left[\sin\theta(x) - C_f\cos\theta(x)\right] dx, \qquad I_2(t) = C_f \int\limits_{x_l(t)}^{x_r(t)} h_{\rm sl}(x,t) K(x)\, dx,$$

and the sought quantity $x_c(t)$ is connected to the solution $S(t)$ of the problem (11), (12) by the relation (4):

$$S(t) = \int\limits_{0}^{x_c(t)} \sqrt{1 + \left[\,h'_{\rm bt}(\xi)\,\right]^2}\, d\xi.$$

Let us note, that the equation (11) is valid for a moving landslide only. A landslide is at rest at the initial time moment, therefore, this equation can be used for $t > 0$ only in the case when a landslide really starts to move.

The equation (11) shows that the sufficient condition of a landslide displacement at the initial time moment is the condition of the positivity of the integral I_1 at $t = 0$:

$$I_1 = \int_{x_l^0}^{x_r^0} h_{\rm sl}^0(x)\left[\sin\theta(x) - C_f\cos\theta(x)\right]dx > 0.$$

Due to the non-linearity of the equation (11) the numerical methods are required for its solution. It is reasonable to rewrite the problem (11), (12) in the form of Cauchy problem for the system of ordinary differential equations of the first order:

$$(\gamma + C_w)\frac{V}{W}\dot{v}_c = (\gamma - 1)gI_1 - \left[(\gamma - 1)I_2 + \frac{C_d}{2}T\right]v_c^2, \qquad v_c(0) = 0, \tag{13}$$

$$\dot{S} = v_c, \qquad S(0) = S_0, \tag{14}$$

where $\gamma = \rho_{sl}/\rho_w$ is the relation of a landslide material density to the water density, $\gamma > 1$. The solution of this problem is sought till the moment of a landslide stop, i.e. till the moment when the value of a landslide velocity v_c turns to be zero for the first time (with a given accuracy).

4 Model Area and Model Landslide

The following model bottom topography with non-zero curvature is chosen (Fig. 1) for investigating the influence of bottom irregularity on surface water waves, generated by an underwater landslide motion. The topography is given by the smooth monotone decreasing function (2):

$$z = h_{bt}(x) = \frac{h_+ + h_-}{2} + \frac{h_+ - h_-}{2}\tanh\left[c(x - \xi)\right], \tag{15}$$

where h_+ and h_- are bottom depths in the right and left infinitely remote points correspondingly, $c = \dfrac{2\tan\theta_0}{h_- - h_+}$, θ_0 is the maximal bottom slope angle, $\xi = \dfrac{1}{2c}\ln\dfrac{h_0 - h_+}{h_- - h_0}$ is the flex point, h_0 is the depth at the point $x = 0$.

The following parameter values are used: $h_+ = -1400$ m, $h_- = -5$ m, $\theta_0 = 6°$, $h_0 = -10$ m. The domain length is 70000 m.

The initial form of the model landslide is:

$$h_{\rm sl}^0(x) = \begin{cases} \dfrac{T}{2}\left[1 + \cos\left(\dfrac{2\pi(x - x_c^0)}{b}\right)\right], & |x - x_c^0| \le b/2 \\ 0, & |x - x_c^0| > b/2 \end{cases}, \tag{16}$$

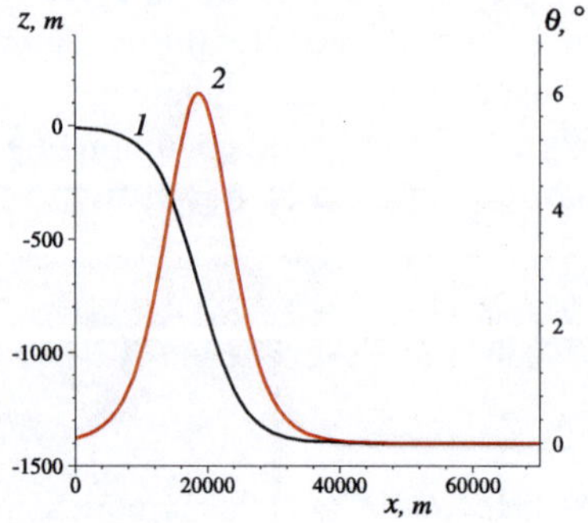

Fig. 1 Model topography $z = h_{bt}(x)$ (*line 1*) and local bottom slope angles $\theta(x)$ (*line 2*)

where x_c^0, T, b are the given abscissa of the landslide vertex at $t = 0$, the landslide width and length (along the axis Ox) correspondingly, $x_l^0 = x_c^0 - b/2, x_r^0 = x_c^0 + b/2 = x_l^0 + b$.

For the calculations the landslide width $T = 25$ meters, the landslide length $b = 5000$ meters. At the initial time moment the landslide vertex is positioned above the bottom point, where the depth is $d_c = -200$ meters. Thus,

$$x_c^0 = \xi - \frac{1}{2c} \ln\left(\frac{d_c - h_+}{h_- - d_c}\right) \approx 12644 \text{ m.}$$

Further, these values together with the values of constants, entering the law of motion ($\gamma = 1.5$, $C_w = 1$, $C_d = 1$, $\theta_* = 1°$) are used in the numerical experiments by default, if something else is not stated.

5 The Results of the Numerical Experiments

For numerical modelling of the surface waves, generated by a landslide motion, the predictor-corrector scheme on adaptive grid [6] is used. This scheme approximates the shallow water equations (1), written down in the new coordinates (q,t):

$$(J\mathbf{u})_t + (\mathbf{f} - x_t\mathbf{u})_q = \mathbf{G}.$$

Here $J(q,t)$ is the Jacobian of the non-degenerate coordinate transformation

$$x = x(q,t), \quad x(0,t) = 0, \quad x(1,t) = L, \tag{17}$$

which at each moment of time t maps the unit interval $[0,\ 1]$ to the interval $[0,\ L]$ in a one-to-one manner. The adaptive grid is generated using the equidistribution method [7]. The method consists in the choice of the transformation from the set of all possible transformations of the form (17), which is the solution of the initial boundary value problem:

$$\begin{aligned} &(w(x,t)x_q)_q = \beta x_t, \quad q \in (0,\,1), \quad t > 0, \\ &x(0,t) = 0, \quad x(1,t) = L, \quad x(q,0) = x^0(q), \end{aligned} \tag{18}$$

where β is the positive parameter, selected experimentally in order to reduce the oscillations of the node trajectories of the moving grid, $w(x,t)$ is the given control function.

5.1 *Test Problem on Surface Wave Generation by Body Movement on Horizontal Bottom*

Let us consider the problem on surface wave generation by the rigid body of the given shape (16), moving with a constant velocity $v_c > 0$ on the horizontal bottom $h_{bt}(x) \equiv -h_0$. Therefore, the function, describing the moving bottom, changes with respect to time according to the following law:

$$h(x,t) = h_0 - h^0_{\mathrm{sl}}(x + x^0_c - x_c(t)),$$

where $x_c(t) = x^0_c + v_c t$.

In this test problem the total depth $H_0(x)$ and the fluid velocity $u_0(x)$ were chosen at the initial time moment in such a way that the Cauchy problem for the equation (1) with these initial data has the solution in the form of the wave, propagating with a constant velocity v_c:

$$u(x,t) = u_0(x - v_c t), \quad H(x,t) = H_0(x - v_c t). \tag{19}$$

Substituting this solution to the equations (1) and requiring that the fluid holds still at infinity, we obtain that the initial functions have to satisfy the following relationships:

$$H_0(x)\Big[u_0(x) - v_c\Big] = -v_c h_0, \qquad \frac{\big[u_0(x) - v_c\big]^2}{2} + g\Big[H_0(x) - h(x,0)\Big] = \frac{v_c^2}{2}. \tag{20}$$

Whence the cubic equation for defining the function $H_0(x)$ follows:

$$\overline{H}_0^3 - \overline{H}_0^2\left(\overline{h}(x,0) + \frac{\overline{v}_c^2}{2}\right) + \frac{\overline{v}_c^2}{2} = 0, \tag{21}$$

where $\overline{H}_0 = H_0/h_0$, $\overline{h} = h/h_0$, $\overline{v}_c = v_c/\sqrt{gh_0}$.

The equation (21) can have no positive roots; or one positive root; or two positive roots, corresponding to subcritical and supercritical flows. It depends on the given values of the body velocity v_c and its width T. Here the case of subcritical flow is considered, which is realized under the condition of subcriticality of the flow over the vertex of moving body:

$$\overline{T} < 1 - \frac{3\overline{v}_c^{2/3} - \overline{v}_c^2}{2},$$

where $\overline{T} = T/h_0$. Fig. 2 (a) shows the upper boundary of the domain of subcritical flow. Setting the values $T > 0$ and $v_c > 0$, corresponding to this domain, we obtain a larger positive root of the equation (21). Then, the initial velocity $u_0(x)$ is defined using the fist formula from (20). After that, the exact solution of the equation (1) is obtained using the formulas (19).

Fig. 3 (a) shows the graphs of free boundary for exact and numerical solutions at the time moment $t = 2000$ s. To this moment the wave in the form of hollow has passed the distance, approximately equal to eight of its lengths. In the calculations the values of quantities x_c^0, b are taken the same as in Section 4, the body velocity is taken equal to 20 m/s.

The lines *1* on the Fig. 3 (a) are obtained for $h_0 = 200$ m, $T = 25$ m (on Fig. 2 the point $(\overline{v}_c, \overline{T})$, corresponding to these data, is depicted by the circle *1*). It can be seen that the sufficiently good accuracy of numerical solution takes place despite of the fact that the computational grid with a small amount of nodes $N = 100$ is used. The reason is the adaptive gird, condensing in the vicinity of hollow (see Fig. 3 (b)) and tracing the wave movement. Let us note, that when the uniform fixed grid is used, the comparable accuracy is reached for the number of nodes by an order more.

The lines *2* on Fig. 3 (a) and the point *2* on Fig. 2 (a) correspond to the body movement on horizontal bottom with the depth $h_0 = 400$ m. It can be seen, that duplication of depth leads to the decrease of the wave amplitude in nearly three times. Non-linear decrease of the wave amplitude is also caused by the decrease of the body width (see lines *3* on the Fig. 3 (a and the point *3* on Fig. 2 (a), obtained for $h_0 = 200$ m, $T = 12.5$ m).

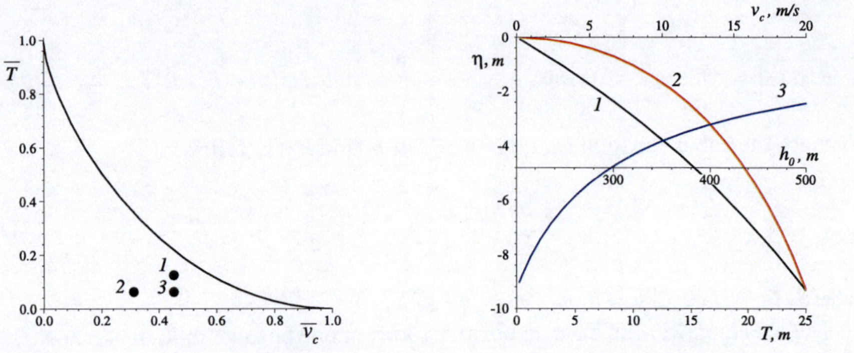

a *b*

Fig. 2 *a* – the boundary of the domain of subcritical flow in the plane of parameters $\overline{T}$ and $\overline{v}_c$; *b* – graphs of the dependencies $\eta_{\min}(T)$ (*1*), $\eta_{\min}(v_c)$ (2), $\eta_{\min}(h_0)$ (*3*)

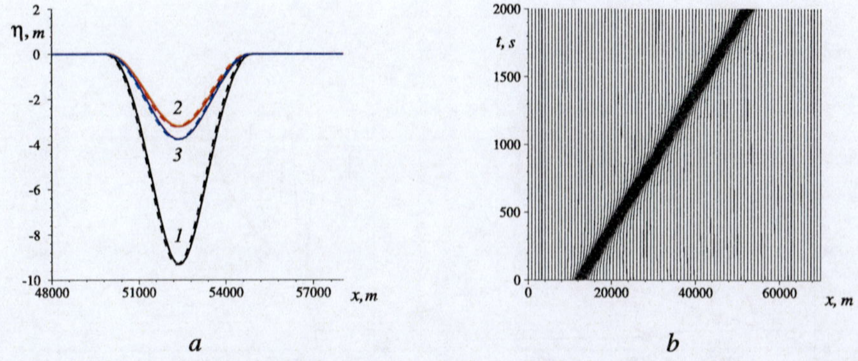

Fig. 3 *a* – graphs of the exact (solid line) and numerical (dashed line) solutions at the time moment $t = 2000$ s; *b* – the trajectories of the nodes of adaptive grid, $N = 100$

The non-linear character of dependencies of minimal value of $\eta_{\min}$ in the wave hollow on the incoming data is shown in Fig. 2 (*b*). The graphs are shown for the variation of the wave amplitude $\eta_{\min}$ at the variation of one of the parameters T, v_c or h_0 and unaltered values of two others, which have the values $T = 25$ m, $v_c = 20$ m/s and $h_0 = 200$ m. Let us note, that the amplitude of generated wave does not depend on the body length, which influences the wave length only.

The good correspondence of the obtained numerical results with the exact solution of the test problem hold out a hope for fidelity of the data provided below for the problem on wave generation by a landslide moving on irregular bottom.

5.2 General Characteristics of Wave Regimes, Generated by Landslide Movement on Irregular Slope

Let us consider the specifics of wave generation due to the landslide movement on irregular bottom (15). The irregular bottom make influence not only on the character of wave transformation during their propagating, but also on the motion law of the landslide.

Fig. 4 (*a*) shows the trajectory of movement of the landslide vertex (line *1*), moving under the action of gravity force, buoyancy force, friction force and surrounding water resistance force, and the graph of velocity variation (line *2*).

On the topography profile (line *3*) two extreme points depict the initial ($t = 0$ s, $x_c = 12644$ m) and final ($t = 1370$ s, $x_c = 37233$ m) positions of the landslide. The middle point depicts the position ($x_c = 22624$ m) of the landslide at the moment ($t = 512$ s) of its maximal velocity ($v_c \approx 37.5$ m/s). At this moment the landslide is positioned below the flex point ($\xi = 18673$ m) of the bottom profile.

As the velocity graph shows, the time interval of the landslide acceleration ($\Delta_1 t = 512$ s) is significantly smaller than the deceleration time interval ($\Delta_2 t = 858$ s). The

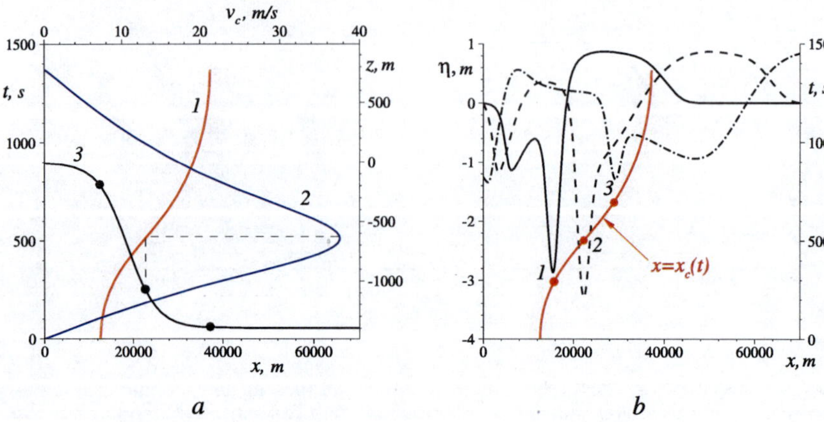

Fig. 4 *a* – graphs of the functions $x = x_c(t)$ *(1)*, $v = v_c(t)$ (2) and $z = h_{bt}(x)$ *(3)*; *b* – graphs of the free boundary at time moments $t = 300$ s *(1)*, $t = 500$ s (2) and $t = 700$ s *(3)*

distances covered by the landslide within these two time intervals are related in the nearly same way.

The velocity graph also shows that the landslide stop happens sharply (at non-zero negative value of acceleration).

Fig. 4 (*b*) shows the profiles of free boundary at time moments, close to the landslide's reaching the maximal velocity. The points show the positions of the landslide vertex at these moments of time (on the line $x = x_c(t)$). It is clear that the maximal deviation of the free boundary from unperturbed state is reached exactly above the landslide vertex, and the "hollow", which is formed here, moves together with the landslide. In contrast to the test problem, the amplitude of the "hollow" changes with time. Before the landslide the hydraulic bore is formed, which is pushed by the landslide to the direction of the depth increase and gradually separated from it with generation of the depression wave on the trailing edge (see line *3* in Fig. 4 (*b*)). On the left side from the "hollow", which accompanies the landslide, another depression wave is generated, moving to the direction of the shore.

The specified peculiarities of the landslide mechanism of wave generation are well illustrated by the Fig. 5 (*a*), which shows the surface $z = \eta(x,t)$. It is clear that the depression wave, which was moving to the shore, is reflected from the vertical wall (which is placed here) in the form of the hydraulic bore, which moves to the open sea.

The investigation of the free boundary dynamics during longer time periods show that the oscillations of the water level on the left wall continue after the reflection of the incoming depression wave. But the maximal vertical climb of wave happens during the interaction of this particular wave with the wall.

It is also seen that during the landslide deceleration the amplitude of the "hollow", going above the landslide, monotonously decreases. When the landslide stops, two oppositely directed waves with small amplitudes are formed.

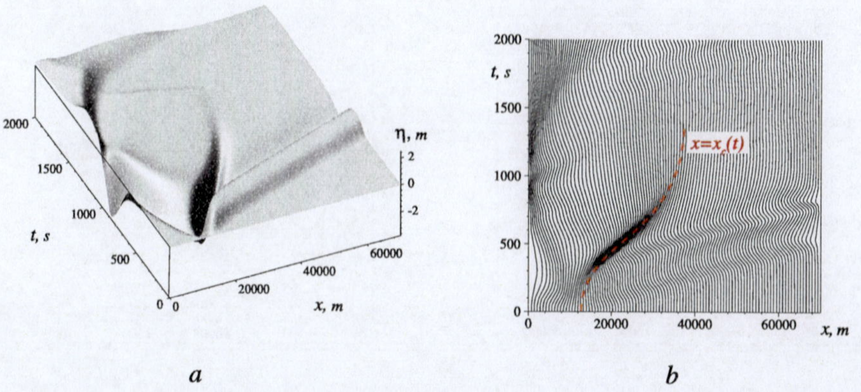

Fig. 5 Dynamics of the free surface (*a*) and trajectories of the nodes of the adaptive grid (*b*)

Fig. 5 (*b*) shows the trajectories of the nodes of the adaptive grid, which is condensed in the zones of maximal values of the quantity $|\eta(x,t)|$ like in the test problem above. It is clearly seen that the initially uniform grid reforms to the non-uniform grid and automatically traces hydraulic bores and depression waves.

The trajectory of the landslide vertex movement is depicted by a dashed line. It is seen that for large landslide velocities the grid has a strong condensation in the vicinity of this trajectory and is adapted to the moving "hollow". When the "hollow" amplitude decreases, the condensation disperses and the grid is adapted to other waves with dominated amplitudes.

5.3 *Dependency on Landslide Dimensions*

In order to investigate the influence of such characteristics as bottom slope, length, width and initial depth of the landslide on the wave processes, the numerical experiments are done, where one parameter is varied while others are kept the same as in Section 4.

Firstly, let us consider the dependency of the generated wave regime on the landslide width T. The volume of the model landslide (16) is defined using the formula $V = WTb/2$, therefore, the quantity T enters as a factor into all summands of the equation (13). Therefore, the landslide law of motion will not be changed when its width changes T (see Fig. 6 (*a*)).

The amplitude of the generated waves significantly depends on the landslide width T. Fig. 6 (*b*) shows the mareograms, written down at two mareograph points. The mareograph A is positioned on the shore and fixes the height of the free surface on the vertical wall (at the point $x = 0$). The moving mareograph B moves together with the landslide, has the abscissa $x = x_c(t)$ and measures the depth of the "hollow" above the landslide. It is seen that this depth monotonously increases with an increase of T, that is in a qualitative correspondence with the solution of the test

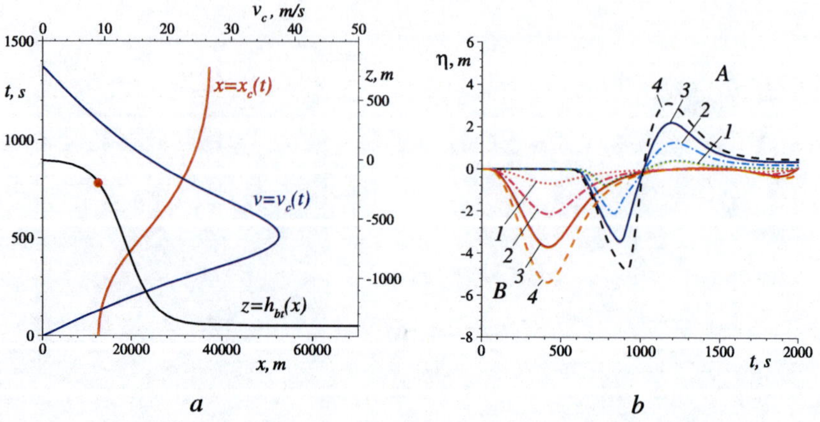

Fig. 6 Influence of the landslide width T on the characteristics of the landslide movement and the generated waves. $b = 5000$ m; $d_c = -200$ m; $\theta_0 = 6°$. a – graphs of functions $x = x_c(t)$, $v = v_c(t)$ and $z = h_{bt}(x)$; b – mareograms A and B. $T = 5$ m (*1*); 15 m (2); 25 m (*3*); 35 m (*4*)

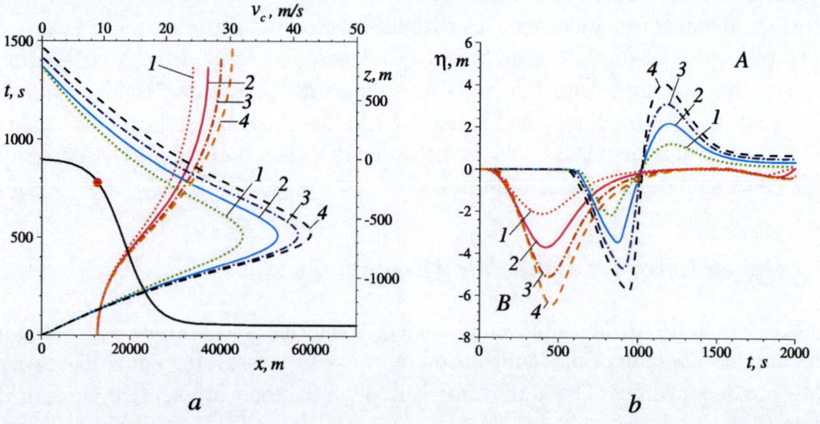

Fig. 7 The influence of the landslide length b on the characteristics of the landslide and generated waves motion. $T = 25$ m; $d_c = -200$ m; $\theta_0 = 6°$. a – graphs of the functions $x = x_c(t)$, $v = v_c(t)$ and $z = h_{bt}(x)$; b – mareograms A and B. $b = 3000$ m (*1*); 5000 m (*2*); 7000 m (*3*); 9000 m (*4*)

problem (see the behaviour of the line *1* in Fig. 2 (*b*)). Let us note, that for every fixed value of T the depth of the "hollow" reaches its maximal value on the acceleration interval, and this happens significantly earlier than the moment when the landslide reaches its maximal velocity.

The increase of the "hollow" depth, as the landslide width increases, contributes to the growth of the depression wave amplitude, coming at the shore, and as a consequence to the growth of the vertical splash (see mareograms *A*). In addition, the nearly linear character of all mentioned dependencies on the value of T is observed.

The increase of the landslide length b leads to the same qualitative changes of the mareograms (see Fig. 7 (*b*)), as the increase of the landslide width T. The law of motion is influenced by the change of b so that a longer landslide moves further and with a greater velocity. As a consequence, it passes a longer distance on a slope than a shorter landslide (see Fig.7 (*a*), where the circle depicts a start point for the landslide).

5.4 *Dependency of Initial Depth of Landslide and Bottom Slope*

In Fig. 8 (*a*) the circles depict the starting points for the landslides with different initial depths d_c. It is seen that for any initial depth the stopping points are practically identical, but movement times and maximal velocities differ significantly: they are larger for "shallow-water" landslides than for "deep-water" landslides. Correspondingly, the amplitudes of generated waves are larger for the landslides starting from a smaller depth (see Fig. 8 (*b*)), that is in qualitative correspondence with the results of the solution of the test problem (see the behaviour of the line *3* in Fig. 2 (*b*)).

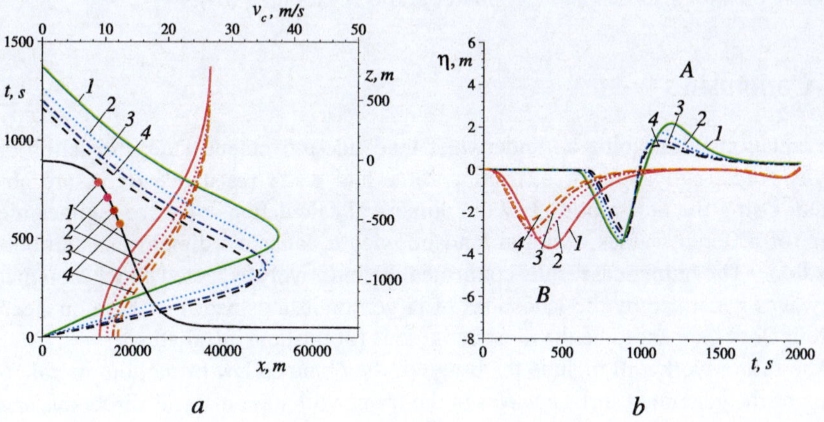

Fig. 8 The influence of the landslide initial depth d_c on the characteristics of the landslide and generated waves motion. $T = 25$ m; $b = 5000$ m; $\theta_0 = 6°$. *a* – graphs of the functions $x = x_c(t)$, $v = v_c(t)$ and $z = h_{bt}(x)$; *b* – mareograms *A* and *B*. $d_c = -200$ m (*1*); -300 m (*2*); -400 m (*3*); -500 m (*4*)

Fig. 9 (*a*) shows the graphs of the trajectories of landslides and their velocities for different values of the slope angle θ_0 of the bottom profile in the flex point. Initial depths are identical, therefore, the landslide starting points (depicted by circles) on more steeper slopes are positioned closer to the shore than on less steeper slopes.

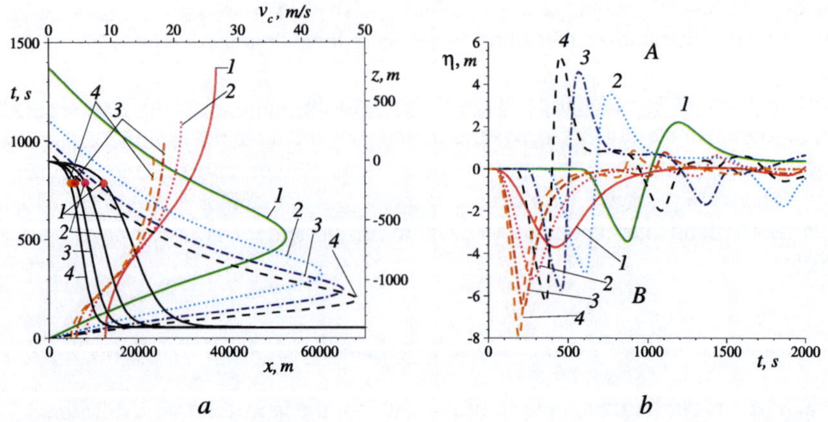

Fig. 9 The influence of the bottom slope on the characteristics of the landslide and generated waves motion. $T = 25$ m; $b = 5000$ m; $d_c = -200$ m. *a* – graphs of the functions $x = x_c(t)$, $v = v_c(t)$ and $z = h_{\rm bt}(x)$; *b* – mareograms *A* and *B*. $\theta_0 = 6°$ (*1*); 9° (*2*); 12° (*3*); 15° (*4*)

A greater slope steepness leads to a greater landslide velocity, but to a smaller acceleration interval and movement time. With increase in the slope steepness the splashes on wall also increase (see mareograms *A* on Fig. 9 (*b*)).

6 Conclusion

The equations, describing an underwater landslide movement under the action of gravity force, buoyancy force, friction force and water resistance force, are obtained. Using the new motion law the numerical calculations of wave regimes are done for different values of initial landslide depth, landslide dimensions and bottom slope. The numerical results confirmed the intuitively expected conclusion that the waves generated by the landslides of large dimensions near the shore on steep bottom slopes appear to be the most dangerous for onshore installations.

The future work will include the usage of the obtained law of motion for calculation of the generated surface waves in the framework of non-linear-dispersive and full hydrodynamic models.

Acknowledgements. This work has been supported by the Russian Foundation for Basic Research (grants 09-05-00294, 10-05-91052-NTSNI), the President's Program "Leading scientific schools of Russian Federation (grant NSH-6068.2010.9) and the Fundamental Research Program of SB RAS (project IV.31.2.1).

References

1. Beisel, S.A., Chubarov, L.B., Shokin, Y.I.: Some features of the landslide mechanism of surface waves generation in real basins. This book
2. Watts, P., Imamura, F., Grilli, S.T.: Comparing model simulations of three benchmark tsunami generation cases. Sci. Tsunami Hazards 18(2), 107–123 (2000)
3. Chubarov, L.B., Eletskii, S.V., Fedotova, Z.I., Khakimzyanov, G.S.: Simulation of surface waves generation by an underwater landslide. Rus. J. Numer. Anal. Math. Model. 20(5), 425–437 (2005)
4. Watts, P., Grilli, S.T., Kirby, J.T., Fryer, G.J., Tappin, D.R.: Landslide tsunami case studies using a Boussinesq model and a fully nonlinear tsunami generation model. Natur. Hazards Earth Syst. Sci. 3(5), 391–402 (2003)
5. Shokin, Y.I., Khakimzyanov, G.S.: Construction of monotonic schemes on the basis of method of differential approximation. In: Krause, E., Shokin, Y.I., Resch, M.M., Shokina, N. (eds.) Computational Science and High Performance Computing II. The 2nd Russian-German Advanced Research Workshop, Notes on Numerical Fluid Mechanics and Multidisciplinary Design (NNFM), Stuttgart, Germany, March 23-27, vol. 99, pp. 13–20 (2006)
6. Shokin, Y.I., Sergeeva, Y.V., Khakimzyanov, G.S.: Predictor-corrector scheme for the solution of shallow water equations. Rus. J. Numer. Anal. Math. Model. 21(5), 459–479 (2006)
7. Khakimzyanov, G.S., Shokin, Y.I., Barakhnin, V.B., Shokina, N.Y.: Numerical Modelling of Fluid Flows with Surface Waves. SB RAS Publishing House, Nobosibirsk (2001) (in Russian)

Simulation of Geophysical Problems with DUNE-FEM

Slavko Brdar, Andreas Dedner, Robert Klöfkorn,
Mirko Kränkel, and Dietmar Kröner

Abstract. In this work we present simulations of different types of geophysical problems using the DUNE and DUNE-FEM software framework. We consider two-phase flow in porous media, a Stokes-Darcy coupled problem, and atmospheric flow problems. The basis of our schemes is the Discontinuous Galerkin discretizations.

1 Introduction

The main aspect of the DUNE and DUNE-FEM project is the development of an interface library for solving partial differential equations, especially non-linear systems of evolution equations. These systems are used in many different fields of applications from astrophysics to industrial problems to model the evolution of physical quantities U, e.g., density, momentum, and energy of a fluid. A very general form of these equations is

$$\partial_t U + \nabla \cdot (F(U) - A(U)\nabla U) = S(U) + \mathscr{T}[U]. \quad (1)$$

This is an advection-diffusion-reaction equation including a non-local operator $\mathscr{T}$; this last term is used to model for example self-gravitation or radiation in astrophysics. Our goal is to develop a general framework for solving this type of equations based on modern software design techniques. Hereby, efficiency and the reuse of code is a major aspect in the design process. Furthermore, the interface should allow for a straight forward modification of both the model and the underlying scheme, so that the development and testing of new numerical schemes becomes an easy task. Also, due to the complexity of the applications mentioned above, parallelization combined with local grid adaptation is a central ingredient for the

Slavko Brdar · Andreas Dedner · Robert Klöfkorn · Mirko Kränkel · Dietmar Kröner
Section of Applied Mathematics, University of Freiburg i. Br., Hermann-Herder-Straße 10, Freiburg i. Br., 79104, Germany
e-mail: dedner, slavko, robertk, kraenkel,
dietmar@mathematik.uni-freiburg.de

E. Krause et al. (Eds.): Computational Sci., & High Performance Computing IV, NNFM 115, pp. 93–106.
springerlink.com

implementation of efficient numerical schemes. To be able to concentrate on the underlying applications and on the design of new numerical schemes, aspects like load-balancing in a distributed computational environment should not be a concern of the researcher working on the project; the underlying software library should handle all these aspects in a transparent manner.

2 The DUNE Interface Library – Recent Development

All the design restrictions mentioned in the introduction require the use of modern software design techniques. These strategies have already been successfully applied during the development of the DUNE grid interface [2, 3, 4]. The grid interface library allows the generic access to the grid structure which forms the basis of most numerical schemes for solving partial differential equations. Lately, version 2.0 of DUNE (consisting of the the DUNE core modules: DUNE-COMMON, DUNE-GRID, DUNE-ISTL) has been released showing that the development on the grid interface and the iterative solver template library has reached a stable state. Furthermore, a new module DUNE-LOCALFUNCTIONS has been added to the DUNE core modules, which provides interfaces for finite element basis functions; available implementations include Lagrange, Raviart-Thomas, and Nedelec type finite elements.

Based on the DUNE core modules we are developing the package DUNE-FEM of which version 1.0 has been released [13]. In this package, interfaces for discrete functions and operators are defined and a wide range of examples are implemented. The structure of the interfaces closely follows the mathematical formalism used to define grid based numerical schemes, e.g., the finite-element framework found, for example, in [9]. The package thus includes interfaces for

- Function spaces and functions
- Discrete function spaces (combining function space and vector valued finite base function set), DG and Lagrange higher order implemented
- Discrete functions (with element wise representation, dof handling)
- Discrete spatial operators (with efficient operations, e.g., $+$ and $\circ$) DG for elliptic and hyperbolic operators, CG for elliptic operators
- Inverse operators (Newton, Krylov methods...)
- IMEX Runge-Kutta methods for time dependent problems
- Automatic handling of grid adaptation, parallelization, and load balancing

During the design process the efficient and easy use of parallel computers has always been a central goal. In the following we present some geophysical flow problems which have been solved using DUNE and in particular DUNE-FEM.

3 Two-Phase Flow in Porous Media

In this section we present results for two-phase flow in porous media using the DG discretizations described in detail in [21]. We consider the following model problem for two-phase flow in porous media (see also [15]). The flow of the wetting phase w (such as water) and the non-wetting phase o (such as oil) is described by Darcy's law and the continuity equation for each phase. We consider the domain $\Omega \subset \mathbb{R}^d$, $d \in \{1,2,3\}$ and assume that the boundary of the domain is divided into three disjoint open sets $\partial\Omega = \Gamma_{\text{in}} \cup \Gamma_{\text{out}} \cup \Gamma_{\text{no}}$. We use a model based on a global pressure formulation.

3.1 Two-Phase Flow Problem

We restrict our consideration to incompressible fluid flow, i.e., the densities $\rho_{w,}\rho_o$ are constant. Furthermore, we make the assumption that the porosity ϕ is constant over the entire domain. Using the wetting phase saturation $s := s_w$ and global pressure p we obtain

$$-\nabla \cdot (\lambda \mathbf{K} \nabla p) = 0, \qquad \text{in } \Omega, \tag{2}$$

$$\mathbf{u} = -\lambda \mathbf{K} \nabla p, \qquad \text{in } \Omega, \tag{3}$$

and $p = p_{\text{in}}$ on Γ_{in}, $p = p_{\text{out}}$ on Γ_{out}, and $(\lambda \mathbf{K} \nabla p) \cdot \mathbf{n} = 0$ on Γ_{no} with the global velocity $\mathbf{u} = \mathbf{v}_w + \mathbf{v}_o$ and $\mathbf{n}$ denoting the outward unit normal to Γ_{no}. For the wetting phase saturation s we obtain

$$\phi\, \partial_t s + \nabla \cdot \mathbf{F}_w(\mathbf{u},s) - \nabla \cdot (\mathbf{D}(s) \nabla s) = 0 \qquad \text{in } (0,T] \times \Omega, \tag{4}$$

$$s(0,\cdot) = s_0(\cdot) \qquad \text{in } \Omega,$$

where $\mathbf{F}_w(\mathbf{u},s) = \mathbf{u}\lambda_w(s)/\lambda(s)$ is the convective flux and $\mathbf{D}(s)\nabla s$ the diffusive flux with $\mathbf{D}(s) := \big((\lambda_w(s)\,\lambda_o(s))/\lambda(s)\big)|p_c'(s)|\mathbf{K}$. On the boundary we have

$$s = s_{\text{dir}} \text{ on } (0,T] \times \Gamma_{\text{in}}, \qquad \mathbf{D}(s)\nabla s \cdot \mathbf{n} = 0 \text{ on } (0,T] \times \Gamma_{\text{out}},$$

and

$$(\mathbf{F}_w(s) - \mathbf{D}(s)\nabla s) \cdot \mathbf{n} = 0 \text{ on } (0,T] \times \Gamma_{\text{no}}.$$

The permeability coefficients $\mathbf{K}$ in equation (3) and (3) can be discontinuous in space and may vary over several orders of magnitude. The coefficients λ_w and λ_o are the mobilities for the wetting and the non-wetting phase, respectively and are given by the ratios of the relative permeabilities k_{r_α} and the viscosities μ_α $\lambda_\alpha = \frac{k_{r\alpha}}{\mu_\alpha}$, $\alpha = w,o,$; the relative permeabilities are non-linear functions of the wetting phase saturation, i.e., $k_{r_\alpha} = k_{r_\alpha}(s_w)$. We follow [15] by choosing the Brooks-Corey model, $k_{rw}(s) = s^{\frac{2+3\theta}{\theta}}$, $k_{ro}(s) = (1-s)^2(1-s^{\frac{2+\theta}{2}})$. The additional parameter $\theta \in [0.2, 3]$ characterizes the inhomogeneity of the medium. The total mobility is $\lambda = \lambda_w + \lambda_o$. In addition the following closure relations must also be satisfied:

$$s_w + s_o = 1, \qquad p_o - p_w = p_c(s_w), \tag{5}$$

where $p_c = p_c(s_w)$ is the capillary pressure, a function of the wetting phase saturation given by $p_c(s) = p_d s^{-1/\theta}$. Here, $p_d > 0$ is the constant entry pressure, needed to displace the fluid from the largest pore.

Equation (4) is in the form of Equation (1) and we apply the DG discretizations presented in [21] (e.g., a stabilized Runge-Kutta Discontinuous Galerkin (RK-DG) method for the saturation equation and the Compact Discontinuous Galerkin (CDG) method for the pressure equation) which are available in the DUNE-FEM package.

3.1.1 Discretization

The global pressure is calculated by using equation (3). This equation is discretized by using the CDG method described in [21, Cha. 9]. With k_p we denote the polynomial order of the basis functions used to construct the Discontinuous Galerkin space for the pressure.

The global velocity (needed in the saturation equation) is calculated as described by equation (3). The discretization of this equations corresponds to the discretization of the auxiliary variable $\sigma_{\mathscr{G}}$ in the CDG method. For the Discontinuous Galerkin space for the global velocity we use the same polynomial degree as for the pressure.

In [5] a post processing procedure is described (and tested for triangular grids) for the global velocity, aimed at avoiding oscillations due to fluctuations in the velocity average used in the flux calculations of the saturation equation. The projection guarantees that the normal components of the velocity are continuous across element intersections. The problem is that this projection is very complicated to implemented for higher order polynomials especially in 3d. Furthermore, investigations in [21] show that small oscillations occur although the projection is applied.

Another, more promising, approach (that has been tested in [21]) is suggested in [17]. The idea is to project the global velocity to a Raviart-Thomas space of the same polynomial order under consideration of an appropriate flux correction. Preliminary 1d results for two-phase flow in porous media are presented in [16] (for more details concerning the implementation see [21]).

The saturation is calculated by discretizing equation (4) using the stabilized RK-DG method (cf. [21, Cha. 6 and 7]) and the CDG method for the diffusive part (cf. [21, Cha. 9]). With k_s we denote the polynomial order of the basis functions used to construct the Discontinuous Galerkin space for the saturation. For the time discretization of the saturation equation an explicit Runge-Kutta method of order $k_s + 1$ is used (see [21, Sec. 6.3]).

The system is decoupled by using an Implicit Pressure – Explicit Saturation (IMPES) scheme. The classical IMPES method is for example described in [8] or [18]. The algorithm is the following, assuming that the initial data for the saturation is given and all boundary conditions are met. For $n \geq 0$:

1. Calculate $p_{\mathscr{G}}^{n+1}$ by using $-\nabla \cdot \left(\lambda(s_{\mathscr{G}}^n)\mathbf{K}\nabla p_{\mathscr{G}}^{n+1}\right) = 0$.
2. Calculate $\mathbf{u}_{\mathscr{G}}^{n+1}$ by using $\mathbf{u}_{\mathscr{G}}^{n+1} = -\lambda(s_{\mathscr{G}}^n)\mathbf{K}\nabla p_{\mathscr{G}}^{n+1}$.
3. Finally proceed to next time step by calculating $s_{\mathscr{G}}^{n+1}$ using $\phi\partial_t s_{\mathscr{G}} = \mathscr{L}_{\text{expl}}[\mathbf{u}_{\mathscr{G}}^{n+1}, s_{\mathscr{G}}^n]$ with $\mathscr{L}_{\text{expl}}[\mathbf{u}_{\mathscr{G}}^{n+1}, s_{\mathscr{G}}^n] = -\nabla \cdot \left(\mathbf{F}_w(\mathbf{u}_{\mathscr{G}}^{n+1}, s_{\mathscr{G}}^n)\right) + \nabla \cdot \mathbf{D}(s_{\mathscr{G}}^n)\nabla s_{\mathscr{G}}^n$.

3.2 Numerical Results for the Two-Phase Flow Problem

In this section we present numerical results for the two-phase flow problem described in the previous section. For a heterogeneous medium we consider the test case presented in [15]. This allows us to compare our results with those obtained in [15]. The setup is the following.

Test Case 1 Heterogeneous flow

Computational domain:	$\Omega_{\mathscr{G}} := [0,100]^d$, $\Omega_1 := [37.5,75] \times [25,75]$, $\Omega_2 := \Omega_{\mathscr{G}} \setminus \Omega_1$, $d = 2$
Time interval:	$t \in [0,550]$
Permeability tensor:	$\mathbf{K} = 10^{-12}\,\mathrm{Id}_d$ in Ω_1, and $\mathbf{K} = 10^{-8}\,\mathrm{Id}_d$ in Ω_2
Brooks-Corey:	$\theta = 3$ and $p_d = 10^3$
Viscosity:	$\mu_w = 10^{-3}$, and $\mu_o = 10^{-2}$
Porosity:	$\phi = 0.2$
Boundary conditions:	$\Gamma_{\mathrm{in}} = \{0\} \times (0,100)$, $\Gamma_{\mathrm{out}} = \{100\} \times (0,100)$, $\Gamma_{\mathrm{no}} = \partial\Omega_{\mathscr{G}} \setminus (\Gamma_{\mathrm{in}} \cup \Gamma_{\mathrm{out}})$, $p_w^{\mathrm{in}} = 3 \times 10^6$, $p_w^{\mathrm{out}} = 10^6$, $s_{\mathrm{dir}} = 0.85$ Using (5) and the definition of the global pressure [21, (11.6)] p_{in} and p_{out} can be calculated from p_w^{in} and p_w^{out}, respectively.
Initial data:	$s_0 = 0.2$ in $\Omega_{\mathscr{G}}$
Velocity projection:	H-div projection from [5].

In Fig. 1(a) and 1(b) we see the pressure and saturation distribution for Test Case 1 at time $t = 550$ for polynomial degrees $(k_p, k_s) = (2,1)$. One can see that the zone of low permeability Ω_1 is captured on all grid levels, in the pressure as well as in the saturation profile. Water is hardly entering the domain Ω_1. In [15] for $\triangle x = \triangle y = 6.25$ the area of Ω_1 is flooded by water, i.e., the lower order scheme in [15] is not able to capture this area, results on finer grids are not presented.

The results for our higher order method with $(k_p, k_s) = (3,2)$ are presented in Fig. 1(c) and 1(d). One can see that the choice of a higher order method increases the approximation quality, the gradient of the saturation around the boundary of Ω_1 is very steep. Furthermore, overshoots and undershoots are kept at a minimum which indicates that the limiter based stabilization technique presented in [12] (see also [21, Cha. 7]) also works well for this kind of problem. Comparing our results with the ones obtained in [15] we see that the resolution of our method is superior to the one obtained with the higher order fully implicit schemes in [15]. Furthermore, the results obtained in [15] show some slight oscillations which do not occur with our method of higher order.

To conclude this section we state that with the considered DG discretizations a higher order alternative for discretizing two-phase flow problems in porous media was presented. The critical point here is the correct approximation of the global velocity. There are preliminary 1d results in [16] showing that an appropriate flux correction is needed. However, for the CDG method these are not yet available and therefore an active part of our ongoing research.

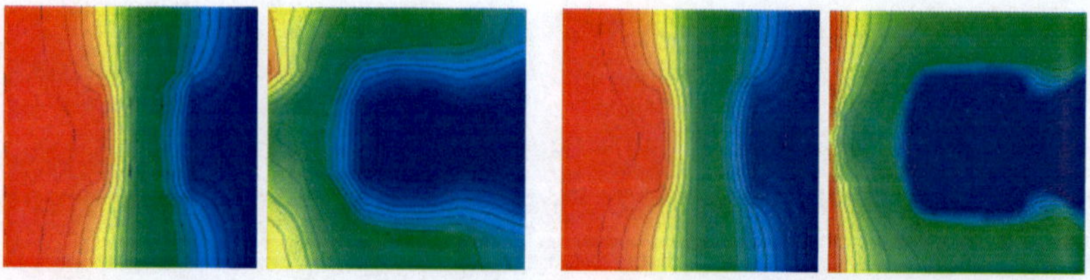

(a) p ($k_p = 2$) and s ($k_s = 1$), $h = 6.25$ (b) p ($k_p = 2$) and s ($k_s = 1$), $h = 1.5625$

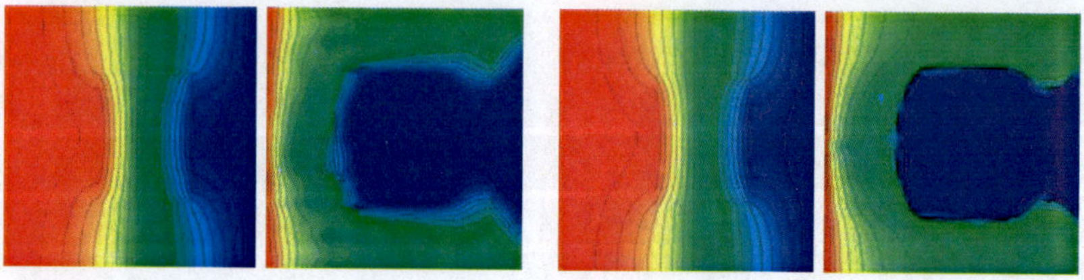

(c) p ($k_p = 3$) and s ($k_s = 2$), $h = 6.25$ (d) p ($k_p = 3$) and s ($k_s = 2$), $h = 1.5625$

Fig. 1 Results for Test Case 1 at time $t = 550$ on grids with $h = \triangle x = \triangle y = 6.25$ (left) and $h = \triangle x = \triangle y = 1.5625$ (right)

4 The Stokes-Darcy Problem

The Stokes-Darcy problem is a model for coupling free fluid and flow through porous media. A typical example for this situation would be river flowing over sand. In this situation the free fluid flow should be governed by the incompressible Navier-Stokes equations or it's simplifications and the flow in the porous medium is described by Darcy's law. This leads to two significantly different PDEs in the two flow regimes. Therefore, the right coupling conditions at the interface must be used. The common interface condition, which seems to be in good agreement with experimental observations are the so called Beaver-Joseph conditions. In [19] the use of these conditions is justified by results based on homogenization theory.

4.1 The Flow Equations

The motion of the fluid in the free flow regime is governed by the Stokes equation as a simplification of the full Navier-Stokes equations:

$$\begin{aligned} \partial_t \mathbf{V}_f - \nabla \cdot \mathbf{T}(\mathbf{V}_f, \mathbf{p}_f) &= g_f \quad \text{in} \quad \Omega_f \quad \text{(conservation of momentum)} \\ \nabla \cdot \mathbf{V}_f &= 0 \quad \text{in} \quad \Omega_f \quad \text{(conservation of mass).} \end{aligned} \tag{6}$$

Here $\mathbf{V}_f$ is the velocity, $\mathbf{p}_f$ the pressure,

$$\mathbf{T}(\mathbf{V}_f, \mathbf{p}_f) := -p_f \mathbf{I} + 2\mu \mathbf{D}(\mathbf{V}_f) \tag{7}$$

denotes the stress tensor where

$$\mathbf{D}(\mathbf{V}_f) = \frac{1}{2}(\nabla \mathbf{V}_f + (\nabla \mathbf{V}_f)^T) \tag{8}$$

is the deformation rate tensor. Furthermore, we have the kinematic velocity μ and the external force g_f.

In the porous medium Darcy's law for the piezometric head ϕ and the Darcy velocity $\mathbf{q}$ is considered, where the Darcy velocity $\mathbf{q}$ is proportional to the fluid velocity $\mathbf{V}_p$.

$$\begin{aligned} S_0 \partial_t \phi + \nabla \cdot \mathbf{q} = g_p \quad & \text{in} \quad \Omega_p \quad \text{(conservation of mass)} \\ \mathbf{q} = -\mathbf{K} \nabla \phi \quad & \text{in} \quad \Omega_p \quad \text{(Darcy's law)} \end{aligned} \tag{9}$$

with S_0 being the specific storage capacity and $\mathbf{K}$ the permeability tensor.

4.2 The Beaver-Joseph Interface Condition

The coupling conditions at the interface Γ are

- Continuity of normal velocities:

$$\mathbf{V}_f \cdot \mathbf{n}_f + \mathbf{V}_p \cdot \mathbf{n}_p = 0, \tag{10}$$

  where $\mathbf{n}_f$ is the outer normal at the free flow regime and $\mathbf{n}_f = -\mathbf{n}_p$.
- Balance of normal forces:

$$-[\mathbf{T}(\mathbf{V}_f, \mathbf{p}_f) \cdot \mathbf{n}_f] \cdot \mathbf{n}_f = \rho_f g \phi. \tag{11}$$

- The Beaver-Joseph condition:
  This condition relates the shear stress to the difference of the tangential velocities at the interface Γ

$$-[\mathbf{T}(\mathbf{V}_f, \mathbf{p}_f) \cdot \mathbf{n}_f] \cdot \tau_i = \frac{\alpha}{\sqrt{\tau_i \cdot \mathbf{K} \cdot \tau_i}} (\mathbf{V}_f - \mathbf{V}_p) \cdot \mathbf{K} \cdot \tau_i, \quad i = 1, \ldots, d-1, \tag{12}$$

  where $\{\tau_i\}$,i=1,...,$d-1$ form an orthonormal basis of the tangential space at Γ.

- The Beaver-Joseph-Saffman condition:
  In the case that the tangential part of the Darcy velocity is small compared to the tangential velocity in the free fluid domain, this term can be neglected and one obtains the Beaver-Joseph-Saffman condition

$$-[\mathbf{T}(\mathbf{V}_f, \mathbf{p}_f) \cdot \mathbf{n}_f] \cdot \tau_i = \frac{\alpha}{\sqrt{\tau_i \cdot \mathbf{K} \cdot \tau_i}} (\mathbf{V}_f) \cdot \tau_i, \quad i = 1, \ldots, d-1. \tag{13}$$

4.3 Discretization

Both the Stokes and the Darcy equations are treated by a local Discontinuous Galerkin method given in [10] resp. [1]. Due to the interface conditions the two problems are coupled and therefore this discretization leads to a large indefinite linear system. Here the system is solved by a GMRES-solver. In the numerical experiments convergence of the linear solver is only obtained, when the GMRES-method is used without restarting, which leads to a large memory consumption and makes further research concerning the preconditioning of the system necessary.

4.4 Numerical Results

In Fig. 2 some numerical results for the Stokes-Darcy problem are shown. The Problem was solved on a triangular grid. In the discretization element-wise polynomial and orthonormal basis functions of degree 2 for $\mathbf{V}_f$ and ϕ resp. 1 for $\mathbf{p}_f$ were used. In order to test the method an analytical solution was constructed, which satisfies the differential equations and the interface conditions. In our test case the right hand side and the boundary data were chosen so that

$$\begin{aligned}
\mathbf{V}_{f1}(x,y) &:= x^2y+x^2+y^2+xy+x \\
\mathbf{V}_{f2}(x,y) &:= \frac{1}{3}x^3-xy^2+\frac{1}{2}x^2-\frac{1}{2}y^2-2xy+2x-y \\
\mathbf{p}_f(x,y) &:= \frac{1}{3}x^3-\frac{1}{3}x^3y-\frac{1}{2}x^2y+\frac{1}{2}x^2+\frac{1}{2}xy^2+\frac{3}{4}y^2-\frac{17}{2}x-4 \\
\phi(x,y) &:= \frac{1}{3}x^3-\frac{1}{3}x^3y-\frac{1}{2}x^2y+\frac{1}{2}x^2+\frac{1}{2}xy^2+\frac{3}{4}y^2-\frac{1}{2}x
\end{aligned}$$

are a solution for the Stokes-Darcy problem.

Table 2 Convergence study for the Stokes-Darcy problem. L^2-error of the piezometric head and the pressure. The decrease in the EOC is due the limitations of the accuracy of the GMRES solver caused by the high memory consumption of this method.

Elements	$\|\|\phi-\phi_h\|\|_2$	EOC	$\|\|\mathbf{p}_f-\mathbf{p}_{f_h}\|\|_2$	EOC
64	0.000237289		0.00769258	
256	3.03249e-05	2.968070429	0.00200995	1.936307929
1024	3.92356e-06	2.950267720	0.000540779	1.894048580
4094	1.96588e-06	0.9969880020	0.000378242	0.5157295599

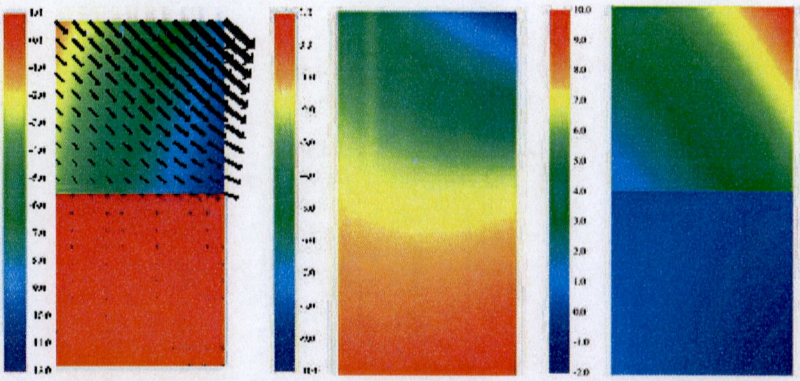

(a) velocity and pressure (b) normal velocity (c) tangential velocity

Fig. 2 Numerical solution of the Stokes-Darcy Problem, where the free fluid flow is in the upper domain, the porous medium flow in the lower part of the figure. The pressure and the tangential part of the velocity are discontinuous across the interface, the normal components of the velocity are continuous

5 Atmospheric Simulations

In this work we apply the DG scheme for simulating two- and three-dimensional atmospheric flow problems. The mathematical model that lies behind these simulations is represented by the fully compressible Euler equations including the gravitational force; atmospheric factors like Coriolis force, viscosity and turbulence are neglected in this stage. We take into account the air moisture (water vapor, cloud water) and the phase change between the moisture components, whereas total mass of the dry air remains constant over time. The presented model is able to give predictions for warm non-precipitating clouds, meaning no ice structures and no rain is assumed. The modeling of phase changes requires special attention. In this work we use *the Kessler scheme* described in [14] to include the phase changes in our model. Due to the design of our scheme more complex models can be easily added.

5.1 The Euler Equations with Moisture

To model the atmospheric flow problems, the evolution of the density of each air component ρ_d, ρ_v, ρ_c, of the velocity vector $\mathbf{v}$, and of one thermodynamic variable Θ have to be tracked. This evolution is described by *the compressible moist Euler equations*

$$\begin{aligned}
\partial_t \rho + \nabla \cdot (\rho \mathbf{v}) &= 0 \\
\partial_t (\rho \mathbf{v}) + \nabla \cdot (\rho \mathbf{v} \otimes \mathbf{v} + p\mathbf{I}) &= -\rho \mathbf{g} \\
\partial_t \Theta + \nabla \cdot (\Theta \mathbf{v}) &= S_\Theta(\rho, \rho_v, \rho_c, p) \\
\partial_t \rho_v + \nabla \cdot (\rho_v \mathbf{v}) &= S_v(\rho_v, \rho_c, T) \\
\partial_t \rho_c + \nabla \cdot (\rho_v \mathbf{v}) &= -S_v(\rho_v, \rho_c, T) \\
p &= p_o \left(\frac{\Theta R_d}{p_0} \right)^{1/(1-\kappa_m)}
\end{aligned} \tag{14}$$

where

$\Theta = \rho T \left(\frac{p_0}{p}\right)^{\kappa_m} \frac{R_m}{R_d}$	moist potential temperature
$R_m = R_d + R_v \frac{\rho_v}{\rho}$	gas constant of moist air
$\kappa_m = \frac{R_m}{c_{pml}}$	
$c_{pml} = c_{pd} + c_{pv} \frac{\rho_v}{\rho} + c_{pl} \frac{\rho_c}{\rho}$	specific heat of moist air at constant pressure
$p = \rho R_m T$	total pressure
$\rho = \rho_d + \rho_v + \rho_c$	total density (*d*=dry, *v*=vapor, *c*=water)
S_v, S_Θ	source functions
c_{pd}, c_{pv}, c_{pl}, R_d, R_v	physical constants

In Eqn. (14) the source term S_v describes the condensation-evaporation process. To derive the S_v function we follow the approach in [14]. Numerous measurements have shown that more than 99% of a cloud is in saturation equilibrium, meaning that only two different thermodynamic states can occur

$$\textit{the saturated cloud} \text{ case: } \rho_v = \rho_{v,sat}(T, p) \text{ and } \rho_c > 0,$$
$$\textit{the subsaturated no-cloud} \text{ case: } \rho_v < \rho_{v,sat}(T, p) \text{ and } \rho_c = 0,$$

where $\rho_{v,sat}(T) = \frac{p_{v,sat}(T)}{R_v T}$ and $p_{v,sat}$ is pressure of water vapor at the saturation point given as

$$p_{v,\mathrm{sat}} = p_{0,\mathrm{sat}} \left(\frac{T}{T_0} \right)^{(c_{pv} - c_{pl})/R_v} \exp\left(\frac{1}{T_0} - \frac{1}{T} \right).$$

Situations occur where we have *subsaturation with cloud water* ($p_v < p_{v,sat}$ and $q_c > 0$) as one case in which a condensation ($\rho_c \to \rho_v$, $S_v > 0$) should take place, or *supersaturation* ($p_v > p_{v,sat}$ and $q_c = 0$) where evaporation ($\rho_v \to \rho_c$, $S_v < 0$) takes place. In either case a source function S_v prescribed as

$$S_v(\rho_v, \rho_c, T) = \alpha_{sat} \left\{ \left(\frac{p_{v,\mathrm{sat}}(T)}{R_v T} - \rho_v \right) + \rho_c - \sqrt{\left(\frac{p_{v,\mathrm{sat}}(T)}{R_v T} - \rho_v \right)^2 + \rho_c^2} \right\}$$

will balance out water vapor and cloud water towards the saturation equilibrium. The constant α_{sat} prescribes the intensity of the phase change, and by comparing our simulation results with the one in [7] we found out that $\alpha_{sat} = 1$ would be a good choice.

The *subsaturation with cloud water* and *supersaturation* cases have to be brought isobarically to the saturation equilibrium by adjusting the temperature T, water vapor ρ_v and cloud water ρ_c. Taking into account latent heat release $l_{\text{heat}}(T) = L_0 + (c_{pv} - c_{pl})T$ associated with the phase changes one can derive the moist potential temperature source function as follows

$$S_\Theta(\rho, \rho_v, \rho_c, p, T) = S_v \cdot \left(-\frac{l_{\text{heat}}(T)}{T c_{pml}} - \kappa_m \log\left(\frac{p}{p_0}\right)\left(\frac{R_v}{R_m} - \frac{c_{pv}}{c_{pml}}\right) + \frac{R_v}{R_m} \right) \tag{15}$$

Notice that when only dry air is present in the atmosphere we get $S_\Theta = 0$, $S_v = 0$ and the equations Eqn. $(14)_4$ and Eqn. $(14)_5$ can be excluded from the mathematical model, leaving us with *the dry Euler equations.* In this case Θ becomes density potential temperature $\rho\theta = \rho T \left(\frac{p_0}{p}\right)^{R/c_{pd}}$.

5.2 Discretization

We adopt the approach of using a high order method on unstructured locally adapting grids. The use of unstructured grids allows for a better resolution of the orography, which in return makes it easier to formulate boundary conditions. One of the high order methods suitable for implementation on such grids is the Discontinuous Galerkin (DG) method. Therefore, we choose the DG method for spatial discretization of Eqn. (14) and combine it with Runge-Kutta (RK) methods for the time integration. In the following the focus is on testing the DG method for atmospheric simulations and we will not discuss the time integration, even though the design of efficient methods are a crucial issue for atmospheric simulations to avoid the severe time step limitation caused by fast acoustic waves (see [22, 20] for an approach to this problem). One other problem in the discretization Eqn. (14) is that some terms in line 2 of Eqn. (14) can be of different scales then the remaining ones, and since we are interested in the effects of the small scale terms, this problem has to be resolved adequately. This requires some form of *well-balancing*. We solve this problem using the *BgFix* scheme described in [11]; the approach is based on splitting the solution vector $\mathbf{u}$ into a large scale part $\bar{\mathbf{u}}$ and a small scale deviation $\tilde{\mathbf{u}}$ from $\bar{\mathbf{u}}$. The large scale part represents an atmosphere at rest and therefore satisfies the stationary form of Eqn. (14). The system Eqn. (14) is then rewritten to describe the evolution of the perturbations $\tilde{\mathbf{u}}$ without losing important aspects of the original system, e.g., the conservation property of the different physical quantities. A further problem arises in atmospheric simulation are the outer boundaries where some form of transparent boundary conditions have to be prescribed. Our approach to this problem is based on the *absorbing layer* technique (see [6]).

5.3 Numerical Results

In the atmospheric community there are several test cases for evaluating a numerical scheme. We show results for the *rising warm thermal bubble* test in [7] for the dry atmosphere, as well as *rising warm thermal bubble with moisture* test in [7] for moist atmosphere.

For the evaluation of the DG scheme for moist atmospheric flow, we use the benchmark problem in [7]. The setting for this problem is as follows: the initial atmosphere is neutrally stable with constant total water components $\rho_c + \rho_v = const$ and with the reversible phase changes: changes in $\rho_{v,sat}$, ρ_v and ρ_c are equal and there is not change in $\rho_v + \rho_c$. Inside a ball within the domain a perturbation from the initial atmosphere is introduced by increasing the dry potential temperature by 2° K. The results shown in Fig. 3 are comparable with [7].

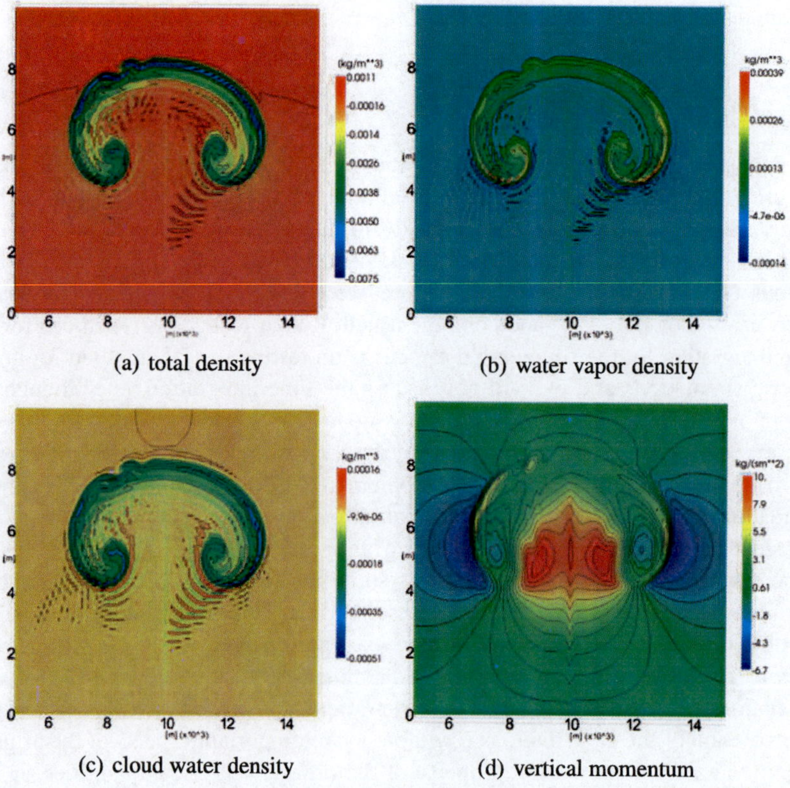

(a) total density (b) water vapor density

(c) cloud water density (d) vertical momentum

Fig. 3 Rising warm moist thermal bubble at t=2400 s

References

1. Arnold, D.N., Brezzi, F., Cockburn, B., Marini, L.D.: Unified analysis of discontinuous Galerkin methods for elliptic problems. SIAM J. Numer. Anal. 39(5), 1749–1779 (2002)
2. Bastian, P., Blatt, M., Dedner, A., Engwer, C., Klöfkorn, R., Kornhuber, R., Ohlberger, M., Sander, O.: A generic grid interface for parallel and adaptive scientific computing. II: Implementation and tests in dune. Computing 82(2-3), 121–138 (2008)
3. Bastian, P., Blatt, M., Dedner, A., Engwer, C., Klöfkorn, R., Ohlberger, M., Sander, O.: A generic grid interface for parallel and adaptive scientific computing. I: Abstract framework. Computing 82(2-3), 103–119 (2008)
4. Bastian, P., Droske, M., Engwer, C., Klöfkorn, R., Neubauer, T., Rumpf, M.: Towards a unified framework for scientific computing. In: Kornhuber, R., Hoppe, R., Keyes, D., Périaux, J., Pironneau, O., Xu, J. (eds.) Proceedings of the 15th Conference on Domain Decomposition Methods. LNCSE, vol. 40, pp. 167–174. Springer, Heidelberg (1976)
5. Bastian, P., Rivière, B.: Superconvergence and $H(div)$ projection for discontinuous Galerkin methods. Int. J. Numer. Methods Fluids 42(10), 1043–1057 (2003)
6. Bonaventura, L.: Numerical simulation of hydrostatic mountain waves. J. Atmos. Sci. 35, 78–107 (1978)
7. Bryan, G.H., Fritsch, J.M.: A benchmark simulation for moist nonhydrostatic numerical models. Mon. Weath. Rev. 130, 2917–2928 (2002)
8. Chen, Z., Huan, G., Ma, Y.: Computational methods for multiphase flows in porous media. Computational Science & Engineering, vol. 2, xxix 531 p. Society for Industrial and Applied Mathematics (SIAM), Philadelphia (2006)
9. Ciarlet, P., Lions, J.: Handbook of numerical analysis. Finite element methods (Part 1), vol. II. North-Holland, Amsterdam (1991)
10. Cockburn, B., Kanschat, G., Schotzau, D., Schwab, C.: Local discontinuous Galerkin methods for the Stokes system. SIAM J. Numer. Anal. 40(1), 319–343 (2002)
11. Dedner, A.: Solving the system of Radiation Magnetohydrodynamics: for solar physical simulations in 3d. Ph.D. thesis, Universität Freiburg (2003), `http://www.freidok.uni-freiburg.de/volltexte/1098/`
12. Dedner, A., Klöfkorn, R.: A generic stabilization approach for higher order Discontinuous Galerkin methods for convection dominated problems. Preprint no. 8 (submitted to SIAM Sci. Comput.), Mathematisches Institut, Unversität Freiburg (2008), `http://www.mathematik.uni-freiburg.de/IAM/homepages/robertk/postscript/dedner_kloefkorn_limiter.pdf`
13. Dedner, A., Klöfkorn, R., Nolte, M., Ohlberger, M.: A generic interface for parallel and adaptive discretization schemes. Abstraction principles and the DUNE-FEM module. Computing (to appear)
14. Doms, G., Förstner, J., Heise, E., Herzog, H.J., Raschendorfler, M., Schrodin, R., Reinhardt, T., Vogel, G.: A description of the nonhydrostatic regional model lm. part i: Physical parametrization. Internal Correspondence at the DWD, 1–126 (2005)
15. Epshteyn, Y., Rivière, B.: Fully implicit discontinuous finite element methods for two-phase flow. Appl. Numer. Math. 57(4), 383–401 (2007), doi:10.1016/j.apnum.2006.04.004
16. Ern, A., Mozolevski, I., Schuh, L.: Accurate velocity reconstruction for discontinuous Galerkin approximations of two-phase porous medium flows. C. R., Math., Acad. Sci. Paris 347(9-10), 551–554 (2009), doi:10.1016/j.crma.2009.02.011
17. Ern, A., Nicaise, S., Vohralík, M.: An accurate **H**(div) flux reconstruction for discontinuous Galerkin approximations of elliptic problems. C. R., Math., Acad. Sci. Paris 345(12), 709–712 (2007), doi:10.1016/j.crma.2007.10.036

18. Helmig, R.: Multiphase flow and transport processes in the subsurface: a contribution to the modelling of hydrosystems. Springer, Berlin (1997)
19. Jäger, W., Mikelic, A., Neuss, N.: Asymptotic analysis of the laminar viscous flow over a porous bed. SIAM J. Sci. Comput. 22(6), 2006–2028 (2001)
20. Klein, R.: Asymptotics, structure, and integration of sound-proof atmospheric flow equations. Theo. Comp. Fluid Dyn. 23(3), 161–195 (2009)
21. Klöfkorn, R.: Numerics for Evolution Equations - A General Interface Based Design Concept. Ph.D. thesis, Universität Freiburg (2009), `http://www.freidok.uni-freiburg.de/volltexte/7175/`
22. Restelli, M., Giraldo, F.: A conservative discontinuous galerkin semi-implicit formulation for the navier-stokes equations in nonhydrostatic mesoscale modeling. J. Sci. Comput. 31, 2231–2257 (2009)

On Modeling of Mechanical Properties of Fibrous Composites

Kseniya S. Golushko, Sergey K. Golushko, and Andrey V. Yurchenko

Abstract. The various models of fibrous composites are considered: the filament model, the model with one-dimensional fibres, the specified model with one-dimensional fibres, the model with two-dimensional fibres. The parametrical analysis of a stressed-deformed state (SDS) of multilayered reinforced toroidal rubber-cord shells and pressure vessels is made. The influence of selection of composite material models on simulation results for composite constructions is investigated.

1 Introduction

Thin plates and shells are common elements of many modern structures and designs within the aircraft and shipbuilding, mechanical engineering, the petroleum, gas and chemical industry. Opportunities of their using have considerably extended with occurrence of a composite materials (CM). Because of CM lightness, strength, rigidity they are essentially better compared to traditional metals and alloys in specific characteristics. Having an opportunity of change of the internal structure, the CM offers to designers great opportunities for tuning the SDS of structures, thus providing the best of those structures and corresponding CM usage. Increase of requirements to durability and reliability of modern constructions results in necessity of using the nonlinear and non-classical theories of shells and plates alongside with the classical linear one.

There are two levels of nonhomogeneity characterizing the fibrous composite materials: the micrononhomogeneity corresponded to the presence of two materials phases (a matrix or binding and a reinforcement or filling material), and the macrononhomogeneity, corresponded to the presence in a material of focused

Kseniya S. Golushko · Sergey K. Golushko · Andrey V. Yurchenko
Institute of Computational Technologies SB RAS, Lavrentiev Ave. 6,
630090 Novosibirsk, Russia
e-mail: golushko@ict.nsc.ru

E. Krause et al. (Eds.): Computational Sci., & High Performance Computing IV, NNFM 115, pp. 107–120.
springerlink.com

micrononhomogeneous layers with different space orientation. The main goal of the micromechanics of composites is the definition of the effective modules of elasticity, i.e. the factors connecting averaged stresses and deformations. There are two basic approaches to the definition of these factors: phenomenological and structural. In the phenomenological approach the microstructure of a composite is practically ignored, the material is considered to be conditionally homogeneous, having some set of experimentally found constants. The structural approach is based on the analysis of CM according to its structure and mechanical properties of components. Considering the second level of nonhomogeneity of CM, the preference should be given to the structural approach as the phenomenological approach in this case cannot be realized. Besides the structural approach allows to obtain the stresses in elements of a composition after the definition of the stressed-deformed state of a construction. It enables to investigate the local effects in binding and reinforcement, on border binding and reinforcing elements, to determine character of destruction, to state and solve problems on rational designing of constructions from CM.

At present time a great number of the micromodels of composites of a different level of complexity is developed which can be divided into the following groups:

- the structural models which take into account physic-mechanical properties and volumetric contents of components, and directions of an arrangement of fibrous;
- the self-conjugated models where the composite is represented as one fibre surrounded with the infinitive environment;
- the models which take into account the form of fibres and suppose their regular arrangement;
- the energy models allowing to receive the top and bottom borders for effective modules of elasticity;
- the statistical models based on the assumption of casual distribution of fibres.

At the solving of direct problems on calculation of SDS of real composite constructions and especially at the solving of inverse problems of their optimum and rational designing the researchers are compelled to be guided by models which, describing the basic properties of composites, have at the same time the simplest kind, i.e. the models of the first type. The structural models which are considered below satisfy such requirements.

2 Structural Models of Composite Materials

For most of CM-models and all that considered here we can write the relations between averaged stresses $\sigma_{\alpha\beta}$, $\tau_{\alpha 3}$ and deformations $e_{\alpha\beta}$, $\gamma_{\alpha 3}$ (generalized Hooke's Law):

$$\sigma_{\alpha\alpha} = a_{\alpha\alpha}e_{\alpha\alpha} + a_{\alpha\beta}e_{\beta\beta} + a_{\alpha 3}\cdot 2e_{\alpha\beta} - a_{\alpha\Theta}\Theta,$$

$$\sigma_{\alpha\beta} = a_{\alpha 3}e_{\alpha\alpha} + a_{\beta 3}e_{\beta\beta} + a_{33}\cdot 2e_{\alpha\beta} - a_{3\Theta}\Theta, \tag{1}$$

$$\gamma_{\alpha 3} = q_{\alpha\alpha}\tau_{\alpha 3} + q_{\alpha\beta}\tau_{\beta 3},$$

where Θ is the increase of temperature. Relation (1) is called the thermoelasticity relations, or, when no temperature influence is considered, simply elasticity relations.

Here we present the generalized relations for the several structural CM-models: the model with 2d fibers (2dFM) [1], the model with 1d fibers (1dFM) and the improved version of it (1dFMI) [2, 3]. The material of reinforced layer considered as elastic with the symmetry regarding to hypersurface shaping the corresponding shell (or plate). Here are basic hypothesis for the considered models [1].

1. The polyreinforced layer (fig. 1) is elastic homogeneous and isotropic binding (matrix) with the regular grid of long elastic, isotropic by themselves fibers, that are implanted in regarding to special directions, called the reinforcement directions.

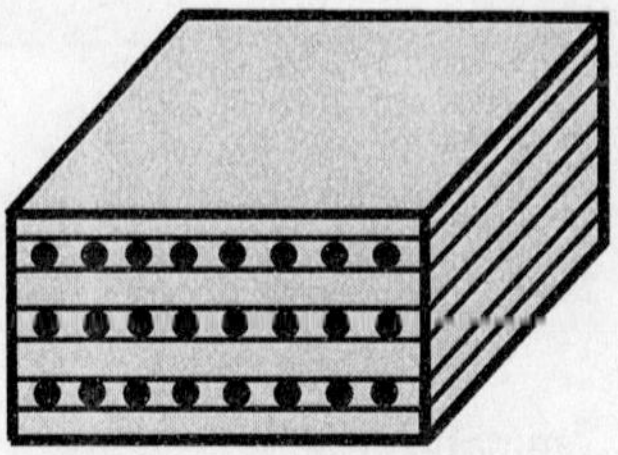

Fig. 1

2. The number of reinforcing fibers is big enough to consider the layer as quasihomogeneous.
3. The fibers implanted can hold the stretching and the compressing stresses.
4. The displacements remains small comparing to unity during all the deformation process.
5. The composite phases junction is ideal, means that there is no sliding between the binding and the reinforcing fibers.
6. The space between fibers is big enough compared to their transverse dimension and the same time small enough compared to shell or plate dimensions, so we can neglect the local effects near the fibers and the deformation's nonregularity in space between fibers.
7. a. Reinforced layer is the net of one-dimensional fibers that holds all the loading; the stresses and the strength of binding are neglected (fig. 2 a).
   b. Reinforced layer is the wire of one-dimensional fibers; the transversal shearing stresses are held by binding sub-layers; the reinforcing layers are absolutely rigid to shearing (fig. 2 b,c).
   c. Reinforcing fibers are considered as rectangular in transversal cross section (fig. 2 d) with corresponding 2d SDS.

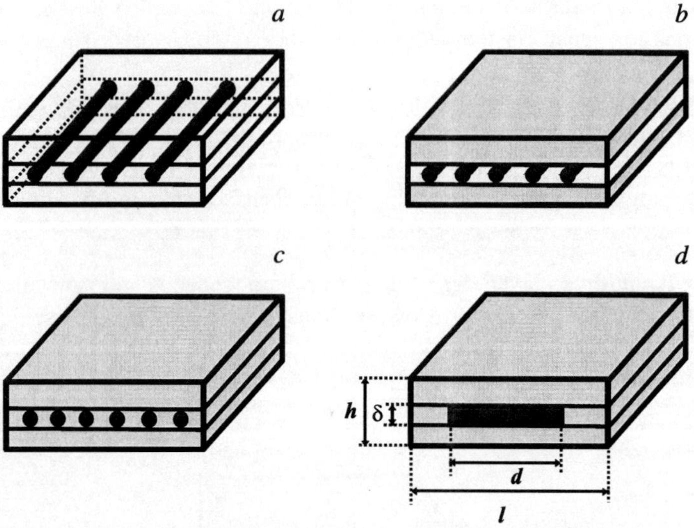

Fig. 2

Assumptions 1 — 6 are common for all considered models and the assumption 7a is for filament model (FM) [4] and the assumptions 7b, 7c are corresponded to 1dFM and 2dFM.

The coefficients in the relations (1) defined by the formulas below [5, 6]:

$$a_{\alpha\alpha} = \bar{a}_z E_c + \sum_{n=1}^{N} a^n_{\alpha\alpha}, \qquad a_{\alpha\beta} = \bar{a}_z \nu_c E_c + \sum_{n=1}^{N} a^n_{\alpha\beta},$$

$$q_{\alpha\beta} = \frac{a_z}{G_c} + \sum_{n=1}^{N} g^n_{\alpha\beta}, \quad a_{\alpha 3} = \sum_{n=1}^{N} a^n_{\alpha 3}, \quad a_{33} = a_z G_c + \sum_{n=1}^{N} a^n_{33},$$

$$a_{\alpha\Theta} = \frac{a_z \alpha_c E_c}{1-\nu_c} + \sum_{n=1}^{N} a^n_{\alpha\Theta}, \quad a_{3\Theta} = \sum_{n=1}^{N} a^n_{3\Theta},$$

$$a^n_{\alpha\alpha} = A^n_{1111} l^4_{\alpha n} + 2(A^n_{1122} + 2A^n_{1212}) l^2_{\alpha n} l^2_{\beta n} + A^n_{2222} l^4_{\beta n},$$

$$a^n_{\alpha\beta} = (A^n_{1111} + A^n_{2222} - 4A^n_{1212}) l^2_{\alpha n} l^2_{\beta n} + A^n_{1122}(l^4_{\alpha n} + l^4_{\beta n}),$$

$$a^n_{\alpha 3} = \left[A^n_{1111} l^2_{\alpha n} - A^n_{2222} l^2_{\beta n} - (A^n_{1122} + 2A^n_{1212})(l^2_{\alpha n} - l^2_{\beta n}) \right] l_{\alpha n} l_{\beta n},$$

$$a^n_{33} = (A^n_{1111} + A^n_{2222} - 2A^n_{1122}) l^2_{1n} l^2_{2n} + A^n_{1212}(l^2_{1n} - l^2_{2n})^2, \qquad (2)$$

$$a^n_{\alpha\Theta} = A^n_{11\Theta} l^2_{\alpha n} + A^n_{22\Theta} l^2_{\beta n}, \quad a^n_{3\Theta} = (A^n_{11\Theta} - A^n_{22\Theta}) l_{1n} l_{2n},$$

$$A_{1111}^n = \omega_z^n \left[\omega^n E_a^n + \omega_c^n E_c + \delta E_c E_a^n (\omega^n \nu_a^n + \omega_c^n \nu_c)^2 \chi_n^{-1}\right],$$
$$A_{1212}^n = \delta \omega_z^n E_c E_a^n \zeta_n^{-1}, \quad A_{1122}^n = \delta \omega_z^n E_c E_a^n (\omega^n \nu_a^n + \omega_c^n \nu_c)^2 \chi_n^{-1},$$
$$A_{2222}^n = \delta \omega_z^n E_c E_a^n \chi_n^{-1}, \quad g_{\alpha\alpha}^n = q_{11}^n l_{\alpha n}^2 + q_{22}^n l_{\beta n}^2, \; g_{\alpha\beta}^n = (q_{11}^n - q_{22}^n) l_{\alpha n} l_{\beta n},$$
$$q_{11}^n = 2\delta \omega_z^n (1+\nu_c)(1+\nu_a^n)\eta_n^{-1}, \quad q_{22}^n = \delta \omega_z^n \zeta_n (E_c E_a^n)^{-1},$$
$$\chi_n = \omega^n [1 - (\nu_a^n)^2] E_c + \omega_c^n [1 - (\nu_c)^2] E_a^n,$$
$$\eta_n = \omega^n (1+\nu_c) E_a^n + \omega_c^n (1+\nu_a^n) E_c,$$
$$\zeta_n = 2\left[\omega^n (1+\nu_a^n) E_c + \omega_c^n (1+\nu_c) E_a^n\right],$$
$$\bar{a}_z = \frac{a_z}{1-(\nu_c)^2}, \quad a_z = 1 - \omega_z,$$
$$\omega_z = \sum_{n=1}^{N} \omega_z^n, \quad l_{1n} = \cos\psi_n, \quad l_{2n} = \sin\psi_n,$$

Here for n-th reinforcing fibers family we have: E_a^n, ν_a^n are the Young's modulus and Poisson's coefficient of the material; ω^n, ω_z^n are the reinforcing intensity within plane and in the normal direction to IT; ψ_n is the angle between the reinforcing direction and the meridional direction of shell; $\omega_c^n = 1 - \omega^n$ for the 2dFM and 1dFMI, and for the 1dFM and FM we have $\omega_c^n = 0$.

We acquire 2dFM when $\delta = 1$; 1dFM and 1dFMI when $\delta = 0$; FM when $\delta = 0$, $E_c = 0$.

After the model have solved and we obtained the averaged stresses and deformation, we can determine the deformations and the stresses for elements of CM.

For the FM the stresses and the deformations of reinforcing fibers derives from [4]:

$$\sigma_{(11)a}^n = E_a^n e_{(11)a}^n, \quad e_{(11)a}^n = e_{11} l_{1n}^2 + 2e_{12} l_{1n} l_{2n} + e_{22} l_{2n}^2. \tag{3}$$

For the 1dFM adds the expressions for the stresses in binding:

$$\sigma_{(\alpha\alpha)c2} = \frac{E_c}{1-(\nu_c)^2}(e_{\alpha\alpha} + \nu_c e_{\beta\beta}), \quad \sigma_{(\alpha\beta)c2} = 2G_c e_{\alpha\beta},$$
$$\tau_{(\alpha 3)c2} = 2G_c \gamma_{(\alpha 3)c}, \quad \gamma_{(\alpha 3)c2} = \gamma_{\alpha 3}/\omega_z. \tag{4}$$

For the 1dFMI we have to add the expressions for the stresses and the deformations of binding between fibers in reinforcing layer:

$$\sigma_{(11)c1}^n = E_c e_{(11)c1}^n, \quad e_{(11)c1}^n = e_{11} l_{1n}^2 + 2e_{12} l_{1n} l_{2n} + e_{22} l_{2n}^2. \tag{5}$$

For the models mentioned above the undescribed components of stresses are equal to zero.

For the 2dFM the stresses in fibers and in binding within reinforcing sublayer and in binding between such sublayers defined by formulas [5]:

$$e_{(\alpha\beta)c2} = e_{\alpha\beta}, \quad \sigma_{(\alpha\alpha)c2} = \frac{E_c}{1-(\nu_c)^2}(e_{(\alpha\alpha)c2} + \nu_c e_{(\beta\beta)c2}),$$

$$\sigma_{(\alpha\beta)c2} = 2G_c e_{(\alpha\beta)c2}, \quad \tau_{(\alpha 3)c2} = \hat{\tau}_{\alpha 3}, \quad \gamma_{(\alpha 3)c2} = G_c^{-1}\tau_{(\alpha 3)c2},$$

$$e^n_{(11)c1} = e^n_{(11)a} = \hat{e}^n_{11}, \quad \gamma^n_{(13)c1} = \gamma^n_{(13)a} = \hat{\gamma}^n_{13},$$

$$\sigma^n_{(22)c1} = \sigma^n_{(22)a} = \hat{\sigma}^n_{22}, \quad \sigma^n_{(\alpha\beta)c1} = \sigma^n_{(\alpha\beta)a} = \hat{\sigma}^n_{\alpha\beta},$$

$$\tau^n_{(23)c1} = \tau^n_{(23)a} = \hat{\tau}^n_{23}, \tag{6}$$

$$\tau^n_{(13)c1} = G_c\gamma^n_{(13)c1}, \quad \gamma^n_{(23)c1} = G_c^{-1}\tau^n_{(23)c1},$$

$$\tau^n_{(13)a} = G^n_a\gamma^n_{(13)a}, \quad \gamma^n_{(23)a} = (G^n_a)^{-1}\tau^n_{(23)a},$$

$$e^n_{(\alpha\beta)c1} = (2G_c)^{-1}\sigma^n_{(\alpha\beta)c1}, \quad e^n_{(\alpha\beta)a} = (2G^n_a)^{-1}\sigma^n_{(\alpha\beta)a},$$

$$e^n_{(22)a} = \frac{1-(\nu^n_a)^2}{E^n_a}\sigma^n_{(22)a} - \nu^n_a e^n_{(11)a}, \quad \sigma^n_{(11)a} = E^n_a e^n_{(11)a} + \nu^n_a\sigma^n_{(22)a},$$

$$e^n_{(22)c1} = \frac{1-(\nu_c)^2}{E_c}\sigma^n_{(22)c1} - \nu^n_c e^n_{(11)c1}, \quad \sigma^n_{(11)c1} = E_c e^n_{(11)c1} + \nu_c\sigma^n_{(22)c1},$$

$$\hat{\sigma}^n_{\alpha\beta} = A^n_{\alpha\beta\lambda\mu}\hat{e}^n_{\lambda\mu}, \quad \hat{\gamma}^n_{\alpha 3} = q^n_{\alpha\beta}\hat{\tau}^n_{\beta 3}, \tag{7}$$

$$\hat{e}^n_{11} = e^n_{11}l^2_{1n} + e^n_{12}l_{1n}l_{2n} + e^n_{22}l^2_{2n}, \quad \hat{e}^n_{22} = e^n_{11}l^2_{2n} - e^n_{12}l_{1n}l_{2n} + e^n_{22}l^2_{1n},$$

$$\hat{e}^n_{12} = (e^n_{22} - e^n_{11})l_{1n}l_{2n} + (l^2_{1n} - l^2_{2n})e^n_{12},$$

$$\hat{\tau}^n_{13} = \tau^n_{13}l_{1n} + \tau^n_{23}l_{2n}, \quad \hat{\tau}^n_{23} = -\tau^n_{13}l_{2n} + \tau^n_{23}l_{1n}.$$

Here λ, $\mu = 1,2$ and a summation convention is mentioned.

3 Influence of CM-Models Selection on Simulation Results for Rubber-Based Toroidal Shell

It is very important to choose the appropriate CM-model when simulating the composite structure or design behaviour. Investigating the influence of CM-model choosing on the results of computer simulation is one of the ways to improve the trust level of the models within specific cases of their usage and to determine their limits. It is especially necessary taking into account the complexity of improved models that doesn't allows us to solve some useful inverse problems of rational designing of CM structures.

Let's consider the toroidal shell, which generatrix is the arc of the circle of radius R_1, situated between angles φ_n and $-\varphi_n$; the circle center is situated at range R_0 from the revolution axe; the thickness of shell is $2h$, and the shell consists of N

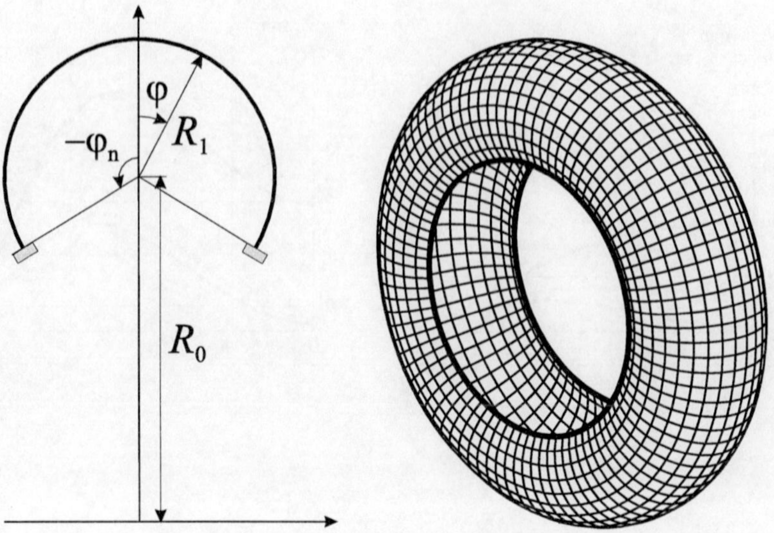

Fig. 3

layers. There is uniformly distributed pressure P loads the shell, that's additionally rigidly jammed on the boundaries.

Below the results of SDS calculation presented for the rubber-based ($E_c = 6.6$ MPa, $\nu_c = 0.49$) toroidal shell ($R_0 = 0.25$ m., $R_1 = 0.1$ m., $2h = 0.024$ m., $\varphi_n = 120° = 2.09$ rad). The shell consist of even number of layers that are two-way reinforced with viscose ($E_a = 1$ GPa, $\nu_a = 0.3$) or with steel ($E_a = 220$ GPa, $\nu_a = 0.3$) cord.

Let's consider the influence of a choice of the CM-model on the results of calculating the SDS of toroidal shells. On the Fig. 4 there are shown the results achieved using different CM-models: the continuous curves corresponds to 2dFM, dashed ones — to Bolotin's model [7], and the dotted-line curves corresponds to 1dFM. The curves that marked with 1 corresponds to $P = 5 \cdot 10^4$ Pa, and the curves that marked with 2 — to $P = 2 \cdot 10^5$ Pa. The Fig. 4 a corresponds to the shell with 2 layers, b — to the shell with 8 layers, both with viscose cord, and the Fig. 4 c corresponds to the shell with 2 layers, reinforced with steel cord. Results are obtained using nonlinear Timoshenko's shells theory [8]. As one can see, the results of using 2dFM and Bolotin's CM-model are very the same. At the same time one can see that involving the 1dFM gives quite different picture. Maximal values of calculated stress intensities may differs from 1dFM to 2dFM for up to four times (curves, marked with 2 on Fig. 4 c). And one can also see that the location of point with maximal deflections can change dramatically. However, there are variants of structures for which the results of 1dFM are close enough to those of 2dFM. For example such results are shown with curves marked by 1 on the Fig. 4 b). But for the same structure one

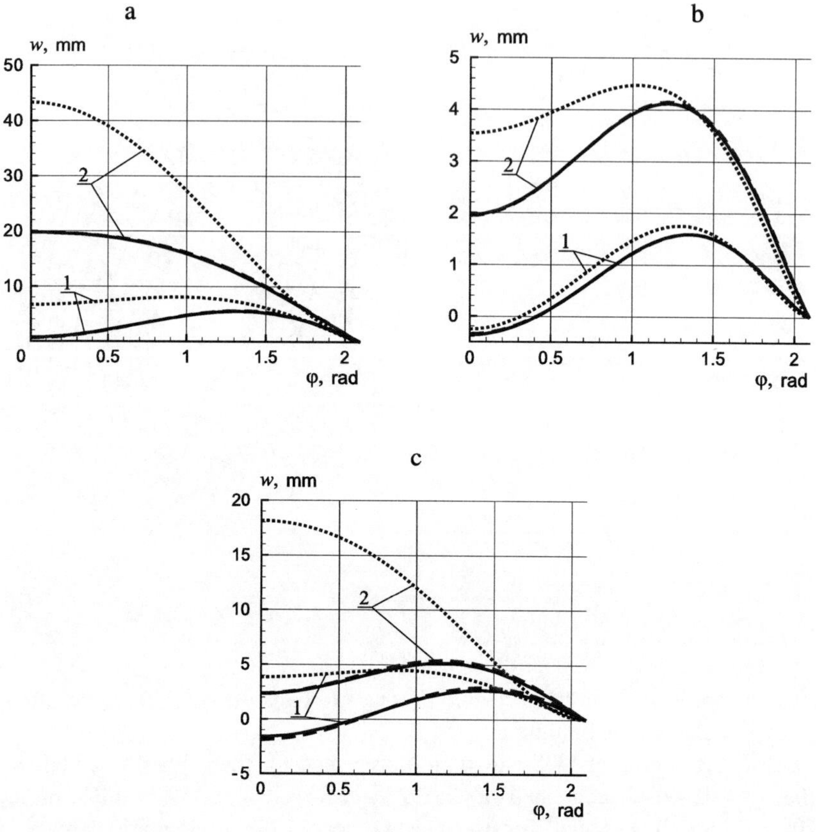

Fig. 4

can see that increase of loading leads to quite different deflection distribution along generatrix of shell.

Let's consider the influence of a choice of the CM-model on simulation results depending on composite microstructure in case of two-layer rubber-based shell reinforced with viscose (Fig. 5 a) or steel (Fig. 5 b, c) cord. There are maximal deflections presented on Fig. 5 a and b; the maximal stress intensities in binding are presented on Fig. 5 c). The presented dependencies are on reinforcing angles. The different marking numbers corresponds to CM-models: 1 — 1dFM, 2 — 2dFM, 3 — Bolotin's model 1, 4 — Bolotin's model 2. The results obtained using linear Timoshenko shells theory [8], in compare there are results obtained using linear Kirchhoff—Love theory of shells [9] in combination with 2dFM presented with firm line.

The results presented shows that for the shells reinforced with viscose simplified model (1dFM) always gives higher estimation for the maximal deflections. But for steel reinforced shell there is no unambiguous interpretation. There is a wide range of reinforcing angles (starting from 0 to 45°) at which the 1dFM underestimates

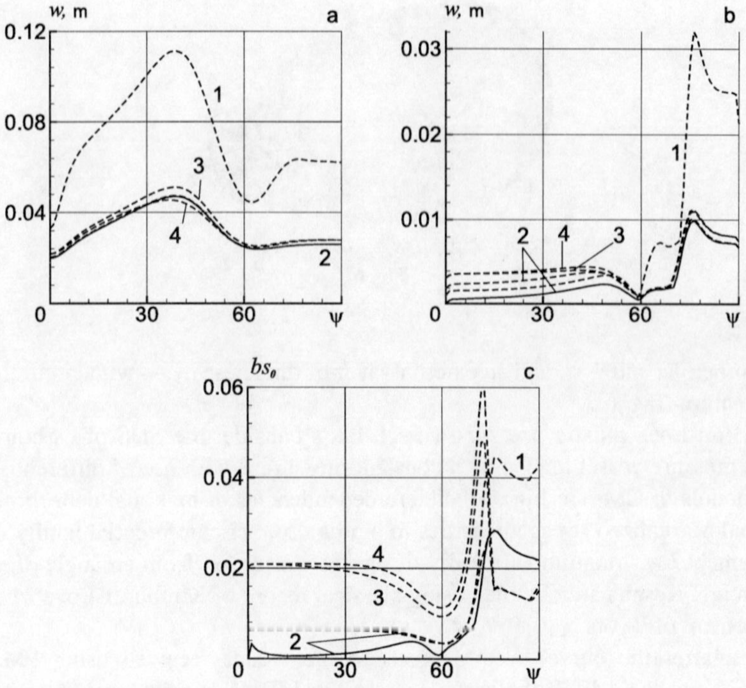

Fig. 5

the deflections in 1.5 times comparing to 2dFM. However in other cases 1dFM can overestimate the deflections in up to 200% compared to other CM-models.

4 Influence of CM-Models Selection on Simulation Results for Combined Pressure Vessels

We consider combined pressure vessel formed by consecutive interface of nodoidal, cylindrical and spherical shells.

In Cartesian coordinates *Oxyz* the equation of generatrix of shell is: $z = r_1\sqrt{1-k'^2\sin^2\phi}$, $x = (2\lambda - r_1)F(k',\phi) + r_1E(k',\phi)$, with $0 \le x \le x_1$; $z = r_1$, with $x_1 \le x \le x_2$; $z = \sqrt{z_1^2 - (x-x_2)^2}$, with $x_2 \le x \le x_3$; here r_1 is the radius of cylindrical shell; $k = (2\lambda - r_1)/r_1$ is the module of elliptic integral; $k' = \sqrt{(1-k^2)}$ is the additional module; *F*, *E* are the elliptic integrals of I and II type; $\phi = arcsin(\sqrt{r_1^2 - x^2}/(k'r_1))$ is the current coordinate; λ is the parameter which characterizes a curve; x_1, x_2 are the coupling points; x_3 is the total length of the vessel. The internal layer of a shell of thickness h_1 is reinforced with longitudinal family of armature, a middle layer of thickness h_2 - with spiral families of armature

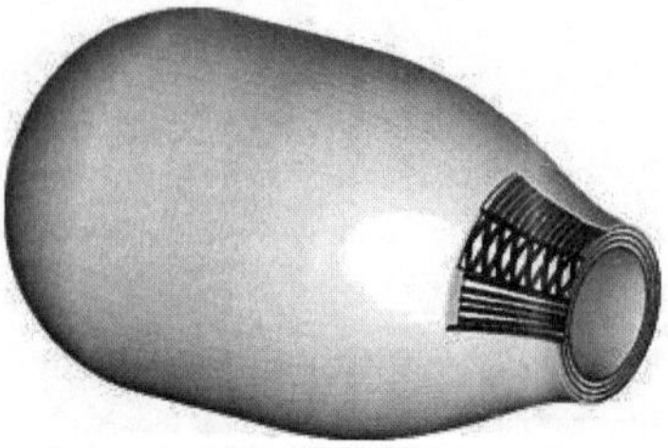

Fig. 6

with corners ψ and $-\psi$, and an external layer of thickness h_3 — with circumferential armature (Fig. 6).

Carbon-fiber plastic pressure vessel. Let's consider the SDS of carbon-fiber plastic pressure vessel loaded by a constant pressure P with use of different structural models of CM. On Fig. 7 there are dependencies of maximal deflections W, maximal normalized stress intensities in binding bs_0, circumferential family of reinforcement bs_2, longitudinal family of reinforcement bs_3 from an angle of spiral reinforcing. Results are obtained using classical theory of Kirchhoff-Love when an arrangement of layers is $(0, 90^o, \psi, -\psi)$.

Hereinafter the curve 1 corresponds to the values received using FM, the curve 2 — to the 1dFM, the curve 3 — to the 1dFMI and the curve 4 — to the 2dFM. To continuous lines there correspond values $h_1 = h_3 = 0.1h$, to dotted lines $h_1 = h_3 = 0.4h$.

From Fig. 7 it is clear that the results obtained using the models with one-dimensional fibers practically coincide. The use of FM makes the distinction with the results received using models with one-dimensional fibers up to 50% for circumferential fibers and 30% for deflections. For spiral family of fibers with $\psi = 70^o$ and $h_1 = h_3 = 0.1h$, stress intensities received with FM are three times higher then those received with the 1dFM. Perceptible difference in the results obtained with the 1dFM and the 2dFM is observed, for example, with $\psi = 60^o$ and $h_1 = h_3 = 0.1h$, and it is 30% for binding, 50% for circumferential family of fibers, 60% for spiral armature, 25% for deflections. However, with $h_1 = h_3 = 0.4h$ and $\psi > 30^o$ the difference between the results obtained with different models of CM does not exceed 10%.

From Fig. 7 one can also see that with $h_1 = h_3 = 0.1h$ and $\psi > 60^o$ initial destruction takes place in the binding, and with $\psi < 60^o$ — in the spiral family of fibers. When decreasing an angle of spiral reinforcing, stress intensities in binding and circumferential armature decrease three times when using the 1dFM or 1dFMI and 1.5 times using the 2dFM. The change of thickness of reinforced layers allows to decrease deflections, stress intensities in binding and spiral armature from 2 to 3 times. Behaviour pattern of maximal stress intensities in circumferential armature and deflections from an angle of spiral reinforcing also changes. From Fig. 7 one can conclude that in this case the best structure of reinforcement, when stress intensities and deflections reach lower-range value is when $h_1 = h_3 = 0.1h$ and $55^o \le \psi \le 60^o$.

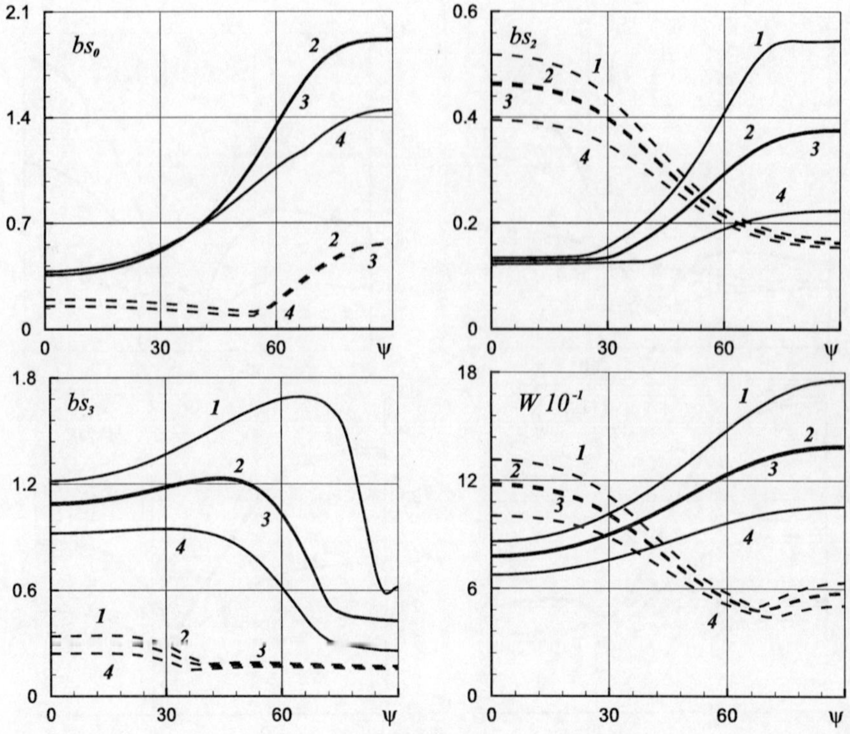

Fig. 7

Fiberglass plastic pressure vessel. Let's investigate the influence of a choice of CM-model on calculated SDS of fiberglass combined pressure vessel when the structure of reinforcement is (90°, ψ, $-\psi$, 0). On Fig. 8 the dependencies of maximal reduced stress intensities in CM and non-dimensional deflections from an angle of spiral reinforcing for fiberglass rigidly jammed combined vessel situated under the action of constant pressure are shown. The results are obtained using the specified theory of Andreev—Nemirovsky [5]. The other parameters have the same values as shown on Fig. 7. To continuous lines there correspond values $h_1 = h_3 = 0.1h$, to dotted lines — $h_1 = h_3 = 0.4h$. From Fig. 8 it follows that stress intensities in the elements of CM and deflections take on lower-range value when $h_1 = h_3 = 0.1h$ and $45^o < \psi < 50^o$. For all that, the values of stress intensities and deflections received with different structural models of CM differ slightly in this area. The increase of an angle of spiral reinforcing up to 60° leads to the increase of stress intensities for 3 times in binding, and for 5 times for fibers in meridian material.

From Fig. 8 one can also see that the results obtained using the models with one-dimensional fibers practically coincide for all area of parameter changing. When $h_1 = h_3 = 0.1h$ and $\psi = 30^o$ distinction between the results received using the 1dFM and the 2dFM makes up to 60% for deflections and stress intensities in binding,

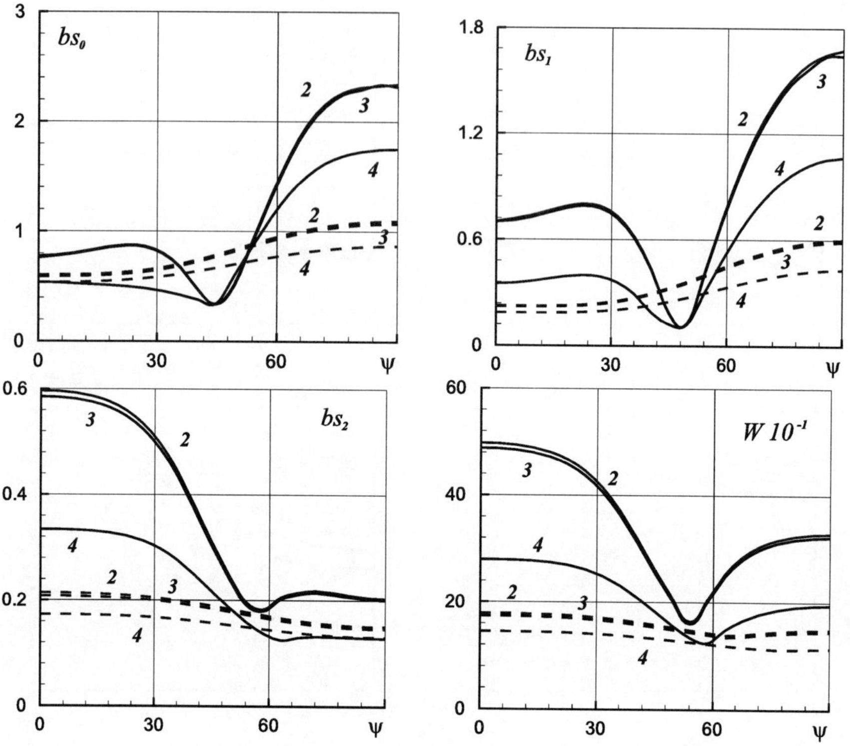

Fig. 8

and up to 80% for stress intensities in longitudinal and circumferential armature. The decrease of thickness of layers with spiral families of fibers allows to decrease stress intensities in circumferential armature and deflections from 1.5 to 2 times when $\psi < 30^o$.

Metal-composite combined structure. Let's consider combined pressure vessel with aluminium matrix and steel fibers when the structure of CM is $(0, 90^o, \psi, -\psi)$. The other parameters have the values for Fig. 7. On Fig. 9 maximal deflections and stress intensities in binding depending on an angle of spiral reinforcing are shown. The results are received using the specified theory of Andreev—Nemirovsky.

From Fig. 9 one can see that the distinction between the results obtained with models with one-dimensional fibers does not exceed 10%. However, for the values obtained with the 1dFM and the 2dFM the distinction is inherent. For example, when $h_1 = h_3 = 0.1h$ and $\psi = 45^o$ the values of stress intensities received using the 1dFM are two times bigger than those received using the 2dFM. From Fig. 9 it also follows that in this case the dependence of stress intensities and deflections from the structure of reinforcement is not so brightly expressed as it is for fiberglass or carbon-fiber plastic vessels.

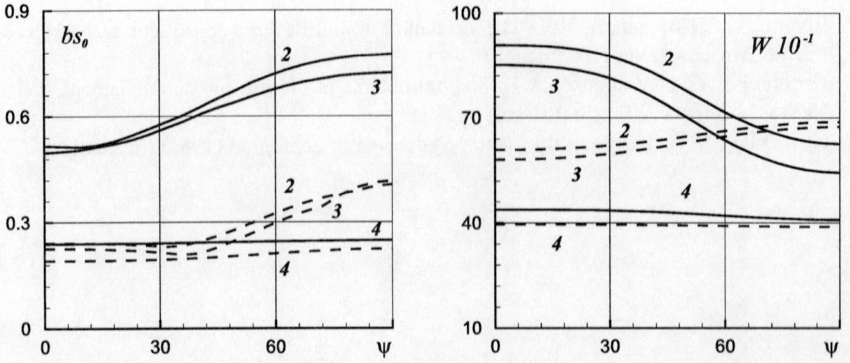

Fig. 9

The degree of influence of structural models selection on SDS of construction appreciably depends on structural and mechanical parameters of CM and can reach from 5% to 80% in the examined cases. We have to note, that in all examined cases stress intensities in the elements of CM and deflections obtained using filament model are higher than corresponding values obtained using models with one-dimensional fibers that are in turn higher than the values obtained using model with two-dimensional fibers. Simplified models underestimate strength properties of CM and therefore durability and stiffness of structures are evaluated "with stock" using such models.

Acknowledgements. Work is done with support of the Russian Foundation for Basic Research (grant 08-01-13509-ofi_c), Integrational project of SB RAS No. 72 and the Program of federal support for scientific schools of Russian Federation (NSh-931.2008.9).

References

1. Nemirovskii, Y.V.: On theory of thermoelastic bend of reinforced shells and plates. Mekhanika polimerov 5, 861–873 (1972) (in Russian)
2. Nemirovsky, Y.V.: About elastoplastic behaviour of the reinforced layer. Prikl. Mekh . Tekhn. Phys. 6, 81–89 (1969) (in Russian)
3. Nemirovsky, Y.V.: The equations of a bend and stability of the reinforced shells and plates from a viscoelastic material. Dinamika sploshnikh sred. Novosibirsk: Institute of hydrodynamics SB RAS 4, 50–63 (1970) (in Russian)
4. Obraztsov, I.F., Vasiliev, V.V., Bunakov, V.A.: Optimum Reinforcing of Shells of Revolution From Composite Materials. Mashinostroenie, Moscow (1977) (in Russian)
5. Andreev, A.N., Nemirovskii, Y.V.: Multilayered Anisotropic Shells and Plates: Bend, Stability, Vibration. Nauka, Novosibirsk (2001) (in Russian)
6. Golushko, S.K., Nemirovskii, Y.V.: Direct and Inverse Problems of Mechanics of Elastic Composite Plates and Shells of Revolution. Fizmatlit, Moscow (2008) (in Russian)

7. Bolotin, V.V., Novitchkov, Y.N.: The mechanics of multilayered constructions. Mashinostroenie, Moscow (1980) (in Russian)
8. Grirorenko, Y.M., Vasilenko, A.T.: Static problems of anisotropic heterogeneous shells. Nauka, Moskow (1992) (in Russian)
9. Novozhilov, V.V.: Theory of thin shells. Sudpromgiz, Leningrad (1951) (in Russian)

Well-Balanced Path-Consistent Finite Volume EG Schemes for the Two-Layer Shallow Water Equations

Michael Dudzinski and Mária Lukáčová-Medviďová

Abstract. We present a new path-consistent well-balanced finite volume method within the framework of the evolution Galerkin (FVEG) schemes. The methodology will be illustrated for two layer shallow water equations with source terms modelling the bottom topography and Coriolis forces. The FVEG methods couple a finite volume formulation with approximate evolution operators. The latter are constructed using the bicharacteristics of multidimensional hyperbolic systems, such that all of the infinitely many directions of wave propagation are taken into account explicitly. We will derive a suitable path in the phase space that is based on the evolution operator and derive the corresponding path-consistent FVEG scheme. The path-consistent FVEG scheme is well-balanced for the stationary steady states as well as for the steady jets in the rotational frame.

1 Mathematical Model

Many types of flows, not necessarily involving water, can be described as shallow water flows. Such flows are all characterized by negligible vertical scales in comparison to horizontal scales. Typical examples are rivers with their flood plains, flows in lakes generated by wind blows, propagation of tsunamis, oceanographic, meteorological and geophysical flows. For smooth flows different methods, such as finite difference schemes, finite element methods, or spectral methods perform quite well. Under some assumptions flows may exhibit discontinuities, such as tidal bores observed in some rivers, the wave resulting from the bursting of a dam or hydraulic jump in jets. In this situation, a moving step front develops, which is comparable to a shock wave in aerodynamics. In this work we consider the two-layer shallow water equations describing motion of immiscible inviscid fluids of constant density. This

Michael Dudzinski · Mária Lukáčová-Medviďová
Institute of Numerical Simulation, Hamburg University of Technology,
Schwarzenbergstrasse 95, 21073 Hamburg, Germany
e-mail: michael.dudzinski, lukacova@tu-harburg.de

E. Krause et al. (Eds.): Computational Sci., & High Performance Computing IV, NNFM 115, pp. 121–135.
springerlink.com

type of flow appears, typically, in oceanographic models when a warm and light upper layer flows over a lower layer of cooler, heavier water with larger salinity.

Taking into account effects of the variable bottom topography and the Coriolis forces due to the earth rotation the system of two-layer shallow water equations can be formulated as follows

$$\left\{\begin{array}{rcl} \partial_t h_1 + \partial_x(h_1 u_1) + \partial_y(h_1 v_1) & = & 0, \\ \partial_t(h_1 u_1) + \partial_x\left(h_1 u_1^2 + \frac{g}{2}h_1^2\right) + \partial_y(h_1 u_1 v_1) & = & -gh_1\partial_x(b+h_2) - fh_1 v_1, \\ \partial_t(h_1 v_1) + \partial_x(h_1 u_1 v_1) + \partial_y\left(h_1 v_1^2 + \frac{g}{2}h_1^2\right) & = & -gh_1\partial_y(b+h_2) + fh_1 u_1, \\ \partial_t h_2 + \partial_x(h_2 u_2) + \partial_y(h_2 v_2) & = & 0, \\ \partial_t(h_2 u_2) + \partial_x\left(h_2 u_2^2 + \frac{g}{2}h_2^2\right) + \partial_y(h_2 u_2 v_2) & = & -gh_2\partial_x(b+rh_1) - fh_2 v_2, \\ \partial_t(h_2 v_2) + \partial_x(h_2 u_2 v_2) + \partial_y\left(h_2 v_2^2 + \frac{g}{2}h_2^2\right) & = & -gh_2\partial_y(b+rh_1) + fh_2 u_2. \end{array}\right. \tag{1}$$

Here h_1, h_2 denote the height of the corresponding layer, u_i and v_i are the vertically averaged velocities in the x- and y-direction, respectively, $i = 1,2$. Furthermore b is the time independent bottom topography, g the gravitational constant, f the Coriolis parameter and $r = \rho_1/\rho_2$ describes the effect of different constant densities $\rho_1 < \rho_2$.

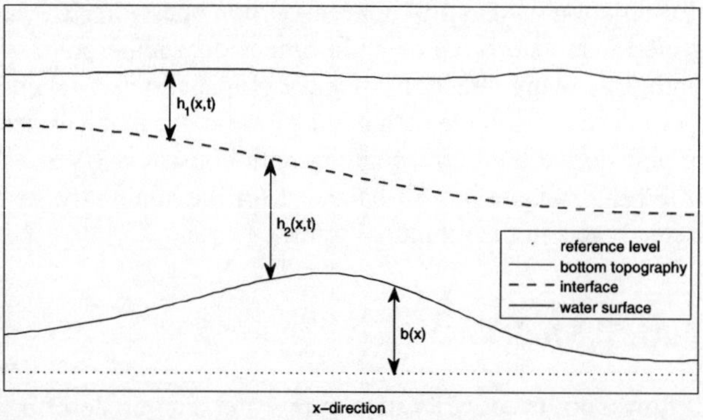

Fig. 1 Two shallow layers

Mathematically there are several challenging problems arising by numerical solution of the system (1). Indeed, we have non-conservative product terms, the system is only conditionally hyperbolic, nonstrictly hyperbolic, and its eigenstructure cannot be obtained in explicit form. In the literature several approaches to attack these problems [1], [2], [7] can be found.

Typically large scale flow in the ocean and the atmosphere is characterized by an approximate balance in the vertical direction between pressure gradient and gravity (hydrostatic balance), and in the horizontal direction between the pressure gradient and the Coriolis forces that arise due to the earth rotation (geostrophic balance). Numerical scheme preserving important physical equilibria are called *well-balanced* schemes. In this paper the following fundamental equilibrium states will be considered

1. rest state (lake at rest):

$$h_1 + h_2 + b = \text{const.}, \; rh_1 + h_2 + b = \text{const.}, \; u_1 = v_1 = 0 = u_2 = v_2; \tag{2}$$

2. geostrophic equilibrium (jet in the rotational frame):

$$\begin{aligned} &g\partial_x(h_1 + h_2 + b) = -fv_1, u_1 = 0, g\partial_y(h_1 + h_2 + b) = 0, \partial_y v_1 = 0, v_1 \partial_y h_1 = 0, \\ &g\partial_x(rh_1 + h_2 + b) = -fv_2, u_2 = 0, g\partial_y(rh_1 + h_2 + b) = 0, \partial_y v_2 = 0, v_2 \partial_y h_2 = 0. \end{aligned} \tag{3}$$

Note that the condition (3) actually states the balance between the Coriolis forces and the pressure gradient in each layer; we assume here additionally that the flow is quasi one-dimensional, thus the y-derivatives are set to zero.

In our previous works [9], [10] we have developed the so-called *well-balanced finite volume evolution Galerkin scheme* for the one-layer shallow water system with bottom topography and the Coriolis forces. The scheme has been shown to be simple, accurate and surprisingly efficient. The finite volume evolution Galerkin schemes can be formulated as two-step predictor-corrector scheme. The first step, called predictor step, evolves the value at a quadrature node to the half-timestep. This has been done by fully multidimensional bicharacteristic theory, cf. [10]. The second step is the standard finite volume update. It approximates the flux integral across the interfaces by a quadrature of the fluxes evaluated at the predicted states at the half time step. Now in order to preserve equilibrium states, it is necessary to satisfy equilibrium conditions at both time steps. Let us note that in [10] we have approximated the right-hand-side source terms by the so-called cell-interface approach that exactly balances flux integrals.

On the other hand, there have been several interesting results presented recently in the literature where the non-conservative character of the right-hand-side source terms of (1) was taken into account. In order to define the concept of the weak solutions the theory developed by Dal Maso et al. [3] has been used. The most important point is to give an interpretation to the nonconservative products as Borel measures that are based on the choice of a family of paths drawn in the phase space. The concept of path-conservative or path-consistent numerical schemes has been introduced by Pares and Castro in [12], see also [11, 4, 6] and the references therein.

The goal of the present paper is two-fold: firstly we will derive a new well-balanced path-consistent finite volume evolution Galerkin method and secondly we apply this method for the two-layer shallow water equations.

2 Operator Splitting Technique

Let us firstly rewrite (1) in the quasi-linear form. Setting $\mathbf{W} = (\mathbf{w}_1, \mathbf{w}_2, b, x, y)^T$ and denoting conservative variables of the corresponding layer by $\mathbf{w}_i = (h_i, h_i u_i, h_i v_i)^T$, $i = 1, 2$, we obtain

$$\mathbf{W}_t + \widetilde{\mathbf{A}}_1(\mathbf{W})\mathbf{W}_x + \widetilde{\mathbf{A}}_2(\mathbf{W})\mathbf{W}_y = 0, \tag{4}$$

where

$$\widetilde{\mathbf{A}}_1 = \begin{pmatrix} \mathbf{A}_1(\mathbf{w}_1) & \mathbf{C}_1(\mathbf{w}_1) & -\mathbf{S}_1(\mathbf{w}_1) & -\mathbf{S}_3(\mathbf{w}_1) & 0 \\ \widetilde{\mathbf{C}}_1(\mathbf{w}_2) & \mathbf{A}_1(\mathbf{w}_2) & -\mathbf{S}_1(\mathbf{w}_2) & -\mathbf{S}_3(\mathbf{w}_2) & 0 \\ 0 & 0 & 0 & 0 & 0 \\ 0 & 0 & 0 & 0 & 0 \\ 0 & 0 & 0 & 0 & 0 \end{pmatrix},$$

$$\widetilde{\mathbf{A}}_2 = \begin{pmatrix} \mathbf{A}_2(\mathbf{w}_1) & \mathbf{C}_2(\mathbf{w}_1) & -\mathbf{S}_2(\mathbf{w}_1) & 0 & -\mathbf{S}_4(\mathbf{w}_1) \\ \widetilde{\mathbf{C}}_2(\mathbf{w}_2) & \mathbf{A}_2(\mathbf{w}_2) & -\mathbf{S}_2(\mathbf{w}_2) & 0 & -\mathbf{S}_4(\mathbf{w}_2) \\ 0 & 0 & 0 & 0 & 0 \\ 0 & 0 & 0 & 0 & 0 \\ 0 & 0 & 0 & 0 & 0 \end{pmatrix},$$

$$\mathbf{S}_1(\mathbf{w}_i) = \begin{pmatrix} 0 \\ -gh_i \\ 0 \end{pmatrix}, \mathbf{S}_2(\mathbf{w}_i) = \begin{pmatrix} 0 \\ 0 \\ -gh_i \end{pmatrix}, \mathbf{S}_3(\mathbf{w}_i) = \begin{pmatrix} 0 \\ fh_iv_i \\ 0 \end{pmatrix}, \mathbf{S}_4(\mathbf{w}_i) = \begin{pmatrix} 0 \\ 0 \\ -fh_iu_i \end{pmatrix}.$$

Further, $\mathbf{A}_1$, $\mathbf{A}_2$ are the usual Jacobian matrices of conservative fluxes in $x-, y-$ directions, respectively, and $\mathbf{C}_1, \mathbf{C}_2, \widetilde{\mathbf{C}}_1, \widetilde{\mathbf{C}}_2$ the corresponding coupling matrices containing the non-conservative terms. For example, we have

$$\mathbf{C}_1(\mathbf{w}_1) = \begin{pmatrix} 0 & 0 & 0 \\ c_1^2 & 0 & 0 \\ 0 & 0 & 0 \end{pmatrix}, \qquad \widetilde{\mathbf{C}}_2(\mathbf{w}_2) = \begin{pmatrix} 0 & 0 & 0 \\ 0 & 0 & 0 \\ rc_2^2 & 0 & 0 \end{pmatrix}.$$

The difficulty of the two-layer system lies in the coupling matrices $\mathbf{C}_i, \widetilde{\mathbf{C}}_i$. Indeed, the matrix pencil

$$\begin{pmatrix} \mathbf{A}_1(\mathbf{w}_1) & \mathbf{C}_1(\mathbf{w}_1) \\ \widetilde{\mathbf{C}}_1(\mathbf{w}_2) & \mathbf{A}_1(\mathbf{w}_2) \end{pmatrix} \cos\theta + \begin{pmatrix} \mathbf{A}_2(\mathbf{w}_1) & \mathbf{C}_2(\mathbf{w}_1) \\ \widetilde{\mathbf{C}}_2(\mathbf{w}_2) & \mathbf{A}_2(\mathbf{w}_2) \end{pmatrix} \sin\theta, \qquad \theta \in (0, 2\pi)$$

may have complex eingenvalues and the hyperbolicity of the system is lost. Even if we would be able to approximate eigenvalues at least in the case when they are real it is too expensive to construct the corresponding eigenvectors. The latter are however necessary in order to apply the characteristic (or bicharacteristic) decomposition needed for the evolution operator in the predictor step of the FVEG scheme. These considerations lead us to a construction of a suitable decomposition of the system (4) into a set of subsystems for which the corresponding eigenstructure is readily available. In the present paper we apply a natural operator splitting that separates each layer, see also [2] for similar approach. Note however, that other splitting techniques are possible [7] and will be reported in our future work.

In what follows the following decomposition will be used:
operator T^1

$$\partial_t \mathbf{w}_1 + \mathbf{A}_1(\mathbf{w}_1)\partial_x \mathbf{w}_1 + \mathbf{A}_2(\mathbf{w}_1)\partial_y \mathbf{w}_1 = \\ -\mathbf{C}_1(\mathbf{w}_1)\partial_x \mathbf{w}_2 - \mathbf{C}_2(\mathbf{w}_1)\partial_y \mathbf{w}_2 + \mathbf{S}_1(\mathbf{w}_1)\partial_x b + \mathbf{S}_2(\mathbf{w}_1)\partial_y b + \mathbf{S}_3(\mathbf{w}_1) + \mathbf{S}_4(\mathbf{w}_1) \tag{5}$$

operator T^2

$$\partial_t \mathbf{w}_2 + \mathbf{A}_1(\mathbf{w}_2)\partial_x \mathbf{w}_2 + \mathbf{A}_2(\mathbf{w}_2)\partial_y \mathbf{w}_2 = \\ -\widetilde{\mathbf{C}}_1(\mathbf{w}_2)\partial_x \mathbf{w}_1 - \widetilde{\mathbf{C}}_2(\mathbf{w}_2)\partial_y \mathbf{w}_1 + \mathbf{S}_1(\mathbf{w}_2)\partial_x b + \mathbf{S}_2(\mathbf{w}_2)\partial_y b + \mathbf{S}_3(\mathbf{w}_2) + \mathbf{S}_4(\mathbf{w}_2) \tag{6}$$

Thus, instead of solving a complex system (4) the second order Strang splitting approach is applied using the operators T^1 and T^2

$$\mathbf{W}^{n+1} = T^1_{\Delta t/2} T^2_{\Delta t} T^1_{\Delta t/2} \mathbf{W}^n. \tag{7}$$

Note that the operators T^1, T^2 can be easily rewritten in the form (4) by setting, respectively, the second and first row in the matrices $\widetilde{\mathbf{A}}_1$ and $\widetilde{\mathbf{A}}_2$ to zero. In fact, we have for the operator T^1

$$\mathbf{W}_t + \widetilde{\mathbf{A}}^1_1(\mathbf{W})\mathbf{W}_x + \widetilde{\mathbf{A}}^1_2(\mathbf{W})\mathbf{W}_y = 0,$$

$$\widetilde{\mathbf{A}}^1_1 = \begin{pmatrix} \mathbf{A}_1(\mathbf{w}_1) & \mathbf{C}_1(\mathbf{w}_1) & -\mathbf{S}_1(\mathbf{w}_1) & -\mathbf{S}_3(\mathbf{w}_1) & 0 \\ 0 & 0 & 0 & 0 & 0 \\ 0 & 0 & 0 & 0 & 0 \\ 0 & 0 & 0 & 0 & 0 \\ 0 & 0 & 0 & 0 & 0 \end{pmatrix}$$

$$\widetilde{\mathbf{A}}^1_2 = \begin{pmatrix} \mathbf{A}_2(\mathbf{w}_1) & \mathbf{C}_2(\mathbf{w}_1) & -\mathbf{S}_2(\mathbf{w}_1) & 0 & -\mathbf{S}_4(\mathbf{w}_1) \\ 0 & 0 & 0 & 0 & 0 \\ 0 & 0 & 0 & 0 & 0 \\ 0 & 0 & 0 & 0 & 0 \\ 0 & 0 & 0 & 0 & 0 \end{pmatrix}.$$

The analogous equations hold for the operator T^2. Clearly, each system (5), (6) now reduces to the one-layer shallow water equations with a generalized bottom topography $h_2 + b$ or $rh_1 + b$. These are known for a given time step using the values from the previous time step. Our next aim is to derive a *well-balanced path-consistent FVEG scheme* for each layer separately.

3 Path-Consistent FVEG Scheme

Let us discretize a computational domain Ω by a regular rectangular mesh consisting of finite volumes $\Omega_{k\ell} = (x_k - \hbar/2, x_k + \hbar/2) \times (y_\ell - \hbar/2, y_\ell + \hbar/2)$, $k,\ell \in \mathbb{Z}$ and denote by $\hbar$ the mesh step and by Δt time step. For each layer the path-consistent FVEG scheme is a predictor-corrector method. The corrector step is a path-consistent FV update, in the predictor step an intermediate solution $\mathbf{W}^*$ on cell interfaces and cell centers is computed using an approximate evolution operator for the related layer, cf. Section 4.

In order to derive a path-consistent FV update we need to construct a suitable path in the phase space

$$\tilde{\Phi}(s, \mathbf{W}^*_{k,\ell'}, \mathbf{W}^*_{k+1,\ell'}) := \begin{cases} \Phi(2s, \mathbf{W}^*_{k,\ell'}, \mathbf{W}^*_{k+1/2,\ell'}) & 0 \le s \le 1/2, \\ \Phi(2s-1, \mathbf{W}^*_{k+1/2,\ell'}, \mathbf{W}^*_{k+1,\ell'}) & 1/2 \le s \le 1, \end{cases}$$

here $\Phi(s, \mathbf{W}^*_a, \mathbf{W}^*_b)$ is a straight path connecting $\mathbf{W}^*_a$ and $\mathbf{W}^*_b$, $\mathbf{W}^*$ is obtained by the evolution Galerkin operator in the predictor step.

The path-consistent finite volume update reads for each operator T^i, $i = 1,2$

$$\begin{aligned} \mathbf{W}^{new}_{k,\ell} = \mathbf{W}^{old}_{k,\ell} &- \frac{\tau}{\hbar} \sum_{\ell' \in L} \alpha_{\ell'} \left(\mathbf{D}^{i,-}_{k+1/2,\ell'} + \mathbf{D}^{i,+}_{k-1/2,\ell'} \right) \\ &- \frac{\tau}{\hbar} \sum_{k' \in K} \beta_{k'} \left(\mathbf{D}^{i,-}_{k',\ell+1/2} + \mathbf{D}^{i,+}_{k',\ell-1/2} \right), \end{aligned} \tag{8}$$

where $\tau = \Delta t$ or $\tau = \Delta t/2$, $L := \{\ell - 1/2, \ell, \ell + 1/2\}$ and $K := \{k - 1/2, k, k + 1/2\}$ are the index sets and $\alpha_{\ell'}$, $\beta_{k'}$ the weights of the Simpson quadrature applied for flux integration along cell interfaces. Now let us define the $\mathbf{D}$ matrices for each $k,\ell \in \mathbb{Z}, k' \in K, \ell' \in L$

$$\mathbf{D}^{i,-}_{k+1/2,\ell'} := \int_0^1 \widetilde{\mathbf{A}}^i_1(\Phi(s; \mathbf{W}^*_{k,\ell'}, \mathbf{W}^*_{k+1/2,\ell'})) \frac{\partial \Phi}{\partial s}(s; \mathbf{W}^*_{k,\ell'}, \mathbf{W}^*_{k+1/2,\ell'})\, ds, \tag{9}$$

$$\mathbf{D}^{i,+}_{k+1/2,\ell'} := \int_0^1 \widetilde{\mathbf{A}}^i_1(\Phi(s; \mathbf{W}^*_{k+1/2,\ell'}, \mathbf{W}^*_{k+1,\ell'})) \frac{\partial \Phi}{\partial s}(s; \mathbf{W}^*_{k+1/2,\ell'}, \mathbf{W}^*_{k,\ell'})\, ds, \tag{10}$$

matrices $\mathbf{D}^{i,+}_{k',\ell+1/2}, \mathbf{D}^{i,-}_{k',\ell+1/2}$ are defined using $\widetilde{\mathbf{A}}^i_2$ in an analogous way, $i = 1,2$.

After a short calculation we obtain for each layer, $i = 1,2$, the following representation of source terms

$$(\mathbf{S}_1(\mathbf{w}^*_i)\partial_x b)_{k+1/2,\ell'} =$$

$$\frac{g}{2} \begin{pmatrix} 0 \\ ((h^*_i)_{k,\ell'} + (h^*_i)_{k+1/2,\ell'})(b^*_{k,\ell'} - b^*_{k+1/2,\ell'}) + ((h^*_i)_{k-1/2,\ell'} + (h^*_i)_{k,\ell'})(b^*_{k-1/2,\ell'} - b^*_{k,\ell'}) \\ 0 \end{pmatrix},$$

$$(\mathbf{S}_2(\mathbf{w}_i^*)\partial_y b)_{k',\ell+1/2} =$$

$$\frac{g}{2}\begin{pmatrix} 0 \\ 0 \\ ((h_i^*)_{k',\ell}+(h_i^*)_{k',\ell+1/2})(b^*_{k',\ell'}-b^*_{k',\ell+1/2})+((h_i^*)_{k',\ell-1/2}+(h_i^*)_{k',\ell})(b^*_{k',\ell-1/2}-b^*_{k',\ell})\end{pmatrix},$$

$$(\mathbf{S}_3(\mathbf{w}_i^*)\partial_x x)_{k+1/2,\ell'} =$$

$$f\frac{\hbar}{4}\begin{pmatrix} 0 \\ (h_i^*)_{k-1/2,\ell'}(v_i^*)_{k-1/2,\ell'}+2(h_i^*)_{k,\ell'}(v_i^*)_{k,\ell'}+(h_i^*)_{k+1/2,\ell'}(v_i^*)_{k+1/2,\ell'} \\ 0 \end{pmatrix},$$

$$(\mathbf{S}_4(\mathbf{w}_i^*)\partial_y y)_{k',\ell+1/2} =$$

$$-f\frac{\hbar}{4}\begin{pmatrix} 0 \\ 0 \\ (h_i^*)_{k',\ell-1/2}(u_i^*)_{k',\ell-1/2}+2(h_i^*)_{k',\ell}(u_i^*)_{k',\ell}+(h_i^*)_{k',\ell+1/2}(u_i^*)_{k',\ell+1/2}\end{pmatrix}.$$

Let us point out that the discretization of source terms obtained above using the path-consistent approach and those derived in [10] are different. Nevertheless both approaches yield well-balanced approximations, cf. [10] and Theorem 1 below.

4 Approximate Evolution Operators

In order to keep the paper self-contained we present the first order approximate evolution operator T^1, the approximate evolution operator T^2 can be written in an analogous way. For the second order version and for the detailed derivation of approximate evolution operators the reader is referred to [10]. The approximate evolution operator $\mathbf{W}^* = EG^1_{\tau/2}\mathbf{W}^{old}$ for the operator T^1_τ reads

$$\begin{aligned}
&h_1(P) = -b(P) - h_2(P) + \\
&\frac{1}{2\pi}\int_0^{2\pi}\left[(h_1(Q)+b(Q)+h_2(Q)) - \frac{\tilde{c}_1}{g}(u_1(Q)\,\mathrm{sgn}(\cos\theta)+v_1(Q)\,\mathrm{sgn}(\sin\theta))\right]\mathrm{d}\theta \\
&+\frac{\tau}{4\pi}\int_0^{2\pi}(\tilde{u}_1(b_x(Q)+(h_2)_x(Q))+\tilde{v}_1(b_y(Q)+(h_2)_y(Q)))\,\mathrm{d}\theta
\end{aligned}$$

$$\begin{aligned}
u_1(P) = \frac{1}{2\pi}\int_0^{2\pi}\Big[&-\frac{1}{\tilde{c}_1}K_1(Q)\,\mathrm{sgn}(\cos\theta)+u_1(Q)\left(\cos^2\theta+\frac{1}{2}\right) \\
&+v_1(Q)\sin\theta\cos\theta\Big]\mathrm{d}\theta
\end{aligned}$$

$$\begin{aligned}
v_1(P) = \frac{1}{2\pi}\int_0^{2\pi}\Big[&-\frac{1}{\tilde{c}_1}L_1(Q)\,\mathrm{sgn}(\sin\theta)+u_1(Q)\sin\theta\cos\theta \\
&+v_1(Q)\left(\sin^2\theta+\frac{1}{2}\right)\Big]\mathrm{d}\theta
\end{aligned}$$

$$h_2(P) = \frac{1}{2\pi}\int_0^{2\pi} h_2(Q)\,\mathrm{d}\theta, \quad b(P) = \frac{1}{2\pi}\int_0^{2\pi} b(Q)\,\mathrm{d}\theta \qquad (11)$$
$$u_2(P) = \frac{1}{2\pi}\int_0^{2\pi} u_2(Q)\,\mathrm{d}\theta, \quad v_2(P) = \frac{1}{2\pi}\int_0^{2\pi} v_2(Q)\,\mathrm{d}\theta$$

Here $\tau = \Delta t/2$, due to the Strang splitting (7) we have $\tau = \Delta t$ for the T^2 operator. Further, K_1 and L_1 are the potential energies in the $x-$ and $y-$ directions defined as $K_1 := g(h_1 + h_2 + b - V_1)$, $L_1 := g(h_1 + h_2 + b + U_1)$, where V_1, U_1 are the primitives to the Coriolis forces, i.e. $\partial_x(V_1) = \frac{f}{g}v_1$, $\partial_y(U_1) = \frac{f}{g}u_1$. In the equations (11) the evolution takes place along the so-called bicharacteristic cone having the peak at $P = (x, y, t^{old} + \tau/2)$ and footpoints Q on the sonic circle, i.e. $Q = (x - \tilde{u}_1\tau/2 - \tilde{c}_1\tau/2\cos\theta, y - \tilde{v}_1\tau/2 - \tilde{c}_1\tau/2\sin\theta, t^{old})$, $\theta \in (0, 2\pi)$; $\tilde{u}_1, \tilde{v}_1, \tilde{c}_1$ are the constant states obtained by a local linearization (e.g. local averaging).

5 Well-Balancing of Path-Consistent FVEG Scheme

The aim of this section is to verify the well-balanced property of the path-consistent FVEG scheme (7), (8).

Theorem 5.1. *Suppose that the solution at time t^n satisfy for all (x,y) the equilibrium conditions (2) and (3). Then the path-consistent FVEG scheme satisfies the same well-balanced conditions for the new time level t^{n+1}. More precisely, the rest at state condition (2) is preserved exactly and the geostrophic equilibrium condition (3) is satisfied for smooth solutions up to at least third order accuracy.*

Proof. *i)* First, it is clear that if the operators T^1 and T^2 are well-balanced the complete path-consistent scheme (7) will be well-balanced as well.

Further, we have shown in [10] that the approximate evolution operator (11) satisfies these conditions exactly for the corresponding generalized bottom topography $h_2 + b$; the same is true for the second layer with the generalized topography $rh_1 + b$.

Thus for each layer, the predicted solutions $\mathbf{W}^*$ satisfy the above conditions (2), (3) exactly.

ii) Now, we need to show that the finite volume update (8) preserves these conditions as well. We will first consider the rest state conditions: using $u_i^* = 0$, $v_i^* = 0$, for $i = 1,2$, and $h_1^* + h_2^* + b = c_1$, $rh_1^* + h_2^* + b = c_2$, $c_1 = $ const., $c_2 = $ const. we get for the first layer for each k, ℓ

$$(h_1)_{k,\ell}^{new} = (h_1)_{k,\ell}^{old}$$

$$\begin{aligned}(h_1u_1)_{k,\ell}^{new} = &- \frac{\Delta t}{\hbar}\sum_{\ell' \in L}\alpha_{\ell'}\frac{g}{2}\Big[\Big((h_1^*)_{k+1/2,\ell'}^2 - (h_1^*)_{k-1/2,\ell'}^2\Big) \\ &+ \big((h_1^*)_{k-1/2,\ell'} + (h_1^*)_{k,\ell'}\big)\Big(B_{k,\ell'}^* - B_{k-1/2,\ell'}^*\Big) \\ &+ \big((h_1^*)_{k+1/2,\ell'} + (h_1^*)_{k,\ell'}\big)\Big(B_{k+1/2,\ell'}^* - B_{k,\ell'}^*\Big)\Big],\end{aligned} \qquad (12)$$

where we set $B = b + h_2$. Now we have on the right hand side of (12)

$$\begin{aligned}
&(h_1^*)^2_{k+1/2,\ell'} - (h_1^*)^2_{k-1/2,\ell'} + \left((h_1^*)_{k-1/2,\ell'} + (h_1^*)^2_{k,\ell'}\right)\left(B^*_{k,\ell'} - B_{k-1/2,\ell'}\right) \\
&\qquad + \left((h_1^*)_{k+1/2,\ell'} + (h_1^*)_{k,\ell'}\right)\left(B^*_{k+1/2,\ell'} - B^*_{k,\ell'}\right) \\
= \; &(h_1^*)_{k+1/2,\ell'} c_1 - (h_1^*)_{k-1/2,\ell'} c_1 + \left((h_1^*)_{k-1/2,\ell'} - (h_1^*)_{k+1/2,\ell'}\right) B^*_{k,\ell'} \\
&\qquad + (h_1^*)_{k,\ell'}\left(B^*_{k+1/2,\ell'} - B^*_{k-1/2,\ell'}\right) \\
= \; &\left((h_1^*)_{k+1/2,\ell'} - (h_1^*)_{k-1/2,\ell'}\right)\left(c_1 - B^*_{k,\ell'}\right) + (h_1^*)_{k,\ell'}\left(B^*_{k+1/2,\ell'} - B^*_{k-1/2,\ell'}\right) \\
= \; &(h_1^*)_{k,\ell'}(c_1 - c_1) = 0.
\end{aligned}$$

The analogous equations hold for the momentum equation in the y-direction. Altogether these imply that $h_1^{new} = h_1^{old}$, $u_1^{new} = 0 = v_1^{new}$. The same relations hold analogously for the second layer $h_2^{new} = h_2^{old}$, $u_2^{new} = 0 = v_2^{new}$, so that together with the results from *i)* we have $h_1^{n+1} + h_2^{n+1} + b = c_1$ and $rh_1^{n+1} + h_2^{n+1} + b = c_2$.

iii) The proof for the geostrophic equilibrium (3) is analogous. The first equation for the conservation of momentum yields $(h_1)^{n+1}_{k\ell} = 0$ and $(h_2)^{n+1}_{k\ell} = 0$. However, the momentum equation in the $x-$ direction yields only $u_i^{n+1} = \mathcal{O}(\hbar^3), \quad i = 1,2.$ Indeed,

$$\begin{aligned}
(h_1 u_1)^{new}_{k\ell} &= -\frac{\Delta t}{\hbar} \sum_{\ell' \in L} \alpha_{\ell'} \left[\frac{g}{2}\left((h_1^*)^2_{k+1/2,\ell'} - (h_1^*)^2_{k-1/2,\ell'}\right)\right. \qquad (13) \\
&\left. + f\frac{\hbar}{4} \quad \left((h_1^*)_{k-1/2,\ell'}(v_1^*)_{k-1/2,\ell'} + 2(h_1^*)_{k\ell'}(v_1^*)_{k\ell'} + (h_1^*)_{k+1/2,\ell'}(v_1^*)_{k+1/2,\ell'}\right)\right] = \\
&-\frac{\Delta t}{\hbar} \quad \sum_{\ell' \in L} \alpha_{\ell'} \left[\frac{g}{2}\left((h_1^*)_{k+1/2,\ell'} + (h_1^*)_{k-1/2,\ell'}\right)\left((K_1^*)_{k+1/2,\ell'} - (K_1^*)_{k-1/2,\ell'}\right) + \mathcal{E}\right],
\end{aligned}$$

here $\mathcal{E}$ is the error term and K_1 is the potential energy, thus setting for simplicity of the presentation $K_1 = h_1 + V_1$. Then $(K_1^*)_{k+1/2,\ell'} - (K_1^*)_{k-1/2,\ell'} = (h_1^*)_{k+1/2,\ell'} - (h_1^*)_{k-1/2,\ell'} + f\frac{\hbar}{2}((v_1^*)_{k+1/2,\ell'} + (v_1^*)_{k-1/2,\ell'}) = 0$. The error term gives

$$\mathcal{E} = \frac{f\hbar}{4}\left((h_1^*)_{k+1/2,\ell'}(v_1^*)_{k-1/2,\ell'} + (h_1^*)_{k-1/2,\ell'}(v_1^*)_{k+1/2,\ell'} - 2(h_1^*)_{k,\ell'}(v_1^*)_{k,\ell'}\right) = \mathcal{O}(\hbar^3),$$

which together with an analogous equation for the second layer concludes the proof.

6 Numerical Experiments

In this section we want to demonstrate the behaviour of new path-consistent FVEG scheme through some numerical experiments.

Problem 1

First, we test accuracy of the path-consistent FVEG scheme. The initial data are as follows

$$\begin{aligned} b(x,y) &= -\frac{1}{2}\exp(\sin^2(\pi x)), \\ h_1(0,x,y) &= 5+\exp(\cos(2\pi x)), \\ h_2(0,x,y) &= 5+\sin^2(\pi x)-b(x), \\ u_1(0,x,y) &= \sin(\cos(2\pi x))/h_1(x,0) \\ u_2(0,x,y) &= 0. \end{aligned}$$

We apply periodic boundary conditions and set $r = 0.98$, $g = 9.812$. The computational domain $[0,1]\times[0,1]$ is divided into $N\times N$ mesh cells, $N = 20,\ldots,160$. In the following table we present behaviour of the relative global L^1 error of the second order well-balanced path-consistent FVEG scheme. The solution was computed until $T = 0.1$ and no limiter has been used in this experiment. The CFL number is chosen to be $CFL = 0.9$.

Table 1 clearly demonstrates the second order accuracy of the FVEG method. Interestingly, we can notice very accurate behaviour of our path-consistent FVEG scheme, for example, in comparison with the well-balanced higher order FV WENO scheme that uses fourth order Runge-Kutta method for time evolution. The latter has been developed by Frings in [5].

Table 1 L_1 errors of the well-balanced path-consistent FVEG method

	FVEG			
N	h_1	EOC	h_2	EOC
20	2.31e-03		1.45e-03	
40	4.28e-04	2.43	4.02e-04	1.85
80	7.73e-05	2.47	8.36e-05	2.27
160	1.59e-05	2.28	1.92e-05	2.12

	FVEG			
N	q_1	EOC	q_2	EOC
20	6.87e-02		9.48e-02	
40	1.32e-02	2.39	1.97e-02	2.27
80	2.29e-03	2.52	3.70e-03	2.41
160	4.61e-04	2.32	7.82e-04	2.24

Problem 2

This is an internal dam-break problem. We have imposed initially a jump at the interface, while still having the total water height constant.

$$h_1(0,x) = \begin{cases} 0.2, & \text{if } x < 5, \\ 1.8, & \text{if } x > 5, \end{cases}$$
$$h_2(0,x) = \begin{cases} 1.8, & \text{if } x < 5, \\ 0.2, & \text{if } x > 5, \end{cases}$$
$$u_1(0,x) = u_2(0,x) = 0,\ b(x) = 0, \qquad x \in (0,10).$$

In the following figures solution obtained by the path-consistent second order FVEG scheme for the case $r = 0.7$ at time $t = 0.1$ and $r = 0.98$ at $t = 5$ is plotted, see Figures 2,3. We can nicely recognize the structure of the solution consisting of three constant states connected by two rarefaction waves and two shocks. No oscillations or smearing of shocks can be seen, that was a problem reported by other authors, cf. [2]. The minmod limiter has been used in the second order reconstruction.

Problem 3

The last problem is a geostrophic adjustment simulation for two-layer shallow water model, analogous problem for the one-layer shallow water has been considered in [6]. The initial conditions are

$$h_1(x,y,0) = 1 + \frac{A_0}{2}\left(1 - \tanh\left(\frac{\sqrt{(\sqrt{\lambda}x)^2 + (y/\sqrt{\lambda})^2} - R_i}{R_E}\right)\right),$$
$$h_2(x,y,0) = 1, \quad u_1(x,y,0) = v_1(x,y,0) = u_2(x,y,0) = v_2(x,y,0) = 0,$$

where the parameters are $A_0 = 0.5$, $\lambda = 2.5$, $R_E = 0.1$, and $R_i = 1$. The gravity and the Coriolis forces parameters are set to $g = 1$, $f = 1$, fraction of layers density is $r = 0.98$. No bottom topography is considered. The computational domain $[-10,10] \times [-10,10]$ has been divided into 400×400 mesh cells. The CFL number was set to 0.6. In Figure 4 we see results of time evolution of initial perturbation of top surface. Results were obtained by the second order path-consistent FVEG scheme using bilinear recovery with minmod limiter. Two circular shock waves propagating in the top surface $h_1 + h_2$ and in the second layer h_2 can be recognized nicely. Behind the shocks there is elevation rotating anti-clockwise. In order to represent absorbing boundary conditions extrapolation has been implemented on the boundary of computational domain.

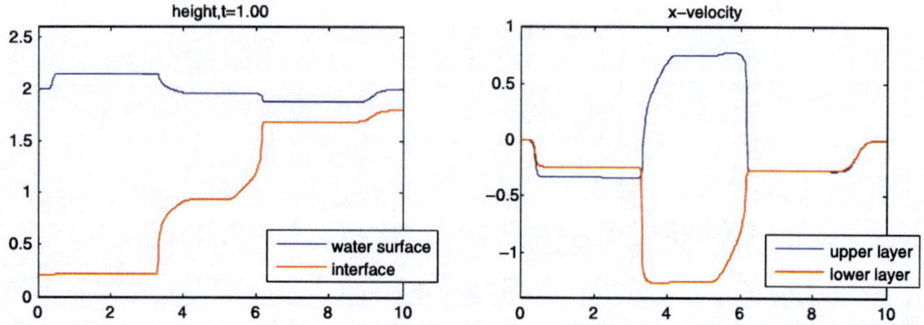

Fig. 2 Solution obtained by the path-consistent FVEG scheme at time $t = 1.0$, $\frac{\rho_1}{\rho_2} = 0.7$

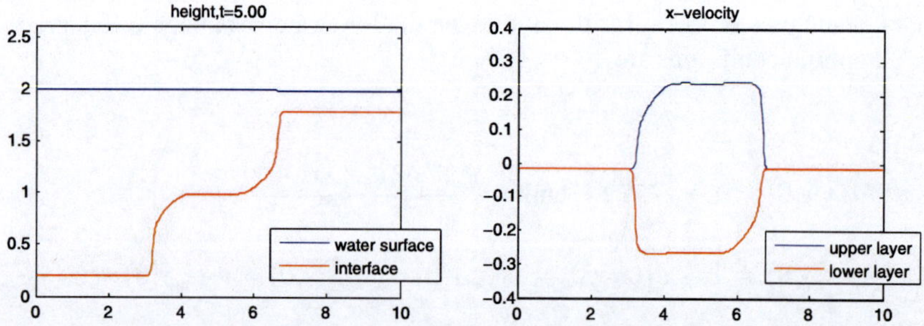

Fig. 3 Solution obtained by the path-consistent FVEG scheme at time $t = 5.0$, $\frac{\rho_1}{\rho_2} = 0.98$

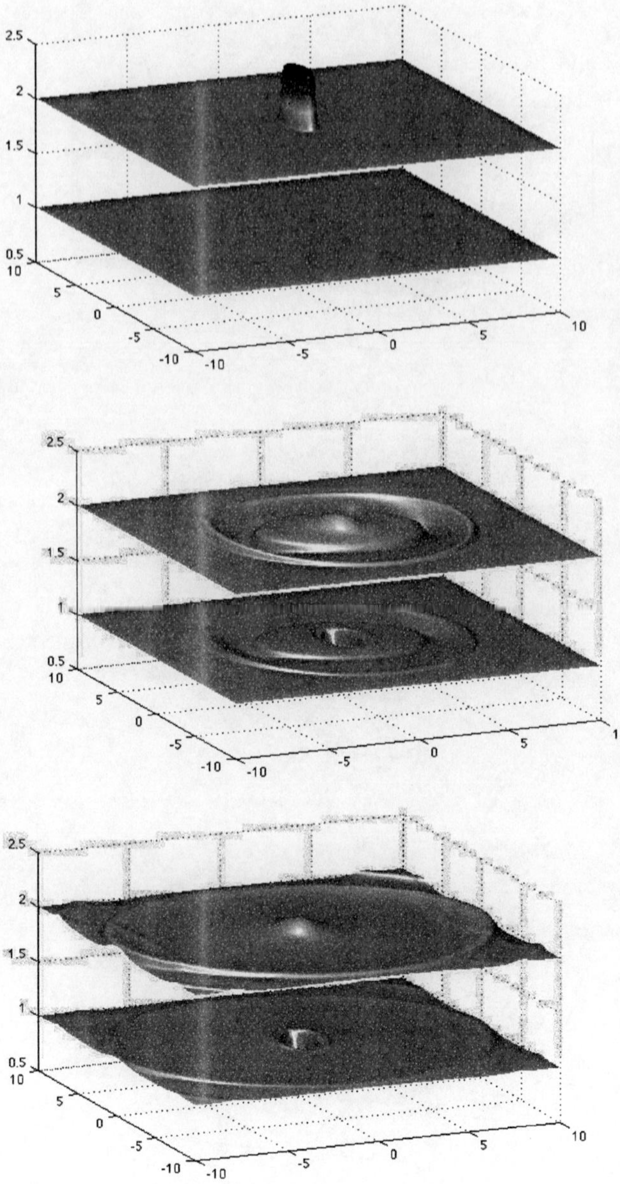

Fig. 4 Effects of Coriolis forces in the two-layer model at $t = 0, 4, 8, 12, 16, 20$

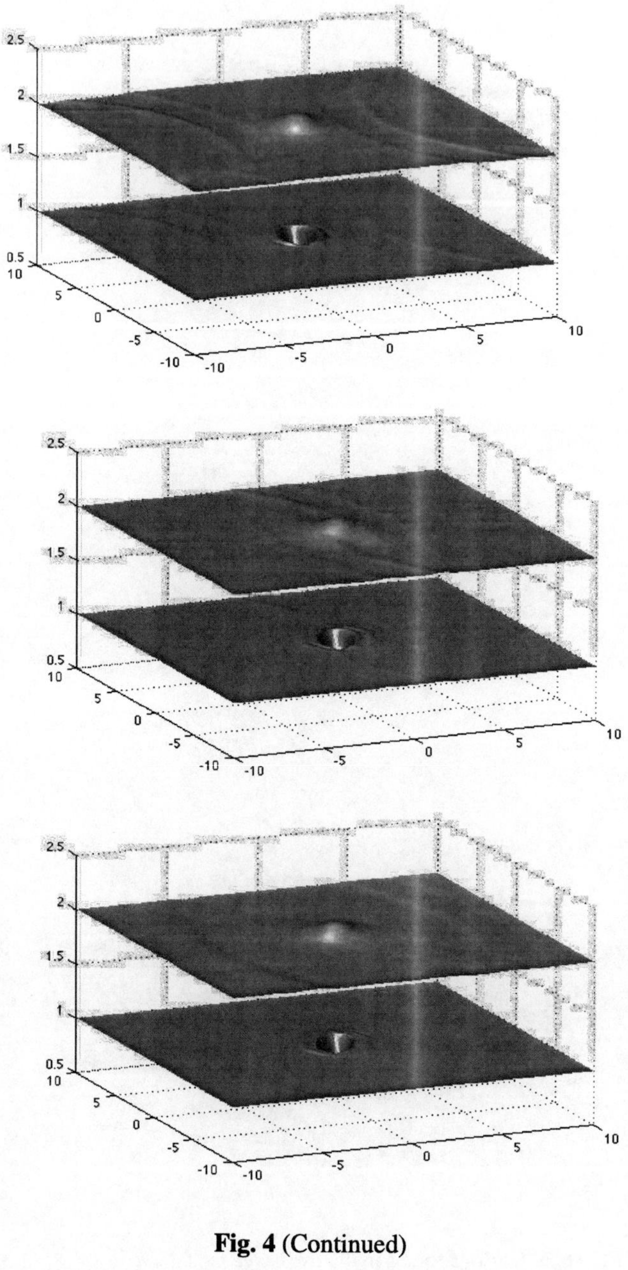

Fig. 4 (Continued)

Acknowledgements. The present research has been partially supported by the Deutsche Forschungsgemeinschaft under the grant LU 1470/2-1 and by the European Graduate School Differential Equations with Applications in Science and Engineering (DEASE), MEST-CT-2005-021122.

References

1. Abgrall, R., Karni, S.: Two–layer shallow water systems: a relaxation approach. SIAM J. Sci. Comput. 31(3), 1603–1627 (2009)
2. Bouchut, F., Morales, T.: An entropy satisfying scheme for two-layer shallow water equations with uncoupled treatment. M2AN 42, 683–698 (2008)
3. Dal Maso, G., LeFloch, P.G., Murat, F.: Definition and weak stability of nonconservative products. J. Math. Pures Appl. 74, 483–548 (1995)
4. Castro, M.J., LeFloch, P.G., Munoz-Ruiz, M.L., Parés, C.: Why many theories of shock waves are necessary: Convergence error in formally path-consistent schemes. J. Comp. Phys. 227(17), 8107–8129 (2008)
5. Frings, J.: Well-Balanced Finite Volumes of Higher Order of Accuracy for Two-Layer Shallow Water Flows. Master Thesis, RWTH Aachen (2007)
6. Castro, M.J., López, J.A., Parés, C.: Finite volume simulation of the geostrophic adjustment in a rotating shallow-water system. SIAM J. Sci. Comput. 31, 444–477 (2008)
7. Kurganov, A., Petrova, G.: Central-upwind schemes for two layer shallow water equations. SIAM J. Sci. Comput. 31, 1742–1773 (2009)
8. Lukáčová-Medvidová, M., Morton, K.W., Warnecke, G.: Finite volume evolution Galerkin (FVEG) methods for hyperbolic problems. SIAM J. Sci. Comput. 26(1), 1–30 (2004)
9. Lukáčová-Medvidová, M., Vlk, Z.: Well-balanced Finite Volume Evolution Galerkin Methods for the Shallow Water Equations with Source Terms. Int. J. Num. Fluids. 47(10-11), 1165–1171 (2005)
10. Lukáčová-Medvidová, M., Noelle, S., Kraft, M.: Well-balanced finite volume evolution Galerkin methods for the shallow water equations. J. Comp. Phys. 221, 122–147 (2007)
11. Parés, C.: Numerical methods for nonconservative hyperbolic systems: a theoretical framework. SIAM J. Numer. Anal. 44, 300–321 (2006)
12. Parés, C., Castro, M.J.: On the well-balance property of Roe's method for nonconservative hyperbolic systems. Applications to shallow-water systems. Math. Model. Numer. Anal. 38, 821–852 (2004)

Some Features of the Landslide Mechanism of Surface Waves Generation in Real Basins

Sofia A. Beisel, Leonid B. Chubarov, and Yurii I. Shokin

Abstract. We present the results of numerical modelling of surface wave generation by the movement of submerged deformable body along the slope, which simulates the real coastal slope. The multiparametric computations are carried out within the shallow water approximation that allowed to determine the dependence of wave pattern on the depth of landslide submergence, length and thickness of the body, relative density, friction coefficient, and slope geometry. The main attention is focused on the effects resulted from the heterogeneity of depths, which simulates specifics of real water basins.

1 Introduction

The interest in the investigations of the landslide mechanism of surface waves generation in nearshore water areas is caused by a number of catastrophic events caused by this mechanism that had happened in the recent years in different basins of the World Ocean. The landslide mechanism of tsunami wave generation is called abnormal in contrast to the conventional seismic one. Here abnormality means inconsistency between significant tsunami wave near the shore and a weak earthquake, which is associated with the generation of this wave and actually can be only a trigger for the landslide mechanism of wave generation. The portion of such events is estimated at about 15% of registered historical tsunamis.

To solve the problem of mathematical modelling of the landslide induced wave generation, both the models of landslide motion and of the ambient fluid should be built. The known approaches to the simulation of landslides include the simulation of the motion of an absolutely rigid body [5, 7] or a set of such bodies [13], simulating a fluid flow of different density, viscosity, etc. [10], or simulating

Sofia A. Beisel · Leonid B. Chubarov · Yurii I. Shokin
Institute of Computational Technologies SB RAS,
Lavrentiev Ave. 6, Novosibirsk, 630090, Russia
e-mail: beisels@gmail.com, chubarov@ict.nsc.ru, shokin@ict.nsc.ru

E. Krause et al. (Eds.): Computational Sci., & High Performance Computing IV, NNFM 115, pp. 137–148.
springerlink.com

the motion of a elasticoplastic medium [4] with or without taking into account the interaction with the ambient fluid. In some situations it appears promising to simulate the phenomenon as a two-layer fluid with layers of various densities and viscosity coefficients [8]. In [7, 12, 14] it is shown that substitution of the real landslide by a model rigid body coupled with an appropriate law of motion gives an adequate description of wave processes in the wide range of such parameters as slope angle, thickness and length of the body, and it's initial submergence.

The specifics of the simulation of landslide generated surface waves is that this waves are generated in the nearshore zone of a small depth, and the duration of the landslide motion is quite long and comparable with the period of the generated wave; the characteristic depth and the vertical size of the landslide are comparable as well. Therefore hydrodynamic aspects of wave processes are being investigated in the framework of approximate models of wave hydrodynamics. Earlier investigations by the authors let us estimate the capabilities of different models by comparing their results with laboratory experiments. These investigations were carried out for the simplest relief consisted of the uniform slope and the flat bottom. The conclusion [2] was that even the simplest shallow water models give an adequate qualitative description of wave regimes within the special range of problem parameters, while nonlinear-dispersive models, that take into account density non-hydrostaticity (or, what is the same, the vertical fluid motion), should be used for detailed investigations.

In previous investigations, when a landslide was represented as a rigid body, only its movement along the uniform slope was considered in the law of motion. For numerical modelling of the landslide motion along such slope [1, 6], a flat bottom was attached at some depth, and for the landslide not to enter this zone, its velocity was being decreased forcibly for the body to stop at the end of the slope. Such artificial deceleration with a jump or a break of the velocity caused stop waves or deceleration waves with significant amplitude.

In the present paper an attempt is made to model the surface waves generation by the landslide moving along the arbitrary curvilinear slope according to the motion law, that takes into account slope irregularity. In particular, the possibility of a landslide to continue its movement along the flat bottom is possible, and its deceleration results from the bottom friction and hydrodynamic drag. Such natural smooth deceleration of the landslide avoids the effects of an abrupt stop noted above.

Our main attention is focused on the comparison of results obtained on the piecewise linear and “real" curvilinear slopes. The results of computational experiments let us determine the base invariable characteristics of wave processes and the parameters, which depend on the slope form. We discuss the results of comparison of wave regimes, generated above different bottom reliefs, analyse the dependence of regime characteristics on geometrical and physical parameters (landslide submergence, its length and thickness, relative density and friction coefficient). It is shown that the character of landslide motion along the irregular slope and surface wave pattern generated by this motion essentially differ from that on the uniform slope.

The modelling of surface waves, generated by the motion of a landslide was made in the framework of the nonlinear model of shallow water over a non stationary bottom. The modified finite-difference MacCormack scheme on a uniform computational grid was utilized [3].

2 Set-Up of Problem

The mathematical set-up is similar to the one presented at the paper of Chubarov L.B., Khakimzyanov G.S. and Shokina N. published at this book, and, therefore, will be skipped here, as well as the law of landslide motion be in use, whose derivation is also given in details at that work. Only model reliefs and the form of landslide will be described below, as they are different from that used in the work noted.

2.1 *Model Reliefs*

Two reliefs were considered to study the influence of bottom irregularity on the surface waves, generated by the motion of underwater landslide. The first one is the digitized vertical section of Mediterranean coast of Israel in the direction normal to the coastal line near Hadera city. Its minimal depth in the coastal zone was 2 m (but it was enlarged to 20 meters to keep full depth positive during the whole period of calculations), and the maximal depth in the deep-water zone was 1450 m. It consisted of the long shelf zone with slope angle about 1 degree, the slope with incline of up to 7 degrees and gentle deep-water slope with incline no more than 2 degrees.

The second relief was the piece-wise linear approximation of the first one and consisted of the flat bottom with depth 20 m and length 3200 m, uniform slope with inclination of 2 angles, and flat bottom with depth 1400 m and length 27300 m. With given parameters the linear slope had the inclination that is equal to the mean inclination of the "real" relief, and had the same depth 535 m at the point of the initial position of the landslide centre of mass ($x = 17900$ m). The total length of both water basins was 70000 m.

To fix the results of calculations 2 virtual gages were used: A – on the "shore" ($x = 0$ m), and B – in the deep-water zone ($x = 60000$ m).

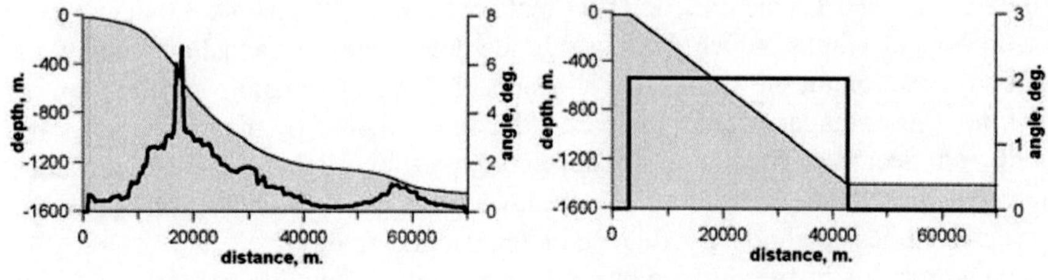

Fig. 1 Characteristics of model bottom reliefs used in numerical experiments: depth (filled contour, axe is at the left) and slope angle (fat line, axe is at the right): on the left – "real" slope, on the right – linear one

2.2 Model Landslide

The shape of the model landslide, which was used in computations, is described by the formula, that is similar to that proposed in [11]:

$$h_{sl}(x,t) = T\frac{[1+\tanh\left((x-x_1(t))\frac{2}{p}\right)][1-\tanh\left((x-x_2(t))\frac{2}{p}\right)]}{[1+\tanh(1)][1-\tanh(-1)]},$$

where $x_1(t) = x_c(t) - \frac{p}{2}$, $x_2(t) = x_c(t) + \frac{p}{2}$, $x_c(t)$ is the x-coordinate of the landslide mass centre.

The maximum landslide thickness T was equal to 25 m, parameter p, that determine landslide length b, was equal to 2500 m (and then b, determined on 10% of maximal landslide thickness, was equal to 5000 m). At the start time the landslide mass centre was at the point of maximum inclination of "real" relief $x_c = 17900$ m, that corresponds to the initial submergence of 535 m.

3 Computational Results

Computational experiments were carried out using the simple effective finite - difference scheme, built on the base of MacCormack scheme of the second order of approximation, on the uniform computational grid. The time step was calculated from the stability condition so that Courant number was equal to 0.9. The total number of computational grid points was equal to 1401.

The following parameter values were used in calculations by default: relative density $\gamma = 1.5$, added mass coefficient $C_w = 1.0$, hydrodynamic drag coefficient $C_d = 1.0$, friction angle $\theta_* = 1°$.

3.1 Dependence on Friction Angle

To study the influence of the friction angle θ_* on the body motion and on the characteristics of waves generated, the following values of θ_* were used: for linear slope – 0.0, 0.5, 1.0 and 1.5 degrees, and for "real" one – 0.0, 1.0, 2.0 and 3.0 degrees.

Analysis of graphs, which demonstrate the dependence of landslide velocity on the body position on the slope, as well as dependence of landslide acceleration on the time, shows that as friction increases, the distance gone by the body and its velocity both decrease. The picture for acceleration differs slightly: its positive values during the acceleration also have evident inverse dependence on friction angle, but its negative ones do almost not depend on the friction angle.

Gauge records of free surface elevations generated by underwater landslide are presented below. The depression wave, recorded near the shore the first, arises at

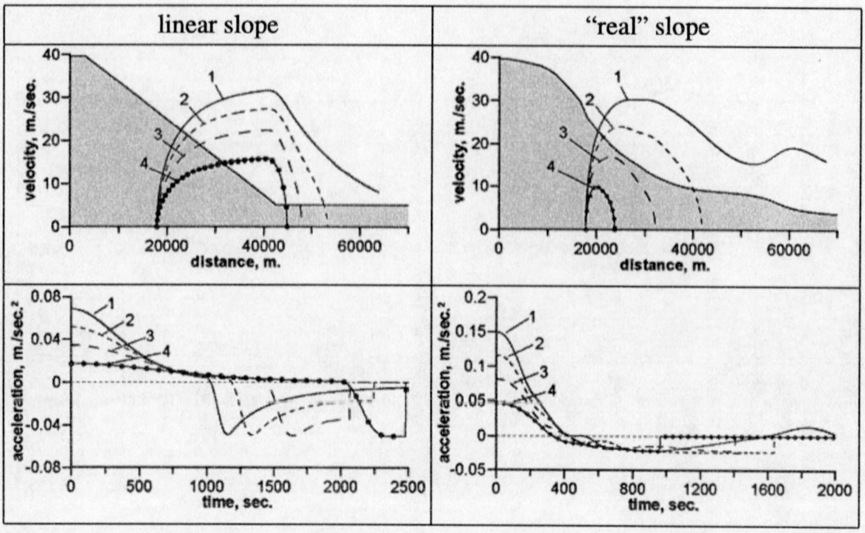

Fig. 2 Characteristics of laws of landslide motion for different values of friction angle θ_*. For linear slope: (1) – 0.0°, (2) – 0.5°, (3) – 1.0°, (4) – 1.5°; for "real" one: (1) – 0.0°, (2) – 1.0°, (3) – 2.0°, (4) – 3.0°

the beginning of landslide motion and goes to the shallow water zone, against the direction of body motion. Its amplitude linearly depends on the initial acceleration, that is seen in big difference between fixed negative amplitudes on gauge A. The first positive wave, registered near the shore, is generated when two negative waves (one is the mentioned above after its reflection off the wall and the second one is the negative wave that goes right over the moving landslide) move over the slope to the zone with larger depths. On the linear slope this wave is seen clearly and its amplitude inversely depends on the friction angle. The second positive wave arises when the body decelerates, and terminates after body finally stops its moving. The amplitude of this wave directly depends on the absolute value of body acceleration just before the stop, therefore this amplitude is the biggest for the widest friction angle.

On the "real" slope the process of deceleration goes faster, therefore both positive waves almost merge, especially for wide friction angle. Here deceleration takes place on different depths: for wider θ_* this depth is smaller and it makes the amplitude of generated wave bigger. As a result, the maximum amplitude of positive wave is the same for all θ_*, its length is smaller for wider θ_* though.

Waves that propagate to the deep-water zone suffer similar changes, when the friction angle varies.

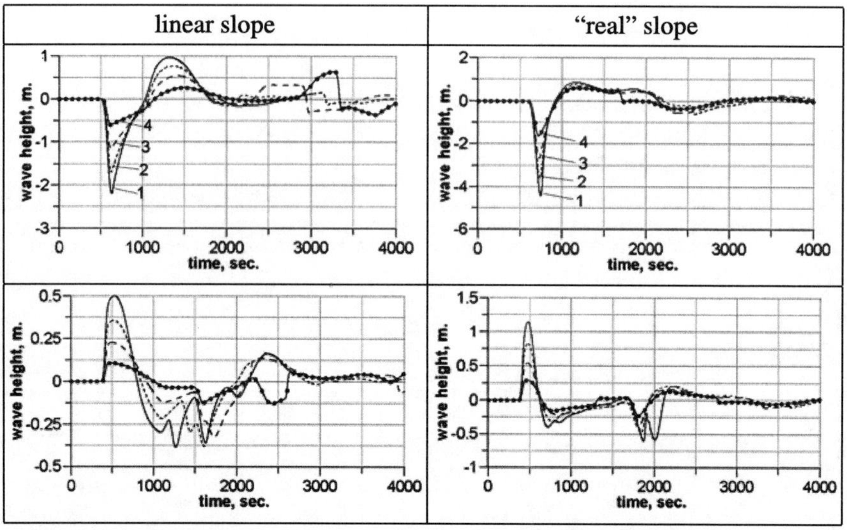

Fig. 3 Gauge records for different friction angles: top row – gauge A, bottom row – gauge B. For linear slope: (1) – 0.0°, (2) – 0.5°, (3) – 1.0°, (4) – 1.5°; for "real" one: (1) – 0.0°, (2) – 1.0°, (3) – 2.0°, (4) – 3.0°

3.2 Dependence on Landslide Size

To investigate the influence of the landslide size on the characteristics of wave regime, the computations with following length and thickness were carried out: $b = 5000$ m, $T = 25$ m (reference landslide), $b = 5000$ m, $T = 50$ m (double thickness); $b = 10000$ m, $T = 25$ m (double length).

As we see on Fig. 4, the law of landslide motion does not depend on body's thickness, while varying of it's length appears differently on different slopes. On the one hand, it is due to the specificity of parameter occurrence in the formula of motion law, and, on the other hand, due to the fact that as body length increases, it occupies new parts of the slope with other inclination. On the "real" slope this effect implies from the very beginning of the motion, and for the linear one – during landslide transition from the inclined part of the slope to the flat one. Thus, on the linear slope initial acceleration does not depend on the length of the body, and its maximal negative value is a little less for bigger length, as the left side of the body lays on the slope for a longer time. On the curvilinear slope both initial acceleration and maximum negative one are smaller for bigger length. The maximum velocity is smaller for longer body on both reliefs.

As for the waves generated (Fig. 5), on the linear slope duplication of size appears in similar ways: in corresponding duplication of positive and negative wave amplitudes. On the "real" slope duplication of length appears weaker.

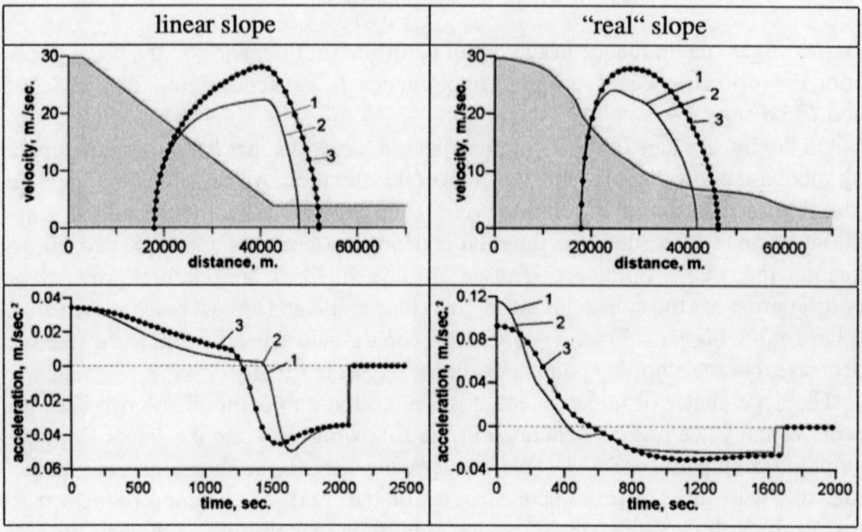

Fig. 4 Characteristics of laws of landslide motion for different landslide sizes: (1) – "reference" landslide, (2) – double thickness, (3) – double length

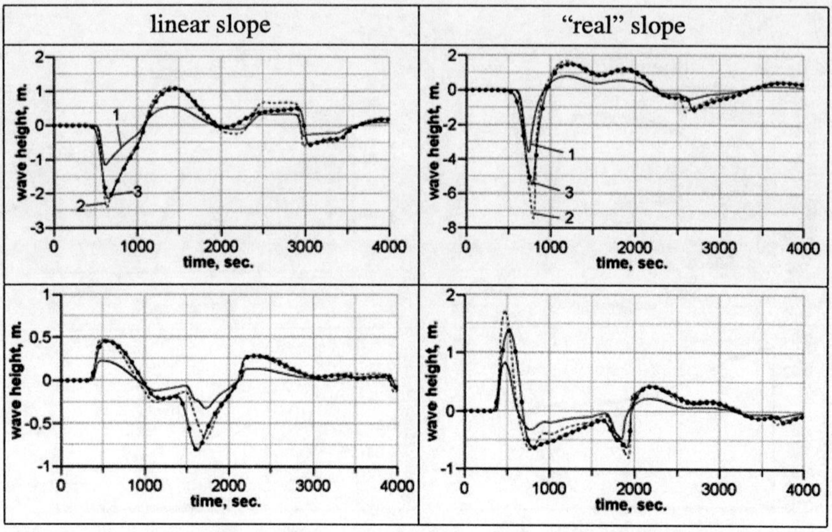

Fig. 5 Gauge records for different landslide sizes: : top row – gauge A, bottom row – gauge B. (1) – "reference" landslide, (2) – double thickness, (3) – double length

3.3 Dependence on Landslide Embedding

To investigate the influence of the initial landslide submergence on the wave generation the following values of initial submergence d_c were considered: 535, 435, 335 and 235 meters.

On the linear slope, as one should expect, the motion character does not depend on submergence, except for the start time of deceleration. As velocities are equal before this deceleration, the stopping point is the same, too. On the curvilinear slope the situation is different. Here, the smaller depth, the smaller slope angle and, consequently, the smaller initial acceleration. But, for small d_c, acceleration grows when body approaches the zone with the largest slope angle and almost reaches the values it has for the bigger submergence. The maximum velocity values increase when d_c decrease, but the stopping points are almost the same for any d_c.

The dependence of underwater landslide motion on the initial submergence appears in the wave regime generated in the following way. On the linear slope the amplitude of the first negative wave, propagating towards the shore, increases monotonically with submergence decreasing, but on the "real" one submergence decreasing results in little decreasing of trough amplitude. Amplitudes of the first positive waves near the shore change equally on both slopes: smaller submergence, bigger amplitude. Amplitudes of the second positive waves, generated by body deceleration, are equal for any submergence on both slopes, and this fact is in good correlation with the equality of landslide velocities during deceleration.

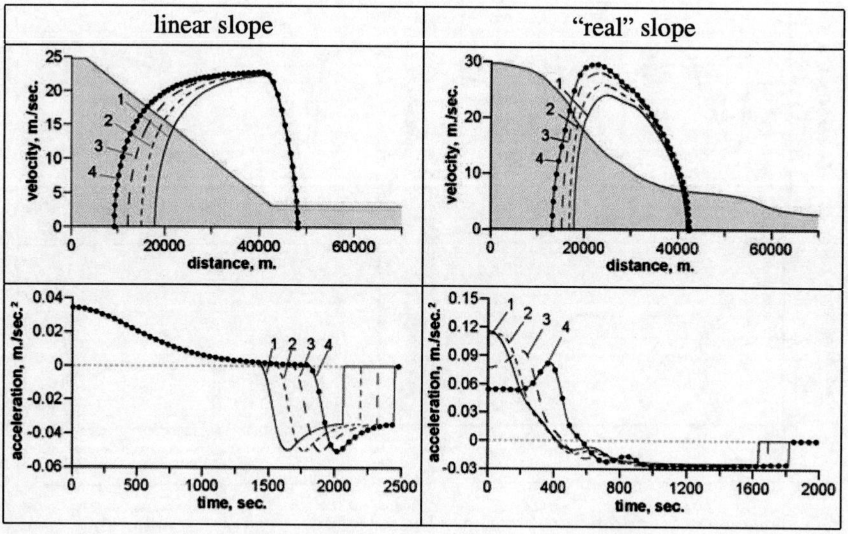

Fig. 6 Characteristics of laws of landslide motion for different landslide submergence d_c: (1) – 535 m, (2) – 435 m, (3) – 335 m, (4) – 235 m

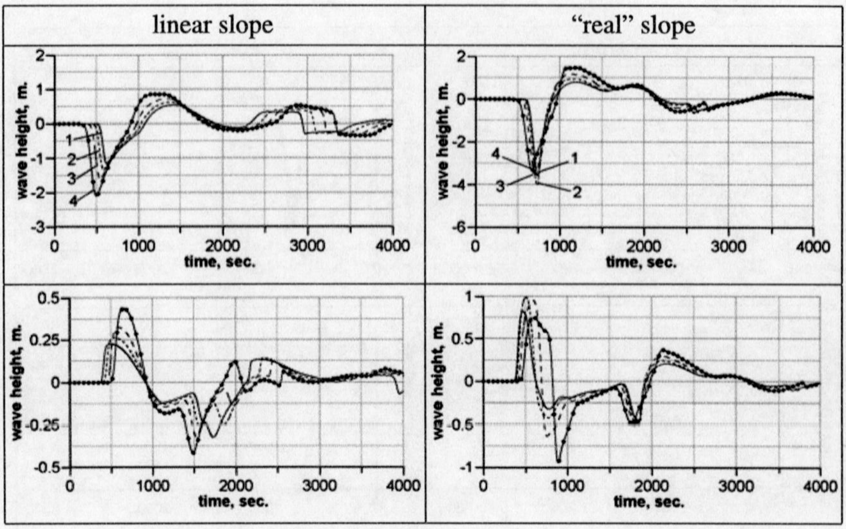

Fig. 7 Gauge records for different landslide submergence d_c: top row – gauge A, bottom row – gauge B. (1) – 535 m, (2) – 435 m, (3) – 335 m, (4) – 235 m

3.4 Dependence on Landslide Density

The influence of the landslide relative density on the body's motion and wave formation appears the simplest. The following values of γ were considered here: 1.5, 1.75, 2.0, 2.25.

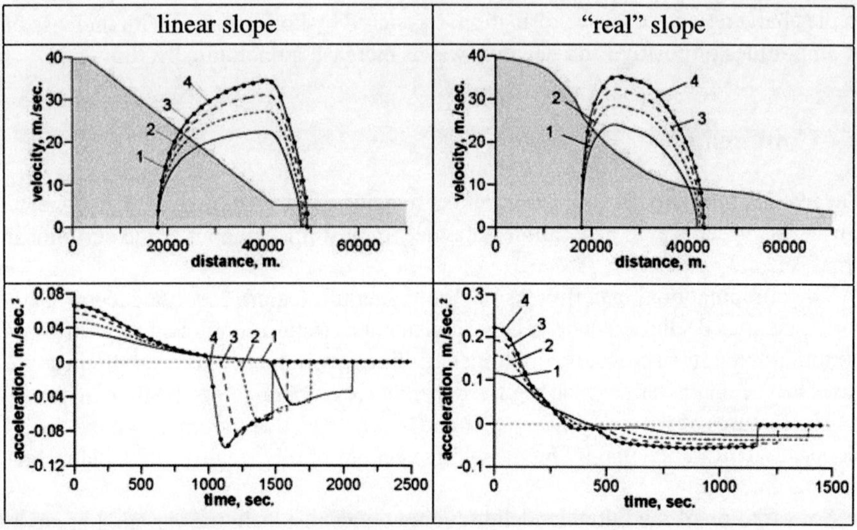

Fig. 8 Characteristics of laws of landslide motion for different relative density γ: (1) – 1.5, (2) – 1.75, (3) – 2.0, (4) – 2.25

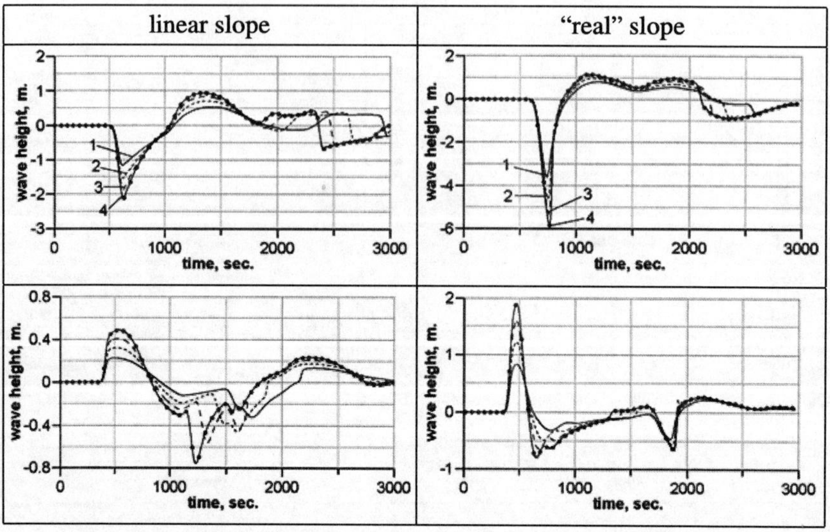

Fig. 9 Gauge records for different relative density γ: top row – gauge A, bottom row – gauge B. (1) – 1.5, (2) – 1.75, (3) – 2.0, (4) – 2.25

On both reliefs with relative density increasing maximum values of acceleration and deceleration increase (approximately in two times with density increasing in 1.5 times), as well as body velocity (in 1.5 times); the duration of motion is in inverse dependence, and the stopping point does not change almost.

The monotonicity of landslide motion dependence on relative density transmits to the characteristics of wave formation, registered by both gauges. With increase of γ, amplitudes of positive and negative waves increase monotonically, too.

4 Conclusions

The results, presented in the article, reflect transition in the study of landslide mechanism of surface waves generation from the simplest linear slopes to the curvilinear "real" ones.

The computational experiments demonstrated the features of influence of landslide size, initial submergence, relative density and friction coefficient over different bottom reliefs. It is shown, that bottom curvilinearity influences essentially in some cases on the character of wave regime dependence on parameters mentioned above.

The features of wave regimes, generated by the landslide motion down the "real" slope, are determined mainly by the taking account of this "reality" in the landslide motion law.

So, variation of the initial landslide submergence results in the variation of angle of the slope, the landslide is situated on, that, in turn, changes the initial acceleration. As a result, for the linear slope the smaller submergence means the bigger wave amplitude, but for the "real" one the most "dangerous" landslide is situated not certainly near the shore.

When landslide length varies, the mean slope angle under the landslide varies too, that has an additional effect on the wave formation.

When the friction force and relative density vary, the characteristics of wave formation change in the same manner over model linear and "real" slopes. Note also, that the friction coefficient is the only considered parameter, which determines the landslide stopping point.

Acknowledgements. The work was supported by the Russian Foundation for Basic Research (grants 09-05-00294, 06-05-72014) and the Program of Federal Support for Scientific Schools of Russian Federation (931.2008.9).

References

1. Chubarov, L.B., Eletskii, S.V., Fedotova, Z.I., Khakimzyanov, G.S.: Simulation of surface waves generation by an underwater landslide. Rus. J. Numer. Anal. Math. Model. 20(5), 425–437 (2005)
2. Eletskij, S.V., Maiorov, Y.B., Maksimov, V.V., Nudner, I.S., Fedotova, Z.I., Kharzhoyan, M.G., Khakimzyanov, G.S., Chubarov, L.B.: Simulation of surface waves generation by a moving part of the bottom down the coastal slope. Comput. Technologies, Special issue 2(part 2), 194–206 (2004) (in Russian)
3. Fedotova, Z.I.: On application of the MacCormack difference scheme for problems of long-wave hydrodynamics. Comput. Technologies, Special issue 11(Part 2), 53–63 (2006) (in Russian)
4. Garagash, I.A., Lobkovskij, L.I.: Geometrical estimation of landslide processes and its monitoring on slope of the Black Sea in view of realization of the "Blue flow" project. In: Proceedings of the Sixth International scientific and technical conference "Up-to-date methods and tools of oceanological investigations", Moscow, pp. 5–15 (2000)
5. Grilli, S.T., Watts, P.: Modeling of waves generated by moving submerged body. Applications to underwater landslide. Engineering Analysis with boundary elements 23, 645–656 (1999)
6. Grilli, S.T., Watts, P.: Tsunami generation by submarine mass failure. I: Modeling, experimental validation, and sensitivity analyses. J. Waterway Port Coast. Ocean Eng. 131(6), 283–297 (2005)
7. Harbitz, C., Pedersen, G.: Model theory and analytical solutions for large water waves due to landslides. In: Preprint Series, Dept. of Mathematics, Univ. of Oslo. No. 4 (1992)
8. Heinrich, P., Piatanesi, A., Hebert, H.: Numerical modeling of tsunami generation and propagation from submarine slumps: the 1998 Papua New Guinea event. Geophys. J. Int. 145, 97–111 (2001)
9. Imamura, F., Imteaz, M.M.A.: Long waves in two-layers: governing equations and numerical model. Sci. Tsunami Hazards 13(1), 3–24 (1995)
10. Jiang, L., LeBlond, P.H.: The coupling of a submarine slide and the surface waves which it generates. J. Geophys. Research 97(C8), 12731–12744 (1992)

11. Lynett, P.J., Liu, P.L.-F.: A numerical study of the runup generated by three-dimensional landslides. J. Geophys. Research 110(C03006) (2005), doi:10.1029/2004JC002443
12. Savage, S., Hutter, K.: The motion of a finite mass of granular material down a rough incline. J. Fluid Mech. 199, 177–215 (1989)
13. Tinti, S., Bortolucci, E., Vannini, C.: A block-based theoretical model suited to gravitational sliding. Natur. Hazards 16, 1–28 (1997)
14. Watts, P., Imamura, F., Grilli, S.T.: Comparing model simulations of three benchmark tsunami generation cases. Sci. Tsunami Hazards 18(2), 107–123 (2000)

Multiscale Analyses for the Shallow Water Equations

Didier Bresch, Rupert Klein, and Carine Lucas

Abstract. This paper explores several asymptotic limit regimes for shallow water flows over multiscale topography. Depending on the length and time scales considered and on the characteristic water depth and height of topography, a variety of mathematically quite different asymptotic limit systems emerges. Specifically, we recover the classical "lake equations" for balanced flow without gravity waves in the single time, single space scale limit (Greenspan, Cambridge Univ. Press, (1968)), discuss a weakly nonlinear and a strongly nonlinear multi-scale version of these wave-free equations involving short-range topography, and we re-derive the equations for long-wave shallow water waves passing over short-range topography by Le Maître et al., JCP (2001).

Didier Bresch
LAMA, équipe EDPs2, Bat. Le Chablais, Campus scientifique,
Université de Savoie, 73376 Le Bourget du Lac, France
e-mail: `Didier.Bresch@univ-savoie.fr`

Rupert Klein
Department of Mathematics and Computer Science, Free University of Berlin,
Arnimallee 6, 14195 Berlin, Germany
e-mail: `Rupert.Klein@math.fu-berlin.de`

Carine Lucas
MAPMO, Université d'Orléans, UFR Sciences,
Bâtiment de mathématiques - Route de Chartres,
B.P. 6759 - 45067 Orléans cedex 2, Université de Orléans, France
e-mail: `carine.lucas@univ-orleans.fr`

E. Krause et al. (Eds.): Computational Sci., & High Performance Computing IV, NNFM 115, pp. 149–164.
springerlink.com

1 Introduction

1.1 Governing Equations and Non-dimensionalization

In this article, we present multiscale analyses for the shallow water equations,

$$\begin{aligned} \partial_t H + \operatorname{div}(Hu) &= 0 \\ \partial_t (Hu) + \operatorname{div}(Hu \otimes u) + g\nabla H^2/2 &= -gH\nabla b \end{aligned}, \tag{1}$$

where $(H,u)(t,x)$ are the depth of the water layer and the flow velocity, g is the gravitational acceleration, $b(x)$ denotes the bottom topography, (t,x) are time and horizontal space coordinates, and $\otimes$ indicates the tensorial product. Non-dimensionalizing (H,u,b,x,t) by characteristic values $H_{\text{ref}}, u_{\text{ref}}, b_{\text{ref}}, L, t_{\text{ref}}$, respectively, we obtain, using the same symbols for the dimensionless variables as used above for the dimensional ones,

$$\begin{aligned} \mathrm{Sr}\,\partial_t H + \operatorname{div}(Hu) &= 0 \\ \mathrm{Sr}\,\partial_t (Hu) + \operatorname{div}(Hu \otimes u) + \frac{1}{\mathrm{Fr}^2}\nabla H^2/2 &= -\frac{\beta}{\mathrm{Fr}^2}H\nabla b \end{aligned}. \tag{2}$$

Here

$$\mathrm{Sr} = \frac{L}{t_{\text{ref}}u_{\text{ref}}}, \qquad \mathrm{Fr} = \sqrt{\frac{gH_{\text{ref}}}{u_{\text{ref}}^2}}, \qquad \beta = \frac{b_{\text{ref}}}{H_{\text{ref}}}, \tag{3}$$

are the Strouhal, and Froude numbers, and the ratio of typical variations of the topography versus the water layer depth.

The asymptotic limit regimes to be analysed in this paper will be defined by particular distinguished limits of these dimensionless parameters, and by multiple spacio-temporal scales. Throughout, we consider low Froude number flows, for which flow velocities are systematically small compared with the speed of the gravity waves, and we introduce the reference asymptotic expansion parameter, ε, via

$$\mathrm{Fr} = \varepsilon^{\alpha} \ll 1, \tag{4}$$

with α depending on the particular flow regime considered.

As discussed in [8], low Froude numbers give rise to multiple length or time scales or both, depending on the particular set-up of initial and boundary conditions, and on the structure of any pertinent source terms. For example, a typical distance $t_{\text{ref}}u_{\text{ref}} = L/\mathrm{Sr}$ which an advected particle traverses during the reference time t_{ref} differs asymptotically from the distance $t_{\text{ref}}\sqrt{gH_{\text{ref}}} = (L/\mathrm{Sr})/\varepsilon^{\alpha}$ traversed by a shallow water gravity during the same time. Similarly, variations of an advected variable over length scales of order $O(L)$ induce temporal variations of that quantity on a time scale $L/u_{\text{ref}} = \mathrm{Sr}\,t_{\text{ref}}$, whereas gravity waves with characteristic length L feature asymptotically shorter time scales of order $L/\sqrt{H_{\text{ref}}} = \varepsilon^{\alpha}\,\mathrm{Sr}\,t_{\text{ref}}$. See also the analogous discussion of low Mach number compressible flows in [7].

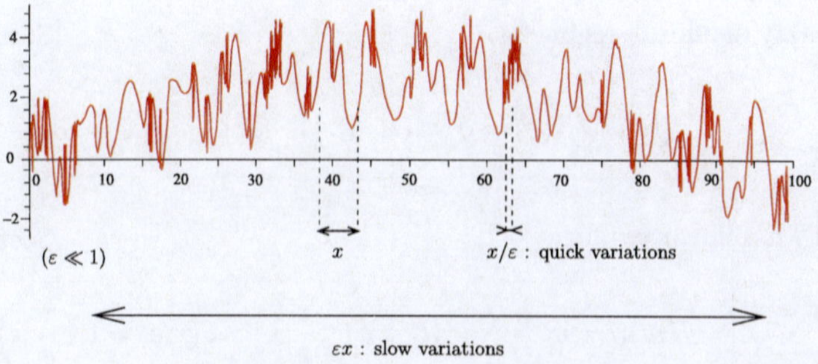

Fig. 1 Slow, classical and fast scales

We are interested here in multiscale topography. The relevant scalings as adopted in the asymptotic analyses to follow below are indicated in Fig. 1. Thus, for $\varepsilon \ll 1$, we refer to a "normal scale", resolved by a dimensionless coordinate x, a short range resolved by $X = x/\varepsilon$, and a long-wave scale represented by $\chi = \varepsilon x$. Depending on the flow regime considered, ε is representative of different powers of the Froude and Strouhal numbers as discussed shortly.

1.2 Asymptotic Flow Regimes and Organization of Paper

Essentially different flow characteristics emerge under the following different distinguished asymptotic limits, all of which we will analyse in some detail below. We will assume

$$\beta \equiv 1 \tag{5}$$

throughout, i.e., we allow for topographical heights comparable with the shallow water depth, and otherwise consider:

- **The "Lake Equations": single-scale, inviscid balanced flow over topography** (Section 2)

  We let $\mathrm{Fr} = \varepsilon$, i.e., $\alpha = 1$, follow the flow over advective time scales so that $\mathrm{Sr} = 1$, and assume

$$\left.\begin{array}{l}\mathrm{Fr} = \varepsilon \\ \mathrm{Sr} = 1\end{array}\right\} \quad (H,u)(t,x;\varepsilon) = \sum_i \varepsilon^i \, (H,u)^i \, (t,x)\,, \quad b(x;\varepsilon) \equiv B(x)\,. \tag{6}$$

- **Inviscid balanced flow over multiscale topography** (Section 3)

  Similar to the previous case, we consider multiscale topography, yet this time we assume its second characteristic scale to be much shorter than L instead of much longer. At the same time, by letting $\mathrm{Fr}(\varepsilon) = o(\mathrm{Sr}(\varepsilon))$ as $\varepsilon \to 0$, we restrict the considered flows to characteristic lengths that are too short to support gravity waves over the given time scales. Moreover, we distinguish a weakly nonlinear and a fully nonlinear regime, characterized by

weakly nonlinear regime (Section 3.1)

$$\left.\begin{array}{r}\mathrm{Fr}^2 = \varepsilon^3 \\ \mathrm{Sr} = \varepsilon^{-1}\end{array}\right\} \quad (H,u)(t,x;\varepsilon) = \sum_i \varepsilon^i (H,u)^i \left(t, \frac{x}{\varepsilon}, x\right) \; ; \; b(x;\varepsilon) = B\left(\frac{x}{\varepsilon}, x\right) \tag{7}$$

fully nonlinear regime: (Section 3.2)

$$\left.\begin{array}{r}\mathrm{Fr} = \varepsilon \\ \mathrm{Sr} = 1\end{array}\right\} \quad (H,u)(t,x;\varepsilon) = \sum_i \varepsilon^i (H,u)^i \left(t, \frac{x}{\varepsilon}, x\right) \; ; \; b(x;\varepsilon) = B\left(\frac{x}{\varepsilon}, x\right). \tag{8}$$

In these balanced, i.e., waveless, flow regimes, gradients of the water height on scale L drive mean flows, which in conjunction with the short-range topography induce a small-scale flow response. Depending on whether we adopt the scalings for strong or weak nonlinearity, the nonlinear advection of momentum by the small-scale flow does or does not affect in turn the mean flow dynamics at leading order, respectively. In case of strong nonlinearity, the result is a nonlinear Darcy-type homogenized equation system, with homogenized nonlinear advective momentum transport replacing the viscous fluxes in the classical Darcy theory.

- **Gravity waves over multiscale topography** (Section 4)

  Again we let $\mathrm{Fr} = \varepsilon$ and $\mathrm{Sr} = 1$, but express $(H,u)(t,x;\varepsilon)$ and $b(x;\varepsilon)$ through

$$(H,u)(t,x;\varepsilon) = \sum_i \varepsilon^i (H,u)^i (t, x, \varepsilon x), \qquad b(x;\varepsilon) = B(x, \varepsilon x). \tag{9}$$

  In this regime we will observe how long-range gravity waves generate localized balanced flow over the normal-scale topography, and how they will be affected by the induced effective average nonlinear momentum transport. Analogous regimes were studied in [7] for weakly compressible flow with small-scale entropy and vorticity in the context of ocean flows in [9], and for near-equatorial atmospheric motions in [12].

We draw conclusions in section 5.

2 Single-Scale Limit: "Lake Equations"

Here we let $\mathrm{Fr} = \varepsilon$ and $\mathrm{Sr} = 1$, so that the dimensionless equations from (2) become

$$\begin{aligned} \partial_t H + \mathrm{div}(Hu) &= 0 \\ \partial_t (Hu) + \mathrm{div}(Hu \otimes u) + \frac{H}{\varepsilon^2} \nabla (H+B) &= 0 \end{aligned}. \tag{10}$$

Single-scale expansion of the solution in terms of ε according to

$O(\varepsilon^{-2})$

$$H^0 \nabla (H^0 + B) = 0, \tag{11}$$

$O(\varepsilon^{-1})$

$$H^1\nabla(H^0+B)+H^0\nabla H^1=0, \tag{12}$$

$O(\varepsilon^0)$

$$\begin{array}{r} \partial_t H^0+\operatorname{div}(H^0u^0)=0 \\ \partial_t(H^0u^0)+\operatorname{div}(H^0u^0\otimes u^0)+H^2\nabla(H^0+B)+H^1\nabla H^1+H^0\nabla H^2=0 \end{array}. \tag{13}$$

From (11), (12) we conclude that

$$H^0+B\equiv c^0(t) \tag{14}$$

and $H^1\equiv c^1(t)$. Then, since the topography, B, is assumed time independent, we have from (13) that

$$\begin{array}{r} \operatorname{div}(H^0u^0)=-\dfrac{dc^0}{dt} \\ \partial_t(H^0u^0)+\operatorname{div}(H^0u^0\otimes u^0)+H^0\nabla H^2=0 \end{array}. \tag{15}$$

The time change of the total water height, $c^0(t)$, follows from integrating (15) over the entire flow domain to be

$$\frac{dc^0}{dt}=-\frac{1}{|\Omega|}\int_{\Omega}H^0u^0\cdot n\,d\sigma, \tag{16}$$

i.e., the change of water height is given by the total flux of water across the domain boundary.

Equations (14)–(16) constitute the classical zero Froude number shallow water or "lake equations". They form a closed system, once appropriate initial and boundary conditions for u^0,c^0 are provided. See [5].

3 Inviscid Balanced Flow over Short-Wave Topography

The second choice for the topography is to consider a bottom that depends on x and on the fast variable $X=x/\varepsilon$, such as the one plotted on Figure 2. For this case, we expand the velocity and the water height as

$$\begin{array}{l} u(t,x;\varepsilon)=u^0(t,X,x)+\varepsilon u^1(t,X,x)+\dots \\ H(t,x;\varepsilon)=H^0(t,X,x)+\varepsilon H^1(t,X,x)+\dots \end{array}, \qquad X=\frac{x}{\varepsilon}, \tag{17}$$

and we replace them in the Shallow Water system from (2), for (X,x) in $\mathbb{T}^2\times\mathscr{D}$.

We begin in section 3.1 with a study of the weakly nonlinear system that arises under the distinguished limit $\mathrm{Sr}=\varepsilon^{-1},\mathrm{Fr}^2=\varepsilon^3$ for the Strouhal and Froude numbers (see (7)): we show that the limit as $\varepsilon\to 0$ leads to a weakly nonlinear limit

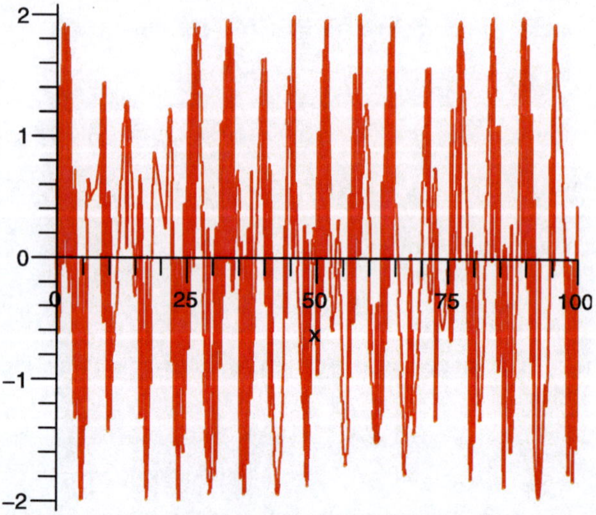

Fig. 2 Example of a topography that depend on the fast variable $X = x/\varepsilon$ and on x

version of the lake equations with oscillatory topography. In section 3.2, we study the Shallow Water equations with a strong nonlinearity and show that the balanced multiscale regime leads to a new nonlinear Darcy-type problem in which cumulative inertial forces of the small-scale flow through the topography replaces the viscous forces in the classical Darcy model.

3.1 Weakly Nonlinear Regime

With $\mathrm{Sr} = \varepsilon^{-1}$ and $\mathrm{Fr}^2 = \varepsilon^3$, the dimensionless shallow water equations from (2) read

$$\begin{aligned} \partial_t H + \varepsilon \mathrm{div}(Hu) &= 0 \\ \partial_t (Hu) + \varepsilon \mathrm{div}(Hu \otimes u) + \frac{H}{\varepsilon^2} \nabla (H + b) &= 0 \end{aligned} . \tag{18}$$

Assuming multiscale topography such that $b(x;\varepsilon) \equiv B(x/\varepsilon, x)$, and adopting the asymptotic expansion scheme from (17), we identify terms multiplied by like powers of ε,

$O(\varepsilon^{-3})$

$$H^0 \nabla_X (H^0 + B) = 0, \tag{19}$$

$O(\varepsilon^{-2})$

$$H^0 \nabla_x (H^0 + B) + H^1 \nabla_X (H^0 + B) + H^0 \nabla_X H^1 = 0, \tag{20}$$

$O(\varepsilon^{-1})$

$$H^1\nabla_x(H^0+B)+H^0\nabla_x H^1+H^2\nabla_X(H^0+B)+H^1\nabla_X H^1+H^0\nabla_X H^2=0\,, \quad (21)$$

$O(\varepsilon^0)$

$$\begin{aligned} \partial_t H^0+\mathrm{div}_X(H^0u^0) &= 0 \\ \partial_t(H^0u^0)+\mathrm{div}_X(H^0u^0\otimes u^0)+H^2\nabla_x(H^0+B)+H^1\nabla_x H^1 & \\ +H^0\nabla_x H^2+H^3\nabla_X(H^0+B)+H^2\nabla_X H^1+H^1\nabla_X H^2+H^0\nabla_X H^3 &= 0 \end{aligned} \quad , \quad (22)$$

$O(\varepsilon^1)$

$$\partial_t H^1+\mathrm{div}_x(H^0u^0)+\mathrm{div}_X(H^1u^0)+\mathrm{div}_X(H^0u^1)=0. \quad (23)$$

Equation (19) enables us to assert that, at leading order, the water height reads: $H^0(t,X,x)=-B(X,x)+c(t,x)$. Then the second term of (20) is equal to zero and, taking the mean value in X, we find that $H^0(t,X,x)+B(X,x)$ does not depend on x. In the same way with $(22)_1$, we have $\partial_t(H^0+B)+\mathrm{div}_X(H^0u^0)=0$, and, computing the mean value in the fast variable, we get that the function c does not even depend on the time: it is a constant in time and space, given by the value of H^0 at the initial time. Consequently we know that

$$H^0(X,x)\equiv -B(X,x)+C \qquad \text{where} \qquad C=B+H^0\big|_{t=0}\,. \quad (24)$$

With this result and (20), we obtain $\nabla_X H^1=0$. Then, (21) integrated in X yields that H^1 does not depend on x, so the first order of the water height does not depend on the space variables either. Averaging (23) in X and integrating in x over the (finite) domain, we find

$$\frac{dH^1}{dt}=-\oint_\Omega \overline{H^0u^0}\cdot n\,d\sigma\,, \quad (25)$$

where here and below, an overbar denotes averaging over the small-scale coordinate.

Equation (25) states that the first-order water height can change in time only due to fluxes through the boundary of the overall flow domain. From here on we assume for simplicity that rigid vertical walls bound the domain, and in that case, $H^1\equiv$ const., and we may assume $H^1\equiv 0$. Equipped with this result, the sublinear growth condition applied to (23) provides

$$\mathrm{div}_x\,\overline{H^0u^0}=0. \quad (26)$$

Inserting the results obtained thus far in (21) we find $\nabla_X H^2=0$, and then (22), (23) yield the following multiscale system of equations for H^2, H^3, and the leading-order velocity, u^0,

$$\begin{aligned}\partial_t(H^0u^0)+\mathrm{div}_X(H^0u^0\otimes u^0)+H^0\nabla_x H^2+H^0\nabla_X H^3 &= 0\\ \mathrm{div}_X(H^0u^0) &= 0\\ \mathrm{div}_x\overline{H^0u^0} &= 0\\ \nabla_X H^2 &= 0\end{aligned}, \tag{27}$$

with $H^0(X,x)=C-B(X,x)$.

We are interested in separately extracting the large and small-scale dynamics represented by this system. Averaging all equations in X eliminates the second and fourth equations, whereas the first and third become the governing equations for the

Non-stationary linear balanced mean flow

$$\begin{aligned}\partial_t\overline{H^0u^0}+\overline{H^0}\nabla_x H^2 &= -\overline{H^3\nabla_X B}\\ \mathrm{div}_x\overline{H^0u^0} &= 0\end{aligned}. \tag{28}$$

This equation describes how the leading-order large-scale flow responds to accumulated small-scale pressure forces on the topography, with its divergence constraint enforced by adjustment of the second order height, H^2, which acts as the usual Lagrangian multiplier.

Subtracting (28) from (27), and observing that $H^0+B\equiv C$ implies $H^0-\overline{H^0}\equiv\widetilde{H^0}=-\widetilde{B}$, we obtain the governing equations for the

Non-stationary nonlinear balanced small-scale flow

$$\begin{aligned}\partial_t\,\widetilde{H^0u^0}+\mathrm{div}_X(H^0u^0\otimes u^0)+\widetilde{H^0\nabla_X H^3} &= \widetilde{B}\,\nabla_x H^2\\ \mathrm{div}_X\,\widetilde{H^0u^0} &= 0\end{aligned}. \tag{29}$$

Here and below, the tilde denote small-scale fluctuations.

Equations (28) and (29) reveal how the small and large-scale flow components interact through their respective Lagrangian multiplier pressure fields. The gradient of the second-order height H^2, which is equivalent to a pressure field and purely large-scale, acts on on fluctuations of the topography to drive the small-scale flow. The divergence of the latter is controlled by the third-order height (or pressure) field, H^3. This field produces an accumulated topographical force, $\overline{H^3\nabla_X B}$, which in turn drives the large-scale flow.

Energy conservation

The multi-scale system derived here observes an energy principle. Multiplying Equation $(27)_1$ by u^0, and integrating in x and X by parts we get

$$\frac{1}{2}\frac{d}{dt}\int\int H^0\left|u^0\right|^2 dx dX - \int\int \left(H^2 \mathrm{div}_x(H^0u^0) + H^3 \mathrm{div}_X(H^0u^0)\right) dx dX = 0. \tag{30}$$

The last term is equal to zero thanks to $(27)_2$, whereas the integrand of the second term may be rewritten as $-H^2\left(\mathrm{div}_X(H^0u^1)\right)$, thanks to Equation (23), and the previous conclusion that $H^1 \equiv 0$. The requirement of sublinear growth of H^0u^1 in X then eliminates this term, and we find

$$\frac{1}{2}\frac{d}{dt}\int\int H^0\left|u^0\right|^2 dx dX = 0. \tag{31}$$

The system in (27) or (28) combined with (29) are obtained here through multiple-scales asymptotics from the full "compressible" Shallow Water equations under a particular distinguished limit of the Strouhal and Froude numbers, and of the asymptotic separation of the topographic scales. In contrast, the same system was obtained by Bresch and Gérard-Varet in [3] through a sequential limit of two small parameters: first they let the Froude number vanish to obtain the balanced "lake equations", and then adopt a distinguished limit between the Strouhal number (their parameter η), and the ratio of the characteristic spacial scales. In this setting, Bresch and Gérard-Varet prove convergence to the limit model with the two-scale method, used, e.g., in [10] for homogenization problems.

3.2 Fully Nonlinear Regime

We are now interested in a strong nonlinear term, that is, we adopt order one Strouhal number, and furthermore we let $\mathrm{Fr} = \varepsilon$. Thus we consider the Shallow Water equations in the form

$$\begin{aligned} \partial_t H + \mathrm{div}(Hu) &= 0 \\ \partial_t (Hu) + \mathrm{div}(Hu \otimes u) + \frac{H}{\varepsilon^2}\nabla(H+b) &= 0 \end{aligned}, \tag{32}$$

with multiscale topography, so that $b(x;\varepsilon) = B(x/\varepsilon, x)$.

We perform the same expansion for our variables as in the weakly nonlinear case, see (17), and we identify like powers of ε in the expanded equations,

$O(\varepsilon^{-3})$

$$H^0\nabla_X(H^0+B) = 0, \tag{33}$$

$O(\varepsilon^{-2})$

$$H^0\nabla_x(H^0+B) + H^1\nabla_X(H^0+B) + H^0\nabla_X H^1 = 0, \tag{34}$$

$O(\varepsilon^{-1})$

$$\begin{aligned} \mathrm{div}_X(H^0u^0) &= 0 \\ \mathrm{div}_X(H^0u^0\otimes u^0) + H^1\nabla_x(H^0+B) + H^0\nabla_x H^1 + H^2\nabla_X(H^0+B) & \\ + H^1\nabla_X H^1 + H^0\nabla_X H^2 &= 0 \end{aligned}, \tag{35}$$

$O(\varepsilon^0)$

$$\partial_t H^0 + \mathrm{div}_x(H^0 u^0) + \mathrm{div}_X(H^0 u^1) + \mathrm{div}_X(H^1 u^0) = 0\,, \tag{36}$$

3.2.1 Sublinear Growth Conditions

Here we derive the effective balanced flow equations for the leading-order solutions from (33)–(36). First, (33) yields

$$\nabla_X(H^0 + B) \equiv 0 \tag{37}$$

as in the weakly nonlinear case. Using this information in (34), we obtain after division by H^0 and averaging in X that

$$\nabla_x(H^0 + B) = \nabla_X H^1 \equiv 0\,. \tag{38}$$

For convenience, we rewrite (35) here taking into account the results just obtained, viz.

$$\mathrm{div}_X(H^0 u^0) = 0\,, \tag{39}$$

$$\mathrm{div}_X(H^0 u^0 \otimes u^0) + H^0 \nabla_x H^1 + H^0 \nabla_X H^2 = 0\,. \tag{40}$$

Equation (40) lends itself to two independent sublinear growth constraints. To obtain the first we simply average in X, to obtain the second we divide by H^0 and then average in X. This yields

$$\overline{H^0}\,\nabla_x H^1 + \overline{H^0 \nabla_X H^2} = 0\,, \tag{41}$$

$$\overline{u^0 \cdot \nabla_X u^0} + \nabla_x H^1 = 0\,. \tag{42}$$

3.2.2 Equations for Small-Scale Flow

Using (39), we conclude from (40) that

$$u^0 \cdot \nabla_X u^0 + \nabla_X H^2 = -\nabla_x H^1\,. \tag{43}$$

We eliminate H^2 from this equation by taking the curl, using $u \cdot \nabla u = \nabla u^2/2 - u \times (\nabla \times u)$, and, using $\omega = \nabla \times u$ with $\mathrm{div}\,\omega \equiv 0$, we have $\nabla \times (u \times \omega) = \omega \cdot \nabla u - \omega\,\mathrm{div}\,u - u \cdot \nabla \omega$ and, taking into account that we have a two-dimensional flow only,

$$u^0 \cdot \nabla_X \zeta^0 + \zeta^0 \mathrm{div}_X u^0 = \mathrm{div}_X(\zeta^0 u^0) = 0\,, \tag{44}$$

where $\zeta = k \cdot \omega = -\partial_{X_2} u_1 + \partial_{X_1} u_2$ and k is the unit vector normal to the flow plane. Combining this result with the divergence condition in (39), we find

$$H^0 u^0 \cdot \nabla_X(\zeta^0 / H^0) = 0\,, \tag{45}$$

so that

$$\zeta^0 = H^0 Q(\psi^{*,0}, x, t)\,, \tag{46}$$

where $\psi^{*,0}$ is a stream function for the mass flux $H^0 u^0$ satisfying

$$\psi^{*,0} = \psi^0 + X^{\perp} \cdot \overline{H^0 u^0} \qquad \text{with} \qquad H^0 u^0 = \nabla_X^{\perp} \psi^{*,0}\,, \tag{47}$$

and $Q(\psi, x, t)$ is a prescribed potential vorticity distribution function. Collecting, we obtain an elliptic determining equation for the fluctuating part of the stream function, ψ^0,

$$H^0 \nabla_X^2 \psi^0 - \nabla_X H^0 \cdot \nabla_X \psi^0 = (H^0)^3 Q(\psi^{*,0}, x, t) - \nabla_X H^0 \cdot \overline{H^0 u^0}^{\perp}\,. \tag{48}$$

This is the cell problem for a stationary vortical flow over variable topography with prescribed vorticity on each of the stream surfaces and with a prescribed large-scale mass flow.

3.2.3 Equations for Large-Scale Flow

In the previous section we have determined the leading-order velocity, u^0, only up to a large-scale mean mass flux, $\overline{H^0 u^0}$. We obtain its governing equation by explicitly introducing the split of the mass flux into mean and fluctuations, i.e.,

$$H^0 u^0 = U + H^0 \widetilde{u} \quad \text{where} \quad U = \overline{H^0 u^0}\,, \quad \widetilde{u} = u^0 - \frac{1}{H^0}\overline{H^0 u^0} = \frac{1}{H^0}\nabla_X^{\perp} \psi^0\,, \tag{49}$$

and averaging (43) in X. This procedure yields

$$U \cdot T + \nabla_x H^1 = -q \tag{50}$$

where

$$T = \overline{\frac{1}{H^0}\nabla_X \widetilde{u}} \qquad \text{and} \qquad q = \overline{\widetilde{u} \cdot \nabla_X \widetilde{u}}\,. \tag{51}$$

This is a nonlinear Darcy-type problem, with the effective mean friction tensor, T, and q the accumulated inertial force from the small-scale flow. The latter being determined by the nonlinear cell problem defined in the previous section.

A determining equation for the first-order pressure is obtained by averaging (36) in X, so that

$$\mathrm{div}_x U = -\mathrm{div}_x \left((\nabla_x H^1 + q) \cdot T^{-1} \right) = -\frac{d\overline{H}^0}{dt}\,. \tag{52}$$

Equations (48)–(52) define an interesting stationary multi-scale problem which, to the best of our knowledge, has not been derived or studied before. A large-scale mean mass flux drives a quasistationary small-scale flow over the topography. The small-scale dynamics is determined by vorticity transport, especially vortex stretching do to the motion in a layer with variable height. The large-scale flow adjusts, in turn, to two accumulated forcings from the small scales both of which are induced by nonlinear advection of momentum. The first results from the advection of small-scale momentum by the mean flow (first term in (50)), whereas the second is the nonlinear average of the nonlinear self-advection of the small-scale momentum (right-hand side of (50)).

4 Gravity Waves over Long-Wave Modulated Topography

Here we let $\mathrm{Sr} = 1$ to focus on advective times for the normal scale L, and $\mathrm{Fr} = \varepsilon$ so as to include gravity wave dynamics on a large scale L/ε. We start again from the dimensionless shallow water equations in (10), and we are now interested in large flow domains with characteristic extension $O(L/\varepsilon)$, and we allow for multi-scale bottom topography with associated long-wave modulations, so that $b(x;\varepsilon) = B(x, \varepsilon x)$, see Fig. 3. These scalings are analogous to those considered in [7] and [9] for weakly compressible flows of a gas and for oceanic motions, respectively. These authors pursued asymptotic analyses to motivate related specialized numerical schemes.

We expand the flow variables as

$$(H,u)(t,x;\varepsilon) = (H,u)^0(t,x,\chi) + \varepsilon\,(H,u)^1(t,x,\chi) + \ldots, \qquad (\chi = \varepsilon x) \tag{53}$$

and, after inserting into (10) collect terms involving like powers of ε as usual,

$O(\varepsilon^{-2})$

$$H^0 \nabla_x (H^0 + B) = 0 \tag{54}$$

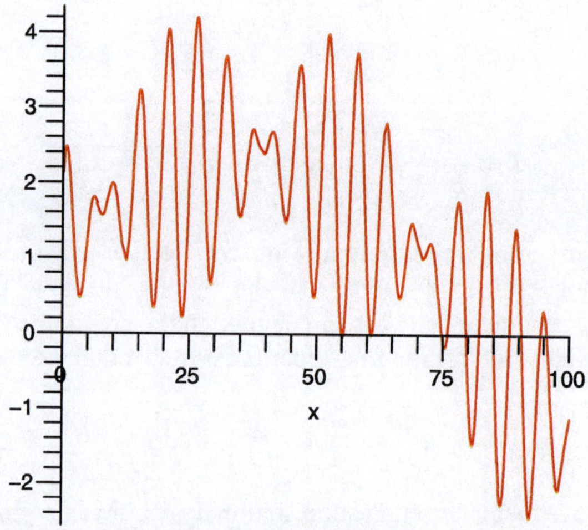

Fig. 3 Example of multiscale topography with leading-order variations on the normal scale represented by the x-coordinate, and long-wave modulations resolved by the slow variable $\chi = \varepsilon x$

$O(\varepsilon^{-1})$

$$H^0\nabla_x H^1 + H^0\nabla_\chi(H^0+B) + H^1\nabla_x(H^0+B) = 0 \tag{55}$$

$O(\varepsilon^0)$

$$\partial_t H^0 + \mathrm{div}_x(H^0u^0) = 0, \tag{56}$$

$$\begin{aligned}\partial_t(H^0u^0) + \mathrm{div}_x(H^0u^0\otimes u^0) + H^0\nabla_x H^2 \\ +H^1\nabla_x H^1 + H^2\nabla_x(H^0+B) + H^0\nabla_\chi H^1 + H^1\nabla_x(H^0+B) = 0,\end{aligned} \tag{57}$$

$O(\varepsilon)$

$$\partial_t H^1 + \mathrm{div}_x(H^0u^1) + \mathrm{div}_x(H^1u^0) + \mathrm{div}_\chi(H^0u^0) = 0. \tag{58}$$

Equation (54) requires H^0+B not to depend on x. Integrating (55) in x, after division by H^0, we find H^0+B to be independent of χ as well through a sublinear growth (or secular) condition, so that the water surface is flat to leading order. Inserting this result back, we conclude that H^1 is independent of x as well.

Next we observe that we may replace $\partial_t H^0 \equiv \partial_t(H^0+B)$ in (56), as the bottom topography is supposed to be time independent. Then, integrating again in x, we find as a sublinear growth condition that H^0+B is time independent. This also implies $\partial_t H^0 \equiv 0$ and we conclude that

$$\mathrm{div}_x(H^0u^0) = 0, \tag{59}$$

i.e., that the leading-order small-scale flow is "incompressible". Equations (57)–(58) then become

$$\partial_t(H^0u^0) + \mathrm{div}_x(H^0u^0\otimes u^0) + H^0\nabla_x H^2 + H^0\nabla_\chi H^1 = 0, \tag{60}$$

$$\partial_t H^1 + \mathrm{div}_x(H^0u^1) + \mathrm{div}_x(H^1u^0) + \mathrm{div}_\chi(H^0u^0) = 0. \tag{61}$$

This is a multiscale system, which we decompose by averaging in x and subtracting the result from the above into separate but coupled descriptions of the long-wave and short-wave components of the flow. Thus we obtain the

Long-wave equations for rough topography

$$\partial_t\left(\overline{H^0u^0}\right) + \overline{H^0}\nabla_\chi H^1 = \overline{H^2\nabla_x H^0}, \tag{62}$$

$$\partial_t H^1 + \mathrm{div}_\chi\left(\overline{H^0u^0}\right) = 0. \tag{63}$$

These are the standard linearized shallow water wave equations, except for the momentum source term on the r.h.s. of (62) which we have derived integrating $H^0\nabla_x H^2$ by parts and using a sublinear growth constraint to eliminate the arising divergence term. This source term represents the net resistance, or large-scale accumulated pressure force, that arises as a result of the small-scale flow through the rough topography. Subtracting the long-wave equations in (62), from the unaveraged equation (60), and using the small-scale divergence constraint from (59), we obtain the determining equations for the small-scale component of the leading-order momentum, $\widetilde{H^0 u^0}$, and for the small-scale structure of the second order height, H^2. With the $\widetilde{\phi} = \phi - \overline{\phi}$ denoting the small-scale component of some field ϕ, we find the

Balanced small-scale flow equations for rough topography

$$\partial_t \widetilde{H^0 u^0} + \mathrm{div}_x (H^0 u^0 \otimes u^0) + H^0 \nabla_x H^2 = -\widetilde{H^0} \nabla_\chi H^1 , \tag{64}$$

$$\mathrm{div}_x \widetilde{H^0 u^0} = 0 . \tag{65}$$

The small-scale flow is driven by the long-wave unbalanced part of the large-scale height gradient, $-\widetilde{H^0} \nabla_\chi H^1$. Since the leading-order momentum is divergence-free on the small scale, the second-order height H^2 assumes the role of a Langrangian multiplier responsible for guaranteeing compliance with this constraint. The nonlinear momentum advection term, $\mathrm{div}_x \left(H^0 u^0 \otimes u^0 \right)$, may be rewritten in terms of the large and small-scale momentum components using

$$\left. \begin{aligned} H^0 u^0 (t,x,\chi) &= \overline{H^0 u^0}(t,\chi) + \widetilde{H^0 u^0}(t,x,\chi) \\ u^0 &= \frac{1}{H^0} \left(\overline{H^0 u^0} + \widetilde{H^0 u^0} \right) \\ \overline{\widetilde{H^0 u^0}} &\equiv 0 \end{aligned} \right. . \tag{66}$$

Equations (62), (63) and (64), (65) are the shallow-water analogue to the single time scale – multiple length scale, variable density low Mach number flow equations derived first in [7], and in the shallow water context in [9].

If the topography does not depend on x, the previous equations may be combined to obtain wave equation with spacially varying signal speed for H^1,

$$\partial_t^2 H^1 - \mathrm{div}_\chi \left((C - B(\chi)) \nabla_\chi H^1 \right) = 0 , \tag{67}$$

where $C = B + H^0 \equiv \mathrm{const.}$

5 Conclusions

Exploring a number of multiscale shallow water regimes using formal multiple-scales asymptotics, we find that there is a range of different possibilities for scale interactions in these flows.

- **Balanced flows.** For flows that are free of surface waves, a multiscale topography can mediate scale interactions through two mechanisms:

  1. **Weakly nonlinear regime.** The large-scale accumulation of net pressure forces on the topography as induced by small-scale flow fluctuations drives the large-scale balanced flow. In turn, the large-scale height gradients that ensure compliance with the large-scale divergence constraint produce

small-scale forces when acting on the topographical fluctuations, thereby driving the small-scale flow.

2. **Strongly nonlinear regime.** Here we find the quasi-steady balanced large-scale flow to follow a Darcy-type equation, yet with the homogenized net forces induced not by viscosity or friction, but rather by the accumulation of small-scale nonlinear momentum fluxes. The small-scale flow in turn is driven by the large-scale mean height gradients, and its detailed structure is determined by the dynamics of vorticity as it gets stretched and compressed when the small-scale flow passes over the variable topography.

- **Long waves passing over multiscale topography.** The interaction mechanism across scales is here similar to that found for the weakly nonlinear balanced flow regime, yet the large-scale flow now involves non-balanced free surface waves.

We understand the present paper as a point of departure for further work providing, on the one hand, asymptotic limit cornerstones against which to measure the performance of general shallow water numerical flow solvers. On the other hand, some of the regimes discussed here are of sufficient practical interest to warrant further work on the rigorous mathematical justification of the model equations derived here through merely formal asymptotic arguments. Another direction of research that we intent to pursue concerns viscous regularizations and rigorous justification of the formal analyses presented here.

Acknowledgements. D. Bresch has been supported by the French "Agence Nationale de la Recherche" project MathOcean (Grant ANR-08-BLAN-0301-01). R. Klein thanks Deutsche Forschungsgemeinschaft for their continued support under grant KL 611/14.

References

1. Ailliot, P., Frénod, E., Monbet, V.: Modeling the coastal ocean over a time period of several weeks. J. Diff. Eq. 246(4), 639–659 (2010)
2. Ailliot, P., Frénod, E., Monbet, V.: Long term object drift forecast in the ocean with tide and wind. Multiscale Model. Simul. 5(2), 514–531 (2006)
3. Bresch, D., Gérard-Varet, D.: On some homogenization problems from shallow water theory. Appl. Math. Letters 20(5), 505–510 (2007)
4. Frénod, E., Mouton, A., Sonnendrücker, E.: Two-scale numerical simulation of the weakly compressible 1D isentropic Euler equations. Numerishe Mathematik 108(2), 263–293 (2007)
5. Greenspan, H.P.: The Theory of Rotating Fluids. Cambridge University Press, London (1968)
6. Hillairet, M.: Propagation of density-oscillations in solutions to the barotropic compressible Navier-Stokes system. J. Math. Fluid Mech. 9(3), 343–376 (2007)
7. Klein, R.: Semi-implicit extension of a Godunov-type scheme based on low Mach number asymptotics i: One-dimensional flow. J. Comput. Phys. 121, 213–237 (1995)
8. Klein, R.: Multiple scales asymptotics for atmospheric flows. In: Proceedings of the 4th European Conference on Mathematics, Stockholm (2004)

9. Le Maître, O., Levin, J., Iskandarani, M., Knio, O.M.: A Multiple pressure splitting of the shallow–water equations. J. Comput. Phys. 166, 116–151 (2001)
10. Lions, P.-L., Masmoudi, N.: Homogenization of the Euler system in a 2D porous medium. J. Math. Pures Appl. 84, 1–20 (2005)
11. Lucas, C.: Cosine effect on shallow water equations and mathematical properties. Quarterly Appl. Math. 67(2), 283–310 (2009)
12. Majda, A.J., Klein, R.: Systematic multi-scale models for the tropics. J. Atmosph. Sci. 60, 393–408 (2003)
13. Marche, F.: Derivation of a new two-dimensional viscous shallow water model with varying topography, bottom friction and capillary effects. European J. Mech. B/Fluids 26(1), 49–63 (2007)
14. Serre, D.: Oscillations non linéaires des systèmes hyperboliques : méthodes et résultats qualitatifs. Ann. Inst. N. Poincaré Anal. Non Linéaire 8(3–4), 351–417 (1991)
15. Serre, D.: Ocillations non linéaires de haute fréquence ; dim = 1. In: Nonlinear partial differential equations and their applications. Collège de France Seminar, vol. XII (Paris, 1991-1993). Pitman Res. Notes Math. Ser., vol. 302, pp. 190–210. Longman Sci. Tech., Harlow (1994)

Shallow Water Equations for Shear Flows

Alexander A. Chesnokov and Valery Yu. Liapidevskii

Abstract. Open channel flows of ideal incompressible fluid with velocity shear are considered in the long wave approximation. Nonlinear integro-differential models of shallow flow with continuous vertical or horizontal velocity distribution are derived. It is shown that mathematically the models are equivalent and, consequently, the obtained early results for two-dimensional open channel flows with a vertical shear can be applied to the 2D flows with horizontally nonhomogeneous velocity field. Stability of shear flows in terms of hyperbolicity of the governing equations is studied. It is shown that the type of the equations of motion can change during the evolution of the flow, which corresponds to the long wave instability for a certain velocity field. A simple mathematical model describing the nonlinear stage of the Kelvin–Helmholtz instability of shear flows is derived. The problem of the mixing layer interaction with a free surface and its transition into a turbulent surface jet is considered.

1 Introduction

The modelling of nonlinear wave motion on the surface of a thin fluid layer is an important fundamental and applied problem which has been the subject of many works, including [9, 14] and others. In the past, studies dealing with water waves concentrated mainly on one-dimensional fluid flows. Classical shallow water model describes flows in an open channel with horizontal velocity components weakly depending on the vertical and transversal coordinates is one of the example. In such model only averaged over depth and cross-section velocities are used in the formulation of the governing equations. It is clear that any applications of the theory to

Alexander A. Chesnokov · Valery Yu. Liapidevskii
M.A. Lavrentiev Institute of Hydrodynamics SB RAS Lavrentiev Ave. 15,
Novosibirsk, 630090, Russia
Novosibirsk State University, Pirogova str. 2,
Novosibirsk, 630090, Russia
e-mail: chesnokov@hydro.nsc.ru, liapid@hydro.nsc.ru

E. Krause et al. (Eds.): Computational Sci., & High Performance Computing IV, NNFM 115, pp. 165–179.
springerlink.com

situations concerning the physical reality more closely should encompass underlying shear flow. The propagation of long gravity waves in shear flows is of considerable interest in various fields of hydrology and meteorology.

Note that while the model governing one-dimensional flows in an open channel forms a hyperbolic system of differential equations, more complicated system of integro-differential equations, which cannot be referred to any classical type, governs the shear flows.

The article focuses on the derivation and theoretical study of the nonlinear equations, describing long waves propagation on shear flows of ideal liquid in an open channel. In semi-Lagrangian frame of reference the shallow water equations for shear flows take form [10, 12]

$$u_t + uu_x + g\int_0^1 H_x\,d\lambda = G, \quad H_t + (uH)_x = 0. \tag{1}$$

In the case $G = 0$ this model describes plane-parallel vertical shear flows of an ideal incompressible liquid with free surface under gravity. The linear waves in shear flow have been considered by Burns [2], who first derived the dispersion relation for normal modes. Some classes of exact solutions to the model were obtained by Freeman [4], Sachdev [8], Teshukov et.al. [12] and others. An infinite system of conservation laws describing long waves propagation was derived by Benney [1], Zakharov [15]. A qualitative analysis of the Eqs. (1) was performed by Teshukov [11] on the base of his generalization of the hyperbolicity concept and the method of characteristics for equations with operator coefficients [10].

In the case

$$G = \frac{gY'(x)}{Y^2}\int_0^1 H\,d\lambda$$

Eqs. (1) describe horizontal shear flows of ideal liquid in an open channel of variable cross-section $Y(x)$. This model has been recently obtained by Chesnokov and Liapidevskii [3].

The stability of shear flows is closely related with the hyperbolicity of governing equations. For high gradient shear flows the type of equations can change, which leads to the long wave instability development. To simulate unstable flows, the tree-layer shallow water approximation taking into account the intermediate turbulent layer formation has been derived in [5]. The main idea here is in using the total conservation laws of mass, momentum and energy to find the mean flow in the turbulent intermediate layer. The boundary layer approximation for the Reynolds equations is based on the hypothesis from [13] about Reynolds stresses expression in free shear turbulent flows. For steady–state flows, the problem on velocity field restoring in the turbulent layer is reduced to the semilinear initial–boundary problem for a hyperbolic system of differential equations. This approach is applied to the problem on the mixing layer evolution at a free surface and its transition into a surface jet.

The plan of the paper is the following. In Sec. 2 we present a brief derivation of the integrodifferential model (1) for shear flows starting from the 3D Euler equations. In Sec. 3 we calculate the perturbation propagation velocity (the model has a discrete and continuous characteristic spectrum) and formulate necessary and sufficient hyperbolicity conditions for the equations of motion (1). We also verify the validity of the hyperbolicity condition for a unsteady exact solution and show that long-wave instability can occur during the evolution of shear flow. In Sec. 4 we consider the development of the mixing layer generated in open channel flow with vertical shear. For this purpose we consider the two-level flow description. As the first step, the three-layer shallow water equations, which describe the mean flow evolution with entrainment of fluid from the potential layer on the turbulent interlayer, are derived. Then the velocity field in the mixing layer is reconstructed with the help of the length scale and intensity of turbulence found at the first step of calculation. Finally, we draw some conclusions.

2 Derivation of Long-Wave Model

We consider the motion of an ideal incompressible fluid with the free boundary $z = h(t,x,y)$ in an open narrow channel with even bottom $z = 0$ in a gravity field. The equations of motion in dimensionless variables are written as

$$u_t + (\mathbf{v}\cdot\nabla)u + p_x = 0, \quad \varepsilon^2(v_t + (\mathbf{v}\cdot\nabla)v) + p_y = 0,$$
$$\varepsilon^2(w_t + (\mathbf{v}\cdot\nabla)w) + p_z = -g, \quad \nabla\cdot\mathbf{v} = 0. \tag{2}$$

On the free boundary $z = h(t,x,y)$, the following kinematic and dynamic boundary conditions should be satisfied:

$$h_t + uh_x + vh_y - w = 0, \quad p = p_0; \tag{3}$$

on the even bottom $z = 0$, the non-penetration condition is imposed:

$$w = 0. \tag{4}$$

Here $u = L^{-1}T\bar{u}$, $v = l^{-1}T\bar{v}$, $w = l^{-1}T\bar{w}$, $p = \rho^{-1}L^{-2}T^2\bar{p}$, $x = L^{-1}\bar{x}$, $y = l^{-1}\bar{y}$, $z = l^{-1}\bar{z}$ and $t = T^{-1}\bar{t}$ are the components of the velocity, pressure, Cartesian coordinates, and time, respectively; $\bar{\mathbf{v}} = (\bar{u},\bar{v},\bar{w})$, $\bar{p}$, $\bar{\mathbf{x}} = (\bar{x},\bar{y},\bar{z})$ and $\bar{t}$ are the corresponding dimensional variables. The quantity L specifies the characteristic scale on the x axis directed along the channel, and the quantity l on the y and z axes; $T = L/\sqrt{al}$ is the characteristic time scale (a has dimension of acceleration); the constants ρ and g are the density of the fluid and the dimensionless acceleration due to gravity, ∇ is a gradient operator, and $\varepsilon = l/L$ is a dimensionless small parameter.

The equations describing long-wave propagation in a narrow open channel are obtained form Eqs. (2) by passing to the limit $\varepsilon \to 0$. In this case, the third equation of system (2) implies a hydrostatic pressure distribution $p = g(h-z) + p_0$, where the constant p_0 is the pressure on the free boundary. Simple transformations of Eqs. (2) taking into account the boundary conditions (3) and (4) lead to the system of equations

$$u_t + uu_x + vu_y + wu_z + gh_x = 0, \quad h_y = 0,$$

$$w = -\int_0^z (u_x + v_y)\,dz, \quad h_t + \Big(\int_0^h u\,dz\Big)_x + \Big(\int_0^h v\,dz\Big)_y = 0, \tag{5}$$

which describes the motion of an ideal fluid with free surface in a narrow open channel in the long-wave approximation.

In what follows, we consider two special classes of fluid motion. The first one corresponds to the plane-parallel fluid motion with vertical shear in a channel with straight lateral boundaries (Fig. 1). This class of flows arises from the model (5) under assumptions $u_y = 0$, $v_y = 0$. The second one corresponds to the horizontal shear flows in a channel with curved lateral walls (Fig. 2). In this case we assume $u_z = 0$, $v_z = 0$ and we also have to add to the model (5) boundary conditions on the lateral walls of the channel of variable cross-section.

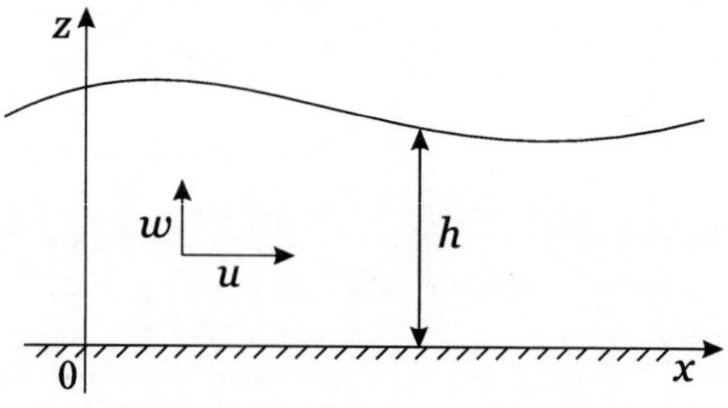

Fig. 1 Plane-parallel vertical shear fluid flow in an open channel with straight lateral walls: section of the channel by the plane $y = \text{const}$

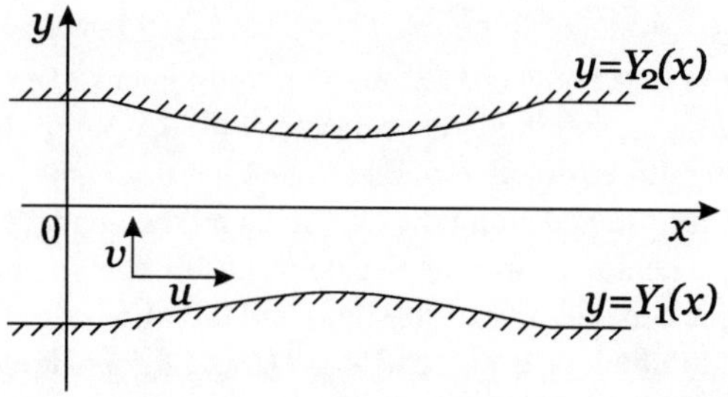

Fig. 2 Horizontal shear fluid flow in an open channel with curved lateral walls: section of the channel by the plane $z = \text{const}$

2.1 Vertical Shear Flows

From Eqs. (5), under assumptions $u_y = v_y = 0$, one obtains the system of equations for the unknown functions $u(t,x,z)$ and $h(t,x)$ describing vertical shear flows in shallow water approximation [1]

$$u_t + uu_x + wu_z + gh_x = 0, \quad h_t + \left(\int_0^h u\,dz \right)_x = 0, \quad w = -\int_0^z u_x\,dz. \tag{6}$$

For convenience, in the following study we need another formulation of the governing equations. First, we formulate (6) using Euler–Lagrangian coordinates [15]. Let us denote by $\lambda \in [0,1]$ a coordinate that identifies the liquid layer. Using the change of variable $y = \Phi(t,x,\lambda)$ $(0 \le \lambda \le 1)$, where the function $\Phi(t,x,\lambda)$ is a solution of the Cauchy problem

$$\Phi_t + u(t,x,\Phi)\Phi_x = w(t,x,\Phi), \quad \Phi(0,x,\lambda) = \lambda h_0(x), \tag{7}$$

one can map the flow region with unknown boundary $z = h(t,x)$ onto the fixed strip $0 \le \lambda \le 1$, $-\infty < x < \infty$. In this domain the functions $u(t,x,\lambda)$ and $H(t,x,\lambda) = \Phi_\lambda$ satisfy the system

$$u_t + uu_x + g\int_0^1 H_x\,d\lambda = 0, \quad H_t + (uH)_x = 0. \tag{8}$$

Transformation to semi-Lagrangian coordinates is a reversible change of variables if $\Phi_\lambda > 0$. Indeed, let the functions $u(t,x,\lambda)$ and $H(t,x,\lambda)$ be found. Then, the fluid depth $h(t,x)$ is known, and the Eulerian coordinate z and the velocity component w can be obtained from formulae

$$z = \Phi(t,x,\lambda) = \int_0^\lambda H\,d\lambda, \quad w = \Phi_t + u\Phi_x.$$

The derived model (8) is a direct consequence of the shallow water equations for shear flows (6). In fact, let us denote by $\tilde{u}$ and $\tilde{w}$ the velocity variables expressed in the new coordinates (t,x,λ). Then one has

$$\tilde{u}(t,x,\lambda) = u(t,x,\Phi(t,x,\lambda)), \quad \tilde{w}(t,x,\lambda) = w(t,x,\Phi(t,x,\lambda)).$$

It follows

$$\frac{\partial \tilde{u}}{\partial t} = \frac{\partial u}{\partial t} + \frac{\partial u}{\partial z}\frac{\partial \Phi}{\partial x}, \quad \frac{\partial \tilde{u}}{\partial x} = \frac{\partial u}{\partial x} + \frac{\partial u}{\partial z}\frac{\partial \Phi}{\partial x}, \quad \frac{\partial \tilde{u}}{\partial \lambda} = \frac{\partial \Phi}{\partial \lambda}\frac{\partial u}{\partial z}; \tag{9}$$

furthermore, $h(t,x) = \Phi(t,x,1) = \int_0^1 H(t,x,\lambda)\,d\lambda$. Therefore

$$\frac{\partial \tilde{u}}{\partial t} + \tilde{u}\frac{\partial \tilde{u}}{\partial x} = \frac{\partial u}{\partial t} + u\frac{\partial u}{\partial x} + \frac{\partial u}{\partial z}\left(\frac{\partial \Phi}{\partial t} + u\frac{\partial \Phi}{\partial x}\right) = \frac{\partial u}{\partial t} + u\frac{\partial u}{\partial x} + w\frac{\partial u}{\partial z} = -g\frac{\partial h}{\partial x},$$

which is the first equation of (8). Differentiating of the first equation in (7) with respect to λ yields

$$H_t + \frac{\partial u}{\partial z}H\Phi_x + uH_x = \frac{\partial w}{\partial z}H.$$

Using continuity equation $w_z = -u_x$ it follows

$$H_t + uH_x + Hu_x + Hu_z\Phi_x = 0$$

and, making use of (9) one has $H_t + \tilde{u}H_x + H\tilde{u}_x = 0$, which is equivalent to the second equation of (8). Thus, integrodifferential system of equations (8) describes plane-parallel shear flows in shallow water approximation.

2.2 Horizontal Shear Flows

Let us consider the class of fluid flows in an open channel of variable cross-section in which the horizontal velocity components u and v do not depend on the vertical coordinate z. The geometry is illustrated in Fig. 2. In this case, the long-wave model (5) becomes

$$\begin{gathered} u_t + uu_x + vu_y + gh_x = 0, \quad h_y = 0, \\ h_t + (hu)_x + (hv)_y = 0, \quad uY_i'(x) - v\big|_{y=Y_i} = 0. \end{gathered} \tag{10}$$

The last equation in (10) corresponds to the non-penetration condition on the lateral channel walls $y = Y_i(x)$, $i = 1,2$.

To study the mathematical properties of Eqs. (10), it is convenient to transform to semi-Lagrangian coordinates by the change of the variable $y = \Phi(t,x,\lambda)$, where the function Φ is a solution of the Cauchy problem [15]

$$\Phi_t + u(t,x,\Phi)\Phi_x = v(t,x,\Phi), \quad \Phi(0,x,\lambda) = \lambda Y_2(x) + (1-\lambda)Y_1(x),$$

The Lagrangian variable $\lambda \in [0,1]$; the values $\lambda = 0$ and $\lambda = 1$ correspond to the lateral boundaries of the channel $y = Y_1(x)$ and $y = Y_2(x)$. In the new variables, the functions $u(t,x,\lambda)$ and $H(t,x,\lambda) = h\Phi_\lambda$ are described by the integrodifferential system of equations [3]

$$u_t + uu_x + gh_x = 0, \quad H_t + (uH)_x = 0, \quad h = \frac{1}{Y}\int_0^1 H\,d\lambda, \tag{11}$$

where $Y(x) = Y_2(x) - Y_1(x)$ is the given channel width. Eqs. (11) are derived as a consequence of the long-wave model (10).

Obviously, the integrodifferential equations (1) are a generalization of the derived above models (8) and (11) for shear flows in an open channel in shallow water approximation. We note that, in the case of straight lateral boundaries ($Y_i = \text{const}$) and zero vorticity ($\omega = u_\lambda / H = 0$), model (1) reduces to the classical one-dimensional shallow water equations [9].

3 Generalized Hyperbolicity of Equations of Shear Flows

The model (1) belongs to the class of equations with operator coefficients for which the notations of characteristics and hyperbolicity were proposed [10]. The system (1) can be presented as

$$\mathbf{U}_t + A\langle \mathbf{U}_x \rangle = \mathbf{G}, \tag{12}$$

where $\mathbf{U} = (u, H)$ is the column vector of the unknown quantities, $\mathbf{G}$ is the right side of Eqs. (12), and $A\langle \mathbf{U}_x \rangle$ is the result of the action of the matrix operator A on the vector $\mathbf{U}_x$. According to [10], the characteristic of (12) is defined by equation $x'(t) = k(t,x)$, where characteristic velocity k is an eigenvalue of the problem

$$(\mathbf{F}, (A - kI)\langle \mathbf{b} \rangle) = 0. \tag{13}$$

The solution of Eq. (13) for the functional $\mathbf{F} = (F_1, F_2)$ is sought in the class of generalized functions. The functional $\mathbf{F}$ acts on the variable λ, and t and x are treated as parameters; I is an identical operator; $\mathbf{b}$ is an arbitrary smooth vector function with the components $b_1(\lambda)$ and $b_2(\lambda)$. Applying the functional $\mathbf{F}$ to Eq. (12), we obtain the following relation on the characteristic:

$$(\mathbf{F}, \mathbf{U}_t + k\mathbf{U}_x) = (\mathbf{F}, \mathbf{G}). \tag{14}$$

We say that system (12) is generalized hyperbolic [10] if all eigenvalues k are real and the set of relations on the characteristics (14) is equivalent to the Eqs. (12), i.e. the system of eigenfunctionals represents a full basis.

The characteristic properties of the model (1) are analysed under the assumption of a monotonic change in the velocity $u(t,x,\lambda)$ with respect to Lagrangian coordinate λ. For definiteness, we set $u_\lambda > 0$. Note, if this condition is satisfied at the initial time $t = 0$, then it is satisfied for all $t > 0$ by virtue of system (1). We also assume that the Jacobian of the transformation to the semi-Lagrangian variables is greater than zero ($H > 0$).

The characteristic equation for the perturbation propagation velocity k has the form

$$\chi(k) = 1 - \frac{g}{S} \int_0^1 \frac{H\, d\lambda}{(u-k)^2} = 0. \tag{15}$$

This equation expresses the condition of the existence of nontrivial solutions of problem (13). Here $S=1$ in the case of vertical shear flows (model (8)) and $S=Y(x)$ in the case of horizontal shear flows (model (11)). Taking into account that $\chi(k)\to 1$ as $|k|\to+\infty$, and $\chi(k)\to-\infty$ as $k\to u_0=u(t,x,0)$ or $u_1=u(t,x,1)$, we notice that Eqs. (15) has two real roots $k_1\in(-\infty,u_0)$ and $k_2\in(u_1,+\infty)$. Moreover, the problem (13) has nontrivial solutions if $k=u(t,x,\lambda_0)$ for all $\lambda_0\in[0,1]$, which forms continuous characteristic spectrum.

The hyperbolicity conditions of the system (1) are formulated in terms of the analytical function $\chi(z)$, or more precisely, its limiting values from the upper χ^+ and lower χ^- complex half-planes on the segment $[u_0,u_1]$:

$$\chi^{\pm}(u(\lambda))=1+\frac{g}{S}\left(\frac{1}{\omega_1(u_1-u(\lambda))}-\frac{1}{\omega_0(u_0-u(\lambda))}-\right.$$

$$\left.-\int_0^1\left(\frac{1}{\omega(\nu)}\right)_\nu\frac{d\nu}{u(\nu)-u(\lambda)}\mp\frac{\pi i}{u_\lambda}\left(\frac{1}{\omega}\right)_\lambda\right). \tag{16}$$

Here $\omega=u_\lambda/H$; the subscripts 0 and 1 correspond to the values of the functions for $\lambda=0$ and $\lambda=1$; i is an imaginary unit. The limiting values of the function $\chi(z)$ are calculated using Sokhotsky–Plemelj formulae. According to [10, 11], the hyperbolicity criterion can be expressed in the form:

For flows with a monotonic velocity profile ($u_\lambda\neq 0$), conditions

$$\chi^{\pm}\neq 0,\quad \Delta\arg\frac{\chi^+(u)}{\chi^-(u)}=0 \tag{17}$$

($\Delta\arg\chi^{\pm}$ is the increment of the complex function $\chi^{\pm}$ as λ changes from 0 to 1 for fixed t and x) are necessary and sufficient for hyperbolicity of Eqs. (1) if the functions u, H and ω are differentiable and the functions u_λ and ω_λ satisfy the Hölder condition with respect variable λ.

3.1 Validation of Hyperbolicity Conditions

An example of verifying the validity of the hyperbolicity conditions (17) is given below. We consider the exact solution of Eqs. (8)

$$u=(x-C(\lambda))t^{-1},\quad H=t^{-1} \tag{18}$$

that describes the fluid spreading in one-dimensional open channel under the pressure action.

Let us show that, for the solution considered, complex roots of the characteristic equation (15) can appear during the evolution of the flow. Let the function $C(\lambda)$ be given implicitly by the equation

$$C^3+9^{-1}C+\lambda-2^{-1}=0.$$

This cubic equation has one real root and two imaginary roots for each $\lambda \in [0,1]$. We note that $C(\lambda)$ is a monotonically decreasing function since $C'(\lambda) = -(3C^2 + 9^{-1})^{-1} < 0$. In addition, $C(1/2) = 0$ and $C_0 = -C_1 \approx 0.75$ (C_0, C_1 are the values of the function $C(\lambda)$ for $\lambda = 0$, $\lambda = 1$). Substitution of solution (18) into (16) yields

$$\chi^{\pm}(C) = 1 + 2gt\left(\frac{(3C_0^2 + 1/9)C_0}{C_0^2 - C^2} - 6C_0 - 3C\ln\frac{C_0 - C}{C_0 + C}\right) \pm 6\pi gCti.$$

In this case, the functions $\chi^{\pm}$ depend only on t and $C(\lambda)$.

It is convenient to verify the hyperbolicity conditions (17) in terms of the functions $\Psi^{\pm} = (C_0^2 - C)\chi^{\pm}(C)$ that do not have poles at the points $C = \pm C_0$. Plots of the function Ψ^{+} for C varying from C_0 to C_1 are given in Fig. 3 (a — $t = 0.1$, b — $t = 0.239$, c — $t = 0.3$). Here the values of $\mathrm{Re}\Psi^{+}(C)$ are plotted on the abscissa, and the values of $\mathrm{Im}\Psi^{+}(C)$ on the ordinate. The plots of the function Ψ^{-} are similar (circulation direction is opposite). For $C = \pm C_0$ and $C = 0$, the imaginary part of the functions $\Psi^{\pm}$ vanish and the functions at these points take the following values:

$$\Psi^{\pm}(C_0) = \Psi^{\pm}(C_1) = 2C_0(3C_0^2 + 9^{-1})t > 0, \quad (t > 0),$$

$$\Psi^{\pm}(0) = C_0^2 - 2C_0(3C_0^2 - 9^{-1})t.$$

At $t = t_* = 2^{-1}C_0(3C_0^2 - 9^{-1})^{-1} \approx 0.239$, the functions $\Psi^{\pm}$ vanish at the point $C = 0$, which leads to the violations of conditions (17). It follows from Fig. 3, that for $t < t_*$, the increment of the functions $\Psi^{\pm}$ is equal to zero and the hyperbolicity conditions (17) are satisfied. In Fig. 3, the curve constructed for $t = t_*$ corresponds to the neutral case. For $t > t_*$, the argument of the functions $\Psi^{\pm}$ gains an increment: $\Delta\arg\Psi^{+}(C) = 2\pi$ and $\Delta\arg\Psi^{-}(C) = -2\pi$. Thus, $\Delta\arg(\Psi^{+}/\Psi^{-}) = 4\pi$, which implies that the characteristic equation (15) has a couple of complex roots. This example shows that the system (8) can change its type in the process of flow evolution, which corresponds to long-wave instability for the considered velocity distribution.

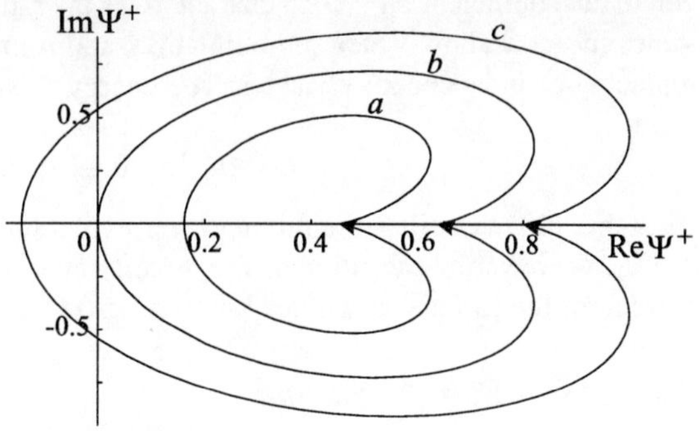

Fig. 3 Parametric representation of real and imaginary parts of the function Ψ^{+} (the arrows show the circulation direction)

4 Mixing Layer at Free Surface

We consider the problem of the formation of a mixing layer in an incompressible homogeneous fluid for open channel flows. Aeration of the flow is ignored, and, hence, the liquid density ρ is constant ($\rho \equiv 1$). We apply the three-layer scheme of flow, in which the mixing layer is considered as an intermediate turbulent layer between two layers with potential flow. For upper and lower layers we use the non-homogeneous shallow water equations in which the entrainment of fluid from the layers into the mixing layer is taken into account. To describe the evolution of the averaged quantities in the mixing layer, we add to the system the total conservation laws of mass, momentum and energy [6]. Under the assumption of hydrostatic pressure distribution in the layers, the governing equations are written as

$$h_t^+ + (h^+u^+)_x = -\kappa^+, \quad h_t^- + (h^-u^-)_x = -\kappa^-$$

$$u_t^+ + \left(0.5u^{+2} + gh\right)_x = 0, \quad u_t^- + \left(0.5u^{-2} + gh\right)_x = 0,$$

$$h_t + \bar{Q}_x = 0, \quad \bar{Q}_t + \left(h^+u^{+2} + \eta\bar{u}^{\,2} + h^-u^{-2} + 0.5gh^2\right)_x = 0,$$

$$(h^+u^{+2} + \eta(\bar{u}^{\,2} + q^2) + h^-u^{-2} + gh^2)_t +$$

$$+(h^+u^{+3} + \eta\bar{u}(\bar{u}^{\,2} + q^2) + h^-u^{-3} + 2gh\bar{Q})_x = -\bar{\varepsilon}. \tag{19}$$

Here t is the time, x is the horizontal coordinate, g is the gravity acceleration, h^+, h^- are the depths, u^+, u^- are the mean horizontal velocities in the upper and lower layer, respectively, η is the thickness and $\bar{u}$ is the mean horizontal velocity in the interlayer, $h = h^+ + \eta + h^-$ is the total depth and $\bar{Q} = h^+u^+ + \eta\bar{u} + h^-u^-$. The entrainment rate $\kappa^\pm$ is supposed to be proportional to the root-mean-square velocity $\bar{q}$ of turbulent flow:

$$\kappa^\pm = \sigma_0\bar{q},$$

where the coefficient $\sigma_0 = 0.15$ characterizes the ratio of the vertical and horizontal scales in the shallow water approximation, and it may be eliminated from (19) by replacing of independent variables. The energy dissipation term $\bar{\varepsilon}$ is taken in the form:

$$\bar{\varepsilon} = \theta\bar{q}^{\,3}, \quad \theta \equiv \text{const}$$

Note that (19) describes the mixing layer evolution for $h^\pm > 0$. If, say, $h^+ = 0$ at $x = x_1$, we have the transition of the mixing layer in a surface jet and (19) will be correct for the surface jet with $h^+ \equiv 0$, $\kappa^+ \equiv 0$ (Fig. 4).

4.1 *Mean Flow Evolution*

A steady-state two-dimensional mixing layer forms when two layers of fluid with depths h_0^+, h_0^- and velocities u_0^+, u_0^- merge at $x = 0$ (Fig. 4). As a consequence of (19) we have the following relations for mean characteristics of the flow

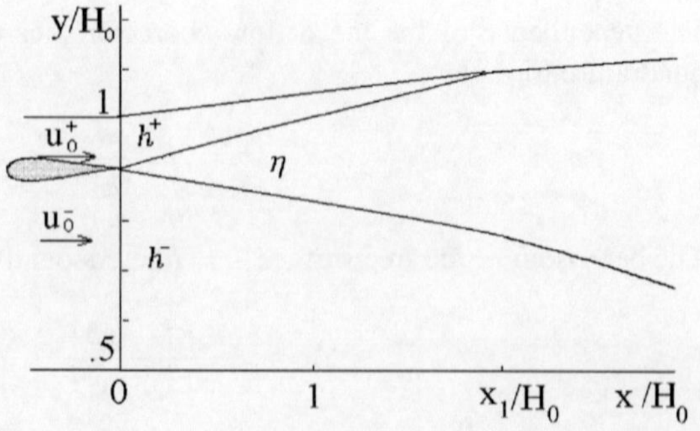

Fig. 4 Mixing layer at a free surface ($u_0^+/\sqrt{gh_0} = 2$, $u_0^-/\sqrt{gh_0} = 0.7$, $h_0^-/h_0 = 0.9$)

$$h^+u^+ + 0.5Q = h_0^+ u_0^+ = Q^+, \quad h^-u^- + 0.5Q = h_0^- u_0^- = Q^-$$

$$0.5u^{+2} + gh = 0.5u_0^{+2} + gh_0 = J^+, \quad 0.5u^{-2} + gh = 0.5u_0^{-2} + gh_0 = J^-$$

$$h^+u^{+2} + \eta\bar{u}^2 + h^-u^{-2} + 0.5gh^2 = h_0^+ u_0^{+2} + h_0^- u_0^{-2} + 0.5gh_0^2 = F \quad (20)$$

Here $Q = \eta\bar{u}$, $h_0 = h_0^+ + h_0^-$ and all unknown variables can be expressed from (20) as functions of Q. For the stationary mixing layer (19) is reduced to the system of ODE

$$\frac{dQ}{dx} = 2\sigma_0\bar{q},$$

$$\frac{d\bar{q}}{dx} = \frac{\sigma_0}{Q}\left(f_l(Q) - (1+\delta)\sigma_0\bar{q}^{\,2}\right), \quad (21)$$

where $\delta = \theta/(2\sigma_0)$, $f_l(Q) = \bar{u}^2 + 0.5u^{+2} + 0.5u^{-2} - \bar{u}(u^+ + u^-)$. Eqs. (21) may be rewritten as a linear ODE with the unknown function $\bar{q}^{\,2} = \bar{q}^{\,2}(Q)$

$$Q\frac{d\bar{q}^{\,2}}{dQ} = \left(f_l(Q) - (1+\delta)\sigma_0\bar{q}^{\,2}\right),$$

which has the bounded solution for $Q > 0$

$$\bar{q}^{\,2}(Q) = Q^{-(1+\delta)} \int_0^Q s^\delta f_l(s)\,ds.$$

Note that $f_l(0) = 0.25(u_0^+ - u_0^-)^2$, $\bar{q}^{\,2}(0) = f_l(0)/(1+\delta)$ and the entrainment in the mixing layer starts at $x = 0$ with the finite rate

$$\bar{q}_0 = (u_0^+ - u_0^-)/\sqrt{1+\delta}.$$

The dependence of the mean flow characteristics on x can be restored from the quadrature formula

$$x = \int_0^Q \frac{ds}{2\sigma_0 \bar{q}(s)}$$

The behaviour of the free surface $h = h(x)$ depends on the sign of the determinant

$$\Delta_l = 1 - \frac{g\eta}{\bar{u}^2} - \frac{gh^+}{u_0^{+2}} - \frac{gh^-}{u_0^{-2}}.$$

In subcritical flows ($\Delta_l > 0$) the total depth $h(x)$ increases and in supercritical flows ($\Delta_l < 0$) the function $h(x)$ decreases. The dependence of the total depth and the boundaries of the mixing layer is shown in Fig. 4. The upper boundary of the mixing layer reaches the free boundary at $x = x_1$.

4.2 Velocity Field in Mixing Layer

The distribution of the mean quantities in the mixing layer have been found above. We use the boundary layer approximation to calculate the horizontal and vertical velocity components $u = u(x,z)$, $w = w(x,z)$ as well as the root-mean-square velocity $q = q(x,z)$ in the free turbulent flow [13]. For steady-state flows the governing equations take the form ($\rho \equiv 1$):

$$uu_x + wu_z + \tau_z = -p_x,$$

$$uq_x + wq_z + \tau u_z = -\varepsilon, \quad u_x + w_z = 0, \tag{22}$$

where the Reynolds stress τ is expressed by the formula

$$\tau = -\sigma \bar{q} q, \quad \sigma = \sigma_0 \mathrm{sgn}(u_z). \tag{23}$$

The hydrostatic assumption gives $p(x) = gh(x)$ and the dissipation rate ε is based on the length scale η and the mean turbulence level $\bar{q}^2$ in the mixing layer

$$\varepsilon = \beta \bar{q} q / \eta, \quad \beta \equiv \mathrm{const}. \tag{24}$$

Note also that $\mathrm{sgn}(u_z) = \mathrm{sgn}(u^+ - u^-)$ in the mixing layer. Outside the mixing layer $u_z = 0$ and, as consequence, $u = u^-(x)$ for $0 < z < h^-(x)$; $u = u^+(x)$ for $h(x) - h^+(x) < z < h(x)$. The function $w(x)$ can be found in this region by the continuity equation from the boundary conditions $w|_{z=0} = 0$ and $w|_{z=h} = u^+ h_x$.

Therefore, at the boundaries of the mixing layer the velocities $u(x,z)$ and $w(x,z)$ are known and $q(x,z) = 0$. It is required to construct a continuous solution of (22) – (24) inside the mixing layer ($h^-(x) < z < h(x) - h^+(x)$).

Let $u_0^+ > u_0^-$ and the velocity profile be monotone ($u_z \geq 0$). Then we have $\sigma \equiv \sigma_0$. It is convenient to use the variables x and ψ as independent variables (ψ is the stream function). In this variables (22) becomes a semilinear system

$$u_x - \sigma \bar{q} q_\psi = -g h_x / u,$$
$$q_x - \sigma \bar{q} u_\psi = -\beta \bar{q} q / (\eta u). \tag{25}$$

A solution of (25) is constructed for $0 \leq x \leq x_1$, $\psi_0^- \leq \psi \leq \psi_0^+$, where

$$\psi_0^- = -\int_0^{h_0} u(0,z)\, dz = -h_0^- u_0^-, \quad \psi_0^+ = \int_{h_0}^{h_0} u(0,z)\, dz = h_0^+ u_0^+.$$

The point A, at which two uniform layers with different velocities merge, corresponds to the origin of coordinates on the (ψ, x)–plane (Fig. 5). The boundaries of the mixing layer are represented by the curves AB and AC. The solution $u = u^-(x)$, $q = \bar{q} = 0$ is known to the left of AB (region I). Similarly, the solution of (25) has the form $u = u^+(x)$, $q = \bar{q} = 0$ to the right of AC (region II).

Note that the curves AB and AC, which are given by the functions $\psi = \psi^-(x)$ and $\psi = \psi^+(x)$, respectively, are the characteristics of (25)

$$\frac{d\psi^\pm(x)}{dx} = \pm \sigma \bar{q}(x).$$

For the semilinear hyperbolic system (25) ($\bar{q}(x) > 0$) we have the Goursat problem, which can be solved by the standard method of characteristics. If $u^-(x) \geq u_{\min} > 0$ in region I ($0 < x < x_1$) the estimate $u(x, \psi) \geq u_{\min} > 0$ holds in BAC for a monotone velocity profile and the solution of (25) is bounded for $0 < x < x_1$. The behaviour of the solution near the origin of coordinates is represented by a self – similar solution of (25) for pressure – gradient – free flows ($p_x \equiv 0$), which can be found in an explicit form [7]. It is shown in [7] that the energy conservation law ($\theta = 0$) may be applied to describe the large billow evolution in mixing layers. Fig. 6 shows the horizontal velocity (a) and the Reynolds stress (b) distribution in the mixing layer at the free surface ($x = x_1$) for the inflow parameters shown in Fig. 4.

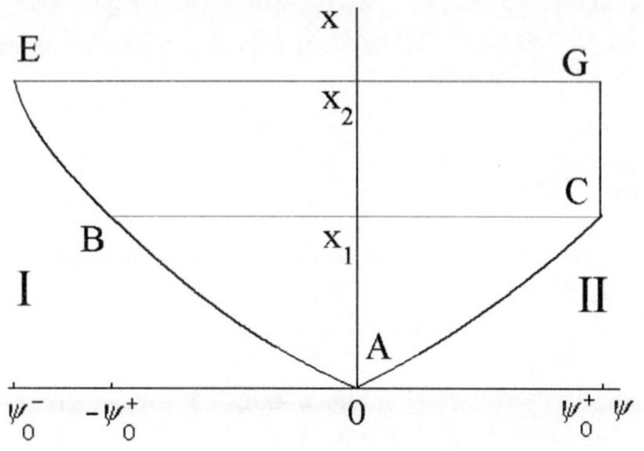

Fig. 5 (ψ, x)–flow diagram

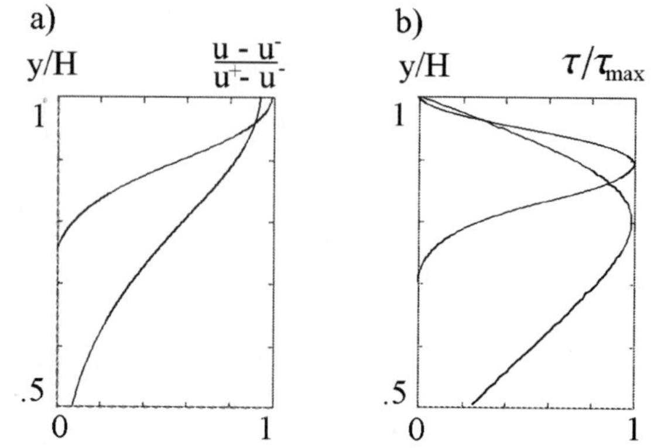

Fig. 6 **a** Non-dimensional horizontal velocity $(u - u^-)/(u^+ - u^-)$ versus z/h at $x = x_1$. **b** Non-dimensional shear stress $\tau/\tau_{\max}$ versus z/h at $x = x_1$. The initial flow parameters are given in Fig. 4 ($\theta = 0$, $\beta = 1.8$)

5 Conclusions

In this paper we have overviewed the nonlinear integro-differential models of shallow flows in an open channel with continuous vertical or horizontal velocity distribution. We have shown that mathematically these models are equivalent and, consequently, the obtained early results for plane-parallel open channel flows with vertical shear can be applied to the 2D horizontally-sheared flows. Based on Teshukov's method of generalized characteristics we have formulated necessary and sufficient conditions of hyperbolicity to the governing equations of flows with horizontally nonhomogeneous velocity field. These conditions allow one to study stability of shear flows. We have found that the type of the governing equations can change during the evolution of the flow, which corresponds to the long wave instability for a certain velocity field. We have also derived mathematical model describing the nonlinear stage of the Kelvin-Helmgoltz instability of shear flows. Using this model we have studied the problem of the mixing layer interaction with a free surface and its transition into a turbulent surface jet.

Acknowledgements. This work was supported by the Russian Academy of Sciences (Grants No. 4.7 and No. 2.14.1) and the Ministry of Education and Science of the Russian Federation (Grant No. 2.1.1/3543).

References

1. Benney, D.J.: Some properties of long nonlinear waves. Stud. Appl. Math. 52, 45–50 (1973)
2. Burns, J.C.: Long waves in running water. Proc. Cambridge Philos. Soc. 49, 695–706 (1953)
3. Chesnokov, A.A., Liapidevskii, V.Y.: Wave motion of an ideal fluid in a narrow open channel. J. Appl. Mech. Techn. Phys. 50, 220–228 (2009)
4. Freeman, N.C.: Simple waves on shear flow: similarity solutions. J. Fluid Mech. 56, 257–263 (1972)

5. Liapidevskii, V.Y.: Blocking of the flow of a two-layer mixible liquid around an obstacle. J. Appl. Math. Mech. 58, 685–690 (1994)
6. Liapidevskii, V.Y.: Structure of a turbulent bore in a homogeneous liquid. J. Appl. Mech. Tech. Phys. 40, 238–248 (1999)
7. Liapidevskii, V.Y.: Mixing layer in homogeneous fluid. J. Appl. Math. Mech. 41, 647–657 (2000)
8. Sachdev, P.L.: Exact self-similar time-dependent free surface flow under gravity. J. Fluid Mech. 96, 797–802 (1980)
9. Stoker, J.J.: Water waves. Mathematical theory and applications. Interscience, New York (1957)
10. Teshukov, V.M.: The hyperbolicity of the long wave equations. Dokl. Akad. Nauk. 284, 555–562 (1985)
11. Teshukov, V.M.: Long waves in an eddying barotropic liquid. J. Appl. Mech. Techn. Phys. 35, 823–831 (1994)
12. Teshukov, V., Russo, G., Chesnokov, A.: Analytical and numerical solutions of the shallow water equations for 2D rotational flows. Math. Models Methods Appl. Sci. 14, 1451–1481 (2004)
13. Townsend, A.A.: The Structure of Turbulent Shear Flow. Cambridge University Press, Cambridge (1956)
14. Witham, J.: Linear and Nonlinear Waves. Wiley, New York (1974)
15. Zakharov, V.E.: Benney equations and quasi-classical approximation in the method of inverse problem. Funct. Anal. Appl. 14, 15–24 (1980)

A Language for the Definition of Fortran Source to Source Transformations

Uwe Küster

Abstract. Based on experiences with imperfect compiler optimization we try to develop a tool which allows to program rules for source to source code transformations. The language provides mechanisms and constructs which facilitate the parsing of codes, the collection of the necessary information, the transformation of the code segments and the definition of the transformed code including all features to get a well defined transformed program. The transformation language can also be used to change source code for other purposes. We describe some features of the language and give some examples.

1 Compiler

Modern computer architectures may deliver very high floating point performance. But they do this on the base of deep implicit and explicit parallelism. And they suffer on restricted memory bandwidth and large memory latencies, which limit the effective performance unless caches, local memories and registers can be reused for providing the necessary data. The unavoidable memory latencies have to be hidden by prefetching the data. Modern compilers are expected to produce code which is effectively using the hardware features of the architecture. There are some well known transformations enabling performance as loop unrolling loop fusion, loop blocking, inlining. But a compiler as an automatic tool can not understand the user intentions. The compiler is forced to generate accurate code under all and even very unlikely circumstances.

It cannot estimate loop iterations counts unless they are disclosed in an obvious way, does not see effective bounds of algorithmically relevant parameters, can not recognize index arrays as injective or as having a limited extend or regular memory access patterns, can not distinguish overlapping and nonoverlapping pointers. Run

Uwe Küster
High Performance Computing Center Stuttgart,
Nobelstraße 19, Stuttgart, 70569, Germany
e-mail: kuester@hlrs.de

E. Krause et al. (Eds.): Computational Sci., & High Performance Computing IV, NNFM 115, pp. 181–191.
springerlink.com

time tests for analysis of the actual situation, for example the calculations of the address range of an index array which will be covered within a loop, would consume computing time and would be contraproductive in most cases. Complementing the languages by constructs ensuring better optimization and parallelization seems to be a way. This is done for the modern HPCS language Chapel [1]. This language allows for the differentiation of programming an algorithm and the later parallel optimization of the source code for performance increase. This important target is not yet reached in the current open source development of Chapel. Even if Chapel shows steps in a reasonable direction, also here user supported optimization will remain important.

The array syntax of Fortran could serve as an example for performance relevant programming support. But the array syntax is build on performance inhibiting semantics as the assumption that the complete right hand side of an array expression has to be evaluated before assigning the data to the left hand side. This forces the compiler to introduce copy operations implying additional expensive memory operations.

More complicated data structures performance supporting language primitives could be linked lists with unique elements or explicit trees without cross references. Handling these kinds of structures could be done efficiently in parallel.

Directives are a generalization of languages enabling the programmer to show semantics of the coded algorithm. Directives expressing that a loop has no inner dependencies, limiting the loop iteration count, enabling prefetching, triggering loop unrolling and loop collapsing may be essential for code optimization. They enable inlining, influence alignments and vectorization and may show the frequency of branch. But syntax and meaning of these directives are not portable between compilers and machines except for the standardized parallelization paradigm OpenMP as an very important example.

Most directives are bound to loops. They influence directly the control constructs of a loop. They do not disclose properties of arrays in a larger code segment or in a data structure. If the programmer knows about some special features of his algorithm (the values of variable `dim` are 1,2 or 3) it will not be possible for him to express this in a way enabling new transformations. He has to implement the transformations himself. These may generate confusing code with many repetitions which is difficult to maintain and may even be contraproductive on other hardware architectures.

2 Self Defined Transformations

A mechanism to define annotating directives to initiate problem dependent code transformations would be helpful for the generation of maintainable and fast code. The programmer would implement some of these directives at some appropriate positions. A transformation tool would execute the transformations for the designated location. The transformed source code would be still readable and be compiled in the next step. The original code remains unchanged except for the additional directives.

We assume that the user knows about the influence of the directives.

For realize this mechanism we have implemented a transformation language which allows for parsing of source code and the new user defined directives within the source code and for formulation of the appropriate transformations.

The transformation language allows for these requirements.

The supported source code language is Fortran. The technology could be also applied to other languages.

In the following we will describe this transformation language and give some examples.

3 Transformation Language

The transformation language allows for an abstraction of program transformation rules. The language is defined in a way that the written transformation programs may transform source code to source code by the implemented rules. We will describe the different elements of the language in the following. The language contains transformation variables, the transformation control constructs, self definable procedures, primitives for parsing support and a library of the intrinsic procedures. The source code is represented internally as a double linked list (not by a tree!) of *tokens* as `,` `(` `subroutine` `myvar` `.and.`. Parts of the source code can be accessed by *segments* which are the intervals of the tokens between a starting token and an ending token. The commands of the transformation language operate on this double linked list by changing tokens and segments.

The transformation languages has three different modi. The language is operating by control constructs. It is possible to extract and insert new tokens or source code elements in the existing source code. Special constructs allow for an environment for parsing and facilitate of the source code.

3.1 Transformation Variables

The value of a transformation variable can accessed by preceding its name by "@". This is similar to scripting languages. Transformation variables can have the following intrinsic types

- **strings**,
- **integers**,
- **logical** values,
- **tokens**,
- **segments**,
- **variables** (Fortran names of variables, array, procedures, modules, ...),
- **variable_sets** which are sets of these source code variables
- **designs** which are themselves collections of *strings*, *integers*, *logicals*, *segments*, *Fortran variables* and recursively *designs*.

Each of these parts is described by a *selection_parameter* which can be used to access the part. Wild cards in the selection_parameters are allowed. Selection_parameters can be combined by "%".

There are no variables to store floating point numbers and no arrays. The design variables take the role of self defined data structures and of arrays. The transformation variables have not to be declared, they get their type by an assignment or as resulting value of an intrinsic procedure.

3.2 *Transformation Constructs*

In the transformation modus the control constructs and the transformation variables are accessible for writing and reading. The control constructs and intrinsics support the formulation of the transformation programs.

- The loop control construct is
  §do(it_var= loop_control_expression) ... **§enddo(** it_var)
  The do loop may include **§cycle(** it_var) to get the next iteration or **§exit(**outer_it_var) to leave the loop. Both include the name of the iteration variable. This allows to leave also an outer loop. The form of the loop_control_expression allows for an appropriate scanning of sets. It may be consist on a start and an end integer or a segment with a statement or token attribute. In the first case all statements of the segment are tested and in the second case all tokens. The loop_control_expression may also be a design with a supplemental expression of selection_parameters.
- The branching construct is
  §if(< logical_expression > **) §then(**if_name)
  ...
  §elseif(if_name)
  ...
  §else(if_name)
  ...
  §endif(if_name)
  with the obvious meaning. The construct names in parenthesis ensure a proper nesting.
- **§set(**variable = set_expression) to define and set transformation variables. This replaces the assignment operator. The assignment operator is reserved for usage in the source code to be defined. set_expression is similar to the loop_control_expression in a do loop. It may also contain.
- **§evaluate(**expression) which evaluates an arithmetic or string expression.
- **§get(**@variable, selection_parameter,movement,result**)** to extract the content from transformation variables of different data types and for the definition of new transformation variables; movement can be composed expressions simplifying the orientation in constructs as Fortran do loop, select case constructs or if then else endif constructs.

- **§add_to**(@design,para,@entity) stores a token or a segment or other design parts from entity in design. The stored content can be found later by using the parameter para as selection_parameter in §set(...) or §get(...) or §do(...).
- **§print**(string, @variable) prints transformation variables of all types in variable.

3.3 Self Defined Procedures in Transformation Code

The transformation language provides self defined procedures which are called via **§exe**(procedure_name, list_of_procedure_arguments). All variables in the procedure are global except these which are explicitly declared as local. Procedures can be called recursively. They may be stored in different files as usual.

3.4 Intrinsic Procedures

The transformation language contains a variety of different intrinsic functions. Some of these may detect and parse Fortran constructs as

- **§is_function_or_array**(...), which is true if an function call or an array element has been found,
- **§is_assignment**(...), which is true if an assignment has been found,
- **§is_do_enddo**(...), true for the begin or the end of a do loop.
- ...

If successful they return a **design** containing the important parameters of the constructs which can be later used for transformations.

- **§parse_all**(...)
  parses a source code construct found by help of parsing primitives which are described in the next section and delivers parameters in a **design** variable. The construct may be a Fortran statement or also a self defined directive.

There are intrinsics for manoeuvring in the Fortran text like

- **§next_active_statement(**@actual_token,new_token),
- **§last_statement(**@actual_token,new_token),

Other intrinsics handle the variables of the source code language itself as

- **§var_decl(**@variable,@varset,declaration_list)
  incorporates the intended Fortran variable in the @varset and defines declarations attributes.
- **§add_declarations(**@insert_token,@varset)
  adds all names collected in the variables set given by @varset as Fortran declarations at the point insert_token. These names have been given all necessary attributes by **§var_decl(**
- **§find_variable_in_varset(...)**
  finds Fortran variables specified in a set of variables within a segment of the Fortran text.

3.5 Parsing Primitives in Parsing Mode

The given Fortran code has to be parsed to 'understand' the different entities, to collect the attributes of variables, to analyse do-loops, to limit a print instruction, to analyse dependencies. We have not used a parser as it is provided in compilers but have defined a mechanism which allows for parsing in interpreting way and for the collection of the important data to be used in the transformation steps. The parsing environment is given by the **§parse_all(** ...) command. This command takes a sequence of parsing primitives as parameters. The parsing primitives all begin with the #-sign and can be combined with transformation constructs. As examples for parsing primitives can be understood

- **#name** parses a Fortran name for variables, arrays, derived types, procedures, e.t.c. and puts the name found in a names database used for further analysis.
- **#bos** and **#eos** parse the begin and the end of a statement.
- **#arithmetic_expression** parses an arithmetic expression like `alpha*(b-c)/d+1`
- **#function_or_array** parses a function call or an array evaluation like `cc(i,j,k)`.
- **#label** parses a label like `1000 continue`.
- **#assignment** parses an assignment like `r=4.*b-d`.
- # parses a blank, and **#,** parses a comma.

A name like **#?var_name** is used to store the result of the subsequent primitive or group of primitives (see below) as entity which may be accessed later by the parameter *var_name* in the resulting **design** in a transformation.
As an example take the following parsing sequence for the begin of a do-loop

```
§if( §parse_all(@start,do_parse_par,
       ' #bos #?begin_construct do #?it #name
       =
         #?it_start #arithmetic_expression
       , #?it_end #arithmetic_expression
         #?incr_switch #{ , #?it_increment #arithmetic_expression #}
       #eos '
   ) §then(if_parse)
       §print(parameters for do loop,@do_parse_par)
§endif(if_parse)
```

If the token in @start points to the begin of the source code statement

```
do j=1-ja,2+jb,jd
```

the intrinsic **§parse_all(...)** is true and delivers the **design** `do_parse_par` which is printed with the result

```
| design at "parameters for do loop"
| design name: "do_parse_par_"
| design number_of_entities= 6
| (1): "begin_construct" is the segment
|                  <
|       80|       do j=1-ja,2+jb,jd
|                    >
| (2): "it" is the segment
|                        <
|       80|       do j=1-ja,2+jb,jd
|                        >
| (3): "it_start" is the segment
|                           <
```

```
|       80|        do j=1-ja,2+jb,jd
|                          >
| (4): "it_end" is the segment
|                          <
|       80|        do j=1-ja,2+jb,jd
|                          >
| (5): "incr_switch" is the integer  1
| (6): "it_increment" is the segment
|                          <
|       80|        do j=1-ja,2+jb,jd
|                          >
| end of design printed at "parameters for do loop"
```

The contents of the variables `#?it`, `#?it_start`, `#?it_end`, `#?it_increment` are shown as segments. In the context of a translation program these variables can be accessed by appending their name to the **design** name (here `do_parse_part`), for example `do_parse_par%it_start`. The variable `incr_switch` is `1` . That means that the content of the succeeding `#{ ... #}` was true for one time.

Primitives and constructs may be combined together via three types of parenthesis. All these primitives and the different parenthesis can be preceded by parsing variable names of kind **#?var_name**.

- **#?vp #(... #)** groups entities. The preceding parsing variable **#?vp** stores the segment of the whole group if the parsing is successful.
- **#?vp #{ ... #}** are handled as true if the enclosed sequence appears at least once. The preceding parsing variable **#?vp** stores the number of successful repetitions of the whole construct if the parsing is successful or 0 otherwise.
- **#?vp #[... #]** are handled as true if one part of the enclosed sequence appears. The preceding parsing variable **#?vp** stores the segment of the first successful entity.

The parenthesis can be nested. This allows for quite complicated parsing constructs. If the parsing of the combined sequence is successful the parsing step delivers a resulting design of all segments and repetition numbers which are accessible via the names of the parsing variables without the prefix **#?** .

4 Examples

The following examples show the applicability of the approach. We show the result of an implemented unrolling transformation, how to parse a self defined directive and the implementation of a mechanism which inserts timing calls at the begin and end of all source code procedures in a file.

4.1 Unrolling of Source Code

The first example shows the original code and the result of a transformation program for unrolling. The transformation program has not be shown because it is too complicated.

```
!directive begin unroll j by 2
 do j = i+1, jmax
   ss = 0.0d0
   do k = i+1, kmax
     ss = ss + aa(j,k) * aa(i,k)
   enddo
   ff=ss/hh
   do k = i+1, kmax
     aa(j,k) = aa(j,k) + ff * aa(i,k)
   enddo
 enddo
!directive end unroll
```

This code will be transformed to the following outer unrolled code

```
!treated!!directive begin unroll j by 2
  j_start=i+1
  j_d=mod(jmax-j_start+1,2) ! case_number
  j_shift=j_start+j_d
  j=j_start ! first it_variable
 select case(j_d)
 case(0)
 case(1)
   ss_j0 = 0.0d0
   do k = i+1, kmax
     ss_j0 = ss_j0 + aa(j+0,k) * aa(i,k)
   enddo
   ff_j0=ss_j0/hh
   do k = i+1, kmax
     aa(j+0,k) = aa(j+0,k) + ff_j0 * aa(i,k)
   enddo
  end select
  do j=j_shift,jmax,2
   ss_j0 = 0.0d0
   ss_j1 = 0.0d0
   do k = i+1, kmax
     ss_j0 = ss_j0 + aa(j+0,k) * aa(i,k)    ! (1a)
     ss_j1 = ss_j1 + aa(j+1,k) * aa(i,k)    ! (1b)
   enddo
   ff_j0=ss_j0/hh
   ff_j1=ss_j1/hh
   do k = i+1, kmax
     aa(j+0,k) = aa(j+0,k) + ff_j0 * aa(i,k)    ! (2a)
     aa(j+1,k) = aa(j+1,k) + ff_j1 * aa(i,k)    ! (2b)
   enddo
  enddo !new
!treated!!directive end unroll
```

This code can be operated quite efficiently because `aa(i,k)` is reused in (1a) and (1b) as well as in (2a) and (2b). If `k` is not too large, `aa(i,k` will reside in the nearest cache for the second loop. Also the multiple reductions in (1a) and (1b) are faster than simple reductions. In general both loops are deeper unrolled than here. Simply the directive has to be changed for this purpose. Not shown are the declarations added for the new variables as `ss_j0`.

4.2 Parsing Directive

For parsing the directive in the first example we use the following sequence in the transformation program. The syntax of the directive and the transformation allow for even more sophisticated unrolling techniques as we see it in usual commercial compilers.

```
§set(str1=' #bos ! @DIRECTIVE begin unroll ')
§set(str2=' #?induction_variable #name ')
§set(str3=' #?keyword #name ')
§set(str4=' #{ #switch_comma #?number #integer #} ')
§set(str5=' #?only_switch #{ only #} #eos ')
    ...
§if( §parse_all(@statement_token,parse_para, @str1,@str2,@str3,@str4,@str5) )
§then(if_parse)
    #* we extract the induction variable from the directive
    §if(§get(@parse_para,induction_variable,string,induction_var)) §then(if_ind)
        §print(++++unroll directive++++-,@statement_token)
    §endif(if_ind)
    §if(§get(@parse_para,keyword,string,unroll_case)) §then()
        §print(got the unroll case ,@unroll_case)
    §endif()
    ...
§endif(if_parse)
```

In the case that a statement will be successfully parsed the **design** `parse_para` will be generated. The first **§get(**@parse_para,induction_variable, ... **)** extracts the string `induction_var` from the segment `parse_para%induction_variable`. If successful the `@statement_toke` will be printed.

4.3 Insertion of Timing Calls

The next transformation procedure inserts timing calls at the begin and at the end of each subroutine. The name of the subroutine is taken as the first parameter of the call. The second argument `actual_time_stack` is additionally declared in the declaration part.

```
§procedure(insert_timing,procedure_body,subroutine_name)

§if( §is_valid_segment(@procedure_body) ) §then(if_valid_procedure_body)
   §get(@procedure_body,end_segment->end_of_last_non_comment_statement->begin_of_statement,test_token)
   §if(§is_call(@test_token,end_token_call,call_para)) §then()
      §get(@call_para,subroutine_name,string,procedure_name)
   §else()
      §set(procedure_name='nonsense_347')
   §endif()

   §if(@procedure_name /= 'timing_end') §then(if_new)
      §get(@procedure_body,begin_segment->end_of_last_statement,temp_token)
      §insert_at(@temp_token,#bos ! # start # timing #eos)
      §insert_at(@temp_token,#bos call # timing_start("@subroutine_name",actual_time_stack) #eos)

      §get(@procedure_body,end_segment->end_of_last_statement,temp_token)
      §insert_at(@temp_token,#bos ! # end # timing #eos)
      §insert_at(@temp_token,#bos call # timing_end("@subroutine_name",actual_time_stack) #eos)

      §get(@declarations,end_segment,temp_token)
      §insert_at(@temp_token,#bos ! # insert # new # declaration # for # timings #eos)
      §insert_at(@temp_token,#bos integer #15 :: # actual_time_stack #eos)
   §endif(if_new)
§endif(if_valid_procedure_body)
§end_procedure(insert_timing)
```

This transformation transforms the following original source code

```
subroutine tt(r,ja,jb,jd)
    integer ja,jb
    integer j,diff
    real r(:)

    do j=1-ja,2+jb,jd
        r(j)=0.
    enddo

end subroutine tt
```

and to the following modified source code

```
subroutine tt(r,ja,jb,jd)
! changed declarations
! from compile in add_missing_declarations_switch
   integer :: ja
   integer :: jb
   integer :: j
   real,dimension(:) :: r
! insert new declaration for timings
integer :: actual_time_stack
! end of declarations

! start timing
call timing_start("tt",actual_time_stack)
   do j=1-ja,2+jb,jd
   r(j)=0.
   enddo
! end timing
call timing_end("tt",actual_time_stack)

end subroutine tt
```

Once the transformation program is defined it can be used for all files containing Fortran source code. This example shows that the transformation language can be used also for other purposes as instrumenting code.

5 Concluding Remarks

We are developing a tool which helps for performance enabling by restructuring the source code. The tool is programmable and can express special source to source transformations which make sense for a broad range of applications or also only for a very specific user code. The tool is not restricted to situations which are completely transparent to a compiler which is forced to generate correct (and potentially slower) code under all circumstances. The produced code is still accessible to the user and could be analysed for correctness. Because the transformation program is open to the user it could be adjusted in the right way. The transformation language is not easy but relatively simple to handle. Up to now we have implemented some transformations which are useful to get higher performance on vector systems. The transformation language is not developed to replace optimizing compilers. It helps for generalization of an existing programming language, here Fortran.

The Rose Compiler project [2] has similar intentions. Rose provides also mechanisms for building source to source transformations. It does this on a sophisticated intermediate inner representation by an abstract syntax tree. On that base automatic transformation programs can be build. The approach might be more general but also more difficult to understand and to handle.

References

1. Chamberlain, B.L., Callahan, D., Zima, H.P.: Parallel programmability and the Chapel language. Int. J. High Perform. Comput. Appl. 21(3), 291–312 (2007)
2. Rose Compiler project, `http://www.rosecompiler.org`

Tsunami Modelling with Unstructured Grids. Interaction between Tides and Tsunami Waves

Alexey Androsov, Jörn Behrens, and Sergey Danilov

Abstract. After the destructive event of December 26, 2004, many attempts have been made to accurately simulate the generation and propagation of tsunami waves in the Indian Ocean. In support of the Tsunami Early Warning System for the Indian Ocean, a very high-resolution finite element model (*TsunAWI*) has been developed for simulations of the wave propagation. It offers geometrical flexibility by working on unstructured triangular grids and is based on finite-element $P_1^{NC} - P_1$ discretization. The paper presents a brief description of the model, with a focus on its verification and validation. The key issue in modelling the tsunami is wetting and drying. The original algorithm to solve this problem is discussed. Full and reduced formulation of the momentum advection for $P_1^{NC} - P_1$ elements and parameterization of horizontal diffusion are presented. Using the well-known Okushiri test case, the influence of nonlinearity on the wave propagation is demonstrated. Numerical experiments simulating the Indian Ocean Tsunami on December 26, 2004 have been conducted. For the whole Indian Ocean, the comparison of simulation results with observational (coast gauge) data is carried out.

A typical tsunami wave is much shorter than tidal waves which are usually neglected in tsunami modelling. However, in coastal areas with strong tidal activity, dynamic nonlinear interaction of tidal and tsunami waves can amplify the magnitude of inundation. To study this effect, water level change due to tide is included in the general scheme.

Alexey Androsov · Sergey Danilov
Alfred Wegener Institute for Polar and Marine Research,
Postfach 12-01-61, 27515 Bremerhaven, Germany
e-mail: alexey.androsov@awi.de, sergey.danilov@awi.de

Jörn Behrens
University of Hamburg, Grindelberg 5,
20144 Hamburg, Germany
e-mail: joern.behrens@zmaw.de

E. Krause et al. (Eds.): Computational Sci., & High Performance Computing IV, NNFM 115, pp. 191–206.
springerlink.com

1 Introduction

In support of the Tsunami Early Warning System for the Indian Ocean, a finite element model *TsunAWI* for simulations of wave propagation has been developed. It is part of the German Indonesian Tsunami Early Warning System (GITEWS) (`www.gitews.de`) serving to predict arrival times and expected wave heights. The finite-element (FE) spatial discretization allows for unstructured triangular meshes of variable resolution.

Finite-element methods are widely used in studies of wave generation and propagation in different fields of fluid dynamics. They are often employed to simulate propagation of long waves such as ocean tides and tsunamis in the ocean in the framework of shallow-water equations [1, 2, 3]. The main reason to prefer FE modelling is that the solution is computed over a mesh that can be adapted to cover basins with complex geometries characterized by irregular bottom topography and coastlines. This is exactly what is needed in the framework of GITEWS.

Although several FE codes dealing successfully with wave propagation in complex geometries exist, the search for numerically efficient and accurate algorithms gives birth for new designs.

The main goal of this paper is twofold. First, it gives a description of the shallow water model used by us and presents comparison results against elementary cases and available observational data. Following [4], the model is built on a continuous linear representation for elevation and discontinuous, nonconforming, linear representation for the velocity. In contrast to [4], it uses a different time stepping scheme and offers three stable versions of momentum advection discretization versus the unstable one in the original approach. The model performance is tested against an analytical solution for a channel, laboratory data (the so-called Okushiri test case) and observational data related to the tsunami event of December 26, 2004.

Second, our investigation addresses the influence of tidal dynamics on tsunami wave propagation in coastal areas. This question has not received its deserved attention so far ([5, 6, 7, 8, 9]). A tsunami wave is much shorter than tidal waves which explains why tidal waves are usually ignored in tsunami modelling. An additional motivation is purely computational. Attaining a periodic regime for tides requires a long simulation. However, the operational models of tsunami wave propagation employ very large meshes, which makes the task of tidal simulation on such meshes too demanding in terms of CPU time.

There are three approaches to the interaction of tsunami waves and tides. The first one presumes that the interaction has linear character and the tide reduces or augments (depending on the phase) the amplitude of an arriving wave [9]. In this case, having simulated tidal patterns in advance, it would be possible to predict arriving waves as a simple superposition of signals. The second approach assumes that the interaction has a nonlinear character caused by the changing fluid layer thickness in the shallow area [6]. Finally, the third approach suggests that the interaction has a nonlinear character due to nonlinear interaction of tidal and tsunami velocities [7].

The studies performed thus far are of theoretical character and do not involve practical examples. In this paper we carry out the analysis of tidal influence for a particular region in the Indian ocean by considering a tsunami wave propagation on the background M_2 tide. Energy evolution is analysed for better understanding of the phenomenon.

The paper is organized as follows. In section 2, the spatial and temporal discretization of the model is presented. Section 3 deals with its verification and validation. The following section 4 discusses tidal-tsunami interactions and section 5 concludes the presentation.

2 Barotropic Shallow Water Model

2.1 Boundary-Value Problem in Cartesian Coordinates

Consider vertically averaged equations of motion and continuity in $\widetilde{\Omega} = \{(x,y) \in \Omega \subset \mathbb{R}^2, 0 \le t \le T\}$, where is Ω a plane domain bounded by boundary $\partial\Omega$,

$$\mathbf{v}_t + (\mathbf{v}\cdot\nabla)\mathbf{v} + g\nabla\zeta = \Phi \equiv f\mathbf{k}\times\mathbf{v} - rH^{-1}\mathbf{v}|\mathbf{v}| + H^{-1}\nabla(K_h H\nabla\mathbf{v}), \qquad (1)$$

$$\zeta_t + \nabla\cdot(H\mathbf{v}) = 0. \qquad (2)$$

Here $\mathbf{v} = (u, v)$ is the velocity vector, $H = h + \zeta$ the total water depth, $H > 0$, h is the unperturbed water depth, and ζ the surface elevation, $\nabla = (\partial/\partial x, \partial/\partial y)$ is the gradient operator, f the Coriolis parameter, $\mathbf{k}$ the unit vector in the vertical direction, r the bottom friction coefficient, and K_h the eddy viscosity coefficient. The set of (1) and (2) is known as the rotating shallow water equations.

On the solid part of the boundary, $\partial\Omega_1$, and on its open part, $\partial\Omega_2$, we impose the following boundary conditions

$$\mathbf{v}_n|_{\partial\Omega_1} = 0, \;\; \Gamma(\mathbf{v},\zeta)|_{\partial\Omega_2} = \Theta_1 \qquad (3)$$

where $\mathbf{v}_n$ is the velocity normal to $\partial\Omega_1$, Γ is the operator of the boundary conditions and Θ_1 is a vector-function determined by the boundary regime and different for inflow and outflow [10]. In practice, when the full necessary information on the open boundary is unavailable, in place of the second condition (3) one commonly imposes the boundary condition on the elevation $\zeta|_{\partial\Omega_2} = \psi(x,y,t)$ or the radiation boundary condition $\mathbf{v}_n = \mathbf{v}\cdot\mathbf{n} = \sqrt{\frac{g}{H}}\zeta$. The latter provides free linear wave passage through the open boundary (when the Coriolis acceleration plays a small role). Here $\mathbf{n}$ is the outer unit normal to $\partial\Omega_2$. The accuracy of the reduced boundary-value formulation with only the sea level assigned at the open boundary, was analysed in [11]. The problem (1)–(3) for the combination $\mathbf{u} = (\mathbf{v},\zeta)$ is solved for given initial conditions:

$$\mathbf{u}|_{t=0} = \mathbf{u}^0. \qquad (4)$$

2.2 Non-conforming Mixed $P_1^{NC} - P_1$ Discretization

2.2.1 Spatial Discretization

The finite element spatial discretization is based on the approach by [4] with some modifications like added viscous and bottom friction terms, corrected momentum advection terms, radiation boundary condition and nodal lumping of mass matrix in the continuity equation. The basic principles of discretization follow the paper of [4] and are not repeated here.

2.2.2 Time Discretization

Simulation of tsunami wave propagation benefits from using an explicit time discretization. Indeed, numerical accuracy requires relatively small time steps, which reduces the main advantage of implicit schemes. Furthermore, modelling the inundation processes usually requires very high spatial resolution in coastal regions (up to some tens of meters) and consequently large number of nodes, drastically increasing necessary computational resources in case of implicit temporal discretization.

The leap-frog discretization was chosen as a simple and easy to implement method. We rewrite Eqs. (1) and (2) in time discrete form,

$$\frac{\mathbf{v}^{n+1} - \mathbf{v}^{n-1}}{2\Delta t} + fk \times \mathbf{v}^n + g\nabla\zeta^n + \frac{r}{H^n}|\mathbf{v}^n|\mathbf{v}^{n+1} - \nabla K_h \nabla \mathbf{v}^{n-1} + (\mathbf{v}^n\nabla)\mathbf{v}^n = 0, \quad (5)$$

$$\frac{\zeta^{n+1} - \zeta^{n-1}}{2\Delta t} + \nabla \cdot (H^n \mathbf{v}^n) = 0. \quad (6)$$

Here Δt is the time step length and n the time index. The leap-frog three-time-level scheme provides second-order accuracy and is neutral within the stability range. This scheme however has a numerical mode which is removed by the standard filtering procedure. Notice that friction and viscosity contributions deviate from the usual leap-frog method.

2.3 Momentum Advection Schemes

Earlier experiments with $P_1^{NC} - P_1$ code reveal problems with spatial noise and instability of momentum advection when the discretization is used in the form described in [4]. Here we suggest a rigorous procedure to derive the FE discretization with discontinuous functions, which guarantees consistency with the original governing equations. Applying this procedure shows that a term is missing from the discretization of momentum suggested in the paper cited above. It serves to penalize discontinuity of normal velocity across the edges.

2.3.1 Consistent P_1^{NC} Discretization

We leave continuous test/basic functions as they are, but augment the discontinuous P_1^{NC} functions by defining them in the following way:

$$\Psi_{ie}(x,y) = \Theta_e \psi_{ie}(x,y). \tag{7}$$

In this definition, $\Theta_e(x,y) = 1$, if point (x,y) is inside element e, 0 if it is outside and $1/2$ if it is on the boundary, and $\psi_{ie}(x,y)$ is the P_1^{NC} basis function due to edge i on element e. The test/basis function Ψ are now defined everywhere so that we can work with them as in the continuous case.

We write

$$(\mathbf{u}\nabla)\mathbf{u} = (\mathbf{ut}\partial_t + \mathbf{un}\partial_n)\mathbf{u} \tag{8}$$

based on tangential (subscript t) and normal (subscript n) directions associated with edge i. Only the second term contains a singularity at the the edge and we will continue by only considering this term. In the vicinity of the edge this term can be written as

$$[(\mathbf{u}\nabla)\mathbf{u}]_s = (\mathbf{u}_f + \Theta(y)(\mathbf{u}_e - \mathbf{u}_f))\mathbf{n}\delta(y)(\mathbf{u}_e - \mathbf{u}_f). \tag{9}$$

Here the coordinates x and y are transformed, so that x is tangential and y is perpendicular to the edge i, subscript s implies that only the singular part is taken into account, and e and f denote elements sharing edge i, with the normal vector pointing to e. This implies that the singular part of

$$\int \hat{\mathbf{u}}(\mathbf{u}\nabla)\mathbf{u}dS \tag{10}$$

due to edge i ($\hat{\mathbf{u}}$ denotes the test function) is the integral over small vicinity of this edge of

$$\int_i (\hat{\mathbf{u}}_f + \Theta(y)(\hat{\mathbf{u}}_e - \hat{\mathbf{u}}_f))(\mathbf{u}_f + \Theta(y)(\mathbf{u}_e - \mathbf{u}_f))\mathbf{n}\delta(y)(\mathbf{u}_e - \mathbf{u}_f)dS. \tag{11}$$

Representing dS as $dxdy$ and using the notation $[\mathbf{u}] = \mathbf{u}_e - \mathbf{u}_f$ and $\langle \mathbf{u} \rangle = \frac{1}{2}(\mathbf{u}_e + \mathbf{u}_f)$ one performs integration over y first to get the following result (and omitting terms that would vanish after integration over x)

$$-\int_i \hat{\mathbf{u}}_f < \mathbf{un} > [\mathbf{u}]dx + \frac{1}{2}\int_i [\hat{\mathbf{u}}]\mathbf{u}_f\mathbf{n}[\mathbf{u}]dx = -\int_i (< \hat{\mathbf{u}} >< \mathbf{un} > -\frac{1}{4}[\hat{\mathbf{u}}][\mathbf{un}])[\mathbf{u}]dx. \tag{12}$$

The last term in this expression can be omitted as it will disappear on performing integration along the edge. The remaining term should be added on all edges to get a consistent discretization of the momentum advection. To see the difference to the approach by [4] we further integrate the elemental integrals by parts to finally get

$$\int \hat{\mathbf{u}}(\mathbf{u}\nabla)\mathbf{u}dS = -\sum_{el}\int_{el}\mathbf{u}\nabla(\hat{\mathbf{u}}\cdot\mathbf{u})dS + \sum_{bnd.ed}\int_{ed}\hat{\mathbf{u}}\mathbf{u}(\mathbf{un})ds$$
$$\sum_{ed}\int_{ed}(<\hat{\mathbf{u}}><\mathbf{u}>[\mathbf{un}]+[\hat{\mathbf{u}}]<\mathbf{u}><\mathbf{un}>+\frac{1}{2}[\hat{\mathbf{u}}][\mathbf{u}][\mathbf{un}])ds. \quad (13)$$

The last term in the integral over edges will disappear after integration and thus can be removed from this expression. If we compare the remaining form with that given in [4] (without upwinding) it can be readily noticed that the term penalizing jump of normal velocity [**un**] is absent there. Upwind terms can be added separately.

2.3.2 P_1 Projection Method

The consistent implementation of the momentum advection involves cycling over edges in the numerical code, in addition to cycling over elements to assemble the elemental (regular) contributions. This is not very convenient. In addition, it was found that while consistent implementation works well, it still requires some viscous dissipation for removing small-scale noise in the velocity field. This lead us to two simpler approaches, which provide some smoothing of the velocity field while removing edge contributions.

According to these approaches, prior to calculating the advection term in the momentum equation we project the velocity from the P_1^{NC} to the P_1 space in order to smooth it. The projected velocity is further denoted as $\bar{\mathbf{u}}$. To make this projection numerically efficient, nodal quadrature (lumped mass matrix) is employed. The projected velocity is then used to estimate the advection term. Finally we proceed as usual by multiplying the result with a P_1^{NC} basis function and integrating over the domain

$$\int \hat{\mathbf{u}}(\mathbf{u}\nabla)\mathbf{u}dS \approx \sum_{el}\int_{el}\hat{\mathbf{u}}(\bar{\mathbf{u}}\cdot\nabla)\bar{\mathbf{u}}dS. \quad (14)$$

This results in a very stable behaviour. In case of the combined $P_1^{NC}-P_1$ discretization of advective transport, only the velocity subject to differentiation is taken from the P_1 space. The consistent non-conforming velocity is used as the advecting one:

$$\int \hat{\mathbf{u}}(\mathbf{u}\nabla)\mathbf{u}dS \approx \sum_{el}\int_{el}\hat{\mathbf{u}}(\mathbf{u}\nabla)\bar{\mathbf{u}}dS. \quad (15)$$

This approach is formally more accurate. Since in both cases one differentiates the continuous velocity, no singular contributions appear. Thus in contrast to the consistent P_1^{NC} advection scheme both ways of using reprojected velocities do not lead to boundary integrals. To distinguish them from the consistent non-conforming implementation we will call them P_1 advection and $P_1^{NC}-P_1$ advection, respectively.

2.4 Other Implementation Details

2.4.1 Wetting and Drying

Wave run-up generated by a tsunami reaching the shoreline may induce devastating flood waves. A tsunami is a long-period wave generated by ocean bottom motion during an earthquake with wave length of about 200 to 350 km. Although the wave amplitude is moderate in deep waters, the tsunami wave slows down and the wave height increases near the shoreline until it breaks. The wave run-up height might reach several metres above the natural sea level and cause significant damage. Seamless simulation of wetting and drying is needed even in far field forecasting, at least to avoid artificial wave reflections from the coast which occur if a "solid boundary" is assumed there.

In our study we adjust the continuity and momentum balance equations in the way that they provide a reasonable solution in the dry regions, i.e. zero velocities and surface elevation equal to land topography. The computational domain is extended over the land. The continuity equation is adjusted by replacing the total depth H with $H_+ = max(H,0)$. For the bottom friction term in the momentum equation we use the standard approach bounding the total depth with some critical value H_{cr}, i.e. H in the friction term denominator is replaced with $max(H,H_{cr})$. The sea surface gradient term needs attention. Special treatment is required for so-called "semi-dry" elements, i.e. elements with one or two dry and, correspondingly two or one wet vertices. For modelling wetting and drying we use a moving boundary technique which utilizes linear least square extrapolation through the wet-dry boundary and into the dry region. We apply the "dry node concept" as developed in [12]. The idea is to exclude dry nodes from the solution and to extrapolate elevation to the dry nodes from their wet neighbours. The number of "wet" points considered for extrapolation in this case should be not less than 4. The aim of this method is to find a plane (representing the elevation or gradient), given by the linear parametric representation

$$f(x,y) = a_0 + a_1 x + a_2 y \tag{16}$$

with (a_0, a_1, a_2) a parameter triplet, which approximates the gradient and extrapolates the elevation into the dry area. With the least squares approach, f can be found by requiring it to be the best fit with respect to the L_2-norm:

$$\|f(x,y) - a_0 - a_1 x - a_2 y\|_2 = min \tag{17}$$

2.4.2 Stabilization by Smagorinsky Viscosity Approach

Because the scheme is neutrally stable it demands horizontal viscosity in places where the gradients of solution are large. Although in general the initial perturbation is smooth, large gradients may form due to nonlinear steepening of the wave front or on reflections from jumps in topography or shallow bathymetry. Using uniform horizontal viscosity on non-uniform grids is a very poor choice leading to strong limitation on time stepping. Using the coefficient depending on the grid size proved not to be efficient either, since high viscosity is only needed when large velocity gradients are observed. For this reason the coefficient of horizontal viscosity is determined according to the Smagorinsky parameterization

$$K_h = cdxdy\sqrt{(\frac{\partial u}{\partial x})^2 + \frac{1}{2}(\frac{\partial v}{\partial x} + \frac{\partial u}{\partial y})^2 + (\frac{\partial v}{\partial y})^2} \tag{18}$$

The product of mesh cell sizes $dxdy$ is replaced by areas of elements in the code while the adjustable coefficient c is set to 0.04 - 0.4. These are about the smallest values that can be used while maintaining numerical stability.

2.4.3 Acceleration by Adaptive Computation

We take advantage of the explicit approach and accelerate the model at the beginning of tsunami wave propagation computations. If one has a local source like an earthquake there is no need to integrate model equations over the whole computational domain especially if the model covers the whole ocean. To avoid unnecessary computations we split the domain in 1 per 1 degree squares and collect the element numbers lying in every square. The integration is carried out only within rectangles with non-zero sea surface elevation. This simple trick applied to the Indian Ocean tsunami situation with a mesh of approx. 2 million nodes accelerated the computational time by a factor of 20 in the first model hour of the wave propagation simulation. Of course the acceleration becomes less and less effective, as the sea surface perturbation propagates into the whole computational domain.

3 Verification and Validation

3.1 Momentum Advection Test

To test the impact of the momentum advection discretization on wave propagation, a test case simulation has been performed. A wave is propagated in a narrow channel with solid boundaries and underwater sill (see Fig. 1a). The initial condition for elevation (10 m of amplitude) and horizontal velocity specify a wave propagating quasi one-dimensionally in the channel. The uniform grid contains 101101 nodes with the horizontal step Δ_{max}=2222 m. When propagating over a flat part of the bottom the wave's shape practically preserves for all types of advection as nonlinearity plays a very limited role there. Yet, on passing over the hill the amplitude of wave increases drastically and nonlinear effects become significant.

Figure 1b compares the maximum velocity as a function of time for two variants of the momentum advection term. As it is apparent from Fig. 1b, there is a very good agreement between the two cases, with $P_1^{NC} - P_1$ advection providing slightly higher amplitudes as expected. The simulation with consistent non-conforming momentum advection showed very similar behaviour too, but required higher values of viscosity for numerical stability. We conclude that all forms of momentum advection can be

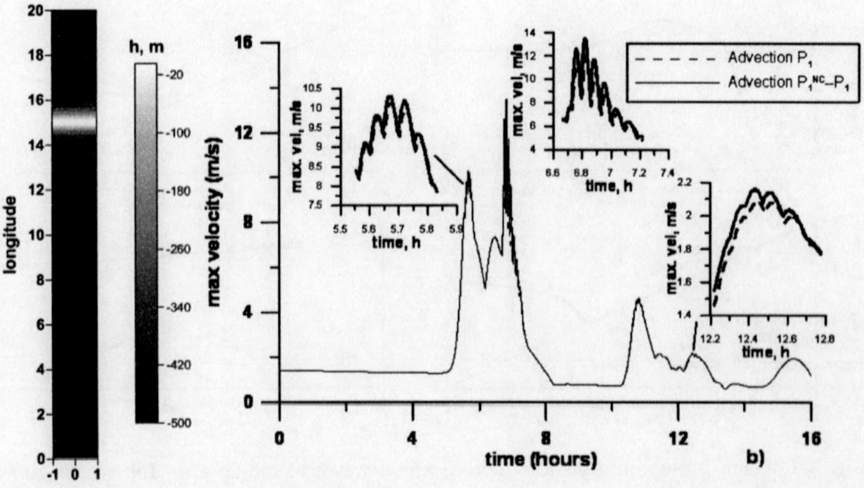

Fig. 1 a) Computational domain with bathymetry. b) Comparison maximal velocity for two representation of advection: correct P_1^{NC} (solid lines) and reduced P_1 (dashed lines)

safely used, but the two with re-projection require less care in tuning friction and viscosity, and less computational effort. They are a preferred choice. Noteworthy the advection scheme of [4] cannot pass this test case.

3.2 Wetting and Drying

The classic analytical solution for wave run-up on a sloping beach was first expressed in [13] and later revisited in [14]. We use this test case to examine the performance of our wetting and drying scheme.

Figure 2 compares the wave form simulated with the numerical model (with non-linear terms switched off) against the analytical solution for several moments of time. Apparently, the numerical solution is in the good agreement with the analytical result. Some difference is observed in the vicinity of the front and can be explained by several factors among which presence of horizontal diffusion (which is absent in the analytical solution) is the most significant one.

3.3 The Okushiri Test Case

The 1993 Okushiri tsunami caused many unexpected phenomena. One of them was the extreme run-up height of 32 m that was measured near the village of Monai in Okushiri Island. This benchmark problem is a simulation of a 1/400 scale laboratory experiment of the Monai run-up, performed in a large-scale tank at Central Research Institute for Electric Power Industry (CRIEPI) in Abiko, Japan.

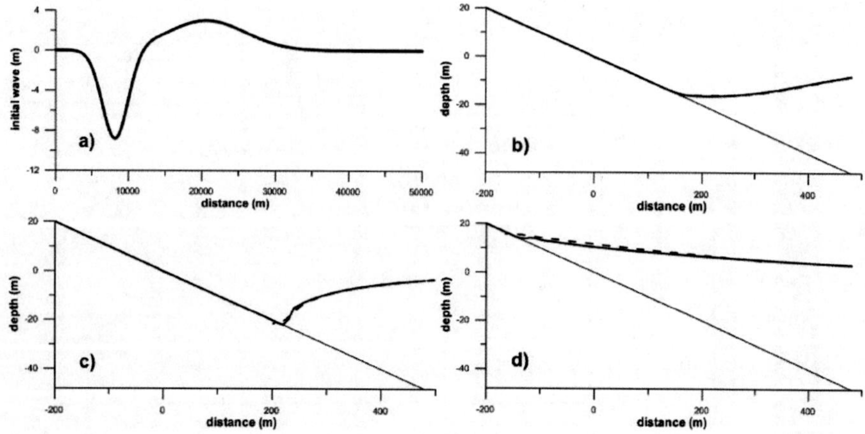

Fig. 2 Tsunami run-up on a plane beach. a) initial wave; b) run-up after 160 sec; c) run-up after 175 sec.; run-up after 220 sec. Dashed and solid lines correspond to analytical and numerical solutions, respectively

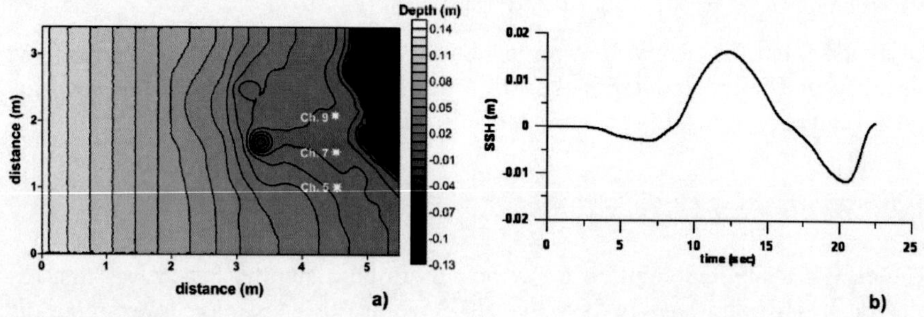

Fig. 3 a) Computational / experimental domain with bathymetry. Stars mark wave gage positions; b) Input wave

The bathymetry and coastal topography used in the laboratory experiment are shown in Fig. 3. Three stations marked by stars in the figure provided the elevation used for the comparison. We note that right, top and bottom boundaries (in the figure plane) are solid and the left boundary is open. The numerical mesh covers the area of the experiment. The initial tsunami wave is imposed at the open boundary and shown in Fig. 3b. Its duration is 22.5 s.

To illustrate the importance of the momentum advection in obtaining good agreement with observations two different simulations were performed: one without (Fig. 4) and the other with the momentum advection term (Fig. 5). It is noteworthy that the first maximum in Fig. 4 agrees well with respect to phase and is slightly overestimated with respect to amplitude. The second maximum leads the observational data for approximately 5 seconds, yet shows reasonable amplitude.

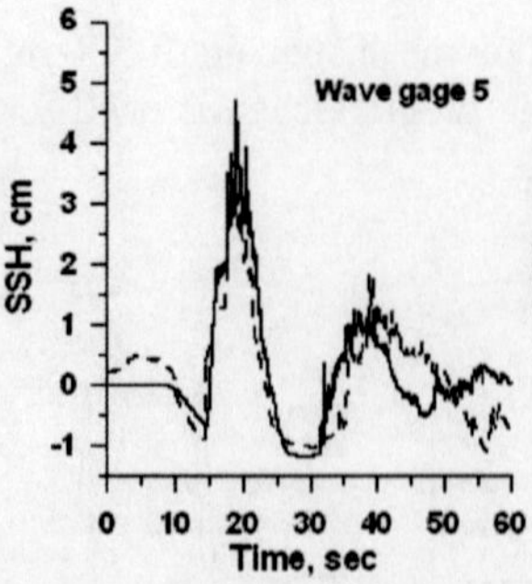

Fig. 4 Comparison experimental data (dashed lines) with numerical simulation (solid lines) for the Okushiri test case in runs without the momentum advection

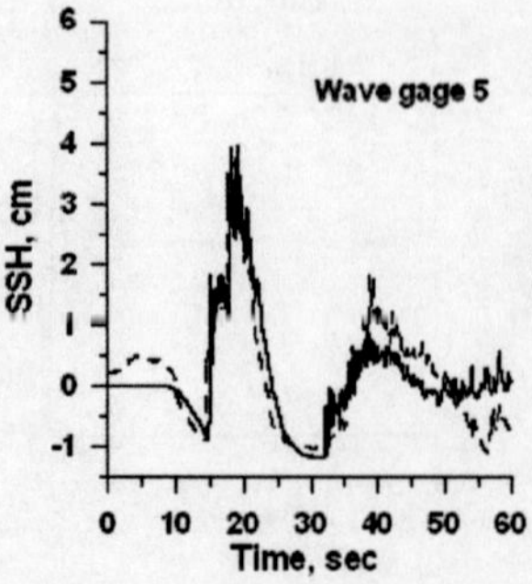

Fig. 5 Comparison experimental data (dashed lines) with numerical simulation (solid lines) for the Okushiri test case in runs with the momentum advection

Including the momentum advection noticeably improves the agreement. In this case, both the first and second maxima are reproducing the observations very well as regards their phase. The amplitude of the first maximum shows almost perfect agreement with observations. It is necessary to note that the laboratory data contain some spurious elevation during the early phase of the experiment (which does not agree with the initial wave shape).

This test case is rather sensitive to the magnitude of viscosity and bottom drag used in numerical simulations and indeed requires the Smagorinsky viscosity to get the agreement in amplitudes.

3.4 Tsunami Simulation

For the purpose of model verification in realistic situations over long time scales we use data from the Great Sumatra-Andaman Tsunami of December 26, 2004. A locally refined mesh of the entire Indian Ocean has been generated, which consists of 2166320 nodes (4304458 elements) and has a minimum mesh spacing of 500 m. The observed elevation at several tide gauges across the Indian Ocean is used for comparison and shown in Fig. 6.

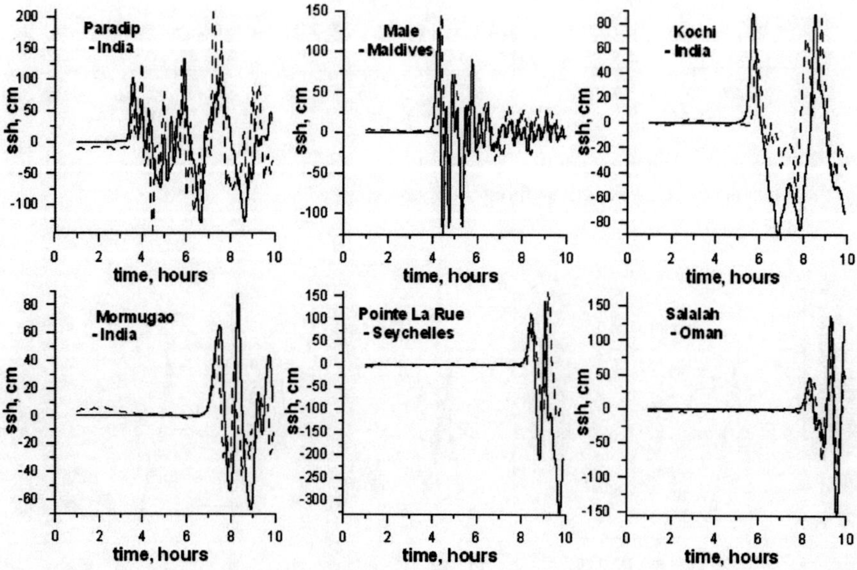

Fig. 6 Numerical experiments simulating the Indian Ocean Tsunami generated by the earthquake of December 26, 2004. Comparison of the computed elevation (solid lines) with the observed one (dashed lines) at different location

Given still only approximately known parameters of the tsunami source the coincidence between the model and observation is indeed good. Not only the arrival time of the first wave is reliably simulated, but the entire shape of the signal is reproduced reasonably well, and with correct amplitude. It should be further noted that correct local bathymetry in harbour areas (where most of the gauges are installed) is not available. Therefore, local phenomena like resonances and reflections cannot be reproduced by the model. This explains the weaker resemblance after the second wave arrival. Overall, these results prove the skill of the model as a tool to simulate tsunami wave propagation. Further results which prove the accuracy of inundation simulations can be found in [15].

4 Tide-Tsunami Interactions

For investigating the influence of tidal motion on tsunami wave propagation in the domain of the Indian ocean, three model runs were carried out: a) pure tidal motion,

b) pure tsunami wave propagation and c) tsunami wave propagation on the background tidal motion. As object of modelling we have chosen the Indonesian coast including south part of Jawa, Bali, Lombok, Sumbawa and Sumba islands.

For calculations, we use the ETOPO 30 sec. morphometry dataset and data on tidal oscillations of the level in the Indian ocean derived from the TPXO6.2 dataset of oceanic tides [16].

The calculation were performed on an unstructured mesh (Fig.7a) with 177132 nodes and 347098 elements with a time step of $\Delta t = 2.0$ sec. The horizontal mesh size varies in the range between Δ_{min}=160 m (in the inundation zone) and Δ_{max}=29 km (deep ocean). The adjustable coefficient c in the Smagorinsky term is taken to be 0.3. The full P_1 projection method is used for advection. Simulations are performed after the tidal motion has reached a stationary state.

The tsunami wave is initialized by a source model as used in the GITEWS scenario database (Fig. 7b). In principle the initial condition is irrelevant for our considerations here.

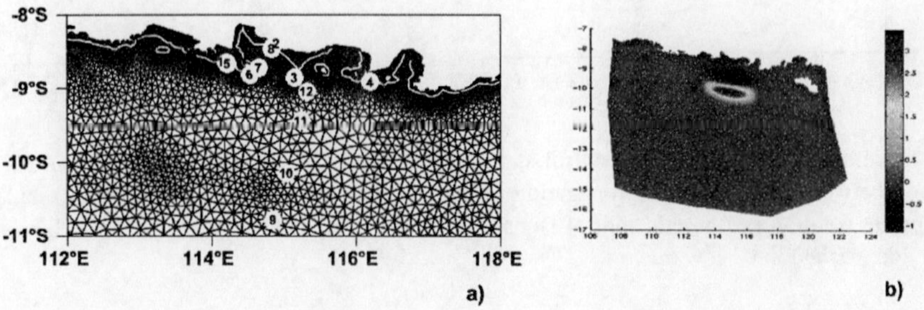

Fig. 7 a) Part of the unstructured mesh with control stations. White curve marks coastal line. b) Initial condition for tsunami wave

The series of experiments tries to shed light on the nonlinear interaction of tidal waves and tsunamis. A full solution (propagation of tsunami wave on the tidal background) and composite solution (an arithmetic sum of tsunami wave and tides computed separately) are compared. It turns out that the difference between these solutions is very significant (Fig. 8) reaching a maximum of 3.5 m with regard to amplitude in the coastal region (St. 3) where nonlinearity is particularly important. It is necessary to notice that the tsunami wave coming first only slightly changes the amplitude $\sim 5-8\%$ (with max. amplitude tsunami wave $\sim$16 m, St.3), but the second wave experiences more significant changes $\sim$ 25% (with max. amplitude of second tsunami wave $\sim$3.5 m).

In search of explanation as to why the impact of tide-tsunami interaction is so significant, we repeated the above cases once more but switching off the momentum advection term in the equation of motion (linear case). Figure 9 clearly shows that the evolution of potential and kinetic energy becomes significantly different from full nonlinear case. Hence we conclude that in the near-shore regime, nonlinear interaction of the wave-induced velocities contributes greatly to the complex

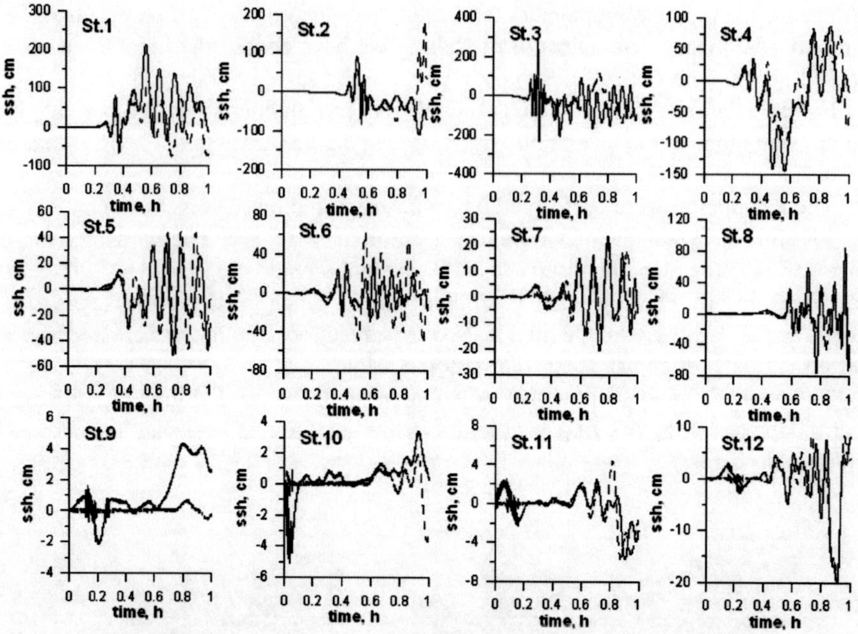

Fig. 8 Nonlinear interaction for control station as numbered in (Fig.7a). Difference in elevation between full solution (propagation of tsunami wave on the tidal background) and composite solution (arithmetic sum of tsunami and tides computed separately). Solid line - high tide; dashed line - low tide

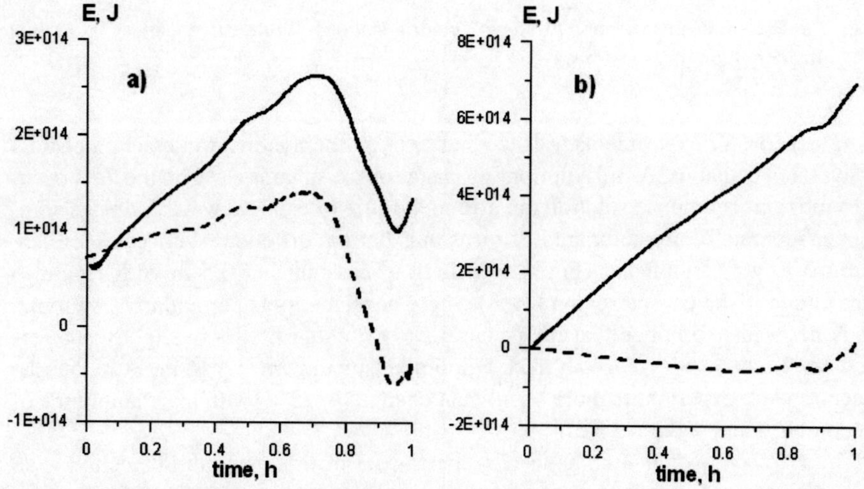

Fig. 9 Difference in potential (a) and kinetic energy (b) between full solution (propagation of tsunami wave on the tidal background) and composite solution (arithmetic sum of tsunami and tides computed separately. Solid line - with advection; dashed line - without advection in momentum equations

behaviour of tide-tsunami wave phenomena. The exact mechanisms still call for a more careful analysis and suggest a topic for future research.

5 Conclusion

The combination of non-conforming finite elements for velocity with conforming linear finite elements for elevation suggests a well-rounded choice for shallow-water modelling on unstructured triangular grids, with a particular focus on simulating tsunami wave propagation. Although our approach was initially inspired by the algorithm proposed in [4] the resulting model is essentially different from it in a number of key directions. First, it is equipped with wetting and drying algorithms and can simulate inundation caused by tsunami. Second, it suggests a choice of stably working discretizations of the momentum advection which all improve over the original method and differ amongst themselves in the degree of smoothing applied. Third, it uses the Smagorinsky horizontal viscosity which is crucial for keeping the dissipation on the level that does not affect the quality of the solution. Finally, the explicit time stepping made possible through the nodal quadrature of the time derivative term in the continuity equation ensures numerically efficient performance while providing a straightforward and easy to implement algorithm.

The accuracy of the model is tested against observational data (the Okushiri test case and the Indian Ocean Tsunami of December 26, 2004) and against an analytical solution given in [13]. We find good agreement of computed and reference solutions, showing the high level of realism necessary for predicting the propagation of tsunami waves in operational environments.

The model can be considered as an easy to use and reliable tool which not only serves the purposes of GITEWS but can be employed for other tasks which can be described in the framework of shallow water equations.

Simulations suggest strong nonlinear interaction between the tsunami and tidal waves. The major difference between tsunami simulations with and without tides occurs in the run-up region. The analysis of energy diagnostics from the simulation results show that the principal cause for the differences is nonlinear interaction between the tidal and tsunami currents. It leads to strong increase in the tsunami amplitude in coastal regions. All these results lead us to infer that the account of tidal dynamics may prove necessary for faithful modelling of tsunami waves.

Acknowledgements. *TsunAWI* is the result of joint developments and collaboration of different groups within AWI, namely, Computing center lead by Wolfgang Hiller and the ocean dynamic section supervised by Jens Schroter. We would like to thank Sven Harig and Claudia Wekerle for technical support, and Dmitry Sein and Dmitry Sidorenko for contributions to early versions of the model.

The authors gratefully acknowledge funding through the GITEWS Project, provided by BMBF under contract no. 03TSU01.

References

1. Kienle, J., Kowalik, Z., Murty, T.S.: Tsunamis generated by eruptions from Mount St. Augustine Volcano, Alaska. Science 236, 1442–1447 (1987)
2. Greenberg, D.A., Murty, T.S., Ruffman, A.: A numerical model for the Halifax Harbor tsunami due to the 1917 explosion. Marine Geodesy 16, 153–167 (1987)
3. Baptista, A.M., Priest, G.R., Murty, T.S.: Field survey of the 1992 Nicaragua Tsunami. Marine Geodesy 16, 1692–1703 (1993)
4. Hanert, E., Le Roux, D.Y., Legat, V., Delesnijder, E.: An efficient Eulerian finite element method for the shallow water equations. Ocean Model 10, 115–136 (2005)
5. Myers, E.P., Baptista, A.M.: Inversion for Tides in the Eastern North Pacific Ocean. Advances Water Resources 24(5), 505–519 (2001)
6. Weisz, R., Winter, C.: Tsunami, tides and run-up: a numerical study. In: Papadopoulos, G.A., Satake, K. (eds.) Proceedings of the International Tsunami Symposium, Chania, Greece, June 27-29, p. 322 (2005)
7. Kowalik, Z., Proshutinsky, T., Proshutinsky, A.: Tide-tsunami interactions. Science of Tsunami Hazards 24(5), 242–256 (2006)
8. Dao, M.H., Tkalich, P.: Tsunami propagation modelling - a sensitivity study. Natur. Hazards Earth System Sci. 7, 741–754 (2007)
9. Mofjeld, H.O., Gonzalez, F.I., Titov, V.V., Venturato, A.J., Newman, J.C.: Effects of Tides on Maximum Tsunami Wave Heights: Probability Distributions. J. Atmos. Oceanic Technol. 24(1), 117–123 (2007)
10. Oliger, J., Sundstrom, A.: Theoretical and practical aspects of some initial boundary value problems in fluid dynamics. SIAM J. Appl. Math. 35, 419–446 (1978)
11. Androsov, A.A., Klevanny, K.A., Salusti, E.S., Voltzinger, N.E.: Open boundary conditions for horizontal 2-D curvilinear-grid long-wave dynamics of a strait. Adv. Water Resour. 18, 267–276 (1995)
12. Lynett, P.J., Wu, T.-R., Liu, P.L.-F.: Modeling wave runup with depth-integrated equations. Coastal Eng. 46, 89–107 (2002)
13. Carrier, G.F., Greenspan, H.P.: Water waves of finite amplitude on a sloping beach. J. Fluid Mech. 4, 97–109 (1958)
14. Siden, G.L.D., Lynch, D.R.: Wave Equation Hydrodynamics on Deforming Elements. Int. J. Numer. Meth. Fluids 8, 1071–1093 (1988)
15. Harig, S., Chaeroni, C., Pranowo, W.S., Behrens, J.: Tsunami simulations on several scales: Comparison of approaches with unstructured meshes and nested grids. Ocean Dynamics 58, 429–440 (2008)
16. Egbert, G.D., Erofeeva, S.Y.: Efficient inverse modeling of barotropic ocean tides. J. Atmos Oceanic Technol. 19(2), 183–204 (2002)

A Geometry Independent Slope Limiter for the Discontinuous Galerkin Method

Vadym Aizinger

Abstract. A simple slope limiting procedure for the discontinuous Galerkin method based on a local constrained optimization approach is applied to the numerical solution of the three-dimensional shallow water equations. The method and its implementation are discussed, followed by numerical examples on several test problems.

1 Introduction

In this paper, a slope limiting scheme for the discontinuous Galerkin (DG) method proposed by Kuzmin in [6] is obtained as the solution to a local constrained optimization problem. The resulting slope limiter is computationally efficient and independent of the element shape, thus also suitable for hybrid meshes containing different element types. The performance of the slope limiter is evaluated using several test problems for the 3D shallow water equations.

The shallow water equations are derived from the incompressible Navier-Stokes equations defined on a domain with a moving free (top) surface. The shallow water assumption reduces the vertical momentum equation to the hydrostatic pressure relation

$$\frac{\partial p}{\partial z} = -\rho g,$$

where p is the hydrostatic pressure, ρ is density and g is gravitational acceleration. In many cases of practical importance, vertical effects are negligible, and one can integrate the horizontal momentum equations and the continuity equation over the depth, applying appropriate boundary conditions at the bottom and the free surface, to obtain the two-dimensional shallow water equations. If the eddy viscosity and/or

Vadym Aizinger
Goethe Center for Scientific Computing (G-CSC),
Goethe University Frankfurt am Main, Kettenhofweg 139,
60325 Frankfurt am Main, Germany
e-mail: vadym.aizinger@gcsc.uni-frankfurt.de

E. Krause et al. (Eds.): Computational Sci., & High Performance Computing IV, NNFM 115, pp. 207–217.
springerlink.com

density variations are important, for example, when salinity and temperature are spatially varying, then the full three-dimensional model should be employed. See [9, 8] for a discussion of shallow water models in both, two and three dimensions.

DG finite element methods are promising because of their ability to handle geometrically complex elements, use of shock-capturing numerical fluxes, adaptivity in polynomial order, and mass conservation properties; see [5] for a historical overview of DG methods. In [2], we investigated DG for the solution of the two-dimensional shallow water equations. Viscosity (second-order derivative) terms are handled in this method through the so-called local discontinuous Galerkin (LDG) framework [3], which involves rewriting the model as a first order system of equations. The present three-dimensional model was described in [1].

The rest of this paper is organized as follows. In the next section, the governing equations and boundary conditions are introduced. In Section 3, we outline the solution strategy, Sections 4 gives a description of the investigated slope limiter. In Section 5, some preliminary numerical results illustrating the performance of the limiter are presented.

2 Mathematical Formulation

The mathematical model of the three-dimensional shallow water flow with constant density includes hydrostatic momentum equations for the horizontal velocity components, a continuity equation as well as boundary and initial conditions. The problem domain has a free surface described by the kinematic boundary condition. The system can be written in conservative form as follows:

$$\frac{\partial \mathbf{u}_{xy}}{\partial t} + \nabla \cdot (\mathbf{u}_{xy} \times \mathbf{u} - D\nabla \mathbf{u}_{xy}) + g\nabla_{xy}\xi - f_c \mathbf{k} \times \mathbf{u}_{xy} = \mathbf{G}, \tag{1}$$

$$\nabla \cdot \mathbf{u} = 0, \tag{2}$$

where $\nabla_{xy} = \left(\frac{\partial}{\partial x}, \frac{\partial}{\partial y}\right)$, ξ is the value of the z coordinate at the free surface, $\mathbf{u} = (u, v, w)$ is the velocity vector, $\mathbf{u}_{xy} = (u, v)$ is the vector of horizontal velocity components, f_c is the Coriolis coefficient, $\mathbf{k} = (0, 0, 1)$ the unit vertical vector, g is acceleration due to gravity, $\mathbf{G} = (G_x, G_y)$ is the vector of body forces (e.g., atmospheric pressure, tidal forcing), and D is the eddy viscosity coefficient.

We specify the following boundary conditions:

- At the bottom: no normal flow

$$\mathbf{u}(z_b) \cdot \mathbf{n} = 0 \tag{3}$$

and no slip for horizontal velocity components

$$u(z_b) = v(z_b) = 0. \tag{4}$$

where z_b is the value of the z coordinate at the sea bed and $\mathbf{n} = (n_x, n_y, n_z)$ is the exterior unit normal to the face.

- At the free surface:

$$\frac{\partial \xi}{\partial t} + u(\xi)\frac{\partial \xi}{\partial x} + v(\xi)\frac{\partial \xi}{\partial y} - w(\xi) = 0, \tag{5}$$

$$\frac{\partial u}{\partial \mathbf{n}} = \frac{\partial v}{\partial \mathbf{n}} = 0. \tag{6}$$

The lateral boundary conditions are similar to those of the 2D system [2].

By integrating equation (2) over the total depth and, taking into account boundary conditions at the top and the bottom boundaries, we obtain the primitive continuity equation for the free surface elevation

$$\frac{\partial \xi}{\partial t} + \frac{\partial}{\partial x}\int_{z_b}^{\xi} u dz + \frac{\partial}{\partial y}\int_{z_b}^{\xi} v dz = 0. \tag{7}$$

Denoting $h = \xi - z_b$ we can rewrite equations (7) and (1) in a compact form

$$\frac{\partial h}{\partial t} + \nabla_{xy} \cdot \mathbf{C_H} = 0, \tag{8}$$

$$\frac{\partial \mathbf{u}_{xy}}{\partial t} + \nabla \cdot (\mathscr{C}_M - D\nabla \mathbf{u}_{xy}) = \mathbf{F}, \tag{9}$$

where

$$\mathbf{C_H} = \begin{pmatrix} \int_{z_b}^{\xi} u\, dz \\ \int_{z_b}^{\xi} v\, dz \end{pmatrix}, \quad \mathscr{C}_M = \begin{pmatrix} u^2 + gh & uv \\ uv & v^2 + gh \\ uw & vw \end{pmatrix}, \quad \mathbf{F} = \begin{pmatrix} G_x + g\frac{\partial z_b}{\partial x} + f_c v \\ G_y + g\frac{\partial z_b}{\partial y} - f_c u \end{pmatrix}.$$

This system consisting of the primitive continuity equation (8), two momentum conservation equations for the horizontal velocity components (9), and the continuity equation (2) forms the basis of our 3D DG discretization.

3 General Issues and Solution Strategy

The 3D shallow water model employs prismatic elements that have a triangular cross-section, strictly vertical lateral faces, and flat but not necessarily parallel top and bottom faces. Such elements can be easily constructed by extending elements of an unstructured triangular 2D mesh in the vertical direction and subdividing the water columns that correspond to each 2D element into several, possibly non-matching, layers. Every few time steps, the free surface geometry and the upper layer of the mesh are updated using computed values of the surface elevation. The frequency of mesh update can be chosen to suit the problem type.

One rather unusual aspect of this model is the treatment of the free surface. Since the free surface elevation is one of the system's primary variables – it is computed using (7) – it lies in a discontinuous polynomial space; thus, the approximate free surface geometry as specified by this variable would generally contain 'holes'. In this situation, consistency problems may arise due to the fact that the problem domain does not have a closed boundary any more. In order to avoid these issues we smooth the free surface mesh (e.g., by means of the least squares fit) and compute all

3D integrals on a grid with a globally continuous free surface. This procedure only affects the mesh geometry and does not change in any way the computed values of state variables or the discontinuous character of the numerical solution.

The time stepping is performed by an explicit Runge-Kutta method of the order corresponding to the order of the space discretization. Within each time sub-step, first, the mass and the momentum conservation equations are solved in a coupled fashion. Then, the vertical velocity component is computed by means of the discrete continuity equation to provide a divergence-free velocity field. The latter computation is done layer-by-layer, starting from the bottom and employing the solution from the layer immediately below or, in the case of the bottommost layer, the bottom boundary condition (3) as an initial value.

4 Space Discretization, LDG Method

The first step of the LDG formulation for the momentum conservation equations is the introduction of an auxiliary flux variable q to obtain a mixed form of (1)

$$\begin{aligned} \frac{\partial \mathbf{u}_{xy}}{\partial t} + \nabla \cdot (\mathscr{C}_M + q) &= \mathbf{F}, \\ q &= -D \nabla \mathbf{u}_{xy}. \end{aligned} \tag{10}$$

Let $\mathscr{T}_h$ be a partition of 3D domain Ω into prisms described above. Multiplying (10) by smooth test functions ϕ, ψ and integrating the equations by parts on each element $\Omega_e \in \mathscr{T}_h$ we arrive at a weak form of (10)

$$\begin{aligned} \int_{\Omega_e} \frac{\partial \mathbf{u}_{xy}}{\partial t} \phi \, d\mathbf{x} + \int_{\partial\Omega_e} (\mathscr{C}_M + q) \cdot \mathbf{n} \, \phi \, ds - \int_{\Omega_e} (\mathscr{C}_M + q) \cdot \nabla \phi \, d\mathbf{x} &= \int_{\Omega_e} \mathbf{F} \, \phi \, d\mathbf{x}, \\ \int_{\Omega_e} q \, \psi \, d\mathbf{x} = - \int_{\partial\Omega_e} D\mathbf{u}_{xy} \, \psi \cdot \mathbf{n} \, ds + \int_{\Omega_e} D\mathbf{u}_{xy} \, \nabla \cdot \psi \, d\mathbf{x}. & \end{aligned} \tag{11}$$

Since all lateral faces of Ω_e are strictly vertical, a weak form of (2) simplifies to

$$\int_{\partial\Omega_{e,top}} w \, n_z \, \sigma \, ds - \int_{\Omega_e} w \left(\frac{\partial \sigma}{\partial z} \right) d\mathbf{x} = \int_{\Omega_e} \mathbf{u}_{xy} \cdot \nabla_{xy} \sigma \, d\mathbf{x} - \int_{\partial\Omega_{e,bot}} w \, n_z \, \sigma \, ds - \int_{\partial\Omega_e} \mathbf{u}_{xy} \cdot \mathbf{n} \, \sigma \, ds, \tag{12}$$

where $\partial\Omega_{e,top}, \partial\Omega_{e,bot}$ denote the top and the bottom element faces.

The weak formulation of the primitive continuity equation is somewhat more involved. Let $\Omega_{e,xy}$ denote the orthogonal projection of Ω_e into the xy-plane. Multiplying (8) by smooth test function $\delta(x,y)$ and integrating by parts we can express the mass balance in the water column corresponding to $\Omega_{e,xy}$ as

$$\int_{\Omega_{e,xy}} \frac{\partial h}{\partial t} \, \delta \, dxdy + \int_{\partial\Omega_{e,xy}} \mathbf{C_H} \cdot \mathbf{n} \, \delta \, ds - \int_{\Omega_{e,xy}} \mathbf{C_H} \cdot \nabla_{xy} \, \delta \, dxdy = 0, \tag{13}$$

As $\mathbf{C_H} \stackrel{\text{def}}{=} \left(\int_{z_b}^{\xi} u dz, \int_{z_b}^{\xi} v dz\right)^T$ and assuming $h > 0$, the equation above can be transformed as follows:

$$\int_{\Omega_{e,xy}} \frac{\partial h}{\partial t} \delta dxdy + \sum_{\Omega_e \in col(\Omega_{e,xy})} \int_{\partial\Omega_{e,lat}} \frac{\mathbf{u}_{xy} h \cdot \mathbf{n}}{h} \delta ds - \sum_{\Omega_e \in col(\Omega_{e,xy})} \int_{\Omega_e} \frac{\mathbf{u}_{xy} h \cdot \nabla_{xy}}{h} \delta d\mathbf{x} = 0, \tag{14}$$

where $\partial\Omega_{e,lat}$ denotes the lateral boundary faces of prism Ω_e, and $col(\Omega_{e,xy})$ is the set of 3D elements in the water column corresponding to $\Omega_{e,xy}$.

This weak formulation is well defined for any $h(t,x,y) \in H^1(0,T;\mathscr{H})$, $\mathbf{u}_{xy}(t,x,y,z) \in H^1(0,T;\ V^{d-1})$, $w(t,x,y,z) \in V$, $q(t,x,y,z) \in Y^{d-1}$, $\delta(x,y) \in \mathscr{H}$, $\phi(x,y,z) \in V^{d-1}$, $\sigma(x,y,z) \in V$, and $\psi(x,y,z) \in Y^{d-1}$, $\forall t \in [0,T]$, where

$$\mathscr{H} \stackrel{\text{def}}{=} L^2(\Omega_{xy}) \bigcap \{h : h_{|\Omega_{e,xy}} \in H^1(\Omega_{e,xy}), \forall \Omega_e \in \mathscr{T}_h\} \tag{15}$$

$$V \stackrel{\text{def}}{=} L^2(\Omega) \bigcap \{u : u_{|\Omega_e} \in H^1(\Omega_e), \forall \Omega_e \in \mathscr{T}_h\}, \tag{16}$$

$$Y \stackrel{\text{def}}{=} L^2(\Omega)^d \bigcap \{\mathbf{q} : \mathbf{q}_{|\Omega_e} \in H^1(\Omega_e)^d, \forall \Omega_e \in \mathscr{T}_h\}, \tag{17}$$

and where Ω_{xy} denotes the orthogonal projection of Ω into the xy-plane.

We seek a semi-discrete solution $(H(t,\cdot), \mathbf{U}_{xy}(t,\cdot), W(t,\cdot), Q(t,\cdot)) \in \mathscr{H}_h \times U_h \times W_h \times Z_h$, for $\mathscr{H}_h \subset \mathscr{H}$, $U_h \subset V^{d-1}$, $W_h \subset V$, and $Z_h \subset Y^{d-1}$, some finite-dimensional subspaces by requiring that for any $t \in [0,T]$, all $\Omega_e \in \mathscr{T}_h$, and for all $(\delta, \phi, \sigma, \psi) \in \mathscr{H}_h \times U_h \times \in W_h \times Z_h$ the following holds:

$$\begin{aligned}
&\int_{\Omega_{e,xy}} \frac{\partial H}{\partial t} \delta dxdy + \sum_{\Omega_e \in col(\Omega_{e,xy})} \int_{\partial\Omega_{e,lat}} \frac{\hat{C}_H}{\xi_s - z_b} \delta ds - \sum_{\Omega_e \in col(\Omega_{e,xy})} \int_{\Omega_e} \frac{\mathbf{U}_{xy} H \cdot \nabla_{xy}}{\xi_s - z_b} \delta d\mathbf{x} = 0,\\
&\int_{\Omega_e} \frac{\partial \mathbf{U}_{xy}}{\partial t} \phi \, d\mathbf{x} + \int_{\partial\Omega_e} (\hat{\mathbf{C}}_{M,\mathbf{n}} + \hat{Q} \cdot \mathbf{n}) \, \phi \, ds - \int_{\Omega_e} (\mathscr{C}_M + Q) \cdot \nabla\phi \, d\mathbf{x} = \int_{\Omega_e} \mathbf{F} \, \phi \, d\mathbf{x},\\
&\int_{\Omega_e} Q \, \psi \, d\mathbf{x} = -\int_{\partial\Omega_e} D\hat{\mathbf{U}}_{xy} \, \psi \cdot \mathbf{n} \, ds + \int_{\Omega_e} D\mathbf{U}_{xy} \, \nabla \cdot \psi \, d\mathbf{x},\\
&\int_{\partial\Omega_{e,top}} W n_z \sigma ds - \int_{\Omega_e} W \left(\frac{\partial \sigma}{\partial z}\right) d\mathbf{x} = \int_{\Omega_e} \mathbf{U}_{xy} \cdot \nabla_{xy} \sigma d\mathbf{x} - \int_{\partial\Omega_{e,bot}} W^- \, n_z \sigma ds - \int_{\partial\Omega_e} \hat{C}_w \sigma ds,
\end{aligned} \tag{18}$$

where $\hat{C}_H, \hat{\mathbf{C}}_{M,\mathbf{n}}$ is a solution to the Riemann problem for the nonlinear boundary flux $\mathbf{U}_{xy} H \cdot \mathbf{n}, \mathscr{C}_M \cdot \mathbf{n}$; $\hat{\mathbf{U}}_{xy}, \hat{Q}$ are the averages of the corresponding variables on both sides of the discontinuity; ξ_s denotes the value of the z coordinate at the free surface of the smoothed mesh; W^- is the value of W taken from the element below (or, for the bottommost element, the boundary condition at the bottom). $\hat{C}_w$ should be set equal to $\frac{\hat{C}_H}{\xi_s - z_b}$ on lateral faces to match the corresponding flux in the primitive continuity equation but can be taken as a central or upwind flux on horizontal faces. This boundary flux formulation transforms integrals over 2D edges in the discrete primitive continuity equation into integrals over lateral faces of 3D elements, thus permitting simultaneous solution of the Riemann problems for the free surface elevation and the momentum. See [1] for a detailed discussion of this approach.

5 Slope Limiter

A slope limiter constitutes an essential part of many DG schemes, particularly of those used for applications that are sensitive to violations of the maximum principle (such as contaminant transport, wetting/drying problems). There exists an extensive literature on slope limiters in the DG context beginning with the classical minmod limiter of Cockburn and Shu [4]. This limiter and variations thereof control the linear components of a DG solution at midpoints of edges by comparing them with the element averages. Another class of limiters is based on the gradient reconstruction techniques developed for the Finite Volume method [7].

However, most of those limiters are still plagued by several serious problems. One of them is the computational cost and additional parallel communication overhead that can be quite significant if several elements are included in the limiting stencil. Another common problem is some degradation of convergence and introduction of spurious slopes into numerical solution. A serious issue in many applications, including the present one, is the fact that very few known limiters can be easily extended to general element shapes or non-matching grids.

A particularly interesting approach that can also be extended to super-linear solution components is proposed in [6]. The main idea of this slope limiter boils down to a simple method to prevent the values of the discontinuous solution at grid nodes from exceeding the limits imposed by the constant solution components on elements containing those nodes. This scheme can be easily generalized by choosing a different set of limiting points on element boundaries, such as edge/face midpoints, hanging nodes, quadrature points, etc.

A DG solution on element Ω_e can be represented as

$$f(\mathbf{x})_{|\Omega_e} = \sum_{k=0}^{n} f_{i,e}\phi_{i,e}(\mathbf{x}), \tag{19}$$

where $\phi_{i,e}, i=0,\dots,n$ are the local basis functions on Ω_e. A hierarchical polynomial basis (e.g., Legendre polynomials in 1D) can be subdivided into the constant $\{\phi_{0,e}\}$, linear $\{\phi_{1,e},\dots,\phi_{d,e}\}$, and super-linear $\{\phi_{d+1,e},\dots,\phi_{n,e}\}$ parts. For a grid node N_i, let $\overline{f}_{i,min}$ and $\overline{f}_{i,max}$ denote the minimum and the maximum values of the constant solution component $f_{0,e}\phi_{0,e}(N_i)$ taken over all elements containing this node. Then we can formulate a local constrained optimization problem on each element Ω_e

$$\begin{gathered}\min_{\substack{(g_{1,e},\dots,g_{d,e})\in \mathbb{R}^d \\ \overline{f}_{0,min} \le f_{0,e}\phi_{0,e} + \sum_{k=1}^{d} g_{k,e}\phi_{k,e}(N_0) \le \overline{f}_{0,max} \\ \vdots \\ \overline{f}_{l,min} \le f_{0,e}\phi_{0,e} + \sum_{k=1}^{d} g_{k,e}\phi_{k,e}(N_l) \le \overline{f}_{l,max}}} \sum_{k=1}^{d} (g_{k,e} - f_{k,e})^2,\end{gathered} \tag{20}$$

where $N_i, i=1,\dots,l$ are the nodes of element Ω_e.

The linear constraints in (20) describe a convex polygonal (polyhedral) set of admissible solutions in a d-dimensional space. This set contains the origin of the coordinate system because the constant solution component on the element itself clearly satisfies all constraints. In Fig. 1, point (a_0, b_0) represents an unlimited DG solution on a triangular element. Classical techniques for obtaining a solution to (20) (point (a_2, b_2) in Fig. 1) involve several iterations of a quadratic optimization procedure in $\mathbb{R}^d$ at a significant computational cost. In addition, the linear components of the limited solution may be quite different from those of the unlimited.

A faster way to find an admissible (however not necessarily the globally optimal) point can be constructed by exploiting the fact that the segment between the linear coefficients of the unlimited solution and the origin of the coefficients space has to contain at least one point satisfying all constraints (the origin itself). This results in a one-dimensional constrained optimization problem (21)

$$\min_{\lambda \in [0,1]} \sum_{k=1}^{d} (\lambda f_{k,e} - f_{k,e})^2, \tag{21}$$

$$\overline{f}_{0,min} \leq f_{0,e}\phi_{0,e} + \textstyle\sum_{k=1}^{d} \lambda f_{k,e}\phi_{k,e}(N_0) \leq \overline{f}_{0,max}$$

$$\vdots$$

$$\overline{f}_{l,min} \leq f_{0,e}\phi_{0,e} + \textstyle\sum_{k=1}^{d} \lambda f_{k,e}\phi_{k,e}(N_l) \leq \overline{f}_{l,max}$$

solution of which ((a_1, b_1) in Fig. 1) is much cheaper to compute and does not produce slopes that are dramatically different from those of the unlimited solution.

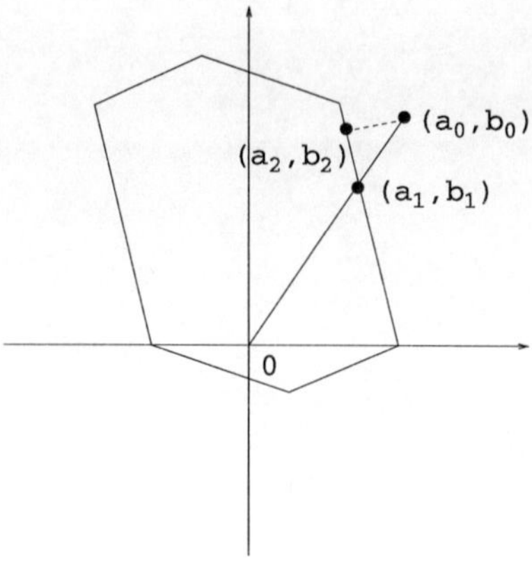

Fig. 1 Illustration of slope limiting for a triangular element

6 Numerical Results

6.1 Supercritical Flow in Constricted Channel

Supercritical channel flows subject to a change in cross-section can lead to the formation of shock and rarefaction waves. Here we take one particular configuration given in Zienkiewicz and Ortiz [10]. The boundary wall is constricted on both sides with constriction angle of 5^0 resulting in a cross-wave pattern. Flow is induced through the left boundary with no flow through the side and bottom boundaries. The inlet Froude number is defined by

$$F_1 = \frac{U_1}{\sqrt{gH_1}}, \tag{22}$$

where U_1 and H_1 are the mean velocity and the depth of the fluid at the inlet, respectively. When $F_1 > 1$, the flow is said to be supercritical. Here we have chosen $F_1 = 2.5$. The domain and the initial finite element mesh are shown in Figure 2. The mesh consists of 3155 triangular elements with no particular orientation. The bathymetry h_b is assumed constant.

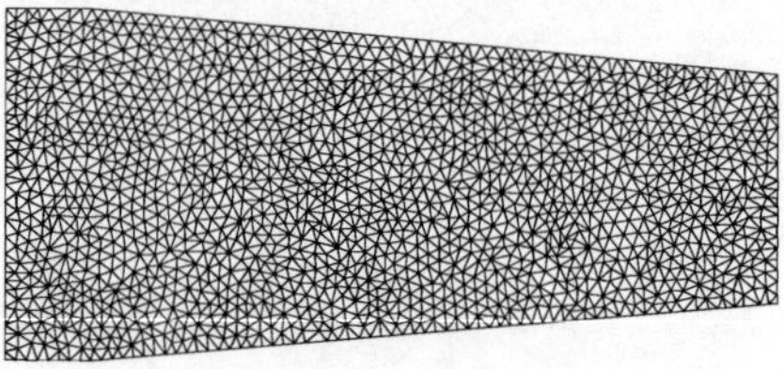

Fig. 2 Grid for the supercritical flow problem

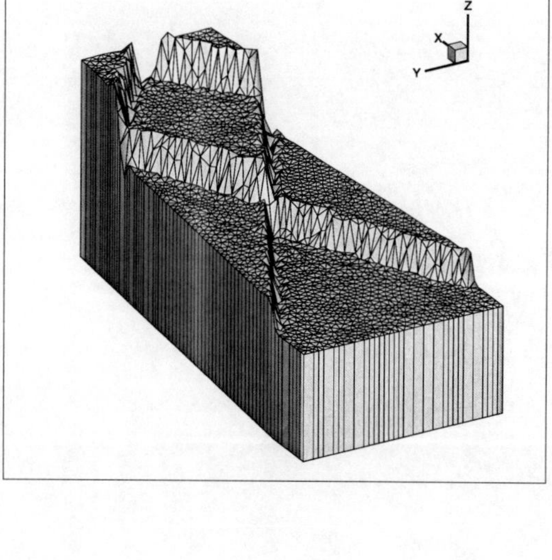

Fig. 3 Supercritical flow problem: no slope limiting

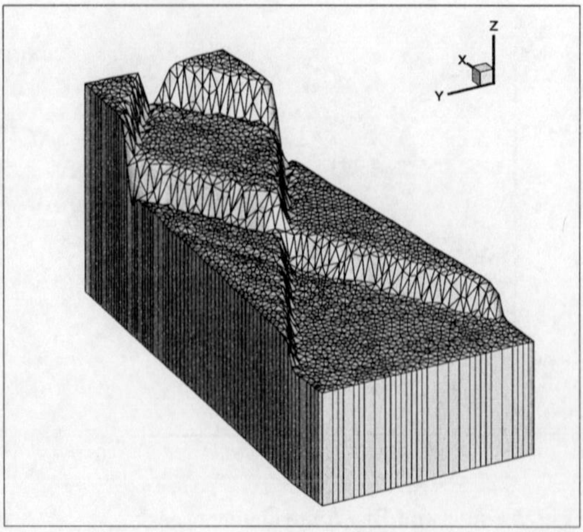

Fig. 4 Supercritical flow problem: slope limited elevation

The plots in Fig. 3 and Fig. 4 illustrate the fact that the slope limiting procedure successfully removes overshoots and undershoots in the free surface elevation without affecting the areas of constant elevation.

6.2 *Desalinization Plant in Galveston Bay*

Our next test case involves a scenario in which a desalinization plant releases saline solution into Galveston Bay in the vicinity of the Houston Ship Channel. A finite element mesh of the bay consisting of 3397 elements is given in Figure 5; Figure 6 illustrates the bathymetry of the bay with all lengths given in meters. The physical domain is quite complicated: A total of seventeen islands are included, and the bathymetry varies sharply from the narrow deep channel to much more shallow regions within the bay.

The following tidal forcing with time(t) in hours was imposed at the open sea boundary:

$$\hat{\xi}(t) = 0.075\cos\left(\frac{2\pi t}{25.82}+3.40\right) + 0.095\cos\left(\frac{2\pi t}{23.94}+3.60\right) + 0.1\cos\left(\frac{2\pi t}{12.66}+5.93\right) + 0.395\cos\left(\frac{2\pi t}{12.42}+0.00\right) + 0.06\cos\left(\frac{2\pi t}{12.00}+0.75\right)\ (meters). \quad (23)$$

At the land and island boundaries, we impose zero normal flux conditions. The Coriolis parameter was set to $7.07\times10^{-5}s^{-1}$. The vertical eddy viscosity was modelled using the k-ε model; the bottom friction coefficient in the quadratic law was taken equal to 0.05, and the bottom roughness was set at 0.01 m. The simulations were run for a total of twelve days with a constant time step fixed at 4 seconds.

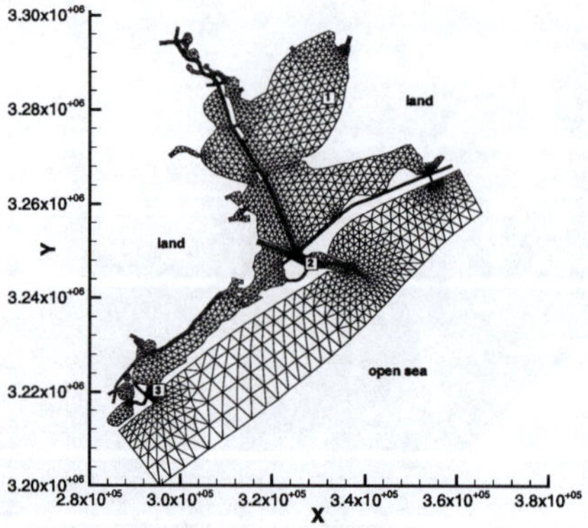

Fig. 5 Galveston Bay finite element mesh

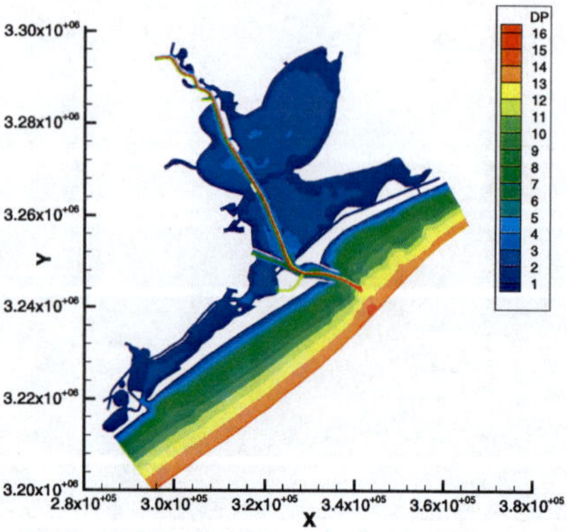

Fig. 6 Galveston Bay bathymetry

Initially, the bay water was set at rest and the salt concentration was constant at 35 ppt. The plant output with a steady rate of 5 m^3/s of 70 ppt salt solution was started at the beginning of the simulation. We compare salinity profiles at the end of the simulation generated using piecewise linear DG approximation without (Figure 7) and with (Figure 8) slope limiter. In these plots, we have zoomed in on the region near the left lower corner of the computational domain.

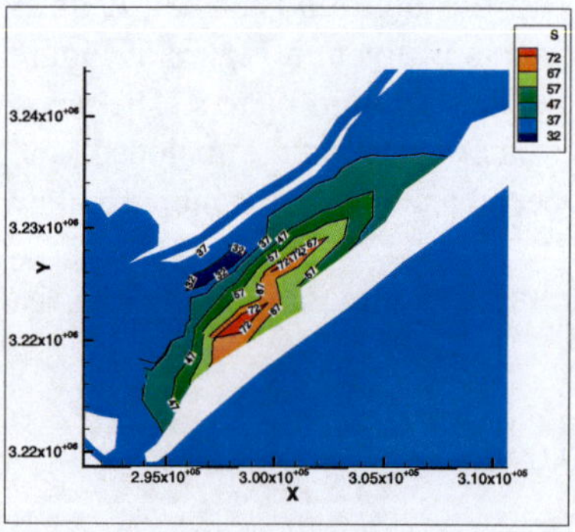

Fig. 7 Salinity at the end of day 12 (unlimited)

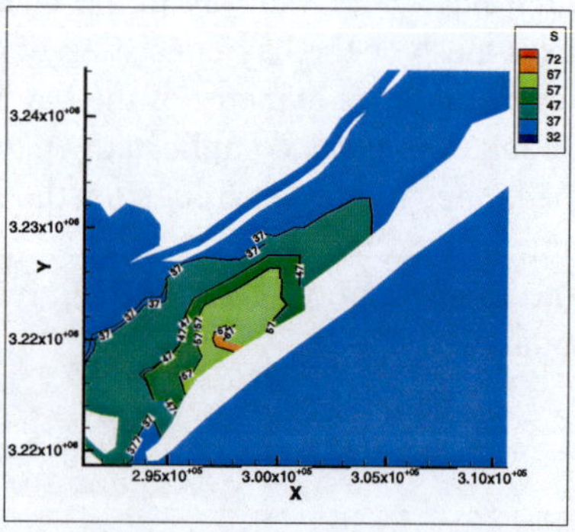

Fig. 8 Salinity at the end of day 12 (limited)

Obviously, the unlimited solution contains significant areas where the salt concentration is significantly over- and undershoots the natural upper and lower bounds of the salt concentration. The limiting postprocessing completely eliminates these problems.

7 Conclusions

The investigated slope limiter shows promising results for problems involving shocks and discontinuities, but the scheme might need some refinements to improve handling of smooth solutions. A combination of the current scheme with a shock detector could be one of the simplest ways to achieve just that.

References

1. Aizinger, V.: A Discontinuous Galerkin Method for Two- and Three-Dimensional Shallow-Water Equations. Ph.D. Dissertation, University of Texas at Austin (2004)
2. Aizinger, V., Dawson, C.: A discontinuous Galerkin method for two-dimensional flow and transport in shallow water. Adv. Water Resources 25, 67–84 (2002)
3. Cockburn, B., Shu, C.-W.: The local discontinuous Galerkin finite element method for convection-diffusion systems. SIAM J. Numer. Anal. 35, 2440–2463 (1998)
4. Cockburn, B., Shu, C.-W.: TVB Runge-Kutta local projection discontinuous Galerkin finite element method for scalar conservation laws II: General framework. Math. Comp. 52, 411–435 (1989)
5. Cockburn, B., Karniadakis, G., Shu, C.-W.: The development of discontinuous Galerkin methods. In: Cockburn, B., Karniadakis, G., Shu, C.-W. (eds.) Discontinuous Galerkin Methods: Theory, Computation and Applications. LNCSE, Part I: Overview, vol. 11, pp. 3–50. Springer, Heidelberg (2000)
6. Kuzmin, D.: A vertex-based hierarchical slope limiter for p-adaptive discontinuous Galerkin methods. J. Comput. Appl. Math. (to appear)
7. Tu, S.Z., Aliabadi, S.: A Slope limiting procedure in discontinuous Galerkin finite element method for gas dynamics applications. Int. J. Numer. Anal. Model. 2, 163–178 (2005)
8. Vreugdenhil, C.B.: Numerical Methods for Shallow-Water Flow. Kluwer Academic Publishers, Dordrecht (1994)
9. Weiyan, T.: Shallow Water Hydrodynamics. Elsevier Oceanography Series, vol. 55. Elsevier, Amsterdam (1992)
10. Zienkiewicz, O.C., Ortiz, P.: A split-characteristic based finite element model for the shallow water equations. Int. J. Num. Meth. Fluids 20, 1061–1080 (1995)

Two-Fluid Barotropic Models for Powder-Snow Avalanche Flows

Yannick Meyapin, Denys Dutykh, and Marguerite Gisclon

Abstract. In the present study we discuss several modelling issues of powder-snow avalanche flows. We take a two-fluid modelling paradigm. For the sake of simplicity, we will restrict our attention to barotropic equations. We begin the exposition by a compressible model with two velocities for each fluid. However, this model may become non-hyperbolic and thus, represents serious challenges for numerical methods. To overcome these issues, we derive a single velocity model as a result of a relaxation process. This model can be easily shown to be hyperbolic for any reasonable equation of state. Finally, an incompressible limit of this model is derived.

1 Introduction

Snow avalanches represent a serious problem for society in mountain regions. The avalanche winter of 1999 attracted a lot of attention to this hazardous natural phenomenon [1, 10]. Further development of mountain regions requires an adequate level of avalanche safety. Therefore, avalanche protective measures (deflecting and catching dams) become increasingly important [9]. During the same winter, several avalanches overran avalanche dams, underlining the need for further research in this field. Proper design of protecting structures necessitates profound understanding of the snow avalanches flow and of the interaction process with dams and other obstacles [6, 13].

Natural snow avalanches are believed to consist of three different layers: a dense core, a fluidised layer and a suspension cloud. Sometimes the surrounding powder cloud is absent and we speak about an avalanche in the flowing regime. Obviously,

Yannick Meyapin · Denys Dutykh · Marguerite Gisclon
LAMA, UMR 5127 CNRS, Université de Savoie,
73376 Le Bourget-du-Lac Cedex, France
e-mail: Yannick.Meyapin@etu.univ-savoie.fr,
Denys.Dutykh@univ-savoie.fr,
Marguerite.Gisclon@univ-savoie.fr

E. Krause et al. (Eds.): Computational Sci., & High Performance Computing IV, NNFM 115, pp. 219–232.
springerlink.com

transition boundaries between these layers are not sharp and this classification is rather conventional.

The dense core consists of snow particles in persistent frictional contact [8]. The density is of the order of 300 kg/m^3 and the depth of this layer does not exceed 3 m. The fluidised regime is characterized by particle's mean-free-paths up to several particle's diameters. This dynamics at microscopic level explains more fluid-like behaviour at large scales. The density of this layer is in the range of 50 - 100 kg/m^3 and the height is about 3 - 5 m. To model successfully this kind of flows it is crucial to know the complex fluid rheology. Finally, these two interior layers can be covered by the powder cloud which is a turbulent suspension of snow particles in the air. The density ranges from 4 to 20 kg/m^3 and an avalanche in aerosol regime can reach the height of 100 m or more [14]. This flow is driven essentially by turbulent advection and particles collisions are unimportant.

In the present study we are concerned with some questions of powder-snow avalanche modelling. Since the interface cannot be defined for this type of flows, we choose the modelling paradigm of two-phase flows. In this approach the governing equations of each phase are spatially averaged to come up with the description of the fluid mixture [7, 15].

It is known [14] that the front of such an avalanche can develop the speed[1] $u_f \approx$ 100 m/s. For comparison, the speed of sound c_0 in the air is about 300 m/s. It means that the local Mach number Ma can reach the value of

$$\mathrm{Ma} := \frac{u_f}{c_0} \approx 0.33.$$

Hence, compressible effects may become important. That is why, we begin our exposition with a compressible model. Then, we gradually simplify it to come up with an incompressible one at the end of the present article. The goal is achieved by taking the limit as the Mach number tends to zero.

The present article is organized as follows. In Section 2 we present a barotropic compressible two-phase model with two velocities. Then, this model is simplified in Section 3 using a velocity relaxation process. The incompressible limit of resulting system is derived in Section 4. Finally, several conclusions and perspectives are drawn out in Section 5.

2 Two-Phase Flow Modelling

Let us consider a domain $\Omega \subseteq \mathbb{R}^3$ where a simultaneous flow of two barotropic fluids occurs. All quantities related to the heavy and light fluids will be denoted by $+$ and $-$ correspondingly. In view of application to snow avalanches, one can consider the heavy fluid of being constituted of snow particles and the light fluid is the air. When the mixing process is extremely complicated and it is impossible to

[1] When we estimate the Mach number magnitude, the particle characteristic velocity should be taken. However, this information is not easily accessible and we took the maximum front velocity. It can lead to some overestimation of the Mach number.

follow the interface between two fluids, the classical modelling procedure consists in applying a volume average operator [7, 15]. Thereby, we make appear two additional variables $\alpha^{\pm}(\mathbf{x},t)$, $\mathbf{x} \in \Omega$ which are called the volume fractions and defined as:

$$\alpha^{\pm}(\mathbf{x},t) := \lim_{\substack{|d\Omega|\to 0 \\ \mathbf{x}\in d\Omega}} \frac{|d\Omega^{\pm}|}{|d\Omega|},$$

the heavy fluid occupies volume $d\Omega^{+} \subseteq d\Omega$ and the light one the volume $d\Omega^{-} \subseteq d\Omega$ (see Figure 1) such that

$$|d\Omega| \equiv |d\Omega^{+}| + |d\Omega^{-}|. \tag{1}$$

From the relation (1) it is obvious that $\alpha^{+}(\mathbf{x},t) + \alpha^{-}(\mathbf{x},t) \equiv 1, \forall \mathbf{x} \in \Omega$.

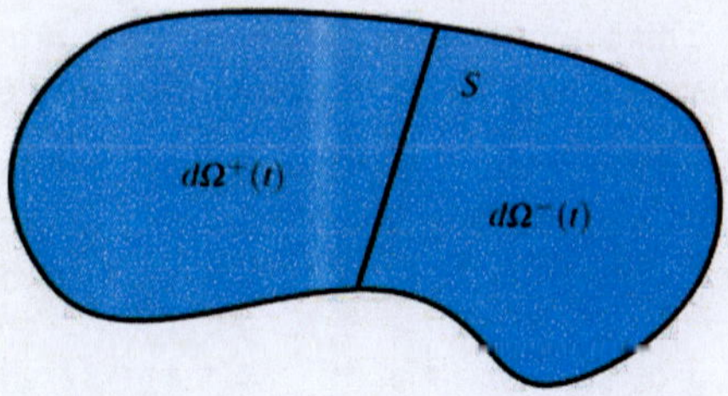

Fig. 1 An elementary fluid volume $d\Omega$ occupied by two phases

After performing the averaging process, one obtains two equations of mass and momentum conservation:

$$\partial_t(\alpha^{\pm}\rho^{\pm}) + \nabla\cdot(\alpha^{\pm}\rho^{\pm}\mathbf{u}^{\pm}) = 0, \tag{2}$$

$$\partial_t(\alpha^{\pm}\rho^{\pm}\mathbf{u}^{\pm}) + \nabla\cdot(\alpha^{\pm}\rho^{\pm}\mathbf{u}^{\pm}\otimes\mathbf{u}^{\pm}) + \alpha^{\pm}\nabla p = \nabla\cdot(\alpha^{\pm}\tau^{\pm}) + \alpha^{\pm}\rho^{\pm}\mathbf{g}, \tag{3}$$

where $\rho^{\pm}(\mathbf{x},t), \mathbf{u}^{\pm}(\mathbf{x},t), \tau^{\pm}(\mathbf{x},t)$ are densities, velocities and viscous stress tensors of each fluid respectively. Traditionally, the vector $\mathbf{g}$ denotes the gravity acceleration. We assume that both fluids share the same pressure[2] $p = p^{\pm}(\rho^{\pm})$ and equations of state of each phase fulfil minimal thermodynamical requirements:

$$p^{\pm}(\rho^{\pm}) > 0, \quad \frac{\partial p^{\pm}(\rho^{\pm})}{\partial \rho^{\pm}} > 0, \quad \text{for} \quad \rho^{\pm} > 0. \tag{4}$$

In order to obtain a well-posed problem, governing equations (2), (3) should be completed by appropriate initial and boundary conditions.

If we assume both fluids to be Newtonian, the viscous stress tensor $\tau^{\pm}$ takes the following classical form:

$$\tau^{\pm} = \lambda^{\pm}\mathrm{tr}\mathbb{D}(\mathbf{u}^{\pm})\mathbf{I} + 2\mu^{\pm}\mathbb{D}(\mathbf{u}^{\pm}), \quad \mathrm{tr}\mathbb{D}(\mathbf{u}^{\pm}) = \nabla\cdot\mathbf{u}^{\pm}, \tag{5}$$

[2] In general, this kind of assumptions is reasonable, since relaxation processes will tend to equilibrate the system when time evolves.

where $\mathbf{I} := (\delta_{ij})_{1\leq i,j\leq 3}$ is the identity tensor, $\mathbb{D}(\mathbf{u}) := \frac{1}{2}\Big(\nabla\mathbf{u}^{\pm} + {}^{t}(\nabla\mathbf{u}^{\pm})\Big)$ is the deformation rate and $\lambda^{\pm}$, $\mu^{\pm}$ are viscosity coefficients. For ideal gases, for example, these coefficients are related by Stokes relation $\lambda^{\pm} + \frac{2}{3}\mu^{\pm} = 0$. In application to powder-snow avalanches, viscosity coefficients $\lambda^{\pm}$, $\mu^{\pm}$ should be understood in the sense of eddy viscosity.

Remark 2.1. From physical point of view, presented here model (2), (3) is far from being complete. For example, one could supplement it by capillarity effects in the Korteweg form. Also we omitted all the terms which model mass, momentum and energy exchange between two phases. Generally, their form is strongly dependent on the physical situation under consideration.

Remark 2.2. Since we do not consider the total energy conservation equation, the fluids are implicitly assumed to be barotropic. In the absence of viscous stresses $\tau^{\pm}$, the flow is isentropic. This simplification can be adopted provided that important energy transfers do not occur. Non-isentropic flows are considered in [11].

Remark 2.3. While considering two-phase flows, it is useful to introduce several additional quantities which play an important role in the description of such flows. The mixture density ρ and mass fractions $m^{\pm}$ are naturally defined as:

$$\rho(\mathbf{x},t) := \alpha^{+}\rho^{+} + \alpha^{-}\rho^{-} > 0, \quad \forall(\mathbf{x},t) \in \Omega \times [0,T],$$

$$m^{\pm} := \frac{\alpha^{\pm}\rho^{\pm}}{\rho}, \quad m^{+} + m^{-} = 1.$$

The total density ρ is assumed to be strictly positive everywhere in the domain Ω. Hence, the void creation is forbidden in our modelling.

Important quantities ρ, $m^{\pm}$ will appear several times below.

In principle, one could use equations (2), (3) to model various two-phase flows. However, this system remains quite expensive for large scale simulations required by real-life applications. The major difficulty comes from the advection operator associated to model (2), (3) which can be non-hyperbolic [3, 15]. In the next section we will derive a simplified two-fluid model which is proposed as a candidate for powder-snow avalanche compressible simulations.

3 Velocity Relaxation

We would like to reduce the number of variables in the system (2), (3). The main idea is to introduce the common velocity field for both phases. For this purpose, we will introduce a relaxation term to the momentum conservation equation (3):

$$\partial_t(\alpha^{\pm}\rho^{\pm}\mathbf{u}^{\pm}) + \nabla\cdot(\alpha^{\pm}\rho^{\pm}\mathbf{u}^{\pm}\otimes\mathbf{u}^{\pm}) + \alpha^{\pm}\nabla p = \nabla\cdot(\alpha^{\pm}\tau^{\pm}) + \alpha^{\pm}\rho^{\pm}\mathbf{g} \pm \frac{\kappa}{\varepsilon}(\mathbf{u}^{+} - \mathbf{u}^{-}), \tag{6}$$

where $\kappa = \mathrm{O}(1)$ is a constant and ε is a small parameter which controls the magnitude of the relaxation term. Physically this additional term represents the friction between two phases. In the following, we are going to take the singular limit as the relaxation parameter $\varepsilon \to 0$. This is achieved with Chapman-Enskog type expansion. In this way, we constrain velocities $\mathbf{u}^{\pm}(\mathbf{x},t)$ to tend to the common value $\mathbf{u}(\mathbf{x},t)$. This technique has been already successfully applied to the Baer-Nunziato model [2] in [12].

The first step consists in rewriting the governing equations (2), (6) in the quasilinear form. To shorten notations, we will also use the material time derivative which is classically defined for any smooth scalar function $\varphi(\mathbf{x},t)$ as

$$\frac{d^{\pm}\varphi}{dt} := \frac{\partial \varphi}{\partial t} + \mathbf{u}^{\pm}\cdot\nabla\varphi.$$

Lemma 3.1. *Smooth solutions to equations (2), (6) satisfy the following system:*

$$\alpha^{\pm}\frac{d^{\pm}p}{dt} + \rho^{\pm}(c_s^{\pm})^2\frac{d^{\pm}\alpha^{\pm}}{dt} + \alpha^{\pm}\rho^{\pm}(c_s^{\pm})^2\nabla\cdot\mathbf{u}^{\pm} = 0, \tag{7}$$

$$\alpha^{\pm}\rho^{\pm}\frac{d^{\pm}\mathbf{u}^{\pm}}{dt} + \alpha^{\pm}\nabla p = \nabla\cdot(\alpha^{\pm}\tau^{\pm}) + \alpha^{\pm}\rho^{\pm}\mathbf{g} \pm \frac{\kappa}{\varepsilon}(\mathbf{u}^{+} - \mathbf{u}^{-}), \tag{8}$$

where $(c_s^{\pm})^2 := \left.\frac{\partial p^{\pm}}{\partial \rho^{\pm}}\right|_{s^{\pm}}$ *represents the sound speed in each phase* $\pm$.

Proof. This result follows from direct calculations. First of all, we remark that the mass conservation equation (2) can be rewritten using the material derivative as follows:

$$\frac{d^{\pm}(\alpha^{\pm}\rho^{\pm})}{dt} + \alpha^{\pm}\rho^{\pm}\nabla\cdot\mathbf{u}^{\pm} = 0. \tag{9}$$

Using equations of state $p = p^{\pm}(\rho^{\pm})$, we can express the density material derivative in terms of the pressure and the sound speed:

$$\frac{d^{\pm}\rho^{\pm}}{dt} = \frac{1}{(c_s^{\pm})^2}\frac{d^{\pm}p}{dt}.$$

Now, it is straightforward to derive equation (7) from (9).

Finally, if we multiply equation (9) by $\mathbf{u}^{\pm}$ and subtract it from the momentum conservation equation (6), we will get desired result (8).

Equations (7), (8) can be also recast in the matrix form which is particularly useful for further developments:

$$\mathbf{A}(V_\varepsilon)\frac{\partial V_\varepsilon}{\partial t} + \mathbf{B}(V_\varepsilon)\nabla V_\varepsilon = \nabla\cdot\mathbf{T}(V_\varepsilon) + \mathbf{S}(V_\varepsilon) + \frac{\mathbf{R}(V_\varepsilon)}{\varepsilon}, \tag{10}$$

where we introduced several notations. The vector V_ε represents four unknown physical variables $V_\varepsilon := {}^t(p, \alpha^{+}, \mathbf{u}^{+}, \mathbf{u}^{-})$ and $\frac{\partial V_\varepsilon}{\partial t} := {}^t(\partial_t p, \partial_t \alpha^{+}, \partial_t \mathbf{u}^{+}, \partial_t \mathbf{u}^{-})$ and $\nabla V_\varepsilon := {}^t(\nabla p, \nabla\alpha^{+}, (\cdot\nabla)\mathbf{u}^{+}, (\cdot\nabla)\mathbf{u}^{-})$. Matrices $\mathbf{A}(V_\varepsilon)$ and $\mathbf{B}(V_\varepsilon)$ are defined as

$$\mathbf{A}(V_\varepsilon) := \begin{pmatrix} \alpha^+ & \rho^+(c_s^+)^2 & 0 & 0 \\ \alpha^- & -\rho^-(c_s^-)^2 & 0 & 0 \\ 0 & 0 & \alpha^+\rho^+\mathbf{I} & 0 \\ 0 & 0 & 0 & \alpha^-\rho^-\mathbf{I} \end{pmatrix},$$

$$\mathbf{B}(V_\varepsilon) := \begin{pmatrix} \alpha^+\mathbf{u}^+ & \rho^+(c_s^+)^2\mathbf{u}^+ & \alpha^+\rho^+(c_s^+)^2\mathbf{I} & 0 \\ \alpha^-\mathbf{u}^- & -\rho^-(c_s^-)^2\mathbf{u}^- & 0 & \alpha^-\rho^-(c_s^-)^2\mathbf{I} \\ \alpha^+\mathbf{I} & 0 & \alpha^+\rho^+\mathbf{u}^+ & 0 \\ 0 & \alpha^-\mathbf{I} & 0 & \alpha^-\rho^-\mathbf{u}^- \end{pmatrix}.$$

In these matrix notations the size of zero entries must be chosen to make the multiplication operation possible.

On the right hand side of (10), the work of viscous forces is denoted by symbol $\nabla\cdot\mathbf{T}(V_\varepsilon) := {}^t(0,0,\nabla\cdot\tau^+,\nabla\cdot\tau^-)$. The source term $\mathbf{S}(V_\varepsilon) := {}^t(0,0,\alpha^+\rho^+\mathbf{g},\alpha^-\rho^-\mathbf{g})$ incorporates the gravity force and $\mathbf{R}(V_\varepsilon) := {}^t(0,0,\kappa(\mathbf{u}^+ - \mathbf{u}^-),-\kappa(\mathbf{u}^+ - \mathbf{u}^-))$ contains the relaxation terms.

Since we expect the limit $V_\varepsilon \to V$ to be finite as $\varepsilon \to 0$, necessary the limiting vector V lies in the hypersurface $\mathbf{R}(V) = 0$. In terms of physical variables, it implies $\mathbf{u}^+ \equiv \mathbf{u}^-$. Consequently, we find our solution in the form of the following Chapman-Enskog type expansion:

$$V_\varepsilon = V + \varepsilon W + \mathrm{O}(\varepsilon^2).$$

After substituting this expansion into (10) and taking into account that $\mathbf{R}(V) \equiv 0$, at the leading order in ε one obtains:

$$\mathbf{A}(V)\frac{\partial V}{\partial t} + \mathbf{B}(V)\nabla V = \nabla\cdot\mathbf{T}(V) + \mathbf{S}(V) + \mathbf{R}'(V)W, \tag{11}$$

where

$$\mathbf{R}'(V) := \begin{pmatrix} 0 & 0 & 0 & 0 \\ 0 & 0 & 0 & 0 \\ 0 & 0 & \kappa\mathbf{I} & -\kappa\mathbf{I} \\ 0 & 0 & -\kappa\mathbf{I} & \kappa\mathbf{I} \end{pmatrix}.$$

Henceforth, we make a technical assumption of the presence of both phases in any point $\mathbf{x} \in \Omega$ of the flow domain. Mathematically it means that $0 < \alpha^+ < 1$. Since $\alpha^+ + \alpha^- = 1$, the same inequality holds for α^-. Otherwise, the relaxation process physically does not make sense and we will have some mathematical technical difficulties.

Under the aforementioned assumption, the matrix $\mathbf{A}(V)$ is invertible. Hence, we can multiply on the left both sides of (11) by $\mathbf{PA}^{-1}(V)$ where the projection matrix $\mathbf{P}$ is to be specified below:

$$\mathbf{P}\frac{\partial V}{\partial t} + \mathbf{PA}^{-1}(V)\mathbf{B}(V)\nabla V = \mathbf{PA}^{-1}(V)\nabla\cdot\mathbf{T}(V) + \mathbf{P}\tilde{\mathbf{R}}'(V)W + \mathbf{PA}^{-1}(V)\mathbf{S}(V), \quad (12)$$

where $\tilde{\mathbf{R}}'(V) := \mathbf{A}^{-1}(V)\mathbf{R}'(V)$ and has the following components

$$\tilde{\mathbf{R}}'(V) = \begin{pmatrix} 0\,0 & 0 & 0 \\ 0\,0 & 0 & 0 \\ 0\,0 & \dfrac{\kappa}{\alpha^+\rho^+}\mathbf{I} & -\dfrac{\kappa}{\alpha^+\rho^+}\mathbf{I} \\ 0\,0 & -\dfrac{\kappa}{\alpha^-\rho^-}\mathbf{I} & \dfrac{\kappa}{\alpha^-\rho^-}\mathbf{I} \end{pmatrix}.$$

The vector of physical variables V has four (in 1D) components ${}^t(p,\alpha^+,\mathbf{u},\mathbf{u})$ and only three are different. In order to remove the redundant information, we will introduce the new vector U defined as $U := {}^t(p,\alpha^+,\mathbf{u})$. The Jacobian matrix of this transformation can be easily computed:

$$\mathbf{J} := \frac{\partial V}{\partial U} = \begin{pmatrix} 1\,0\,0 \\ 0\,1\,0 \\ 0\,0\,\mathbf{I} \\ 0\,0\,\mathbf{I} \end{pmatrix}.$$

In new variables equation (12) becomes:

$$\mathbf{PJ}\frac{\partial U}{\partial t} + \mathbf{PA}^{-1}(U)\mathbf{B}(U)\mathbf{J}\nabla U = \mathbf{PA}^{-1}(U)\nabla\cdot\mathbf{T}(U) + \mathbf{P}\tilde{\mathbf{R}}'(U)W + \mathbf{PA}^{-1}(U)\mathbf{S}(U). \quad (13)$$

Now we can formulate two conditions to construct the matrix $\mathbf{P}$. First of all, the vector W is unknown and we need to remove it from equation (13). Hence, we require $\mathbf{P}\tilde{\mathbf{R}}'(V) = 0$. Then, we would like the governing equations to be explicitly resolved with respect to time derivatives. It gives us the second condition $\mathbf{PJ} = \mathbf{I}$. The existence and effective construction of the matrix $\mathbf{P}$ satisfying two aforementioned conditions

$$\mathbf{P}\tilde{\mathbf{R}}'(V) = 0, \quad \mathbf{PJ} = \mathbf{I},$$

are discussed below. Presented in this section results follow in great lines [12].

We will consider a slightly more general setting. Let be vector $V \in \mathbb{R}^n$ and its reduced counterpart $U \in \mathbb{R}^{n-k}$, $k < n$. In such geometry, $\tilde{\mathbf{R}}'(V) \in \mathrm{Mat}_{n,n}(\mathbb{R})$, $\mathbf{J} \in \mathrm{Mat}_{n,n-k}(\mathbb{R})$ and, consequently, $\mathbf{P} \in \mathrm{Mat}_{n-k,n}(\mathbb{R})$. Here, the notation $\mathrm{Mat}_{m,n}(\mathbb{R})$ denotes the set of $m \times n$ matrices with coefficients in $\mathbb{R}$. We have to say also that from algebraic point of view, matrices $\tilde{\mathbf{R}}'(V)$ and $\mathbf{R}'(V)$ are completely equivalent. Thus, for simplicity, in the following propositions we will reason in terms of $\mathbf{R}'(V)$.

Lemma 3.2. *The columns of the Jacobian matrix* $\mathbf{J}$ *form a basis of* $\ker\big(\mathbf{R}'(V)\big)$.

Proof. If we differentiate the relation $\mathbf{R}(V) = 0$ with respect to U, we will get the identity $\mathbf{R}'(V)\mathbf{J} = 0$. It implies that $\mathrm{range}\big(\mathbf{J}\big) \subseteq \ker\big(\mathbf{R}'(V)\big)$. By direct computation one verifies that $\dim \mathrm{range}\big(\mathbf{R}'(V)\big) = k$. From the well-known identity $\mathrm{range}\big(\mathbf{R}'(V)\big) \oplus \ker\big(\mathbf{R}'(V)\big) = \mathbb{R}^n$, one concludes that $\dim \ker\big(\mathbf{R}'(V)\big) = n-k$. But in the same time, the rank of $\mathbf{J}$ is equal to $n-k$ as well. It proves the result.

Theorem 3.1. *We suppose that for all* V, $\mathrm{range}\big(\mathbf{R}'(V)\big) \cap \ker\big(\mathbf{R}'(V)\big) = \{0\}$ *then there exists a matrix* $\mathbf{P} \in \mathrm{Mat}_{n-k,n}(\mathbb{R})$ *such that* $\mathbf{P}\mathbf{R}'(V) = 0$ *and* $\mathbf{P}\mathbf{J} = \mathbf{I}_{n-k}$.

Proof. Hypothesis $\mathrm{range}\big(\mathbf{R}'(V)\big) \cap \ker\big(\mathbf{R}'(V)\big) = \{0\}$ implies that $\mathrm{range}\big(\mathbf{R}'(V)\big) \oplus \ker\big(\mathbf{R}'(V)\big) = \mathbb{R}^n$. From Lemma 3.2 it follows that $\mathrm{range}\big(\mathbf{J}\big) = \ker\big(\mathbf{R}'(V)\big)$. Thus, the space $\mathbb{R}^n$ can be also represented as a direct sum $\mathrm{range}\big(\mathbf{R}'(V)\big) \oplus \mathrm{range}\big(\mathbf{J}\big)$. We will define $\mathbf{P}$ to be the projection on $\ker\big(\mathbf{R}'(V)\big) \equiv \mathrm{range}\big(\mathbf{J}\big)$. Since obviously $\mathbf{R}'(V) \in \mathrm{range}\big(\mathbf{R}'(V)\big)$ and $\mathbf{J} \in \mathrm{range}\big(\mathbf{J}\big)$, we have two required identities: $\mathbf{P}\mathbf{J} = \mathbf{I}_{n-k}$ and $\mathbf{P}\mathbf{R}'(V) = 0$.

Now, in order to compute effectively the projection matrix $\mathbf{P}$, we will construct an auxiliary matrix $\mathbf{D}(V) = [J^1, \dots, J^{n-k}, I^1, \dots, I^k]$, where J^i is the column i of the matrix $\mathbf{J}$ and $\{I^1, \dots, I^k\}$ are vectors which form a basis of $\mathrm{range}\big(\mathbf{R}'(V)\big)$. We remark that $\mathbf{P}\mathbf{D}(V) = [\mathbf{I}_{n-k}, 0]$. Lemma 3.2 implies that the matrix $\mathbf{D}(V)$ is invertible. Thus, the projection $\mathbf{P}$ can be computed by inverting $\mathbf{D}(V)$:

$$\mathbf{P} = [\mathbf{I}_{n-k}, 0] \cdot \mathbf{D}^{-1}(V).$$

Let us apply this general framework to our model (12), where $n = 4$ and $k = 1$. The matrix $\mathbf{D}(V)$ and its inverse $\mathbf{D}^{-1}(V)$ take this form:

$$\mathbf{D}(V) = \begin{pmatrix} 1 & 0 & 0 & 0 \\ 0 & 1 & 0 & 0 \\ 0 & 0 & \mathbf{I} & \dfrac{\kappa}{\alpha^+\rho^+}\mathbf{I} \\ 0 & 0 & \mathbf{I} & -\dfrac{\kappa}{\alpha^-\rho^-}\mathbf{I} \end{pmatrix}, \quad \mathbf{D}^{-1}(V) = \begin{pmatrix} 1 & 0 & 0 & 0 \\ 0 & 1 & 0 & 0 \\ 0 & 0 & m^+\mathbf{I} & m^-\mathbf{I} \\ 0 & 0 & \dfrac{m^+m^-\rho}{\kappa}\mathbf{I} & -\dfrac{m^+m^-\rho}{\kappa}\mathbf{I} \end{pmatrix},$$

where $m^\pm$ are mass fractions defined in Remark 2.3.

Now, the projection matrix $\mathbf{P}$ can be immediately computed:

$$\mathbf{P} = \begin{pmatrix} 1 & 0 & 0 & 0 \\ 0 & 1 & 0 & 0 \\ 0 & 0 & m^+\mathbf{I} & m^-\mathbf{I} \end{pmatrix}.$$

Finally, after computing all matrix products $\mathbf{PA}^{-1}(U)\mathbf{B}(U)\mathbf{J}$, $\mathbf{PA}^{-1}(U)\nabla\cdot\mathbf{T}(U)$, $\mathbf{PA}^{-1}(U)\mathbf{S}(U)$ present in equation (13), we obtain the desired single velocity model:

$$\frac{\partial p}{\partial t} + \mathbf{u}\cdot\nabla p + \rho c_s^2 \nabla\cdot\mathbf{u} = 0, \tag{14}$$

$$\frac{\partial \alpha^+}{\partial t} + \mathbf{u}\cdot\nabla\alpha^+ + \alpha^+\alpha^-\delta\nabla\cdot\mathbf{u} = 0, \tag{15}$$

$$\rho\frac{\partial \mathbf{u}}{\partial t} + \rho(\mathbf{u}\cdot\nabla)\mathbf{u} = \rho\mathbf{g} + \nabla\cdot\tau, \tag{16}$$

where $\rho = \alpha^+\rho^+ + \alpha^-\rho^-$ is the mixture density and c_s^2 is the sound velocity in the mixture which is determined by this formula:

$$\rho c_s^2 := \frac{\rho^+\rho^-(c_s^+)^2(c_s^-)^2}{\alpha^-\rho^+(c_s^+)^2 + \alpha^+\rho^-(c_s^-)^2},$$

and δ is given by

$$\delta := \frac{\rho^+(c_s^+)^2 - \rho^-(c_s^-)^2}{\alpha^-\rho^+(c_s^+)^2 + \alpha^+\rho^-(c_s^-)^2}.$$

Finally, $\tau := \lambda \operatorname{tr}\mathbb{D}(\mathbf{u})\mathbf{I} + 2\mu\mathbb{D}(\mathbf{u})$ is the viscous stress tensor of the mixture. Viscosity coefficients λ, μ are naturally defined as

$$\lambda := \alpha^+\lambda^+ + \alpha^-\lambda^-, \quad \mu := \alpha^+\mu^+ + \alpha^-\mu^-.$$

Equations (14) – (16) can be recast in the conservative form which is more convenient for numerical computations and theoretical analysis. To achieve this purpose, we replace the pressure p in (14) by $\rho^\pm$ using the equation of state:

$$\frac{\partial \rho^\pm}{\partial t} + \mathbf{u}\cdot\nabla\rho^\pm + \frac{\rho c_s^2}{(c_s^\pm)^2}\nabla\cdot\mathbf{u} = 0.$$

The last equation is multiplied by $\alpha^\pm$, the second equation (15) is multiplied by $\rho^\pm$ and we sum them to come up with two mass conservation equations. Transformation of the momentum conservation equation (16) is straightforward. The resulting conservative system takes this form:

$$\partial_t(\alpha^\pm\rho^\pm) + \nabla\cdot(\alpha^\pm\rho^\pm\mathbf{u}) = 0, \tag{17}$$

$$\partial_t(\rho\mathbf{u}) + \nabla\cdot(\rho\mathbf{u}\otimes\mathbf{u}) + \nabla p = \nabla\cdot\tau + \rho\mathbf{g}. \tag{18}$$

These equations represent a barotropic version of the four-equations model proposed in [4, 5].

It can be shown that the advection operator of the model (17), (18) is hyperbolic for any reasonable equation of state (4). Moreover, this system contains fewer variables which allow more efficient computations required in practice.

4 Incompressible Limit

The main scope of this paper is certainly around compressible two-fluid models. However, we decided to derive an incompressible limit of the single velocity model (17), (18) for the case when acoustic effects should be filtered out. The presence of acoustic waves represent, for example, a major restriction for the time step, if an explicit scheme is used.

For the sake of simplicity, we will neglect dissipative effects which do not affect the acoustic wave propagation. Thus, in this section we consider the following system of equations:

$$\partial_t(\alpha^{\pm}\rho^{\pm}) + \nabla\cdot(\alpha^{\pm}\rho^{\pm}\mathbf{u}) = 0, \tag{19}$$

$$\rho\partial_t\mathbf{u} + \rho(\mathbf{u}\cdot\nabla)\mathbf{u} + \nabla p = \rho\mathbf{g}. \tag{20}$$

For convenience, we rewrite equation (18) in nonconservative form.

In order to estimate the relative importance of various terms, we introduce dimensionless variables. The characteristic length, time, and velocity scales are denoted by ℓ, t_0 and U_0 respectively. For example, ℓ may be chosen as the diameter of the fluid domain Ω, t_0 is the biggest vortex turnover time and U_0 is the typical flow velocity. The density and the sound velocity scales are chosen to be those of the heavy fluid, i.e. ρ_0^+ and c_{0s}^+ correspondingly. Since we are interested in acoustic effects, the natural pressure scale is given by $\rho_0^+(c_{0s}^+)^2$. If we summarize these remarks, dependent and independent dimensionless variables (denoted with primes) are defined as:

$$\mathbf{x}' := \frac{\mathbf{x}}{\ell}, \quad t' := \frac{t}{t_0}, \quad \mathbf{u}' := \frac{\mathbf{u}}{U_0}, \quad (\rho^{\pm})' := \frac{\rho^{\pm}}{\rho_0^+}, \quad p' := \frac{p}{\rho_0^+(c_{0s}^+)^2}.$$

Remark 4.4. There is nothing to do for the volume fractions $\alpha^{\pm}$, since this quantity is dimensionless by definition.

After dropping the tildes, nondimensional system of equation becomes:

$$\mathrm{St}\,\partial_t(\alpha^{\pm}\rho^{\pm}) + \nabla\cdot(\alpha^{\pm}\rho^{\pm}\mathbf{u}) = 0, \tag{21}$$

$$\mathrm{St}\,\rho\partial_t\mathbf{u} + \rho(\mathbf{u}\cdot\nabla)\mathbf{u} + \frac{1}{\mathrm{Ma}^2}\nabla p = \frac{1}{\mathrm{Fr}^2}\rho\mathbf{g}, \tag{22}$$

where several scaling parameters have appeared:

- Strouhal number $\mathrm{St} := \dfrac{\ell}{U_0 t_0}$. In this study we will assume the Strouhal number to be equal to one, i.e. $t_0 = \dfrac{\ell}{U_0}$.
- Mach number $\mathrm{Ma} := \dfrac{U_0}{c_{0s}^+}$ which measures the relative importance of the flow speed and the sound speed in the medium.

- Froude number $\mathrm{Fr} := \frac{U_0}{\sqrt{g\ell}}$ compares inertia and gravitational force. This parameter will not play an important rôle in the present study.

All physical variables $\alpha^{\pm}$, $\rho^{\pm}$, p and $\mathbf{u}$ are expanded in formal series in powers of the Mach number:

$$\varphi = \varphi_0 + \mathrm{Ma}\,\varphi_1 + \mathrm{Ma}^2\varphi_2 + \ldots, \qquad \varphi \in \{\alpha^{\pm}, \rho^{\pm}, p, \mathbf{u}\}. \tag{23}$$

Formal expansion (23) is then substituted into the system (21), (22). At the orders Ma^{-2} and Ma^{-1}, we obtain

$$\nabla p_0 = \nabla p_1 = 0.$$

In other words, $p_0 = p_0(t)$ and $p_1 = p_1(t)$ are only functions of time. At the order Ma^0 we get the following system of equations:

$$\partial_t(\alpha_0^{\pm}\rho_0^{\pm}) + \nabla\cdot(\alpha_0^{\pm}\rho_0^{\pm}\mathbf{u}_0) = 0, \tag{24}$$

$$\rho_0\partial_t\mathbf{u}_0 + \rho_0(\mathbf{u}_0\cdot\nabla)\mathbf{u}_0 + \nabla\pi = \frac{1}{\mathrm{Fr}^2}\rho_0\mathbf{g}, \tag{25}$$

where by π we denote p_2.

Using the same asymptotic expansion (23), one can show that at the leading order we keep usual relations between densities and volume fractions:

$$\alpha_0^+ + \alpha_0^- = 1, \quad \rho_0 = \alpha_0^+\rho_0^+ + \alpha_0^-\rho_0^-. \tag{26}$$

In order to investigate the behaviour of $\rho_0^{\pm}$, we will invert the equation of state[3] $\rho^{\pm} = \rho^{\pm}(p) = (p^{\pm})^{-1}(p)$ and expand it in powers of Ma:

$$\rho^{\pm}(p) = \rho^{\pm}(p_0) + \mathrm{Ma}\left.\frac{\partial\rho^{\pm}}{\partial p}\right|_{p_0} p_1 + \mathrm{Ma}^2\left(\left.\frac{\partial\rho^{\pm}}{\partial p}\right|_{p_0} p_2 + \left.\frac{\partial^2\rho^{\pm}}{\partial p^2}\right|_{p_0} p_1^2\right) + O(\mathrm{Ma}^3)$$

On the other hand, from (23) we know that

$$\rho^{\pm} = \rho_0^{\pm} + \mathrm{Ma}\rho_1^{\pm} + \mathrm{Ma}^2\rho_2^{\pm} + \ldots$$

Matching these expansions at two lowest orders shows that $\rho_{0,1}^{\pm}$ are functions only of the time variable:

$$\rho_0^{\pm} = \rho^{\pm}(p_0(t)) =: r_0^{\pm}(t), \quad \rho_1^{\pm} = \left.\frac{\partial\rho^{\pm}}{\partial p}\right|_{p_0(t)} p_1(t) =: r_1^{\pm}(t).$$

It is possible to show that $\rho_{0,1}^{\pm}$ are just constants. Consider the Gibbs relation which reads

$$T^{\pm}ds^{\pm} = de^{\pm} - \frac{p}{(\rho^{\pm})^2}d\rho^{\pm}.$$

[3] The function $p = p^{\pm}(\rho^{\pm})$ is invertible since it is a strictly increasing function $\frac{\partial p}{\partial\rho^{\pm}} > 0$.

Since we consider isentropic flows, $ds^{\pm} \equiv 0$ and, consequently, the Gibbs relation takes a much simpler form:

$$de^{\pm} = \frac{p}{(\rho^{\pm})^2} d\rho^{\pm}. \tag{27}$$

It can be shown by considering the total energy conservation equation [11], that the internal energy $e^{\pm}$ naturally scales with U_0^2. After dividing (27) by dt and switching to dimensionless variables, equation (27) takes the following form (after dropping the primes):

$$\frac{de^{\pm}}{dt} = \frac{p}{\mathrm{Ma}^2(\rho^{\pm})^2} \frac{d\rho^{\pm}}{dt}.$$

Expanding $e^{\pm}$ in the series (23) and looking at two leading terms, leads to the desired result:

$$\frac{d\rho_{0,1}^{\pm}}{dt} = 0 \quad \Rightarrow \quad \rho_{0,1}^{\pm} = \text{const.}$$

The incompressibility condition $\nabla \cdot \mathbf{u}_0 = 0$ is obtained by summing up mass conservation equations (24) and taking into account relation (26).

If we summarize all developments made above and switch back to dimensional variables, the resulting incompressible system will become:

$$\partial_t \alpha^{\pm} + \nabla \alpha^{\pm} \cdot \mathbf{u} = 0, \tag{28}$$

$$\nabla \cdot \mathbf{u} = 0, \tag{29}$$

$$\rho \partial_t \mathbf{u} + \rho (\mathbf{u} \cdot \nabla) \mathbf{u} + \nabla \pi = \rho \mathbf{g} + \nabla \cdot \tau, \tag{30}$$

where we dropped the index 0 and added again dissipative effects. Viscous stress tensor τ is still defined by expression (5), as in compressible case. In this case, we can speak about two-fluid Navier-Stokes equations. This system of equations (28) – (30) is much easier to solve numerically than its compressible analogue (17), (18). In particular, this simplification is due to removed stiffness of acoustic waves.

5 Conclusions and Perspectives

In this study we presented several barotropic two-fluid models which can be used for numerical simulation of powder-snow avalanche flows. One of the main objectives of this paper was to reveal the connection between barotropic models with single and two velocities. The extension to more general fluids is in progress [11].

Our exposition began with compressible two-phase model (2), (3) possessing two velocity variables. Then, using a relaxation process, we constrained the system to have a common velocity for both phases. Mathematically it was achieved with a Chapman-Enskog type expansion. Resulting model (17), (18) is hyperbolic for any reasonable equation of state (4). Finally, two-fluid Navier-Stokes equations (28) – (30) were derived as an incompressible limit of the single velocity model (17), (18).

Hence, we presented three different two-fluid models which are related by formal derivation procedures. Simplifications made above, represent a good trade-off between accuracy and computational complexity. The final choice should be made after determining the flow regime and main goals of the simulation.

We did not incorporate yet any turbulence modelling. In this study we were focused essentially on the advection operators. However, it is obvious that the physical flow under consideration is fully turbulent in its aerosol part [14]. As the first physical approximation, turbulence effects can be taken into account by adding eddy viscosity terms and, thus, by modifying the viscous stress tensor τ. It will be done in future studies.

Acknowledgement. The authors would like to acknowledge the University of Savoie for the PPF grant linked to the project: "Mathématiques et avalanches de neige, une rencontre possible?". The support from the Research network VOR (Professors Jacky Mazars and Denis Jongmans) and Cluster Environment through the program "Risques gravitaires, sismes" is also acknowledged.

We would like to thank Professor Carmen de Jong for interesting discussions around snow avalanches. Special thanks go to our colleagues and friends Didier Bresch and Céline Acary-Robert for their continuous help and support. Finally, the second author thanks Professors Jean-Michel Ghidaglia and Frédéric Dias for introducing him to the beautiful field of two-phase flows.

References

1. Ancey, C., Bain, V., Bardou, E., Borrel, G., Burnet, R., Jarry, F., Kolbl, O., Meunier, M.: Dynamique des avalanches. Presses polytechniques et universitaires romandes (Lausanne, Suisse) (2006)
2. Baer, M., Nunziato, J.: A two-phase mixture theory for the deflagration-to-detonation transition (DDT) in reactive granular materials. Int. J. Multiphase Flow 12(6), 861–889 (1986)
3. Bresch, D., Desjardins, B., Ghidaglia, J.M., Grenier, E.: On global weak solutions to a generic two-fluid model. To appear in Arch. Ration. Mech. Anal. (2009)
4. Dias, F., Dutykh, D., Ghidaglia, J.M.: A two-fluid model for violent aerated flows. Submitted to Comput. Fluids (2008)
5. Dutykh, D.: Mathematical modelling of tsunami waves. Ph.D. thesis, École Normale Supérieure de Cachan (2007)
6. Dutykh, D., Acary-Robert, C., Bresch, D.: Numerical simulation of powder-snow avalanche interaction with an obstacle. Submitted to Appl. Math. Model. (2009)
7. Ishii, M.: Thermo-Fluid Dynamic Theory of Two-Phase Flow. Eyrolles, Paris (1975)
8. Issler, D.: Experimental information on the dynamics of dry-snow avalanches. In: Hutter, K., Kirchner, N. (eds.) Dynamic Response of Granular and Porous Materials Under Large and Catastrophic Deformations, vol. 11. Springer, Berlin (2003)
9. Johannesson, T., Gauer, P., Issler, P., Lied, K.: The design of avalanche protection dams. Tech. rep., European Commission (2009)
10. Lied, K.: Satsie: Avalanche studies and model validation in Europe. Tech. rep., European Commission (2006)

11. Meyapin, Y., Dutykh, D., Gisclon, M.: Velocity and energy relaxation in two-phase flows. Submitted to Eur. J. Mech. B / Fluids (2009)
12. Murrone, A., Guillard, H.: A five equation reduced model for compressible two phase flow problems. J. Comput. Phys. 202, 664–698 (2005)
13. Naaim-Bouvet, F., Naaim, M., Bacher, M., Heiligenstein, L.: Physical modelling of the interaction between powder avalanches and defence structures. Nat. Hazards Earth Syst. Sci. 2, 193–202 (2002)
14. Rastello, M., Hopfinger, E.: Sediment-entraining suspension clouds: a model of powder-snow avalanches. J. Fluid. Mech. 509, 181–206 (2004)
15. Rovarch, J.M.: Solveurs tridimensionnels pour les écoulements diphasiques avec transferts d'énergie. Ph.D. thesis, Ecole Normale Supérieure de Cachan (2006)

Hierarchy Analysis and Reduction of Reacting Flow Systems

Viatcheslav Bykov and Ulrich Maas

Abstract. The qualitative system analysis and model reduction for reacting flows has gained an increasing interest during the last years. Nowadays, simulations based on sophisticated algorithms implemented on powerful workstations turn out to be a prevailing tool of the system analysis. However, modelling of realistic systems of technical importance leads to an extreme growth of mathematical models both in complexity and in dimension, i.e. they become not treatable even by modern computational facilities. Recently, the concept of invariant, slow/fast -, attractive and stable manifolds, which appear in the system state space as a manifestation of a restricted number of degrees of freedom exhibiting by the system, has proven to be an efficient tool of system analysis and model reduction. In the current work questions about the specific low-dimensional manifolds' identification, the analysis of their properties, their approximation and the application to model reduction of complex reacting flow systems is discussed.

1 Introduction

In view of the rapid progress in computational ability of new powerful workstations, supercomputers and integration software packages, the role of careful analysis of a reacting flow for accurate model formulation considerably increases. There is a constant trend of considering more and more detailed descriptions because simple models can not satisfy needs of industry and science any more. As a result more detailed and sophisticated descriptions are in use which leads to an increase of the system dimension and its complexity. This trend is crucial particularly in modelling of reacting flows of combustion, biochemical processes and atmospheric chemistry.

Viatcheslav Bykov · Ulrich Maas
Institute for Technical Thermodynamics, Karlsruhe Institute of Technology,
Engelbert-Arndold-Straße 4, Geb. 10.91, 76131 Karlsruhe, Germany
e-mail: viatcheslav.bykov@kit.edu,
ulrich.maas@kit.edu

E. Krause et al. (Eds.): Computational Sci., & High Performance Computing IV, NNFM 115, pp. 233–252.
springerlink.com

Novel software packages significantly simplify not only the actual integration part (CHEMKIN, COSILAB, KINTECH etc. see e.g [51, 21, 33]), the development stage of the detailed mathematical model has become almost automatic as well (see e.g. [22, 12, 47] additionally). Typically, kinetic mechanisms comprising hundreds of chemically reacting species [48] and thousands of elementary reactions are considered nowadays. Besides the dimensionality problem, the detailed description of chemical reactions introduces a range of characteristic time scales that cover about ten orders of magnitude (the time scales may differ by many orders of magnitude from 1 s to 10^{-12} s). The drastic disparity in time-scales results in a so-called stiffness of the system of governing equations of a reacting flow and causes scale problems also in space. These scaling problems together with high dimensionality and the non-linearity of the chemical source term create serious difficulties for numerical solution procedures and make the practical problems beyond the capacity of even today's workstations and super-computers.

Obviously, there is a strong need for algorithms that handle automatically the enormous dimension, stiffness and non-linearity of reacting flow systems. It is of great interest to find a compromise between complexity and accuracy of the resulting system. There are two standard ways to deal with this problem. The physical model can be simplified first, meaning that problems of academic interest e.g. small length scales, simple geometries, symmetric flows etc. might be considered in this level. Secondly, methods of model reduction can be developed and used to describe the chemical kinetics by means of reduced systems, while still retaining the essential dynamics of the detailed system without significant damage to the overall qualitative and quantitative properties of the reduced model by comparison with the original one [38, 36]. Hence, reduction procedures providing automatically with reduced descriptions for practical engineering problems would represent an ideal compromise between computability and accuracy. In this respect the current work is aimed to introduce a hierarchical system analysis and present recent progress in formulation of model reduction by manifold methods.

1.1 Multi-scale Hierarchy and Manifolds Method

A large number of practical reacting flow systems involve sub-processes with essentially different time scales defined as reciprocal values of certain eigenvalue problems (see Fig. 1, where typical scales are sketched). These drastically different time scales are the origin of the stiffness of the system of governing equations on one hand, but on the other hand they allow to assume that some fast modes or processes are relaxed quickly and then only the slow processes govern the overall system dynamics. As a result, the long term system evolution is represented primarily by the dynamics of the slow motions along a stable invariant geometrical attractor in the detailed state space. Figure 2 illustrates the geometrical idea and shows the systems' solutions trajectories in the system state space.

Hence, the main idea of model reduction, is to consider thermo-chemical states of the system to be confined only to this slow manifold and the whole state

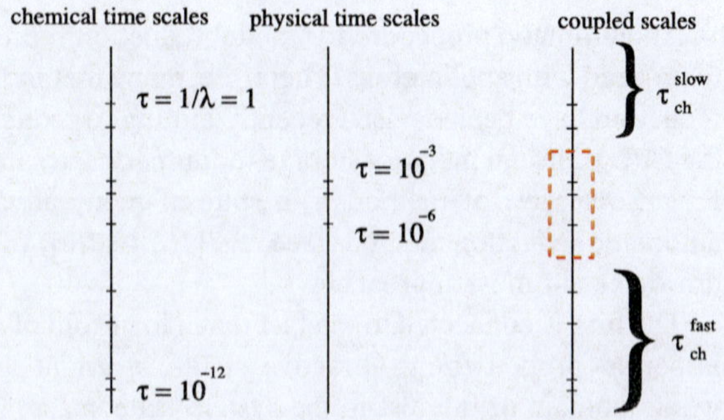

Fig. 1 Sketch of time scales of chemical and physical processes

vector can be redefined by using the manifold. In this respect, although the existence of multiple scales complicates the numerical treatment of the reacting flow system enormously it also allows to construct low-dimensional approximations of the detailed model describing the so-called short- and long-term dynamics as well as the rate limiting processes accurately. Accordingly, model reduction, in our view, means finding this manifold along with a subset or combination of system parameters that consistently represents the full detailed system on this manifold in terms of a sufficiently smaller number of variables/parameters used to parametrize the manifold. As byproduct of such reduction methodology, i.e. reducing the dimension by removing the fast modes, it decreases stiffness of the system of governing differential equations. This makes the existence of multi-scales the most important feature of chemical kinetic mechanisms and of reacting flow systems.

The main problem of approaches dealing with multi-scales hierarchy lies within its implicit nature, i.e. for complex realistic models this structure is typically hidden in the original mathematical model. Sometimes, it is possible to represent the multi-scales explicitly (even analytically) for comparatively simple models. In the general case, however, the transition from the original system with the hidden multi-scale structure to its explicit representation is not a trivial task. It requires a deep understanding of physics, chemistry and properties of the mathematical model of the phenomenon under consideration. A large amount of time, human resources and experience is required to find a suitable system representation in a strictly decomposed form discussed below.

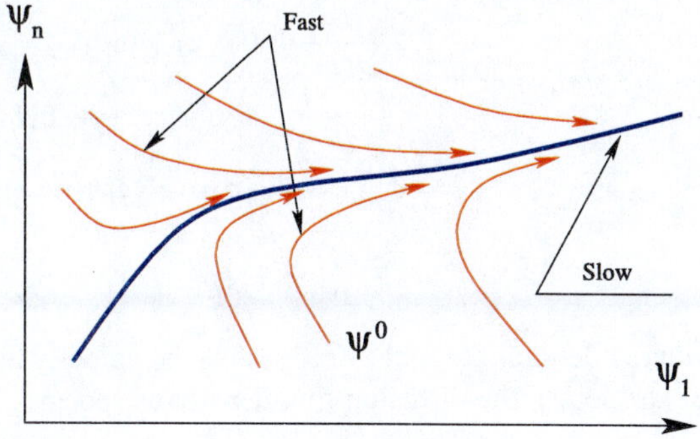

Fig. 2 State space geometry

An automatic procedure for model reduction represents a rather alternative way to proceed with applications. Therefore, many methods performing automatic model reduction have been devised recently aiming to reduce the stiffness, the dimension, the CPU time and memory storage for numerical treatments (see e.g. [22, 47, 36, 20] for an overview of methods). In spite of many advantages of available currently automatic reduction methods (see e.g. [15, 44, 26]) there are principal, fundamental drawbacks of most approaches.

The first is connected to the fact that a manifold of a fixed dimension with certain important properties e.g. attractive, stable, invariant does not exist everywhere in the whole domain of interest in the system state space. Most approaches have a local character and, therefore, can not provide such information, while, typically their results are extended and used automatically everywhere. This inevitably leads to inaccuracies of the produced reduced models or the needed to increase the reduced system's dimension.

The next difficult and fundamental problem of model reduction is the coupling of molecular transport with thermo-chemical processes [25, 39, 40]. It is a very difficult problem which is normally solved by either neglecting such an interaction globally or by applying operator splitting methodologies (see e.g. [45]), which implies neglecting of the coupling locally either in time or in space. In both cases, the interaction is not taken into account and excluded from the reduced model. An algorithm which overcomes these drawbacks: allows automatic analysis of stability, dimension and existence of the manifolds and avoids oversimplification of the coupling processes between chemical reaction and transport in most general case is presented and discussed in the current work.

1.2 Dynamics of Reacting Flows

In our previous work [30] questions of the strongly coupled description and interaction of reaction, turbulence and transport processes were in the focus of discussion by considering the system in most general form as the closed set of the standard conservation equations:

$$\begin{aligned}
\frac{\partial \rho}{\partial t} + \mathrm{div}\,(\rho\,\mathrm{v}) &= 0 \\
\frac{\partial \rho_i}{\partial t} + \mathrm{div}\,(\rho_i \mathrm{v}) + \mathrm{div}\,(\mathrm{j_i}) &= \rho M_i F_i \\
\frac{\partial (\rho\,\mathrm{v})}{\partial t} + \mathrm{div}\,(\rho\,\mathrm{v} \otimes \mathrm{v}) + \mathrm{div}\,(\overline{\overline{\mathrm{p}}}) &= 0 \\
\frac{\partial (\rho\,\mathrm{u})}{\partial t} + \mathrm{div}\,(\rho\,\mathrm{u}\,\mathrm{v} + \mathrm{j}_q) + \overline{\overline{\mathrm{p}}} : \mathrm{grad}\,(\mathrm{v}) &= 0
\end{aligned} \tag{1}$$

$$p = (\rho, \mathrm{u}, \rho_1, \ldots, \rho_{n_s})$$

where t denotes the time, v represents the velocity field, ρ the density, ρ_i the density of specie i, $\mathrm{j_i}$ the diffusion flux density of specie i, M_i the molar mass of specie of specie i, F_i the molar scale rate of formation of specie i, $\overline{\overline{\mathrm{p}}}$ the pressure tensor, u the specific inner energy, j_q the conductive heat flux density, n_s number of different chemical species and p the pressure.

By introducing the state vector ψ - ($n = n_s + 2$)-dimensional vector

$$\psi = \left(h, p, \frac{w_1}{M_1}, \ldots, \frac{w_{n_s}}{M_{n_s}} \right)^T, \tag{2}$$

where $w_1, \ldots, w_{n_s}$ the n_s species mass fractions $w_i = \frac{\rho_i}{\rho}$ the typical system of PDEs (1) for modelling of reacting flows in general symbolic vector notation reads

$$\frac{\partial \psi}{\partial t} = F(\psi) - \mathrm{v} \cdot \mathrm{grad}(\psi) - \frac{1}{\rho} \mathrm{div}(D \cdot \mathrm{grad}(\psi)) \equiv \Phi(\psi,) \tag{3}$$

h denotes the enthalpy, p the pressure. Here D is the (n by n)-dimensional matrix of the transport coefficients and $F(\Psi)$ is the n-dimensional vector of thermo-chemical source terms [48] both are very complicated functions of of the thermo-kinetic state vector ψ (see e.g. [23, 2, 13]).

Taking into account the fact that most problems with scaling originate from the very stiff reaction source term (see Fig. 1) the analysis of the source term is a very important step of model reduction.

$$\frac{d\psi}{dt} = F(\psi). \tag{4}$$

At present there are several methods for an analysis of the system's hierarchy. These methods have been improved constantly producing variations of methods designed for special tasks (see e.g. [22, 47, 36, 20]). In order to design a sophisticated tool of hierarchy system analysis and model reduction it has to be considered in most general formulation by developing appropriate framework and by introducing basic notations and concepts. After basics of model reduction are established, understood and well studied, i.e. the language of model reduction is formed; a straightforward and most efficient way of implementation has to be found.

In our investigation, both analysis and implementation stages are based on the ILDM method, which is a kind of reference approach for the study (see e.g [27] for definitions and detail). It has a rigorous mathematical background established in a number of works [43, 24, 7], where it has been shown that the ILDM provides a good approximation of the invariant manifold of slow motions (up to the second order [24]). What is more, it combines both the system hierarchy analysis and a complete method of automatic reduced model generation, which provides an efficient subsequent implementation scheme into CFD codes [31, 29]. The reason for this choice is simple, at present, the ILDM method is efficient and robust, it is realized in the FORTRAN codes HOMREA and INSFLA [32, 28]. This work is aimed to develop a general mathematical background for model reduction, to fill a gap between model analysis and applications and justify of the model reduction method for reacting flow modelling and simulation. The theoretical basics are revised and then, on the basis of better understanding of the model reduction background an improved implementation scheme as well as superior methods are emerge.

2 Analysis of Chemical Kinetic Mechanisms

In this section we explain basic concepts and problems of model reduction. Because the representation of complex systems by simpler sub-systems using multi-scale nature of different subsystems is almost universally used for modelling in physics and engineering, it is naturally to start with introducing the original mathematical model of multi-scale behaviour as a standard Singularly Perturbed System (SPS) [14, 46, 16, 50]. According to this concept, two sub-systems describe the so-called fast and slow motions of the system solution explicitly, moreover, the discrepancy in time scales are controlled by defined small system parameter. Therefore, in the case of special representation of the system for a considered phenomenon is known and asymptotic limit specified precisely, one can apply the singular perturbation theory in order to obtain peculiarities of the reduced model as it is shown below [5].

2.1 SPS as Mathematical Model of Multi-scales Behaviour

A simple mathematical model for the decomposition of motions leads to a very special system representation of the system as a Singularly Perturbed System (SPS)

$$\varepsilon \frac{dU}{dt} = F_f(U,V)$$

$$\frac{dV}{dt} = F_s(U,V) \tag{5}$$

with the so-called fast variables

$$U = \left(U_1, ..., U_{m_f}\right),$$

which are assumed to change fast compared to slow ones

$$V = (V_1, ..., V_{m_s}), \quad m_f + m_s = n,$$

due to the presence of the system small parameter $0 < \varepsilon \ll 1$, where $F_s(U,V) \sim O(1)$ and $F_f(U,V) \sim O(1)$. The SPS theory can systematically be used as a natural mathematical construction for the system hierarchy analysis and its reduction. A short outline of the main notations and definitions are provided below. More detail on theoretical background can be found in the literature (see e.g. [14, 16, 4]).

Accordingly, one has the fast motions/manifolds estimated to the leading order of magnitude of the system small parameter ε describing the difference in time scales by

$$M_f^0 = \{(U,V) : V = V_0\}, \tag{6}$$

They obviously depend on the initial system state (U_0, V_0) because the slow variables during the fast transient period are estimated to be "frozen" (Fig. 3 shows typical phase space structure of the SPS).

The slow system manifold is then approximated to the leading order, when $\varepsilon \to 0$, by

$$M_s^0 = \left\{(U,V) : F_f(U,V) = 0\right\}, \tag{7}$$

Geometrically, it means that in the ε vicinity of the slow manifold M_s^0 the variable right hand sides (RHS) of (5) become comparable, i.e. variables change with the same "speed". Note, however, that higher order approximations of the invariant manifold of slow motions follow from the application of the SPS theory in a straightforward way [16]. For instance, the first order equation in the implicit form is given by differentiation of the slow manifold (7) along the system (5) vector field $F(U,V) = \left(F_f(U,V), F_s(U,V)\right)^T$:

$$M_s^1 = \left\{(U,V) : D_{U,V}\left(F_f(U,V)\right) F(U,V) = 0\right\}. \tag{8}$$

The stability analysis of the slow manifold is related to the following eigenvalue problem for points (U^*,V^*) on the slow manifold (7)

$$Re\left(\lambda\left(D_U F_f(U^*,V^*)\right)\right)|_{(U^*,V^*)\in M_s} < 0, \tag{9}$$

where $D_U F_f(U^*,V^*)$ is the matrix $m_f \times m_f$ of partial derivatives of the fast subsystem with respect to the fast variable and $\lambda(*)$ are eigenvalues of this matrix. It means, in particular, that knowledge about global fast manifolds is very important in the stability analysis of a reduced model described by a slow manifold. Now, depending on the investigated dynamical regime one can use manifolds defined by (6) or (7) as a manifold equation and proceed with the reduction procedure. Furthermore, in this formulation the problem of reduction reduces to the problem of how to identify this special representation. Clearly, by assuming only one particular asymptotic limit and performing special non-dimensionalization, it might be possible to obtain a representation (5) of the system analytically. But, in the general case, for more complex and practical systems (with large dimension, non-linearity, stiffness etc.) it is either very time/human/computational resources consuming or even prohibitive in some cases.

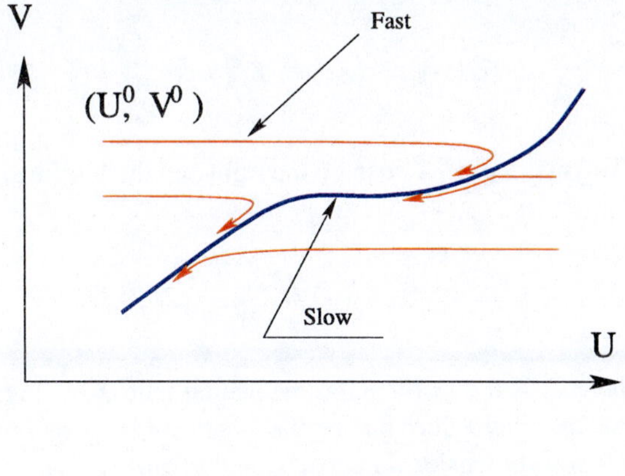

Fig. 3 Standard SPS phase space geometry

Hence, if the original variables are not appropriate for this special form and the system is presented in the general form (4), one has to deal with the following questions in quite general formulation:

1. Is there a multi-scales character exhibited by the system (4)?
2. If yes, then what is a possible/minimal dimension of reduced model?
3. If the dimension is fixed, how to find out slow and fast manifolds as directions of fast relaxations processes and to define properties of the slow manifold: existence, stability, attractiveness, basin of attraction etc.?
4. What is an explicit form of the decomposition? Remark: By the explicit form the standard Singularly Perturbed System is meant here.
5. Is it possible to single out limits of application of a particular asymptotic?

In our view any successful model reduction concept should be able to answer on questions above. It should provide us not only with the approximation of the manifold of slow motions, but with the information of how the slow manifold is embedded into the state space.

2.2 ILDM Method - Local Analysis

Although the mathematical model of the ILDM method is described in details in [31, 29], a short repetition to outline the meaning of the ILDM method with respect to the line above is needed. A concept of generalized coordinates in the context of the standard ILDM shall be presented as well. The ILDM suggests to analyse the (n by n)-dimensional Jacobi matrix F_ψ of the source term

$$(D_\psi F)_{ij} = (F_\psi)_{ij} = \frac{\partial F_i}{\partial \psi_j}.$$

One identifies locally (depending on the state ψ) fast/slow chemical processes by the Schur decomposition with subsequent solution of the Sylvester equation to obtain invariant subspaces of the Jacobian matrix as follows

$$F_\psi(\psi) = (Z_s(\psi)\, Z_f(\psi)) \cdot \begin{pmatrix} N_s(\psi) & 0 \\ 0 & N_f(\psi) \end{pmatrix} \cdot \begin{pmatrix} \tilde{Z}_s(\psi) \\ \tilde{Z}_f(\psi) \end{pmatrix}. \tag{10}$$

The matrices $Z, \tilde{Z}$ span up the right and the left invariant subspaces correspondingly,

$$\tilde{Z} = Z^{-1} = (Z_s\, Z_f)^{-1} = \begin{pmatrix} \tilde{Z}_s \\ \tilde{Z}_f \end{pmatrix}, \tag{11}$$

here Z_s is the (n by m_s)-dimensional matrix belonging to the invariant subspace of the m_s eigenvalues having the smallest real parts (defined by N_s), while Z_f is the (n by m_f)-dimensional matrix related to the m_f eigenvalues (defined by N_f) having the largest real parts respectively, where m_s and m_f denotes the number of eigenvalues according to slow and fast processes i.e.

$$\begin{gathered} i=1,\dots,m_s \quad k=m_s+1,\dots,m_s+m_f \quad m_s+m_f=n \quad a,b>0 \\ \left|Re\left(\lambda_i\left(D_\psi F\right)\right)\right| < a \ll b < \left|Re\left(\lambda_k\left(D_\psi F\right)\right)\right|, \quad Re\left(\lambda_k\left(D_\psi F\right)\right)<0 \end{gathered} \tag{12}$$

The general assumption that the fast processes have already relaxed define a m_s-dimensional manifold in the state space. This sub-space is composed of points where the reaction rates in direction of the m_f fast processes vanish, therefore, they can be expressed by

$$M_s^{ILDM} = \left\{\tilde{Z}_f(\psi)F(\psi)=0\right\}, \tag{13}$$

where $\tilde{Z}_f$ is the (m_f by n)-dimensional matrix of left Schur vectors corresponding to the fast relaxing processes. One can see easily the connection between (13) and (7). If conditions (12) are locally valid, then one has the transformation to the standard SPS form which is defined by invariant subspaces with $\varepsilon = \frac{a}{b} \ll 1$

$$\begin{gathered} U = \tilde{Z}_f\left(\psi_0\right)\psi \\ V = \tilde{Z}_s\left(\psi_0\right)\psi \end{gathered} \tag{14}$$

unfortunately, the transformation is valid only in the vicinity of the slow manifold and cannot be used to find out how the slow manifold is embedded into the state space, even the stability is assumed only locally due to (12). Thus one needs to construct the global tool for system analysis answering the main questions above and providing with information left. For this reason the framework of Singularly Perturbed Vector Fields (SPVF) has been introduced in a number of works [6].

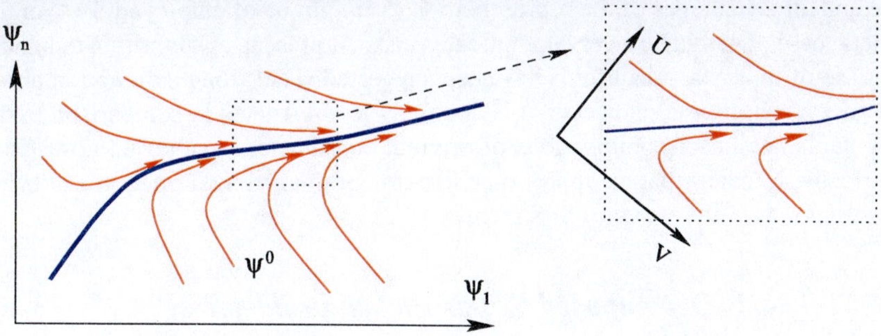

Fig. 4 Standard SPS versus SPVF phase space geometry

2.3 Coordinate Free Singular Perturbations

Strictly speaking concepts of the SPS and SPVF are equivalent locally. The latter can be understood as an analogue of the decomposed form (5), but valid only locally if appropriate coordinate system is specified, while keeping same dimension structure all over the domain. In order to account for changed structure of fast manifolds (fast dynamics of relaxation) and perform the global analysis the so-called fiber bundles are used, which is a natural substitute to a coordinate system. Now, everything depends on whether the fiber structure can be defined in advance or not. Figure 4 illustrates the relation between the standard SPS and SPVF cases. It is shown how the structure of fiber bundles of fast motions can be used to define local coordinates explicitly decomposing the system's motions.

The standard strategy of the SPS theory can be viewed within the SPVF method as follows. First, one tries to use certain scaling in original coordinate system such that it brings to the standard SPS explicit form in the original coordinates. The well known Quasi-steady States Approach (QSSA) is a typical representative of the initial stage of system hierarchy analysis [3, 4, 49]. Of course an assumption that the original variables already appropriate to the decomposition (5) or for definition of the slow manifold is a very strong one. This limits considerably both an application range and the achieved reduced dimension by the QSSA method.

Then, on the next level the local analysis is performed that can be used to efficiently approximate the slow manifold (up to second order). Although it is suggested that the real fast relaxation does not follow the hyper-planes of the constant original coordinates, the fibers of fast motions are still unidentified but assumed to be not important for the reduced model formulation. It means the slow manifold is found, while the way it is embedded into the state space through the fast relaxation is unknown. Obviously, the information on a slow manifold only cannot be efficient for construction of model reduction technique, especially for modelling of the transient system behaviour.

The final step would be complete description of the fast manifolds/fibers structure and all advantages of the representation (5) might be of employed. But, in the general case, it would be a very complicated task. At present, an algorithm to handle the case of linear fast manifolds has been suggested. The strong assumption about linearity has been used; however, it is definitely less restricted in comparison to the first step since linear combinations of original coordinates are allowed now. Geometrically, it means that an appropriate linear coordinate transformation can bring the system (4) to the standard SPS form.

2.4 Linearly Decomposed Vector Fields, Problems of Implementation

Under an assumption of linearly decomposed vector field, it is natural to use a kind of linearization (linear interpolation) of the original vector field that provides with the transformation and fiber structure. The so-called Global-Quasi-Linearization

(GQL) procedure has been suggested in [9]. The main steps of the algorithm are outlined shortly in the following.

Suppose $F(\psi)$ is a SPVF that depends on a "hidden" small parameter ε. The proposed modification of GQL is the following procedure. Choose n linearly independent points $\psi_1, ..., \psi_n$ in such a way that vectors $F(\psi_1), ..., F(\psi_n)$ are also linearly independent. Because a small parameter δ is unknown it is impossible to check linear independence for any comparatively small δ. By using the reference set $\psi = \{\psi_1, ..., \psi_n\}$ and its image by the vector field $F(\psi_1), ..., F(\psi_n)$ as rows for matrices we define a new matrix T as follows

$$T := [F(\psi_1), ..., F(\psi_n)]\, [\psi_1, ..., \psi_n]^{-1}. \tag{15}$$

This matrix has a simple geometrical interpretation. It is the matrix of linear mapping that transforms points $\psi_1, ..., \psi_n$ to $F(\psi_1), ..., F(\psi_n)$. Of course, efficiency of this procedure essentially depends on choice of $\psi_1, ..., \psi_n$. Practical recommendations for the choice are the following. Points $\psi_1, ..., \psi_n$ can not be close one to another, because it can lead to degeneration of the matrix $[\psi_1, ..., \psi_n]^{-1}$ as well as the vector field $F(\psi)$ should have essentially "different" behaviour for different points $\psi_1, ..., \psi_n$. It is obvious that this is a crucial point of the algorithm and must be adapted to every particular model. The problem with choice of the reference sequence and the GQL can be solved by introducing the scale invariant sorting procedure [9, 8].

2.5 *Scale Invariant Form*

Scale invariance is a very important feature of the reduction method, it is a basic property to be independent on the preliminary knowledge and to be automatically applicable. In order to make the method self-consistent and independent of additional dimension analysis some geometrical invariants of the original vector field define the final GQL form and, consequently, the chosen reference set.

For this, the following invariants of the vector field based on an integral of the characteristic polynomial coefficients of the differential $J_F(\psi) = D_\psi F(\psi)$ of the vector field $F(\psi)$ are studied and compared to particular GQLs:

$$I_j(F;\Omega) = \int_\Omega a_j(\psi;F)\, d\psi,\ j = 1, ..., n, \tag{16}$$

where $a_i(\psi;F):\ f_{J_F(\psi)}(t) = t^n + a_1(\psi;F)\, t^{n-1} + ... + a_n(\psi;F)$.

The same invariants for the vector field produced by the GQL procedure are the following:

$$I_j(T;\Omega) = a_j(T)\, |\Omega|,\ j = 1, ..., n. \tag{17}$$

There is a clear interpretation of the integral of the last invariant giving the volume of the image of the domain of interest Ω: $|Im(F(\Omega))| \equiv |I_n(F;\Omega)|$.

Other integrals do not have such clear geometrical interpretations, but certainly have connection to the geometrical structure of the vector field. Hence, the final choice of the GQL is performed on a basis of the following functional

$$\Pi(T;\Omega) = \sum_j \left(a_j(T) - \frac{I_j(F;\Omega)}{|\Omega|} \right)^2 . \tag{18}$$

The chosen GQL minimizes the functional above over the control sequence

$$T^* : \min_T (\Pi(T;\Omega)) = \Pi(T^*;\Omega) . \tag{19}$$

The rest is standard, the spectrum of T^* answers the question about existing hierarchy, a gap condition can be used for definition of the decomposition dimension and the small system parameter, similar to (10)

$$T^* = (Z_s\, Z_f) \cdot \begin{pmatrix} N_s & 0 \\ 0 & N_f \end{pmatrix} \cdot \begin{pmatrix} \tilde{Z}_s \\ \tilde{Z}_f \end{pmatrix} . \tag{20}$$

but invariant subspaces of two groups of eigenvalues defines the fast and slow subspaces and projections

$$Pr_s = Z_s \tilde{Z}_s ,$$

and

$$Pr_f = Z_f \tilde{Z}_f .$$

In this way the system (4) is decomposed explicitly via

$$\begin{aligned} U &= \tilde{Z}_f\, \psi \\ V &= \tilde{Z}_s\, \psi \end{aligned} \tag{21}$$

and the SPS theory can be fully applied to analysis and reduction, for instance, the equation of the slow manifold equation reads

$$M_s^0 = \left\{ \tilde{Z}_f F(\psi) = 0 \right\} . \tag{22}$$

2.6 Computational Tool of Manifold Analysis and Reduction - Tabulation

Approaches described above should be implemented in numerical simulations, therefore, from computational point of view methods and algorithms of efficient use of constructions above are of great importance. In our case adaptation of the tabulation strategy of the ILDM makes the method almost universally applicable to the

CFD modelling and practical simulations in a more robust and efficient way. The reduced manifold is found in terms of generalized coordinates θ by means of a unit grid which is built up during manifold generation. The parametrization is locally adapted in the state space such that it follows the tangent sub-space of the manifold.

As a result, reduced mechanisms are obtained, where the thermochemical state of the system is represented by a small number of variables only, which are denoted by θ. The numerical implementation is done by a path-following calculation algorithm that solves either (13) or (22) with predictor-corrector procedure starts on a known initial point ψ_p (e.g. the equilibrium point), generates an m_s-dimensional mesh on the nodes of the mesh cell by cell [35] by using the direction vectors of the neighbouring cells (see Fig. 5)

$$\psi_{\theta_i} = \psi_p(\theta) - \psi_p(\theta - \delta_i),\ \delta_i = (\delta_{i,1}, ..., \delta_{i,m_s})^T, \tag{23}$$

where $\psi_p(\theta)$, $\psi_p(\theta - \delta_i)$ are cell points, $\delta_{i,j}$ is the Kronecker delta and $\psi_\theta = (\psi_{\theta_1}, ..., \psi_{\theta_{m_s}})$ is an approximation of the (n by m_s)-dimensional gradient of ψ based on cell vertices with respect to θ of a neighbouring cell. Thus, the m_s-dimensional domain of the manifold is described by a mesh, which consists of m_s-dimensional hypercubes. As a result, the state ψ is obtained as known functions of the m_s-dimensional vector of mesh coordinates θ describing completely the manifold of the reduced model as a function in the explicit form

$$M_s = \{\psi : \psi = \psi(\theta)\}. \tag{24}$$

After the manifold is found in the explicit form (24) and properties are studied it can be applied to actual model reduction. Now, it is supposed that the first stage of model reduction of finding a low dimensional manifold has been completed, moreover, the second stage which is the analysis of the overall dynamics on the manifold and its properties such as existence, slowness, attractiveness, boundaries, optimal dimensions etc. has been performed successfully as well.

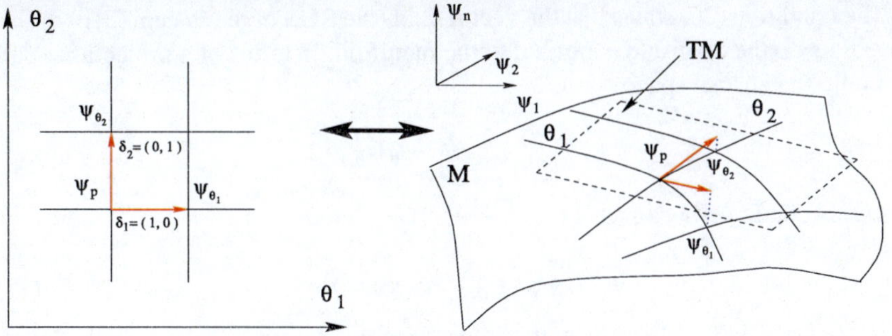

Fig. 5 Tabulation procedure

3 Improvement and Coupling with Diffusion - REDIM

In this section the final stage of model reduction - the implementation of the reduced model together with projection of the system of governing equations on the constructed manifold, i.e. the transformation to the new coordinates, and improvement of the reduced model to handle diffusion term is discussed.

3.1 Invariant Manifolds Concept

In order to determine rigorously the low dimensional reduced manifold that really takes into account the transport properties of the system (1) or (3) and overcome the problem with strongly coupled chemical and transport processes, the well developed framework of invariant system manifolds shall be applied (see e.g. [35, 18, 19, 20] for extensive exhibition of the concept). It will be developed and applied systematically by assuming the detailed system solution in the state space is restricted to an m_s-dimensional manifold defined by an explicit function $\psi(\theta)$ as the tabulation (24) provides. The assumption is natural because of differences in time scales, which lead to a decomposition of the system into fast and slow sub-systems and restricts the states accessed by the system

$$M = \{\psi : \psi = \psi(\theta), \psi : R^{m_s} \to R^n\}. \tag{25}$$

As before θ is an m_s-dimensional vector parameterizing the manifold. Remind that M as an invariant m_s-dimensional manifold if at any point $\psi \in M$ the vector field of (3) defined by the Right Hand Side (RHS) belongs to the tangent space $T_\psi M$ of M (see [18] for references and more definitions). This means for each point on the manifold the following condition has to be valid

$$\left(\psi_\theta^\perp(\theta)\right)^T \cdot \Phi(\psi) \equiv 0, \tag{26}$$

where $\psi_\theta^\perp$ defines the normal space to the manifold as $\left(\psi_\theta^\perp\right)^T \cdot \psi_\theta \equiv 0$. (see Fig. 6).

The condition (26) states that the vector field (the RHS of the system (3)) - $\Phi(\psi)$ for points on the manifold is normal to the manifold. In terms of a projection operator onto the normal space

$$P_{(TM)^\perp} = I - \psi_\theta \psi_\theta^+,$$

of M this condition becomes

$$\left(I - \psi_\theta \psi_\theta^+\right) \cdot \Phi(\psi) = 0. \tag{27}$$

Here ψ_θ^+ is the Moore-Penrose pseudo-inverse of ψ_θ (see e.g. [34, 37, 17]). It always exists when columns in the matrix representation of ψ_θ are linearly independent. This can always be guaranteed by a suitable choice of the local coordinates.

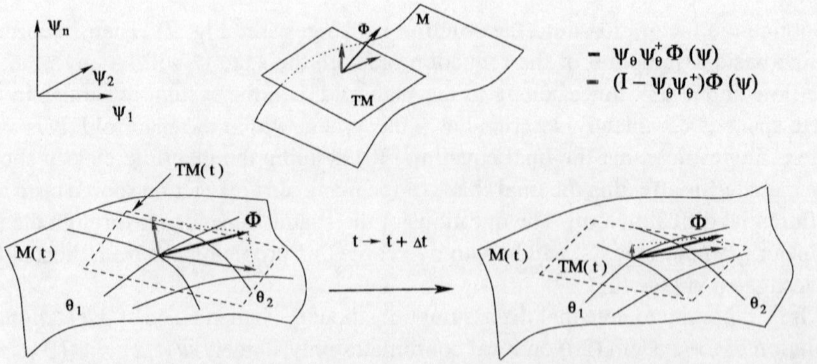

Fig. 6 Evolution of the slow manifold due to the invariance equation

3.2 REDIM Method and Implementation Scheme

In order to understand the properties of the model and present the suggested REDIM method the system (3) on the manifold (25), i.e. calculate the RHS consisting of the source, convection and diffusion terms on the manifold

$$\Phi(\psi(\theta)) = F(\psi(\theta)) - \mathrm{v} \cdot \psi_\theta \,\mathrm{grad}(\theta) - \frac{1}{\rho}\,\mathrm{div}(D \cdot \psi_\theta \,\mathrm{grad}(\theta)). \quad (28)$$

The equation system (27) is a key relation in our work and it is used as a basis to find an approximation for the reduced low dimensional manifold. By rewriting the diffusion term, for simplicity only assume equal diffusivities for the diffusion matrix i.e. $D = d \cdot I$, additionally, the convection term in (28) cancel out (see [41, 42] for a detailed discussion). Thus, after simplifying and applying the invariance condition (26) the invariance equation (27) can be written as following because of $\psi_\theta^+ \psi_\theta = I$

$$(I - \psi_\theta(\theta)\,\psi_\theta^+(\theta)) \cdot \left(F(\psi(\theta)) - \frac{d}{\rho}\,\psi_{\theta\theta} \circ \mathrm{grad}(\theta) \circ \mathrm{grad}(\theta)\right) = 0. \quad (29)$$

This equation describes main terms (i.e. reaction diffusion interaction) influencing the detailed system's solution on the manifold and driving it away from the defined manifold (25) unless the projection onto normal subspaces vanishes. There are different ways to find solutions of the equation (27) or (29) (see e.g. [19, 20]). The one that suggested here is the use of a reformulation into a PDE system for $\psi = \psi(\theta, t)$

$$\frac{\partial \psi(\theta)}{\partial t} = (I - \psi_\theta(\theta)\,\psi_\theta^+(\theta)) \cdot \left(F(\psi) - \frac{d}{\rho}\,\psi_{\theta\theta} \circ \mathrm{grad}(\theta) \circ \mathrm{grad}(\theta)\right), \quad (30)$$

such that a stationary set $\psi(\theta, \infty)$ of the PDEs (30) yields the desired manifold. The system (30) is solved starting from an initial guess for the invariant manifold and continue the integration until the solution converges (see Fig. 7). Then, according to the basic assumption of the reduction method, the state ψ will, at any point of the flow and at any time, belong to the manifold, i.e., the system dynamics in the state space is completely described as a movement within the manifold. It is very interesting to note that the final equation (30) defining the manifold clearly shows the main terms affecting the final shape of the manifold i.e. reaction source term and differential diffusion term. The questions of the spatial dependence through the parameter gradient $\text{grad}(\theta)$ and how to overcome this problem have been thoroughly investigated in [10, 11].

It is important to note that the assumption about the dependence of the stationary solution of the system (30) on local coordinates only, namely $\psi(t,x) = \psi(\theta(t,x))$, is a crucial point of the proposed method. However, in the implicit form, this assumption is universally used by all known standard procedures of model reduction and development.

The reason why many reduction procedures that do not take into account the influence of the transport term on the reduced models in detail nevertheless produce reasonable approximations is partly illustrated by the derived equation (30). In a major part of the domain of interest in the state space the source term is dominating and, therefore, reduced models which are based on an analysis of the source term only are sufficient to approximate the dynamics (see for more details and references [42]).

After the relaxation process of solution (30) has been performed and an approximation of the invariant manifold has been found, all thermochemical states everywhere in chemically reacting system are supposed to be elements of the manifold. The sum of the chemical source term and the transport term does, by definition, not lead to movements in the state space perpendicular to the manifold [28]. In order

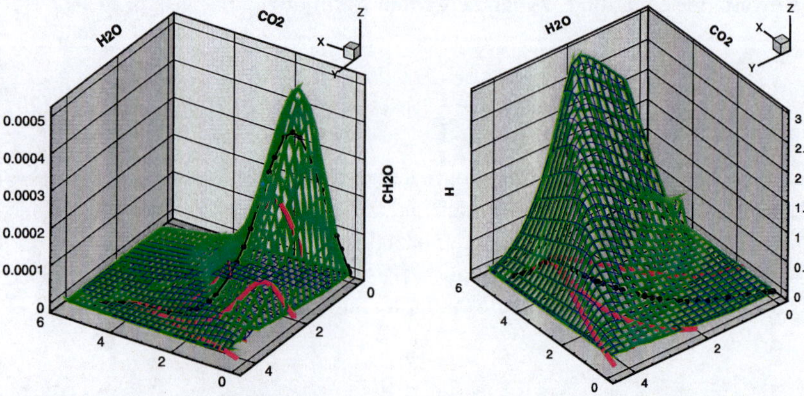

Fig. 7 2D-Relaxation: the projection onto minor (left) and major (right) specie's molar numbers, extended ILDM is shown by flat mesh, REDIM is the grid approaching solid lines, which are the set of stationary solutions that have same element composition and enthalpy in the fresh mixture

to complete the reduction procedure, a projection of the full system is required and can be performed by a transformation of the system into local coordinates of the invariant manifold.

3.2.1 Projection of Governing Equations onto REDIM

Under the assumption that the system (3) defines the full n-dimensional state ψ, but restricts the evolution to a movement tangential to the manifold, the n-dimensional governing equation system (3) can be projected onto the m_s-dimensional REDIM manifold

$$\psi_\theta \frac{\partial \theta}{\partial t} = F(\psi) - \mathrm{v} \cdot \psi_\theta \cdot \mathrm{grad}(\theta) - \frac{1}{\rho} \mathrm{div}(D \cdot \psi_\theta \cdot \mathrm{grad}(\theta)). \tag{31}$$

This is an n-dimensional equation system for m_s unknowns and, therefore, it is overdetermined. In order to obtain the governing equation system in terms of the reduced variables, the (m_s by n)-dimensional pseudo-inverse matrix ψ_θ^+ is used locally again. The column vectors of ψ_θ correspond to parametrization directions in the state space which have to be linearly independent, therefore, existence of ψ_θ^+ is ensured. Multiplication of (31) with ψ_θ^+ yields the equation system in terms of θ only

$$\frac{\partial \theta}{\partial t} = \psi_\theta^+ \cdot \left(F(\psi(\theta)) - \frac{1}{\rho} \cdot \mathrm{div}(D \cdot \psi_\theta(\theta) \cdot \mathrm{grad}(\theta)) \right) - \mathrm{v} \cdot \mathrm{grad}(\theta). \tag{32}$$

The final implementation scheme can be simplified and roughly summarized as follows. The REDIM table has been generated first, i.e. the state vector ψ, the chemical source term $F(\psi)$ and an array containing several physical quantities (temperature, density etc.) are calculated at each mesh point. The next step is the calculation of the term

$$\Xi(\theta) = D \cdot \psi_\theta(\theta),$$

which is needed for the solution of (32), at each point of the generated relaxed ILDM table. This is possible, because ψ and ψ_θ are functions of θ only. The gradient of ψ on the manifold - ψ_θ is approximated by central differencing, and its pseudo-inverse ψ_θ^+ is calculated. Then the m_s-dimensional projected source and diffusion terms $S(\theta)$ are coupled

$$S(\theta) = \psi_\theta^+(\theta) \cdot \left(F(\psi(\theta)) - \frac{1}{\rho} \cdot \mathrm{div}(\Xi(\theta) \cdot \mathrm{grad}(\theta)) \right).$$

Thus, the reduced equation system can be rewritten as

$$\frac{\partial \theta}{\partial t} = S(\theta) - \mathrm{v} \cdot \mathrm{grad}(\theta). \tag{33}$$

3.2.2 Implementation Scheme of REDIM

The full implementation scheme includes the following main steps

- construction of the source term based manifold either (13) or (22);
- extension of the initial guess - $\psi^0(\theta)$ to the whole domain of interest in the state space (see [10] and Fig. 7 left - blue mesh);
- relaxation (30) yields an approximation for the invariant system manifold - $\psi(\theta)$ (see Fig. 7 right and left green grid);

The conservation equations of the reacting flow are solved by a finite difference method on an adaptive mesh (see [32]). On each spatial grid point, the following steps are performed:

- calculation of the m_s-dimensional vector grad(θ);
- interpolation in the REDIM table depending on the reduced coordinates θ gives $\Xi(\theta), S(\theta)$;
- calculation of the right hand side (33).

Finally, at the end of each integration step, a back-transformation of the solution in terms of reduced coordinates θ to detailed variables $\psi(\theta)$ is accomplished by performing another interpolation in the REDIM table. This means that the primary results are spatial profiles of the generalized coordinates, and the species profiles can be recovered by a back-transformation (see e.g. [1]).

4 Conclusions

In the present article, the automatic method of analysis and reduction of detailed models of reacting flows is discussed. It is mainly based on the manifolds approach following the ideas of the ILDM standard technique of model reduction. There are two main questions are touched in the paper: the problem of global analysis of detailed chemical kinetic mechanisms and influence of coupling of the transport processes on the defined low-dimensional approximation of the kinetic mechanism. It is shown that the first problem can be handled by implementing of a coordinate free singular perturbation technique with the help of scale invariant realization strategy.

The second problem is overcome by an accurate approximation of the invariance equation of the PDEs system for the manifold that defines the reduced model. The main emphasis is made on the computational part with detailed description of technical tools and numerical problems arise in the implementation of the methods in modelling and simulations of reacting flows. Even on benchmark examples one can easily see the potential of systematic approach that has a clear geometrical picture in comparison to methodologies of model reduction conventionally used in combustion engineering.

References

1. Bauer, J., Bykov, V., Maas, U.: Proc. of European Conference on Computational Fluid Dynamics, ECCOMAS CFD (2006)
2. Bird, R., Stewart, W., Lightfoot, E.: Transport Phenomena. John Wiley and Sons, New York (1960)
3. Bodenstein, M., Lind, S.C.: Z. Phys. Chem. 27, 168–175 (1906)
4. Bowen, J.R., Acrivos, A., Oppenheim, A.K.: Chem. Eng. Sci. 18, 177–188 (1963)
5. Bykov, V., Goldfarb, I., Gol'dshtein, V.: J. Physics: Conference Series 22, 1–29 (2005)
6. Bykov, V., Goldfarb, I., Gol'dshtein, V.: J. Physics: Conference Series 55, 28–44 (2006)
7. Bykov, V., Goldfarb, I., Gol'dshtein, V., Maas, U.: IMA J. Appl. Math. 71(3), 359–382 (2006)
8. Bykov, V., Gol'dshtein, V.: J. Physics: Conference Series 138, 12003 (2008)
9. Bykov, V., Gol'dshtein, V., Maas, U.: Combust. Theory Model. 12(2), 389–405 (2008)
10. Bykov, V., Maas, U.: Proc. Comb. Inst. 31, 465–472 (2007)
11. Bykov, V., Maas, U.: Proc. Comb. Inst. 32(1), 561–568 (2009)
12. Chevalier, C., Pitz, W.J., Warnatz, J., Westbrook, C.K.: Proc. Comb. Inst. 24, 93–101 (1992)
13. Ern, A., Giovangigli, V.: Multicomponent Transport Algorithms. In: Lecture Notes in Physics, vol. 24. Springer, Berlin (1994)
14. Fenichel, N.: J. Diff. Equat. 31, 53–98 (1979)
15. Fraser, S.J.: J. Chem. Phys. 88, 4732–4738 (1988)
16. Gol'dshtein, V., Sobolev, V.: In: Gindikin, S.G. (ed.) Singularity Theory and Some Problems of Functional Analysis, vol. 153(2), pp. 73–92. American Mathematical Society, Translations, Providence (1992)
17. Golub, G.H., van Loan, C.F.: Matrix Computation. The Hopkins University Press, Baltimore (1989)
18. Gorban, A.N., Karlin, I.V.: Chem. Eng. Sci. 58, 4751–4768 (2003)
19. Gorban, A.N., Karlin, I.V., Zinovyev, A.Y.: Physica A 333, 106–154 (2004)
20. Gorban, A.N., Karlin, I.V., Zinovyev, A.Y.: Physics Reports 396(4-6), 197–403 (2004)
21. Green, W.H., Barton, P.I., Bhattacharjee, B., et al.: Ind. Eng. Chem. Res. 40(23), 5362–5370 (2001)
22. Griffiths, J.F.: Prog. Energy Combust. Sci. 21, 25–107 (1995)
23. Hirschfelder, J., Curtiss, C.: Molecular Theory of Gases and Liquids. John Wiley & Sons, Inc.S, New York (1964)
24. Kaper, H.G., Kaper, T.J.: Argonne National Lab, preprint ANL/MCS-P912-1001 (2001)
25. Lam, S.H.: Combust. Sci. Tech. 179, 767–786 (2007)
26. Lam, S.H., Goussis, D.M.: Int. J. Chem. Kinetics 26, 461–486 (1994)
27. Maas, U.: Mathematische Modellierung instationaerer Verbrennungsprozesse unter Verwenung detaillieter Reaktionsmechanismen. PhD thesis, Naturwissenschaftlich-Mathematische Gesamtfakultät, Ruprecht-Karls-Universitaet Heidelberg (1988)
28. Maas, U.: Appl. Math. 3, 249–266 (1995)
29. Maas, U.: Comput. Visualiz. Sci. 1, 69–82 (1998)
30. Maas, U., Bykov, V., Rybakov, A., Stauch, R.: Hierarchical Modelling of Combustion Processes. In: High Performance Computing on Vector Systems 2008. Springer, Heidelberg (2009)
31. Maas, U., Pope, S.B.: Combust. Flame 88, 239–264 (1992)
32. Maas, U., Warnatz, J.: Combust. Flame 74, 53–69 (1988)
33. Miyoshi, KUCRS software library: version (May 2005) beta, available from the author `http://www.frad.t.u-tokyo.ac.jp/~iyoshi/KUCRS/`

34. Moore, E.H.: Bulletin Amer. Math. Soc. 26, 394–395 (1920)
35. Nafe, J., Maas, U.: Combust Theory Model 6(4), 697–709 (2002)
36. Okino, M.S., Mavrovouniotis, M.L.: Chem. Rev. 98(2), 391–408 (1998)
37. Penrose, R.: Proc. Cambridge Philos. Soc. 51, 406–413 (1955)
38. Mauss, F., Peters, N.: In: Peters, N., Rogg, B. (eds.) Reduced Kinetic Mechanisms for Applications in Combustion Systems. Springer Lecture Notes in Physics, vol. 15, pp. 58–75. Springer, Heidelberg (1993)
39. Ren, Z., Pope, S.B.: Combust. Flame 147(4), 243–261 (2006)
40. Ren, Z., Pope, S.B.: Combust. Theory Model. 11(5), 715–739 (2007)
41. Ren, Z., Pope, S.B., Vladimirsky, A., Guckenheimer, J.M.: J. Chem. Physics 124, 114111 (2006)
42. Ren, Z., Pope, S.B., Vladimirsky, A., Guckenheimer, J.M.: Proc. Comb. Inst. 31, 473–481 (2006)
43. Rhodes, C., Morari, M., Wiggins, S.: Chaos 9, 108–123 (1999)
44. Roussel, M.R., Fraser, S.J.: Chaos 11, 196–206 (2001)
45. Singer, M.A., Pope, S.B., Najm, H.N.: Combust. Theory Model. 10(2), 199–217 (2006)
46. Strygin, B.B., Sobolev, V.A.: Decomposition of Motions by the Integral Manifolds Method. Nauka, Moscow (1988) (in Russian)
47. Tomlin, A.S., Turanyi, T., Pilling, M.J.: In: Pilling, M.J. (ed.) Comprehensive Chemical Kinetics. Low-temperature Combustion and Autoignition, vol. 35, Elsevier, Amsterdam (1997)
48. Warnatz, J., Maas, U., Dibble, R.W.: Combustion, 4 edn. Springer, Heidelberg (2004)
49. Williams, F.A.: Combustion Theory, 2nd edn. Princeton University Press, Princeton (1994)
50. Zagaris, A., Kaper, H.G., Kaper, T.J.: Multiscale Model. Simul. 2, 613–638 (2004)
51. http://www.ca.sandia.gov/chemkin/

Two-Dimensional Nonstationary Mathematical Model of Water Catchment

Dmitry A. Burakov, Evgeniya D. Karepova, and Vladimir V. Shaidurov

Abstract. In the paper the mathematical model, describing the influence of snowmelt and rainfall run-off of given inflow and outflow of a part of a river system being studied in two-dimensional in space and nonstationary in time formulations, is presented. The Navier-Stokes equations for viscous incompressible liquid are taken as the starting point. Then, by height averaging, the shallow water equations serving as the basis for the mathematical model are obtained. Stable combinations of boundary and initial conditions for problems with controlled run-off of water are considered.

1 Introduction

The modelling, calculation, and forecast of snowmelt and rainfall run-off and the formation of a channel are among the most important problems of hydrology. Such calculations are performed every day with variable success but with constant tendency to simplify the techniques being used which is caused by the fact that actual data for more complicated models, including on-line monitoring, are not sufficiently available [1-3].

Most of modern practical approaches are based on the statistical of large amount of long-standing meteorological and hydrological data observations. However, such approaches are less and less valid for forecasting because of natural and man-made changes [4-5].

Dmitry A. Burakov
Krasnoyarsk State Agricultural University, Mira Av. 90, Krasnoyarsk, 660049, Russia
e-mail: daburakov@yandex.ru

Evgeniya D. Karepova · Vladimir V. Shaidurov
Institute of Computational Modeling SB RAS, Akademgorodok, Krasnoyarsk, 660036, Russia
e-mail: jane@icm.krasn.ru, shidurov@icm.krasn.ru

E. Krause et al. (Eds.): Computational Sci., & High Performance Computing IV, NNFM 115, pp. 253–266.
springerlink.com

More complex physical and mathematical models are rarely used since they are not provided with empirical data. But the development of modern observation tools, including satellite survey, and the use of detailed geoinformation relief data gives more and more grounds for application of such models [6-7].

The aim of this paper is to describe the mathematical model based on the shallow water equations. Contrary to well-developed one-dimensional models (the Saint-Venant equation, the kinematic wave, etc.), this model more thoroughly takes into account channel as well as basin and drain surface relief which is of prime importance in flood forecast. Of course, for practice many water drain parameters for various surfaces should be improved. But this can be done in the process of their application. From the mathematical point of view the problem is closed and have stable solution due to nonincreasing energy of flow.

The three-dimensional nonstationary Navier-Stokes equations for viscous incompressible liquid with a free boundary of a flow surface are taken as the starting point of the mathematical model. Then by height averaging with the use of hydrostatic approximation of pressure the shallow water equations are obtained [5], [8].

They serve as the basis for mathematical model: a river run-off taking into account snowmelt and rainfall run-off on a basin area founded by watershed and hydroelectric power stations.

In the paper combinations of boundary and initial conditions providing existence and uniqueness of a solution for problem.

For this model, the application of a difference scheme implicit in time with iteration at each time level is considered.

2 Construction of Mathematical Model

2.1 Original Formulation of Problem

We take the Cartesian coordinate system such that the Oz axis is directed upwards (in the sense of a plumb bob under gravity). Assume that a river bed and an adjacent channel are described by a positive function $z = H(x,y)$ counted from the "sea level" conditional surface along the gravity direction (1).

We take the following model of continuous medium as the basis of the further modelling. Water is considered as viscous barotropic incompressible liquid. Assume also that motion is caused by a single mass force, namely, gravity directed downwards.

Then, to describe water flow in the downslope direction with high accuracy, we can use the following Navier-Stokes system of equations for viscous incompressible liquid [8]:

$$\frac{\partial u}{\partial t} + u\frac{\partial u}{\partial x} + v\frac{\partial u}{\partial y} + w\frac{\partial u}{\partial z} = -\frac{1}{\rho}\frac{\partial p}{\partial x} + \nu \Delta u, \tag{1}$$

$$\frac{\partial v}{\partial t} + u\frac{\partial v}{\partial x} + v\frac{\partial v}{\partial y} + w\frac{\partial v}{\partial z} = -\frac{1}{\rho}\frac{\partial p}{\partial y} + \nu \Delta v, \tag{2}$$

$$\frac{\partial w}{\partial t}+u\frac{\partial w}{\partial x}+v\frac{\partial w}{\partial y}+w\frac{\partial w}{\partial z}=-g-\frac{1}{\rho}\frac{\partial p}{\partial z}+\nu\Delta w, \tag{3}$$

$$\frac{\partial u}{\partial x}+\frac{\partial v}{\partial y}+\frac{\partial w}{\partial z}=0. \tag{4}$$

Here ν is the kinematic viscosity coefficient, ρ is density.

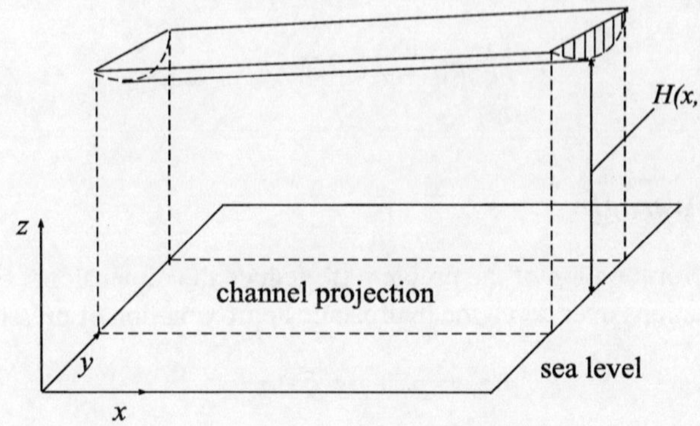

Fig. 1 River channel

For a boundary value problem usually the following boundary conditions are considered:

$$u_b=v_b=0, \qquad w_b=-I \quad \ldots \quad z=H(x,y), \tag{5}$$

$$\frac{\partial \xi}{\partial t}+u_0\frac{\partial \xi}{\partial x}+v_0\frac{\partial \xi}{\partial y}=w+R \quad \ldots \quad z=\xi(t,x,y), \tag{6}$$

where (u_0,v_0) and (u_b,v_b) are horizontal component of velocity on a free surface and on a bed, respectively; I is soil infiltration rate; R is a function accounting snow melt and water inflow. The kinematic condition (6) is given on an unknown moving boundary defined by the function ξ. From the physical point of view the condition $\xi(t,x,y)\geq H(x,y)$ at each instant t defines the closure $D(t)\subset R^2$ of a simply connected domain which is referred to as "a river channel" at instant t. The quotation marks point to distinction from the technical term which defines a channel as $D(t)$ at a steady-state interflood period. Since we are interested, in particular, in a flood period, the term "channel" and the set $D(t)$ are assumed to be time-dependent.

Thus, at each instant the conditions (5), (6) are defined on $D(t)$.

In this terms at each instant t the domain of the functions u,v,w,p is defined as the set $\Omega(t)\subset R^3$

$$\Omega(t)=\{(x,y,z):\quad (x,y)\in D,\quad z\in[H(x,y),\xi(t,x,y)]\}.$$

As initial conditions at the instant $t = t_s$ we take

$$u(t_s,x,y,z) = u_s(x,y,z), \qquad v(t_s,x,y,z) = v_s(x,y,z),$$

$$w(t_s,x,y,z) = w_s(x,y,z) \quad \forall (x,y,z) \in \Omega(t_s). \tag{7}$$

Notice that there is no initial condition for pressure. To determine it uniquely, usually the following condition is imposed:

$$\int_{\Omega(t)} p(t,x,y,z)d\Omega = 0. \tag{8}$$

2.2 Depth Averaging

To simplify the formulation of the problem (to reduce the dimensional by one), we average the equations over depth the hydrostatic approximation of pressure:

$$p(z) = p_a + \rho g(\xi - z). \tag{9}$$

Pressure p_a on a surface usually is assumed to be equal to atmospheric one. Attempts to take into account difference of pressures caused by other reasons, for instance, by rainfall [1], give insignificant distinctions.

When averaging the motion equations (1), (2) over depth, one should apply the rule for differentiation of an integral with variable limits and take into account the boundary conditions (5), (6) and continuity equation (4). As a result, we arrive at the equations [1], [5], [8]

$$\frac{\partial}{\partial t}\int_H^\xi u\,dz + \frac{\partial}{\partial x}\int_H^\xi u^2\,dz + \frac{\partial}{\partial y}\int_H^\xi uv\,dz - u_0 R = -\frac{1}{\rho}\int_H^\xi \frac{\partial p}{\partial x}dz + \nu\int_H^\xi \Delta u\,dz, \tag{10}$$

$$\frac{\partial}{\partial t}\int_H^\xi v\,dz + \frac{\partial}{\partial x}\int_H^\xi uv\,dz + \frac{\partial}{\partial y}\int_H^\xi v^2\,dz - v_0 R = -\frac{1}{\rho}\int_H^\xi \frac{\partial p}{\partial y}dz + \nu\int_H^\xi \Delta v\,dz, \tag{11}$$

$$\frac{\partial \xi}{\partial t} + \frac{\partial}{\partial x}\int_H^\xi u\,dz + \frac{\partial}{\partial y}\int_H^\xi v\,dz = R - I. \tag{12}$$

We denote total depth by $h = \xi - H$ and pass to averaged velocities

$$\bar{u}(t,x,y) = \frac{1}{h}\int_H^\xi u\,dz \quad \text{and} \quad \bar{v}(t,x,y) = \frac{1}{h}\int_H^\xi v\,dz. \tag{13}$$

Rewrite the equations (10)-(12) in the following form:

$$\frac{\partial}{\partial t}(h\bar{u})+\frac{\partial}{\partial x}(h\bar{u}^2)+\frac{\partial}{\partial y}(h\bar{u}\bar{v})+gh\frac{\partial h}{\partial x}=Ru_0+F_x-gh\frac{\partial H}{\partial x}, \tag{14}$$

$$\frac{\partial}{\partial t}(h\bar{v})+\frac{\partial}{\partial x}(h\bar{u}\bar{v})+\frac{\partial}{\partial y}(h\bar{v}^2)+gh\frac{\partial h}{\partial y}=Rv_0+F_y-gh\frac{\partial H}{\partial y}, \tag{15}$$

$$\frac{\partial h}{\partial t}+\frac{\partial}{\partial x}(h\bar{u})+\frac{\partial}{\partial y}(h\bar{v})=R-I, \tag{16}$$

where F_x, F_y are approximations of averaged viscosity forces. In is clear that when replacing quadratic terms of the form

$$\frac{1}{h}\int_H^\xi u^2 dz \approx \bar{u}^2 \tag{17}$$

an error is assumed which is ensured to be zero only in the case of $u = \mathrm{const}(z)$ along the vertical direction. In the remaining cases from the Cauchy-Bunyakovskii inequality it follows that

$$\int_H^\xi u^2 dz \geq h\bar{u}^2. \tag{18}$$

Therefore some authors ([1], [2]) propose to take into account nonuniformity of velocity distribution with depth in more detail and instead of the assumption (17) to use the approximation

$$\int_H^\xi u^2 dz \approx \alpha_x^2 h\bar{u}^2 \tag{19}$$

with some (empirical) coefficient $\alpha_x^2 \geq 1$. In a similar way

$$\int_H^\xi v^2 dz \approx \alpha_y^2 h\bar{v}^2, \quad \alpha_y^2 \geq 1. \tag{20}$$

Then we obtain

$$\int_H^\xi uv\, dz \approx \alpha_x \alpha_y h\bar{u}\bar{v} \tag{21}$$

for the energy relations to be valid.

In principle, in the presented models such relations can be taken into account. But, by doing so, we involve the problems of semiempirical definition of these coefficients which certainly depend on depth, temperature, bed surface etc. Because of this, to simplify the presentation of main mathematical ideas, we leave aside these problems assuming $\alpha_x = \alpha_y = 1$. From the subsequent relations we can see that, in so doing, we somewhat understate kinetic energy of flow.

2.3 Viscosity and Resistance Forces

To approximate the last terms in the equations (10) and (11), we can put [8]

$$\nu \int_H^{\xi} \Delta_2 u \, dz \approx \nu \Delta_2(h\bar{u}) \quad . \quad \nu \int_H^{\xi} \Delta_2 v \, dz \approx \nu \Delta_2(h\bar{v}), \tag{22}$$

where

$$\Delta_2 = \frac{\partial^2}{\partial x^2} + \frac{\partial^2}{\partial y^2}$$

and

$$\nu \int_H^{\xi} \frac{\partial^2 u}{\partial z^2} dz = \nu \frac{\partial u}{\partial z}\Big|_H^{\xi}, \quad \nu \int_H^{\xi} \frac{\partial^2 v}{\partial z^2} dz = \nu \frac{\partial v}{\partial z}\Big|_H^{\xi}. \tag{23}$$

In the subsequent considerations on the basis of practical conclusions [9], we neglect horizontal viscosity of flow in comparison with its vertical component. As a result, we take viscosity forces in the following form:

$$F_x = \nu \frac{\partial u}{\partial z}\Big|_{\xi} - \nu \frac{\partial u}{\partial z}\Big|_H = \frac{1}{\rho}(\tau_{x_0} - \tau_{x_b}), \; F_y = \nu \frac{\partial v}{\partial z}\Big|_{\xi} - \nu \frac{\partial v}{\partial z}\Big|_H = \frac{1}{\rho}(\tau_{y_0} - \tau_{y_b}), \tag{24}$$

where τ_{x_0}, τ_{y_0} are tangent stresses on a water surface and τ_{x_b}, τ_{y_b} are those on a bottom.

To describe these stresses in more detail, additional assumptions taking into account specific conditions are involved. For example, an a water surface we can take into account wind stresses and resistance caused by rainfall. But here we omit these effects which have only slight influence on lengthy flows.

Bottom friction is of greater importance. In hydraulics the ratio of flow friction force and a unit of area of a bottom surface is assumed to be proportional to squared flow velocity. The proportion coefficient C is called Shesi's velocity coefficient. For one-dimensional steady-state uniform flow at a depth h Manning derived the empirical formula

$$C = h^{1/6} \big/ n_r, \tag{25}$$

where n_r is Manning's roughness coefficient tabulated for main characteristics of bed.

In [1], [8] components of tangent stress at each point of two-dimensional flow are described by the same dependence as in the one-dimensional case but with velocity direction at this point into account:

$$\tau_{x_b} = \frac{\rho g \bar{u}\sqrt{\bar{u}^2+\bar{v}^2}}{C^2}, \qquad \tau_{y_b} = \frac{\rho g \bar{v}\sqrt{\bar{u}^2+\bar{v}^2}}{C^2}. \tag{26}$$

Thus, all parameters of the equations (14)-(16) are defined.

3 Mathematical Model of Water Catchment

This model is useful in analysis of the forecast of flow of spring and rainfall floods. Here we reason that the domain D of the problem does not depend on time and is defined only by the boundary of watershed of a river system. For example, for the dynamic estimate of basin for Sayano-Shushenskaya hydroelectric power station it is required to take the domain D bounded by the watershed Γ_{ws} and the hydroelectric power station Γ_{es}. On the map this domain is shown by a thick line and marked by numerical II (Fig. 2). In Fig. 3 an idealized scheme of the site of Sayano-Shushenskaya hydroelectric power station is presented.

Fig. 2 A map of two domains of basin

In principle, with the schedule of flow and evacuation of water at Sayano-Shushenskaya hydroelectric power station being available, for Krasnoyarsk and its suburb we should take other water catchment area D which is shown on the map as well (Fig. 2) and is marked by numerical I.

Thus, the domain D in this model is defined and does not depend on time. The nearer are channels of rivers and feeders, the higher must be accuracy with relief

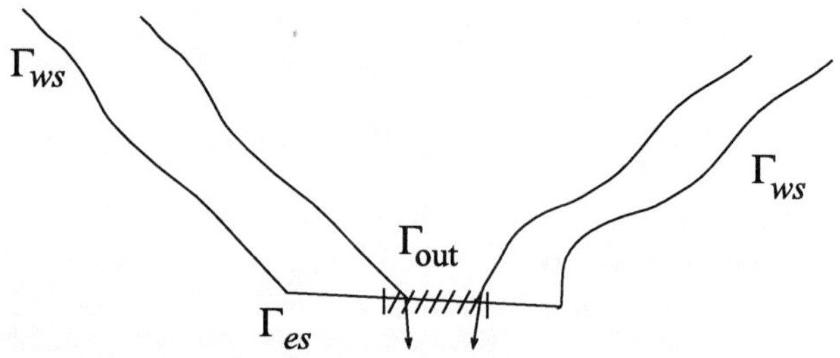

Fig. 3 Outflow segment (issuing characteristics for the domain II)

(the function $H(x,y)$) is defined. Special high accuracy is required on channels and near them.

For the numerical solution of the problem we rearrange the equations. We multiply the equation (16) by $\bar{u}/2$ and subtract it from the equation (14). As a result, we arrive at

$$\sqrt{h}\frac{\partial}{\partial t}(\sqrt{h}\bar{u})+\frac{1}{2}\left(h\frac{\partial \bar{u}^2}{\partial x}+\bar{u}\frac{\partial}{\partial x}(h\bar{u})\right)+\frac{1}{2}\left(\bar{u}\frac{\partial}{\partial y}(h\bar{v})+2h\bar{v}\frac{\partial \bar{u}}{\partial y}\right)+gh\frac{\partial h}{\partial x}=$$

$$=R(u_0-\bar{u}/2)+I\bar{u}/2+F_x-gh\frac{\partial H}{\partial x}, \tag{27}$$

where $F_x=\frac{g\bar{u}}{C^2}\sqrt{\bar{u}^2+\bar{v}^2}$.

In a similar way we subtract the equation (16) multiplied by $\bar{v}/2$ from the equation (15). This gives

$$\sqrt{h}\frac{\partial}{\partial t}(\sqrt{h}\bar{v})+\frac{1}{2}\left(2h\bar{u}\frac{\partial \bar{v}}{\partial x}+\bar{v}\frac{\partial}{\partial x}(h\bar{u})\right)+\frac{1}{2}\left(h\frac{\partial \bar{v}^2}{\partial y}+\bar{v}\frac{\partial}{\partial y}(h\bar{v})\right)+gh\frac{\partial h}{\partial y}=$$

$$=R(v_0-\bar{v}/2)+I\bar{v}/2+F_y-gh\frac{\partial H}{\partial y}, \tag{28}$$

where $F_y=\frac{g\bar{v}}{C^2}\sqrt{\bar{u}^2+\bar{v}^2}$.

The equation (16) remains unchanged.

Now we impose boundary conditions. On the boundary Γ_{ws} of watershed we impose the boundary condition

$$\bar{u}_n=0 \quad \text{on} \quad \Gamma_{ws}, \tag{29}$$

where $\bar{u}_n$ is velocity along a normal to the boundary:

$$\bar{u}_n=\bar{u}\cdot\cos(n,x)+\bar{v}\cdot\cos(n,y),$$

where n is an outer normal vector, $\cos(n,x), \cos(n,y)$ are cosines of the angles between n and the x and y axes, respectively.

It is easily verified that the condition (29) does not allow the characteristics of the equations (27), (28), (16) to enter and to issue [10]. From the theoretical point of view it is sufficient to impose these conditions on Γ_{ws}. As for outflow mode in the site of a hydroelectric power station, we consider the scheduled discharge mode where a known volume of water defined by the relation

$$\int_{\Gamma_{es}} h\bar{u}_n d\Gamma = d_{\text{out}}(t) \tag{30}$$

flows through the site.

Since boundary conditions are not required where characteristics issue out of the domain, we regulate the outflow (30) by the change of passing width in the following way (Fig. 3). On the shaded part Γ_{out} boundary conditions are not imposed and the functions u, v, h are to be determined. Discharge turns out to be equal to the left-hand side of (30).

It is clear that a priori it is not equal to the required function $d_{\text{out}}(t)$. Hence, at each time step t we regulate the length of Γ_{out} (the lock width) such that the equality (30) is achieved. On the remaining part of Γ_{es} we suppose that

$$\bar{u}_n = 0 \quad \text{on} \quad \Gamma_{es} \setminus \Gamma_{\text{out}}. \tag{31}$$

To close the problem, we define initial conditions at some instant t_s:

$$\bar{u}(t_s,x,y) = \bar{u}_s(x,y), \quad \bar{v}(t_s,x,y) = \bar{v}_s(x,y), \quad h(t_s,x,y) = h_s(x,y) \quad \text{on} \quad D. \tag{32}$$

It is obvious that u_s, v_s must satisfy the conditions (29)-(31).

Due to (29), (31), from physical reasoning all characteristics have their origin on the plane $t = t_s$ and issue out of the domain of the problem in the site Γ_{out} at an instant $t \geq t_s$. Hence, the boundary and initial conditions are sufficient to determine the functions in the three-dimensional domain

$$\Omega = [t_s, T] \times D. \tag{33}$$

Now we consider the second version of domain D with inflow on the segment Γ_{in} for the same hydroelectric power station (Fig. 4). Now the characteristics of all the functions u, v, h have their origin on Γ_{in}. Because of this, on Γ_{in} the following boundary conditions should be taken

$$u(t,x,y) = u_{in}(t,x,y), \quad v(t,x,y) = v_{in}(t,x,y),$$

$$h(t,x,y) = h_{in}(t,x,y) \quad \text{on} \quad [0,T] \times \Gamma_{in}. \tag{34}$$

On the remaining part of the boundary the conditions (29)-(31) are taken. The functions u_{in}, v_{in}, h_{in} must satisfy the discharge condition

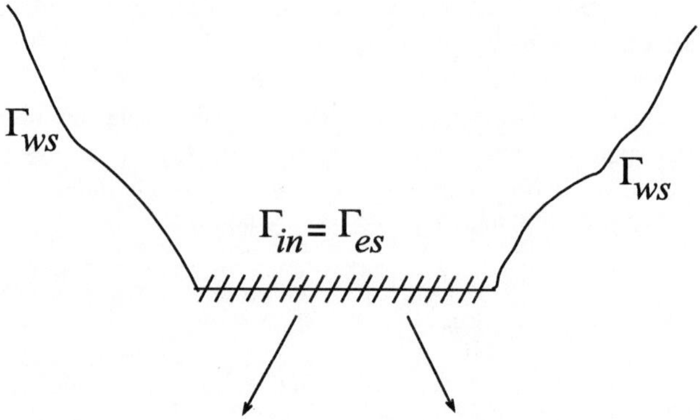

Fig. 4 Segment of entering characteristics for the domain I

$$\int_{\Gamma_{in}} h\bar{u}_{in}\, d\Gamma = d_{in}(t), \qquad t \in [0,T], \tag{35}$$

with the inflow function $d_{in}(t)$ equal to the outflow function $d_{\text{out}}(t)$ for the previous computational domain. It is obvious that to define (34), a large volume of experimental data is required.

Finally, we consider the conditions of flowing a river into a water area such as lake, sea, or ocean through the boundary Γ_Δ (Fig. 5).

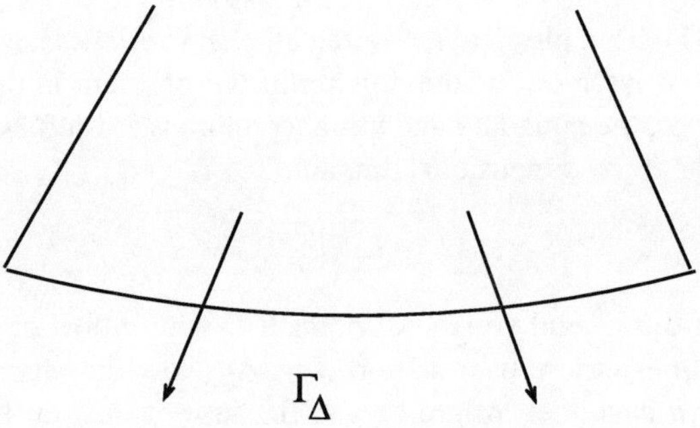

Fig. 5 Delta of a river

In the numerical modelling "liquid bottom" with the function $H(t,x,y)$ equal to the "undisturbed" level of a water area is used. Time-dependence may be required when taking into account possible influence of tide effects. The "undisturbed level" is defined by experimental data and, when combined with the choice of Shesi's coefficient C, it may provide reasonable results. From the theoretical point of view no additional condition on Γ_Δ is required in the case that all characteristics are issuing, i.e., there is no reverse flow.

4 Energy Balance in Basin Model

To derive an a priori estimate, we fix t, multiply the equation (27) by $\bar{u}$ and integrate over D. Applying the first Green formula, we arrive at the equality

$$\frac{1}{2}\int\limits_D \frac{\partial}{\partial t}(h\bar{u}^2)dD + \frac{1}{2}\int\limits_\Gamma hu^2 u_n d\Gamma + \frac{1}{2}\int\limits_D gh\bar{u}\frac{\partial(h+H)}{\partial x}dD =$$

$$= \int\limits_D \Big(R(u_0 - \bar{u}/2) + I\bar{u}/2 + F_x\Big)\bar{u}\,dD. \tag{36}$$

Multiply the equation (28) by $\bar{v}_T$ and integrate over D, applying the first Green formula:

$$\frac{1}{2}\int\limits_D \frac{\partial}{\partial t}(hv^2)dD + \frac{1}{2}\int\limits_\Gamma hv^2 u_n d\Gamma + \frac{1}{2}\int\limits_D ghv\frac{\partial(h+H)}{\partial x}dD =$$

$$= \int\limits_D \Big(R(v_0 - \bar{v}/2) + I\bar{v}/2 + F_y\Big)\bar{v}\,dD. \tag{37}$$

Finally, multiply (16) by $g(h+H)$:

$$g\int\limits_D \frac{\partial}{\partial t}\Big(\frac{h^2}{2} + hH\Big)dD + g\int\limits_D (h+H)\frac{\partial}{\partial x}(h\bar{u})dD + g\int\limits_D (h+H)\frac{\partial}{\partial y}(h\bar{v})dD =$$

$$= g\int\limits_D (R-I)(h+H)\,dD. \tag{38}$$

Sum up the equations (36)-(38) and apply the first Green formula. As a result, we get the equality

$$\frac{1}{2}\frac{\partial}{\partial t}\int\limits_D (h\bar{u}^2 + h\bar{v}^2 + gh^2 + 2ghH)dD + \frac{1}{2}\int\limits_\Gamma h(\bar{u}^2 + \bar{v}^2)u_n\,d\Gamma +$$

$$+ \int\limits_\Gamma gh(h+H)u_n d\Gamma = \int\limits_D \Big(R(u_0 - \bar{u}/2) + I\bar{u}/2 + F_x\Big)\bar{u}\,dD + \tag{39}$$

$$+ \int\limits_D \Big(R(v_0 - \bar{v}/2) + I\bar{v}/2 + F_y\Big)\bar{v}\,dD + \int\limits_D g(R-I)(h+H)dD.$$

The first term defines variation of total energy (kinetic and potential) at instant t. In the second term the expression under the integral sign vanishes on the all boundary Γ_{ws} of watershed and differs from zero only on Γ_{out} and Γ_{in}. On the first segment the integral

$$\frac{1}{2}\int\limits_{\Gamma_{\text{out}}} h(\bar{u}^2+\bar{v}^2)u_n\,d\Gamma$$

defines kinetic energy loss (due to positive sign of u_n).

On the second segment the integral

$$\frac{1}{2}\int\limits_{\Gamma_{in}} h(\bar{u}^2+\bar{v}^2)u_n\,d\Gamma$$

defines acquiring kinetic energy (due to negative sign of u_n).

The third term behaves similarly. The expression under the integral sign vanishes on Γ_{ws}; on Γ_{out} it defines potential energy loss at outflow of D, and on Γ_{in} it defines potential energy.

The role of the terms $F_x\bar{u}$ and $F_y\bar{v}$ in energy loss due to their negative sign is clearly seen. The function $R(t,x,y)$, whose value varies in the range from 0 to $h(t,x,y)$ on a considerable area of basin, is of special importance. The quantity $I(t,x,y)$ (excepting the channel) is of similar order. Thus, from the quantitative point of view, the maximal values of all terms are of the same order.

5 Iterative Solution of Problem

We describe the general scheme of the solution. Assume that a solution of the problem and a segment $\Gamma_{\text{out}}(t-\tau)$ are known at instant $t-\tau$ for small $\tau>0$. We determine a solution and a segment $\Gamma_{\text{out}}(t)$ instant t. Time derivatives are approximated in the following form:

$$\sqrt{h}\frac{\partial}{\partial t}(\sqrt{h}\bar{u})|_t \approx \sqrt{h}|_t\frac{(\sqrt{h}\bar{u})_t-(\sqrt{h}\bar{u})_{t-\tau}}{\tau},$$

$$\sqrt{h}\frac{\partial}{\partial t}(\sqrt{h}\bar{v})|_t \approx \sqrt{h}|_t\frac{(\sqrt{h}\bar{v})_t-(\sqrt{h}\bar{v})_{t-\tau}}{\tau}, \tag{40}$$

$$\left.\frac{\partial h}{\partial t}\right|_t \approx \frac{h|_t-h|_{t-\tau}}{\tau}.$$

The remaining terms are taken at level t such that we obtain a completely implicit scheme.

At an instant t we have a quasilinear stationary problem. The previous time level $t-\tau$ provides a good approximation for its coefficients, so we need a small number of iteration steps to achieve required accuracy.

In iteration the equality (30) is achieved in the following way. If the value of $\int_{\Gamma_{es}} h\bar{u}_n\,d\Gamma$ turns out to be greater than $d_{\text{out}}(t)$, then the "lock" width Γ'_{out} should be decreased to achieve the equality. In the opposite situation, where flow is less than $d_{\text{out}}(t)$, the "lock" width Γ'_{out} should be increased. Then at the next iteration step new Γ'_{out} is taken as a part of the boundary for the problem to be solved at time level t.

For the solution of the stationary problem at time level t, the finite element method with linear elements on triangles is applied. Stability of a solution of a discrete problem is provided by a grid analogue of the energy conservation law of the form (39). For energy to be a weighted Hilbert norm, depth h may not be less than zero over the solution. Nevertheless, due to discretization errors we do not have such a guarantee. Hence, additional efforts are required to provide positive sign of h. We consider two approaches.

The first approach being the most popular is as follows. For approximation of the equation (16) a monotone difference scheme providing nonnegativity of h for nonnegative initial and boundary conditions and the right-hand side $R-I$ is taken. By the Godunov theorem this scheme can be first-order consistent only. In the absence of rainfall, due to zero or small weight h on a large area, the functions $\bar{u}$ and $\bar{v}$ become in essence uncontrolled on this area. This results in uncontrollable errors which usually are suppressed with the help of additional computational viscosity.

The second approach is in passing from the unknown functions $\bar{u}, \bar{v}, h$ in the equations (27), (28), (16) to the functions

$$\tilde{u} = \sqrt{h}\bar{u}, \qquad \tilde{v} = \sqrt{h}\bar{v}, \qquad \tilde{\xi} = \sqrt{g}(h+H). \tag{41}$$

In this case, in spite of a large area where the function h may vanish in the absence of rainfall, energy is expressed in the following form

$$\frac{1}{2}\int_D (\tilde{u}^2 + \tilde{v}^2 + \tilde{\xi}^2)dD. \tag{42}$$

Then we can apply the finite element method for discretization and various iterative schemes to obtain (30). As mentioned above, a previous time level provides an appropriate initial guess for such iterative methods.

In this approach, even though $h(t,x,y)$ takes small negative values, the norm (42) remains a Hilbert one when approximately solving the problem and provides an estimate for stability of a solution of a grid problem.

One more advantage of this approach is that unknown solution are of different smoothness. For example, the relief $H(x,y)$ usually is a piecewise smooth function with bounded but discontinuous first-order derivatives. Second-order derivatives may be infinitely large. Never the less, the flow surface $\xi(t,x,y)$ is a more smooth function. As a consequence, $h(t,x,y) = \xi(t,x,y) - H(x,y)$ is a piecewise smooth function with infinitely large second-order derivatives. When approximating first-order derivatives by divided differences, the values of second-order derivatives become of crucial importance for the truncation error. Therefore along breaks of relief derivatives of h are poorly approximated. Derivatives of the more smooth function $\xi(t,x,y)$ are approximated rather better.

6 Conclusion

In the paper the mathematical model is proposed which is intended for calculation and forecast of snowmelt and rainfall run-off with influence of controlled discharged at a hydroelectric power station taken into account. In this model a large area, whose relief and data on run-off must be formed taking into account satellite data as well as ground-based monitoring, is considered [1-3]. In addition, flood conditions are developed starting from the last year's autumn when contact of soil and snow cover, which has influence on the magnitude of infiltration and soil water reserve, is formed. Development of appropriable geographical information systems provide information of this kind as well as possibility to use it for the solution of the problems formulated above.

Acknowledgements. The work is supported by Russian Foundation of Fundamental Researches (grant 08-01-00621-a) and by State Contract No. 02.740.11.0621 of Federal Program "Scientific and education personnel of innovative Russia".

References

1. Kuchment, L.S.: Models of Formation of River Run-off. Gidrometeoizdat, Leningrad (1980) (in Russian)
2. Vinogradov, Y.B.: Modeling of Formulation of Run-off. Gidrometeoizdat, Leningrad (1998) (in Russian)
3. Koren', V.I.: Mathematical Models for Forecast of River Run-off. Gidrometeoizdat, Leningrad (1991) (in Russian)
4. Burakov, D.A.: Calculation of river floods with the use of analogues. Meteorology and hydrology 10, 79–88 (1982) (in Russian)
5. Burakov, D.A., Karepova, E.D., Shaidurov, V.V.: Mathematical modeling of run-off: theoretical basis, modern state, and outlook. Bulletin of Krasnoyarsk state university. Physics and Mathematics 4, 3–19 (2006) (in Russian)
6. Burakov, D.A., Avdeeva, Y.V.: Technology of on-line forecast of everyday water discharge (level) on the basis of satellite information on snow-covering (using Nizhnyaya Tunguska river as an example). Meteorology and hydrology 10, 75–87 (1996) (in Russian)
7. Burakov, D.A., Kashkin, V.B., Sukhinin, A.I., Romas'ko, V.Y., Ratnenko, I.V.: Methods for determining snow-covering of a eiver basin from satellite data for on-line run-off forecast. Meteorology and hydrology 8, 100–109 (1996) (in Russian)
8. Johnson, R.W. (ed.): The Handbook of Fluid Dynamics. CRC Press/ Springer, USA/Germany (1998)
9. Dronkers, I.: Calculation of Tides in Rivers and Coastal Water. Gidrometeoizdat, Leningrad (1967) (in Russian)
10. Rozhdestvenskii, B.L., Yanenko, N.I.: Systems of Quasilinear Equations and Their Applications in Gas Dynamics. Nauka, Moscow (1978) (in Russian)

Central-Upwind Schemes for Boussinesq Paradigm Equations

Alina Chertock, Christo I. Christov, and Alexander Kurganov

Abstract. We develop a new accurate and robust numerical method for the Boussinesq paradigm equation (BPE). To design the method we first introduce a change of variables, for which the BPE takes the form of a nonlinear wave equation with the global pressure, and rewrite the wave equation as a system of conservation laws with a global flux. We then apply a Godunov-type central-upwind scheme together with an efficient FFT-based elliptic solver to the resulting system. Making use of the new scheme, we investigate the propagation of one- and two-dimensional solitary waves of BPE and identify their solitonic behaviour.

1 Introduction

One of the most important events in mathematical physics was the discovery in the 1960s [16] that the nonlinear waves can behave as particles. The localized waves which can retain their identity during interaction appear to be a very adequate model for particles, especially if some mechanical quantities (such as mass, momentum, energy) are conserved by the governing system of equations. In the case when the system has an infinite number of conservation laws, the "permanent" localized waves are called *solitons*. In the past five decades, a multitude of deep mathematical results were obtained for solitons, and a plethora of analytical solutions found for

Alina Chertock
Department of Mathematics, North Carolina State University, Raleigh, NC 27695, USA
e-mail: chertock@math.ncsu.edu

Christo Christov
Department of Mathematics, University of Louisiana, Lafayette, LA 70504, USA
e-mail: christov@louisiana.edu

Alexander Kurganov
Mathematics Department, Tulane University, New Orleans, LA 70118, USA
e-mail: kurganov@math.tulane.edu

E. Krause et al. (Eds.): Computational Sci., & High Performance Computing IV, NNFM 115, pp. 267–281.
springerlink.com

systems of interacting solitons. As it should have been expected, most of the physical systems are not fully integrable, and for them (even in one space dimension) only numerical approaches can lead to unearthing the pertinent physical mechanisms of the interactions (see, e.g., [4, 6] and references therein).

The overwhelming majority of the analytical and numerical results obtained so far are for the one-dimensional (1-D) case, while in the multiple number of space dimensions, much less can be achieved analytically, and almost nothing is known about the unsteady solutions that involve interactions, especially when full-fledged Boussinesq equations are involved. In the case when one of the fourth-order spatial derivatives can be neglected (the Kadomtsev-Petviashvili equation), two-dimensional (2-D) waves have been found. Being periodic in one of the spatial dimensions, the latter are not truly 2-D waves, localized in both direction. Thus the problem for the Boussinesq equation with fourth-order derivatives in both spatial directions, still awaits it solution.

The difficulties in the multidimensional cases are connected with the unboundedness of the region, and with the slow decay of the solution at infinity. For this reason, the first case to undergo investigation is when the wave profile is stationary in a frame moving with a prescribed phase speed. The shape of the stationary moving wave was reliably computed in [2] using a specialized Galerkin spectral technique, which showed that the decay at infinity is algebraic rather than exponential. A perturbation technique recently developed in [5] confirmed the findings of [2] and put the stationary problem on more rigorous analytical ground.

Upgrading the numerical approach to the unsteady 2-D case is non-trivial, and several different techniques are to be developed, in order to obtain reliable description of the mechanics and physics of the processes involved. The essential properties of the numerical method in this case is to be very efficient computationally, because of the sheer size of the problem: the computational box must be large enough to allow for a good approximation of the localized solution, and the grid must be fine enough to reduce the truncation error. The conservative schemes used in [4, 6] (and in the works of other authors) are not easily extended to 2-D cases, and require an excessive amount of computational time.

The purpose of the present paper is to apply the "machinery" of modern finite volume methods, Godunov-type central-upwind schemes [8, 9, 11, 13], to the BPE and to numerically study their solitary wave solutions.

We first outline the scheme for the 1-D BPE and then proceed to trace the evolution of a 2-D profile described in [5]. This is the first and utmost verification of the "solitonic" properties of the 2-D shapes, before they can be used in computations aimed at the investigation of their collision properties.

2 Boussinesq Paradigm Equation

When Boussinesq derived his famous equation [1], he demonstrated that the nonlinearity can balance dispersion and lead to the existence of waves of permanent shape that propagate in quite similar fashion as the profiles of the d'Alembert solution of the wave equation. Unfortunately, Boussinesq did some additional (and as it turns out) unnecessary assumptions, which rendered his equation incorrect in the sense of Hadamard. An overview of the different Boussinesq equations can be found in [6] (see also references therein). The accurate derivation of the Boussinesq system can be found, e.g., in [4]. In nutshell, Boussinesq's's Boussinesq Equation (BBE) lacks the mixed fourth derivative, which was the cause of it incorrectness. In order to distinguish the equation to which a consistent application of Boussinesq idea would have led, the latter is called Boussinesq Paradigm Equation (BPE). The essential mathematical difference between BBE and BPE is that the former is fully integrable (although incorrect) and the latter is not (although, it is the physically pertinent and mathematically correct model). The BPE furnishes one of those examples, which necessitates an efficient numerical treatment as a precondition for a deeper physical understanding of the problem.

The BPE is a 2-D amplitude equation:

$$w_{tt} = \Delta\left[w - \alpha w^2 + \beta_1 w_{tt} - \beta_2 \Delta w\right], \tag{1}$$

where w is the surface elevation, $\beta_1, \beta_2 > 0$ are two dispersion coefficients, and α is an amplitude parameter. As it has been already mentioned, the main difference between equation 1 and the BBE is that now $\beta_1 \neq 0$.

It has been recently shown in [2, 5] that the 2-D BPE admits stationary soliton solutions as well. Even though no analytical formula for these solutions is available, they can be accurately constructed using either a perturbation or a Galerkin spectral method. However, the stability of the 2-D solitons has still been an open problem, and will be addressed in the present work.

The situation is much more amenable with the 1-D BPE:

$$w_{tt} = \left[w - \alpha w^2 + \beta_1 w_{tt} - \beta_2 w_{xx}\right]_{xx}, \tag{2}$$

for which it has been shown (see [4] for references) that it admits a one-parameter family of soliton solutions given by

$$w^s(x,t;c) = -\frac{3}{2}\frac{c^2-1}{\alpha}\operatorname{sech}^2\left(\frac{x-ct}{2}\sqrt{\frac{c^2-1}{\beta_1 c^2 - \beta_2}}\right), \tag{3}$$

where c is the phase speed of the stationary propagating 1-D localized wave. Although, equation 2 is not fully integrable, the extensive numerical investigation from [6] based on a conservative difference scheme, succeeded to show that in its main aspects, it behaves as its fully integrable "cousin", the BBE. Note that a solution of the above type exists for $|c| > \max\{1, \sqrt{\beta_2/\beta_1}\}$ or $|c| < \min\{1, \sqrt{\beta_2/\beta_1}\}$. The first case comprises of the so-called "supercritical" solitons, while the latter encompasses the "subcritical" ones. In the latter case, the physically relevant phase speeds are those that are significantly smaller than the threshold $\min\{1, \sqrt{\beta_2/\beta_1}\}$, because close to it, the soliton support becomes very short, and the main Boussinesq assumption about waves being long is violated. Moreover, it was shown in [3, 4] that the subcritical solitons are less stable and tend not to preserve their shape upon collisions. For this reason, in the present parer, which is devoted to the development of a new numerical approach, we consider the supercritical shapes in 1-D, which exhibit remarkable stability as their identity is preserved through nonlinear interactions.

The goal of this paper is to develop an efficient and accurate numerical method for the 1-D and 2-D BPEs as well as to numerically study the stability and other properties of their soliton solutions. Our numerical method is designed in several steps. First, we introduce a new variable $\rho := w - \beta_1 w_{xx}$ (in 1-D) or $\rho := w - \beta_1 \Delta w$ (in 2-D), for which the BPE takes the form of a nonlinear wave equation. We then introduce another new variable, the "momentum" m (in 1-D), or two new variables, the x- and y-"momenta" m and n (in 2-D), and rewrite the obtained wave equation as a system of conservation laws with global fluxes. The resulting system is numerically solved by a second-order central-upwind scheme, proposed in [11] and further improved in [9, 13] and [8], together with an efficient FFT-based elliptic solver used to recover w from available values of ρ. The selection of the central-upwind scheme is motivated by its superb properties of this Riemann-problem-solver-free method—efficiency, accuracy and robustness, demonstrated in a wide variety of applications (see, e.g., [8] and references therein), including the models with global fluxes: the incompressible Navier-Stokes equations, [9, 12], and the traffic flow model with Arrhenius look-ahead dynamics, [10].

The rest of the paper is organized as follows. In Sect. 3, we describe the 1-D central-upwind scheme and numerically study stability of the 1-D solitons for different values of c (in 3) as well as their interactions and emergence out of general initial data. In Sect. 4, we use the 2-D extension of the central-upwind scheme to simulate the time evolution of 2-D solitons and to numerically study their stability.

3 Scheme and Algorithm for 1-D BPE

We start by making a change of variables and defining a new variable ρ as

$$\rho = w - \beta_1 w_{xx}. \tag{4}$$

In terms of this new variable, the BPE 2 takes the form of a nonlinear wave equation:

$$\rho_{tt} = p(\rho, w)_{xx}, \quad p(\rho, w) := \frac{\beta_2}{\beta_1}\rho + \left(1 - \frac{\beta_2}{\beta_1}\right) w - \alpha w^2, \tag{5}$$

in which the dependence of the "pressure" p on the "density" ρ is global. Next, we introduce another new variable, the "momentum" m, and rewrite the nonlinear wave equation 5 as a system of conservation laws for the variable $\mathbf{U} := (\rho, m)^T$ with a global flux function $\mathbf{F}(\mathbf{U}) = (m, p(\rho, w(\rho)))^T$:

$$\begin{cases} \rho_t + m_x = 0, \\ m_t + p(\rho, w)_x = 0, \end{cases} \tag{6}$$

where p is defined in 5 and the dependence of w on ρ is given by the modified Helmholtz equation 4.

We then design a numerical method for the system 6, 4 using the following approach: the system 6 is solved by the second-order central-upwind scheme from [8], while the modified Helmholtz equation 4 is solved using the FFT-based elliptic solver. The details on our numerical method are provided in Sect. 3.1.

3.1 Description of 1-D Numerical Method

We assume that the system 6 is augmented with (smooth) initial conditions and periodic boundary conditions (in 1-D, an extension of the proposed numerical method to the nonperiodic case is quite straightforward).

We introduce a spatial grid $\{C_j\}$, where $C_j := (x_{j-\frac{1}{2}}, x_{j+\frac{1}{2}})$ are the cells of a uniform size Δx and denote the cell averages of $\mathbf{U}$ at time t by

$$\overline{\mathbf{U}}_j(t) := \frac{1}{\Delta x}\int_{C_j} \mathbf{U}(x,t)\,dx.$$

According to the semi-discrete central-upwind scheme from [8], the cell averages are evolved in time by solving the following system of ODEs:

$$\frac{d}{dt}\overline{\mathbf{U}}_j(t) = -\frac{\mathbf{H}_{j+\frac{1}{2}}(t) - \mathbf{H}_{j-\frac{1}{2}}(t)}{\Delta x}, \tag{7}$$

where $\mathbf{H}_{j+\frac{1}{2}}$ is a numerical flux given by

$$\mathbf{H}_{j+\frac{1}{2}} = \frac{a^+_{j+\frac{1}{2}}\mathbf{F}(\mathbf{U}^-_{j+\frac{1}{2}}) - a^-_{j+\frac{1}{2}}\mathbf{F}(\mathbf{U}^+_{j+\frac{1}{2}})}{a^+_{j+\frac{1}{2}} - a^-_{j+\frac{1}{2}}} + a^+_{j+\frac{1}{2}}a^-_{j+\frac{1}{2}}\left[\frac{\mathbf{U}^+_{j+\frac{1}{2}} - \mathbf{U}^-_{j+\frac{1}{2}}}{a^+_{j+\frac{1}{2}} - a^-_{j+\frac{1}{2}}} - \mathbf{q}_{j+\frac{1}{2}}\right]. \tag{8}$$

Here, the built-in "anti-diffusion" term (which corresponds to the reduced, compared with the original semi-discrete central-upwind scheme from [9], numerical dissipation) is

$$\mathbf{q}_{j+\frac{1}{2}} = \text{minmod}\left(\frac{\mathbf{U}^+_{j+\frac{1}{2}} - \mathbf{U}^*_{j+\frac{1}{2}}}{a^+_{j+\frac{1}{2}} - a^-_{j+\frac{1}{2}}}, \frac{\mathbf{U}^*_{j+\frac{1}{2}} - \mathbf{U}^-_{j+\frac{1}{2}}}{a^+_{j+\frac{1}{2}} - a^-_{j+\frac{1}{2}}}\right), \tag{9}$$

where

$$\mathbf{U}^*_{j+\frac{1}{2}} = \frac{a^+_{j+\frac{1}{2}}\mathbf{U}^+_{j+\frac{1}{2}} - a^-_{j+\frac{1}{2}}\mathbf{U}^-_{j+\frac{1}{2}} - \left\{\mathbf{F}(\mathbf{U}^+_{j+\frac{1}{2}}) - \mathbf{F}(\mathbf{U}^-_{j+\frac{1}{2}})\right\}}{a^+_{j+\frac{1}{2}} - a^-_{j+\frac{1}{2}}}, \tag{10}$$

and the minmod function,

$$\operatorname{minmod}(z_1,z_2,\ldots,z_m) := \begin{cases} \min(z_1,z_2,\ldots,z_m), & \text{if } z_i>0 \ \forall i=1,\ldots,m,\\ \max(z_1,z_2,\ldots,z_m), & \text{if } z_i<0 \ \forall i=1,\ldots,m,\\ 0, & \text{otherwise,} \end{cases} \tag{11}$$

is applied in a componentwise manner.

In the numerical flux 8–10, $\mathbf{U}^{\pm}_{j+\frac{1}{2}} := \widetilde{\mathbf{U}}(x_{j+\frac{1}{2}})$ are used to denote the point values of a non-oscillatory interpolant $\widetilde{\mathbf{U}}$,

$$\widetilde{\mathbf{U}}(x,t) = \overline{\mathbf{U}}_j + (\mathbf{U}_x)_j(x-x_j), \qquad x_{j-\frac{1}{2}} < x < x_{j+\frac{1}{2}}, \quad \forall j, \tag{12}$$

which is reconstructed at each time step from the previously computed cell averages $\{\overline{\mathbf{U}}_j(t)\}$. The second-order accuracy of the interpolant (12) is guaranteed provided $(\mathbf{U}_x)_j$ is (at least) first-order approximation of the derivative $\mathbf{U}_x(x_j,t)$. In all our numerical experiments, we have computed $(\mathbf{U}_x)_j$ using the minmod limiter (see, e.g., [14, 15]):

$$(\mathbf{U}_x)_j = \operatorname{minmod}\left(2\frac{\overline{\mathbf{U}}_j-\overline{\mathbf{U}}_{j-1}}{\Delta x}, \frac{\overline{\mathbf{U}}_{j+1}-\overline{\mathbf{U}}_{j-1}}{2\Delta x}, 2\frac{\overline{\mathbf{U}}_{j+1}-\overline{\mathbf{U}}_j}{\Delta x}\right), \tag{13}$$

where the minmod function is defined in 11.

Finally, $a^{\pm}_{j+\frac{1}{2}}$ are the local speeds of propagation, which can be estimated using the largest and the smallest eigenvalues of the Jacobian, see [8, 9]. Since the flux in the system 6 is global, we calculate the local speeds as follows:

$$a^{\pm}_{j+\frac{1}{2}} = \pm\max\left(\sqrt{p'(\rho^{+}_{j+\frac{1}{2}})}, \sqrt{p'(\rho^{-}_{j+\frac{1}{2}})}\right), \tag{14}$$

where

$$p'(\rho) = \frac{\beta_2}{\beta_1} + \left(1-\frac{\beta_2}{\beta_1}\right)w'(\rho) + 2\alpha w(\rho)w'(\rho),$$

and the derivative $w'(\rho)$ is calculated using the chain rule and equation 4:

$$w'(\rho) = \frac{w_x}{\rho_x} = \frac{w_x}{w_x - \beta_1 w_{xxx}}.$$

It should be observed that the values of w and its derivatives at each time step are obtained from the given values of ρ by solving the modified Helmholtz equation 4. Thanks to the periodic boundary conditions, equation 4 can be efficiently and highly accurately solved using a standard FFT-based elliptic solver.

Remark 3.1. Since the local speeds in 14 are symmetric, namely,

$$a^{+}_{j+\frac{1}{2}} = -a^{-}_{j+\frac{1}{2}} =: a_{j+\frac{1}{2}}, \quad \forall j,$$

the flux formulae 8–10 are simplified to:

$$\mathbf{H}_{j+\frac{1}{2}} = \frac{1}{2}\left\{\mathbf{F}(\mathbf{U}^-_{j+\frac{1}{2}}) + \mathbf{F}(\mathbf{U}^+_{j+\frac{1}{2}})\right\} - \frac{a_{j+\frac{1}{2}}}{2}\left[\mathbf{U}^+_{j+\frac{1}{2}} - \mathbf{U}^-_{j+\frac{1}{2}} - \mathbf{q}_{j+\frac{1}{2}}\right], \tag{15}$$

$$\mathbf{q}_{j+\frac{1}{2}} = \mathrm{minmod}\left(\mathbf{U}^+_{j+\frac{1}{2}} - \mathbf{U}^*_{j+\frac{1}{2}}, \mathbf{U}^*_{j+\frac{1}{2}} - \mathbf{U}^-_{j+\frac{1}{2}}\right), \tag{16}$$

$$\mathbf{U}^*_{j+\frac{1}{2}} = \frac{1}{2}\left(\mathbf{U}^+_{j+\frac{1}{2}} + \mathbf{U}^-_{j+\frac{1}{2}}\right) - \frac{1}{2a_{j+\frac{1}{2}}}\left\{\mathbf{F}(\mathbf{U}^+_{j+\frac{1}{2}}) - \mathbf{F}(\mathbf{U}^-_{j+\frac{1}{2}})\right\}. \tag{17}$$

Remark 3.2. The ODE system 7 has to be solved by a stable and sufficiently accurate ODE solver. In our numerical experiments, we have used the third-order strong stability preserving (SSP) Runge-Kutta method from [7].

3.2 1-D Numerical Experiments

We apply the central-upwind scheme 7, 15–17 to the system 6. Our goal is to demonstrate stability of the soliton solutions 3 and to numerically verify whether they are attractors for general initial data. In all the 1-D numerical experiments below, we used the same set of parameters α, β_1 and β_2, namely,

$$\alpha = -3, \ \beta_1 = 1.5, \ \beta_2 = 0.5.$$

This particular set was shown to correspond to the flow in a shallow layer [4], but in a paradigmatic sense, any other set of values for the dispersion can be used. If the ratio is kept $\beta_1/\beta_2 = 3$, the relevance to the shallow-layer case can still be shown, but for some stretched spatial coordinates.

Example 1—Single Solitons

We first consider the following initial datum:

$$w(x,0) = w^s(x,0;1.2), \tag{18}$$

which represents a single soliton at the initial time moment, which is then allowed to evolve according to the BPE. In this case, the soliton is expected to travel to the right with the constant speed of $c = 1.2$. We apply the proposed central-upwind scheme to the initial value problem (IVP) 2, 18. The snapshots of the solution, computed at different times on the uniform grid with $\Delta x = 100/512$, are plotted in Fig. 1(a). As one can see, the soliton is numerically stable as predicted in [3, 4].

Next, we take a different initial datum:

$$w(x,0) = w^s(x,0;5), \tag{19}$$

for which the soliton is still expected to travel to the right but with a much higher speed (since $c = 5$). Capturing the evolution of such fast soliton is quite a

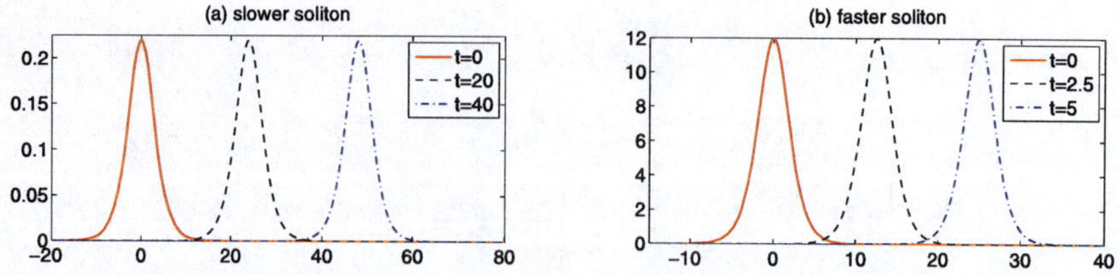

Fig. 1 Numerical evolution of isolated solitons as solutions of (2)

challenging task since the energy functional is not positive definite. For example, the numerical scheme, presented in [6], failed on similar sets of initial data with $c > 2.1$ developing a nonlinear blowup when collision of solitons was considered. The solution, computed by the proposed central-upwind scheme at different times on the uniform grid with $\Delta x = 100/512$, is presented in Fig. 1(b). As one can observe, the scheme captures the evolution of the solitary waves correctly, no instabilities have developed and the quality of the solution is as good as in the previous case.

Example 2—Interaction of Two Solitons

In this example, we simulate a nonlinear interaction of two solitons with different initial speeds. The initial profile is taken as the following superposition of two solitons:

$$w(x,0) = w^s(x+40,0;1.2) + w^s(x-50,0;-1.5), \tag{20}$$

on the interval $[-150,120]$. The initial condition 20 and the solution at times $t = 15,30,40,60$ and 100, computed by the central-upwind scheme with $\Delta x = 270/2048$, are plotted in Fig. 2. As one can see, the higher soliton (initially located on the right) that travels with a negative phase velocity ($c = -1.5$) passes through the lower soliton which travels to the right ($c = 1.2$) after going through a nonlinear interaction. Evidently, the proposed method is capable of capturing the nonlinear soliton interaction with a high resolution. Juxtaposing the profiles before and after the interaction shows that they indeed have the solitonic property of shape preservation. The height, the support and the energy of each soliton are preserved with a high accuracy.

Example 3—Arbitrary Initial Data

Finally, we consider equation 2 subject to arbitrary initial conditions:

$$w(x,0) = \operatorname{sech}^2(0.5x), \quad w_t(x,0) \equiv 0. \tag{21}$$

We call this solution "arbitrary" despite of its "sech" shape, because the phase speed is equal to zero, and the soliton $w^s(x,t;0)$ that corresponds to $c = 0$ has nothing to do with the particularly chosen "sech" shape 21. In this case, according to the analytical results for fully integrable systems, we expect two (or more) soliton solitons of type 3 to emerge out of this initial data. In Fig. 3, we plot the solution at times

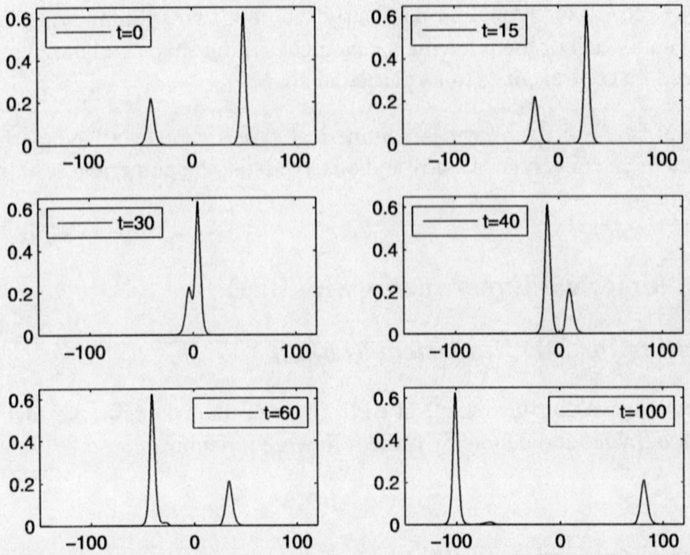

Fig. 2 Evolution of the numerical solution of (2), (20): interaction of two solitons

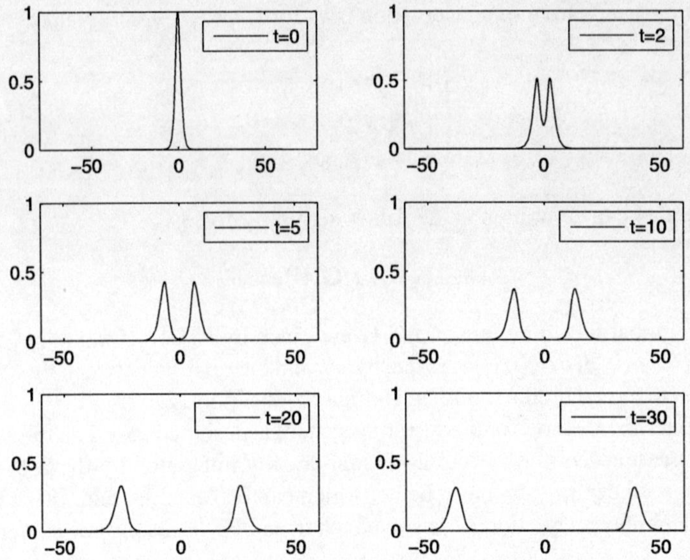

Fig. 3 Two solitons emerging out of the initial data (21)

$t = 0,2,5,10,20$ and 30, computed by the central-upwind scheme on a uniform grid with $\Delta x = 120/1024$. What can be clearly seen are two solitons branching from the initial data, each of them having a canonical soliton shape and moving with the phase speed that corresponds to its particular shape.

Remark 3.3. In all of the above 1-D numerical experiments, we have performed a mesh refinement convergence study and observed that the numerical solutions have been convergent.

4 2-D Numerical Experiments with BPE

4.1 Setting for 2-D Numerical Method

In this section, we consider the 2-D BPE 1. As in the 1-D case, we first make a change of variables and denote by ρ the following function:

$$\rho = w - \beta_1 \Delta w. \tag{22}$$

This allows one to rewrite the BPE 1 in the form of a nonlinear wave equation:

$$\rho_{tt} = \Delta p(\rho, w), \quad p(\rho, w) := \frac{\beta_2}{\beta_1}\rho + \left(1 - \frac{\beta_2}{\beta_1}\right) w - \alpha w^2. \tag{23}$$

Introducing two new variables, the x- and y-"momenta" m and n, and rewriting the last equation as a system of conservation laws for $\mathbf{U} := (\rho, m, n)^T$ yields:

$$\begin{cases} \rho_t + m_x + n_y = 0, \\ m_t + p(\rho, w)_x = 0, \\ n_t + p(\rho, w)_y = 0. \end{cases} \tag{24}$$

The system 24 can be written in the following vector form:

$$\mathbf{U}_t + \mathbf{F}(\mathbf{U})_x + \mathbf{G}(\mathbf{U})_y = 0, \tag{25}$$

where the global flux functions $\mathbf{F}$ and $\mathbf{G}$ are given by $\mathbf{F}(\mathbf{U}) = (m, p(\rho, w(\rho)), 0)^T$ and $\mathbf{G}(\mathbf{U}) = (n, 0, p(\rho, w(\rho)))^T$, respectively, and the relation between the variables ρ and w is governed by the modified Helmholtz equation 22.

The numerical method for the resulting system is designed based on the same (as in the 1-D case) idea: the system 24 is numerically integrated by the 2-D version of the proposed central-upwind scheme, which can be found in, e.g., [8], while the modified Helmholtz equation 22 is solved using an FFT-based elliptic solver.

4.2 2-D Numerical Experiments

For definiteness, in the numerical experiments reported in this section, we consider equation 1 with $\alpha = 1, \beta_1 = 3, \beta_2 = 1$ subject to the initial data

$$w(x,y,0) = w^s(x,y-\eta,t;c) \tag{26}$$

which correspond to a soliton moving along the y-axis with velocity c. Because of the rotational symmetry of the operators in the BPE, any other case can be reduced to a case with only one nontrivial component of the phase speed. In [5], a comprehensive semi-analytical/semi-numerical solution was summarized as best-fit functions of the governing parameters. For $\beta_2 = 1$, the following best fit approximation given in [5] was presented for the shape of the 2-D soliton:

$$\begin{aligned} w^s(x,y,t;c) = f(x,y) &+ c^2[(1-\beta_1)g_a(x,y) + \beta_1 g_b(x,y)] \\ &+ c^2[(1-\beta_1)h_1(x,y) + \beta_1 h_2(x,y)\cos(2\theta)], \end{aligned}$$

where

$$f(x,y) = \frac{2.4(1+0.24r^2)}{\cosh(r)(1+0.095r^2)^{1.5}},$$

$$g_a(x,y) = -\frac{1.2(1-0.177r^{2.4})}{\cosh(r)|1+0.11r^{2.1}|}, \quad g_b(x,y) = -\frac{1.2(1+0.22r^2)}{\cosh(r)|1+0.11r^{2.4}|},$$

$$h_i(x,y) = \frac{a_i r^2 + b_i r^3 + c_i r^4 + v_i r^6}{1 + d_i r + e_i r^2 + f_i r^3 + g_i r^4 + h_i r^5 + q_i r^6 + w_i r^8}.$$

Here, $r = \sqrt{x^2+y^2}$, $\theta = \arctan(y/x)$, and the coefficients are: $a_1 = 1.03993$, $a_2 = 31.2172$, $b_1 = 6.80344$, $b_2 = -10.0834$, $c_1 = -0.22992$, $c_2 = 3.97869$, $d_1 = 12.6069$, $d_2 = 77.9734$, $e_1 = 13.5074$, $e_2 = -76.9199$, $f_1 = 2.46495$, $f_2 = 55.4646$, $g_1 = 2.45953$, $g_2 = -12.9335$, $h_1 = 1.03734$, $h_2 = 1.0351$, $q_1 = -0.0246084$, $q_2 = 0.628801$, $v_1 = 0.0201666$, $v_2 = -0.0290619$, $w_1 = 0.00408432$, $w_2 = -0.00573272$.

We first test our scheme on the case $c = 0$, when the solution has an exponential decay at infinity and hence is tightly localized. In this case, we take a computational domain $[-15,15] \times [-15,15]$, which is sufficiently large and hence the localized solution does not interact with the computational boundaries. The profile in this case is a standing soliton, and apart from the sheer validation significance, the computations are also purported to answer an important question: can the nonlinearity without motion keep this soliton from being annihilated by the diffusion?

We present our results in Fig. 4 for a sequence of time moments from 5 to 20, and on two different grids. The upper row of panels contains the result, obtained on a uniform grid with $\Delta x = \Delta y = 30/128$. One can see that the nonlinearity is not strong enough without motion, and the initial elevation is transformed into a propagating cylindrical wave, similar to the one generated on a water surface when an object is dropped onto it. The lower row of panels contains the same solution, but computed on a finer grid with $\Delta x = \Delta y = 30/256$. The differences between the profiles for

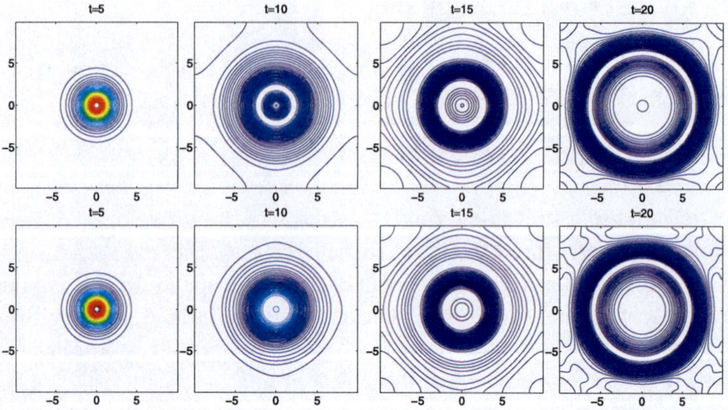

Fig. 4 Evolution of the numerical solution of (1), (26) with $c = 0$ and $\eta = 0$, computed on two different grids

the same time moments are small and compatible with the second-order truncation error. We have thus answered one of the pertinent physical questions.

We now consider the case of a moderate phase speed $c = 0.3$ and compute the solution up to $t = 16$. The results are presented in Fig. 5. The top row of panels shows the lowest resolution ($\Delta x = \Delta y = 30/128$) solution, the plots in the middle row correspond to the medium resolution ($\Delta x = \Delta y = 30/256$), and the lowest row presents the results for the finest grid with $\Delta x = \Delta y = 30/512$. Clearly, the lowest resolution means that more numerical diffusion and dispersion are present, and as a result, the nonlinearity could not keep the shape in the form as prescribed by the initial condition. The interesting result obtained thanks to the numerical technology developed here, is that the soliton not only moves, but also evolves in a similar fashion as the standing soliton, resembling concentric cylindrical waves propagating outwards. In this case, however, the motion does not have a cylindrical symmetry because of the propagation of the solitary wave. The result is again a solitary structure, but of a different symmetry and with some kind of "aging", in the sense that the amplitude decreases in time, and the support increases.

In the middle row of panels in Fig. 5, we present the results obtained on a finer grid with $\Delta x = \Delta y = 30/256$. In this case, the numerical diffusion and dispersion are about four times smaller, and the nonlinearity was able to maintain the shape, prescribed in the initial conditions. This means that the shapes obtained in [5] may qualify as solitons. Unfortunately, the nonlinearity of BPE can lead to blowup even in the 1-D case (see the literature cited in [6]), if the initial condition has negative energy. There are no investigations concerning the possible blowup in 2-D, but in generic sense, one should expect a more stringent conditions on the initial energy in order to keep the shape from blowing up. The blowup can be clearly seen in the bottom row of Fig. 5, where the finest grid solution is shown, as well as in Fig. 6, where the maximum, $u_{\max}$, of the soliton and its position, $y_{\max}$, are traced in time. One can see that at small times the maximum (as presented in the upper panel of

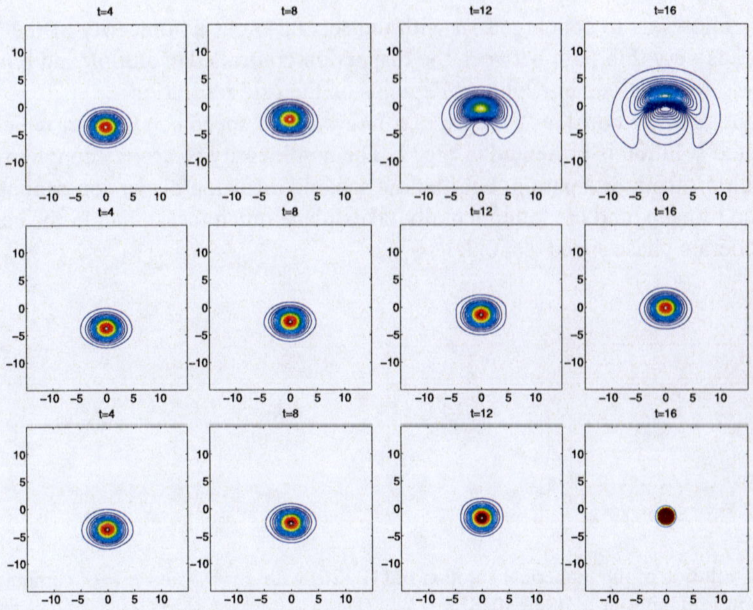

Fig. 5 Evolution of the numerical solution of (1), (26) with $c = 0.3$ and $\eta = -5$

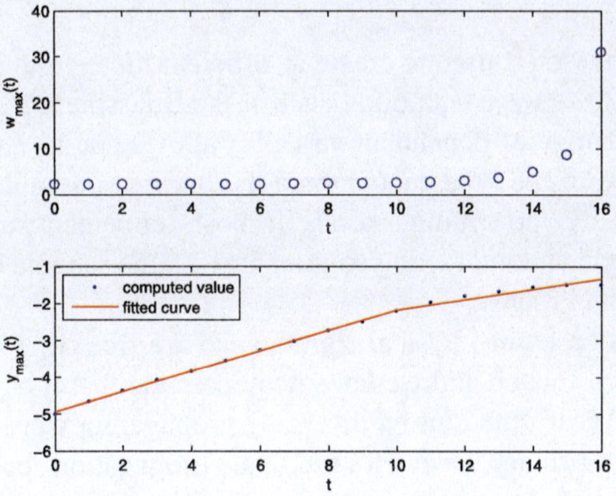

Fig. 6 Evolution of the maximum of the solitary wave and its trajectory for the numerical solution shown in the bottom row of Fig. 5

Fig. 6) almost does not change and the slope of the trajectory (the lower panel of Fig. 6) is 0.28, which is within 5% from the expected theoretical value of 0.3. The disturbing news here is that at large times the maximum increases dramatically and a blowup takes place before $t = 17$.

This allows us to conclude that with the second-order nonlinearity of the BPE, one treads very thin path between the dispersion controlled evolution and blowup, and the selected mode of evolution depends on the grid resolution.

At the end, we consider the case of a larger phase speed $c = 0.7$, for which the computed solution is presented in Fig. 7. The nonlinearity is strong enough to keep the solution from dissipating, but the low scheme diffusion/dispersion cannot prevent the blowup, and the latter actually takes place much faster than in the case of the moderate phase speed $c = 0.3$.

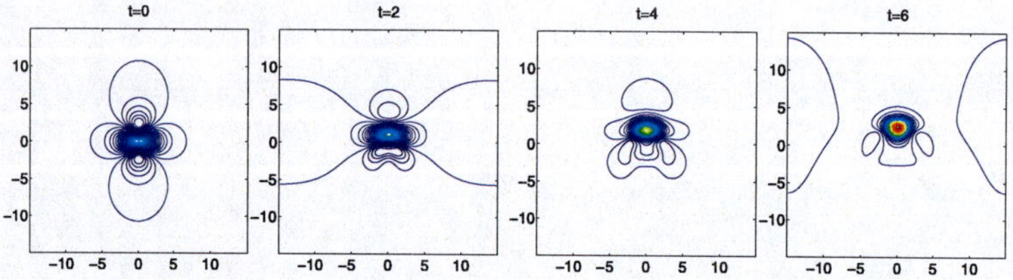

Fig. 7 Evolution of the numerical solution of (1), (26) with $c = 0.7$ and $\eta = 0$, computed on a uniform grid with $\Delta x = \Delta y = 30/512$

5 Conclusion

In this paper, a novel idea is used to create an efficient finite volume scheme for the fourth-order generalized wave equations, such as the Boussinesq Paradigm Equation (BPE). A special change of dependent variables allows one to render the original problem to a system of the type encountered in compressible fluid dynamics. The scheme is validated by performing a series of mesh refinement studies. In treating several cases of single and interacting solitons in 1-D, the scheme has been demonstrated to be stable and robust.

In 2-D, different cases of physical significance are treated. *First*, the standing soliton has been investigated and we have demonstrated that it cannot survive and eventually will be transformed into a diverging propagating wave with cylindrical symmetry. *Second*, a solitary wave with moderate propagation speed is imposed as an initial condition and its evolution is numerically investigated. On the coarsest grid (largest numerical diffusion and dispersion), the propagating wave was transformed into a diverging wave skewed by the propagation. The medium grid allowed to trace the steady translation of the wave for long time period, but eventually the solutions blows up, signifying that the quadratic nonlinearity of BPE is not adequate for modelling permanent soliton-like waves. This conclusion is further confirmed by the computations on the finest grid, on which the blowup takes place earlier in time. *Third*, a case of large phase speed is investigated, and the same behaviour is observed as for the moderate phase speed with the only difference that the time scale of the blowup is much shorter.

Acknowledgements. The work of A. Chertock was supported in part by the NSF Grant DMS-0712898. The work of C. I. Christov was supported, in part, by an ASEE/ONR Summer Faculty Fellowship. The work of A. Kurganov was supported in part by the NSF Grant DMS-0610430.

References

1. Boussinesq, J.V.: Théorie des ondes et des remous qui se propagent le long d'un canal rectangulaire horizontal, en communiquant au liquide contenu dans ce canal des vitesses sensiblement pareilles de la surface au fond. J. Mathématiques Pures et Appliquées 17, 55–108 (1872)
2. Christou, M.A., Christov, C.I.: Localized waves for the regularized long wave equation via a Galerkin spectral method. Math. Comput. Simul. 69, 257–268 (2005)
3. Christov, C.I.: Conservative difference scheme for Boussinesq model of surface waves. In: Morton, W.K., Baines, M.J. (eds.) Proc. ICFD V, pp. 343–349. Oxford University Press, Oxford (1995)
4. Christov, C.I.: An energy-consistent Galilean-invariant dispersive shallow-water model. Wave Motion 34, 161–174 (2001)
5. Christov, C.I., Choudhury, J.: Perturbation solution for the 2D shallow-water waves. Mech. Res. Commun. (2009) (submitted)
6. Christov, C.I., Velarde, M.G.: Inelastic interaction of Boussinesq solitons. J. Bifurcation & Chaos 4, 1095–1112 (1994)
7. Gottlieb, S., Shu, C.W., Tadmor, E.: Strong stability-preserving high-order time discretization methods. SIAM Rev. 43, 89–112 (2001)
8. Kurganov, A., Lin, C.T.: On the reduction of numerical dissipation in central-upwind schemes. Commun. Comput. Phys. 2, 141–163 (2007)
9. Kurganov, A., Noelle, S., Petrova, G.: Semi-discrete central-upwind scheme for hyperbolic conservation laws and hamilton-jacobi equations. SIAM J. Sci. Comput. 23, 707–740 (2001)
10. Kurganov, A., Polizzi, A.: Non-oscillatory central schemes for a traffic flow model with Arrhenius look-ahead dynamics. Networks Heterogeneous Media (to appear)
11. Kurganov, A., Tadmor, E.: New high resolution central schemes for nonlinear conservation laws and convection-diffusion equations. J. Comput. Phys. 160, 241–282 (2000)
12. Kurganov, A., Tadmor, E.: New high-resolution semi-discrete central scheme for hamilton-jacobi equations. J. Comput. Phys. 160, 720–742 (2000)
13. Kurganov, A., Tadmor, E.: Solution of two-dimensional riemann problems for gas dynamics without riemann problem solvers. Numer. Meth. Part. Diff. Eq. 18, 584–608 (2002)
14. Lie, K.A., Noelle, S.: On the artificial compression method for second-order nonoscillatory central difference schemes for systems of conservation laws. SIAM J. Sci. Comput. 24(4), 1157–1174 (2003)
15. Nessyahu, H., Tadmor, E.: Nonoscillatory central differencing for hyperbolic conservation laws. J. Comput. Phys. 87(2), 408–463 (1990)
16. Zabusky, N.J., Kruskal, M.D.: Interaction of 'solitons' in collisionless plasma and the recurrence of initial states. Phys. Rev. Lett. 15, 240–243 (1965)

Parallel Lattice-Boltzmann Simulation of Transitional Flow on Non-uniform Grids

Maik Stiebler, Sören Freudiger, Manfred Krafczyk, and Martin Geier

Abstract. Transitional flows are difficult to address by Reynolds Averaged Navier-Stokes (RANS) simulations as the spectrum is typically not fully developed. In this work the suitability of the lattice Boltzmann method is evaluated for the simulation of transitional flows. Special measures are taken to reduce the computational cost without sacrificing the accuracy of the method. A large eddy simulation turbulence model is employed to allow efficient simulation of the resolved flow structures on relatively coarse computational meshes. In the vicinity of solid walls, where the flow is governed by the presence of a thin boundary layer, local grid-refinement is employed in order to capture the fine structures of the flow. The lattice Boltzmann code is run on an Opteron cluster. In the considered test case, the pressure distribution and the drag force on a sphere are computed in the Reynolds number range 1000 to 10000 and a parallel efficiency of 80% is obtained.

1 Introduction

In recent years, the lattice Boltzmann method (LBM) has been established as a method for solving weakly compressible and incompressible flow problems. The LBM was found to be competitive with traditional methods based on the direct discretization of the Navier-Stokes equations in a wide range of applications with complex flow physics like multiphase flows, suspensions in fluids and flows in porous media. LBM has also been demonstrated to be an efficient simulation tool for laminar flows [7]. However, the efficiency of LB methods in flow problems in the turbulent or transitional regime is not well investigated yet. Efficient simulation of turbulent engineering problems requires the use of non-uniform meshes to resolve

Maik Stiebler · Sören Freudiger · Manfred Krafczyk · Martin Geier
Institute for Computational Modeling in Civil Engineering,
Technische Universität Braunschweig, Mühlenpfordtstraße 4-5,
Braunschweig, 38106, Germany
e-mail: {stiebler, freud, kraft, geier}@irmb.tu-bs.de

E. Krause et al. (Eds.): Computational Sci., & High Performance Computing IV, NNFM 115, pp. 283–295.
springerlink.com

the high velocity gradients in the boundary layer. As a direct numerical simulation (DNS) of technically relevant turbulent flows is typically prohibitively expensive, thus turbulence modelling is required. Two main types of turbulence models have previously been introduced into the LBM context: the Reynolds averaged Navier Stokes (RANS) type models [17] and large eddy simulation models (LES) [10]. In the RANS models, a time- or ensemble-averaged velocity is computed and the turbulent features of the flow have to be modelled including all scales. LES models, on the other hand, apply a spatial low-pass filter to the velocity field. The optimal filter would eliminate only high frequencies in the flow field while keeping large eddies untouched. However, the optimal filter could only be defined in Fourier space and would hence require prohibitively large stencils. A distinguishing property of the lattice Boltzmann model is the local availability of the shear stress without the need for numerical derivatives of the velocity field. This information can be employed to implement a local low-pass filter at very low computational cost and at the same time increase parallel efficiency as compared to simulations without turbulence model due to the increased local amount of computations.

In this article, an efficient implementation of a lattice Boltzmann method with LES turbulence modelling on non-uniform grids is introduced. The first part of this article will shortly recall some basic features of the Lattice Boltzmann multiple-relaxation-times method with subgrid-stress model. The second part will deal with extensions of the method for non-uniform grids. In the third part the method will be applied to calculate some distinct flow features of a sphere in a channel in the transitional regime.

2 Lattice Boltzmann Method with Subgrid Stress Model (LBM-SGS)

Unlike traditional numerical methods that apply some discretization technique to the considered differential equation, the lattice Boltzmann method is an asymptotical method shown to converge to the solution of the incompressible Navier-Stokes equation [14]. The starting point for the derivation of the lattice Boltzmann method is an assumed momentum distribution $f_{i,l}$. The distribution is discrete in space, momentum, and time. In the lattice Boltzmann method the discretization of space, called the lattice, corresponds exactly to the discretization of momentum so that a virtual particle sitting on a node in the lattice moves to a neighbouring node in a discrete time step. Interpolation is hence not required. The Navier-Stokes equation is found to be the equation of motion for the first moment (expected value) of the momentum distribution $f_{i,l}$ provided that a sufficiently isotropic velocity set is used. A common choice for the lattice Boltzmann velocity set is the D3Q19 (three dimensions, 19 speeds) model. Its microscopic velocity components correspond to the third, fifth and seventh row of the matrix given in Appendix 2. The evolution of the lattice Boltzmann model can be split into two steps: free streaming or propagation of the distribution $f_{i,l}$ according to its respective speed; and collision on lattice nodes. In order to retain the Navier-Stokes equation the collision operator has to fulfil some conditions concerning conservation laws, Galilean invariance and dissipation. Still, the collision operator is not uniquely defined. Different choices might have different stability characteristics. In this paper a multiple-relaxation-time collision operator is employed [8].

The lattice Boltzmann model can be written as

$$f_{i,l}(t+\Delta t_l,\mathbf{x}+\mathbf{e}_i\Delta t_l) = f_{i,l}(t,\mathbf{x}) + \Omega_{i,l}, \quad i=0,\dots,b-1, \tag{1}$$

where Δt is the time step and the collision operator is given by

$$\Omega = \mathsf{M}^{-1}\mathsf{S}\,[(\mathsf{M}f) - m^{eq}]. \tag{2}$$

Matrix M is the transformation matrix given in appendix 2, composed of the 19 orthogonal basis vectors $\{\Phi_i, i=0,\dots,b-1\}$ given in appendix 1. They are orthogonal with respect to a weighted inner product, $\langle\Phi_i,\Phi_j\rangle_w = \sum_k \Phi_{ik}\Phi_{jk}w_k = 0$, if $i\neq j$ (in contrast to [8], where $\langle\Phi_i,\Phi_j\rangle = \sum_k \Phi_{ik}\Phi_{jk}\neq 0$, if $i\neq j$). The vector w is composed of the weights $\{w_i, i=0\dots,b-1\}$:

$$w = (\frac{1}{3},\frac{1}{18},\frac{1}{18},\frac{1}{18},\frac{1}{18},\frac{1}{18},\frac{1}{18},\frac{1}{36},\frac{1}{36},\frac{1}{36},\frac{1}{36},\frac{1}{36},\frac{1}{36},\frac{1}{36},\frac{1}{36},\frac{1}{36},\frac{1}{36},\frac{1}{36},\frac{1}{36}).$$

The moments $m = \mathsf{M}f$ are labelled as

$$m = (\rho, e, \varepsilon, j_x, q_x, j_y, q_y, j_z, q_z, 3p_{xx}, 3\pi_{xx}, p_{ww}, \pi_{ww}, p_{xy}, p_{yz}, p_{xz}, m_x, m_y, m_z).$$

m^{eq} is the vector composed of the equilibrium moments given in Eqs. (5) and $\mathsf{S} = \{s_{i,i}, i=0,\dots,b-1\}$ is the diagonal collision matrix. The nonzero collision parameters $s_{i,i}$ (the eigenvalues of the collision matrix $\mathsf{M}^{-1}\mathsf{S}_l\mathsf{M}$) are:

$$\begin{aligned}
s_{1,1} &= s_a\\
s_{2,2} &= s_b\\
s_{4,4} = s_{6,6} = s_{8,8} &= s_c\\
s_{10,10} = s_{12,12} &= s_d\\
s_{9,9} = s_{11,11} = s_{13,13} = s_{14,14} = s_{15,15} &= -\frac{\Delta t}{\tau} = s_\omega\\
s_{16,16} = s_{17,17} = s_{18,18} &= s_e,
\end{aligned} \tag{3}$$

The relaxation time τ is

$$\tau = 3\frac{\nu}{c^2} + \frac{1}{2}\Delta t, \tag{4}$$

where ν is the kinematic viscosity. The parameters s_a, s_b, s_c, s_d and s_e can be freely chosen in the range $[-2,0]$ and tuned to improve stability [11]. While the optimal values for these parameters depend on the specific system under consideration (geometry, initial and boundary conditions), reasonable values are given in [8]. We choose $s_a = s_b = s_c = s_d = s_e = -1.0$. The non-zero equilibrium distribution functions $\{m_i^{eq}, i=0,\dots,18\}$ are given by

$$m_0^{eq} = \rho, \quad m_3^{eq} = \rho_0 u_x, \quad m_5^{eq} = \rho_0 u_y, \quad m_7^{eq} = \rho_0 u_z, \tag{5a}$$

$$m_1^{eq} = e^{eq} = \rho_0 (u_x^2 + u_y^2 + u_z^2), \tag{5b}$$

$$m_9^{eq} = 3p_{xx}^{eq} = \rho_0 (2u_x^2 - u_y^2 - u_z^2), \tag{5c}$$

$$m_{11}^{eq} = p_{zz}^{eq} = \rho_0 (u_y^2 - u_z^2), \tag{5d}$$

$$m_{13}^{eq} = p_{xy}^{eq} = \rho_0 u_x u_y, \tag{5e}$$

$$m_{14}^{eq} = p_{yz}^{eq} = \rho_0 u_y u_z, \tag{5f}$$

$$m_{15}^{eq} = p_{xz}^{eq} = \rho_0 u_x u_z, \tag{5g}$$

where ρ_0 is a constant density and ρ a density variation and the velocities are derived from the moments representing momenta: $u_\alpha = j_\alpha / \rho_0$.

The spatially filtered Navier-Stokes equation for the filtered velocity $\bar{\mathbf{u}}$ is formulated according to the turbulent-viscosity hypothesis:

$$\frac{\partial \bar{u}_\alpha}{\partial t} + \frac{\partial \bar{u}_\alpha \bar{u}_\beta}{\partial x_\beta} = -\frac{\partial P/\rho}{\partial x_\alpha} + \frac{\partial}{\partial x_\beta}\left[(\nu_0 + \nu_\mathrm{T})\left(\frac{\partial \bar{u}_\alpha}{\partial x_\beta} + \frac{\partial \bar{u}_\beta}{\partial x_\alpha}\right)\right] \tag{6}$$

The turbulent viscosity according to the Smagorinsky model depends on the strain rate:

$$\nu_\mathrm{T} = (C_S \Delta x)^2 \|\mathbf{S}\| \tag{7}$$

with the strain rate tensor

$$S_{\alpha\beta} = \frac{1}{2}\left(\frac{\partial \bar{u}_\alpha}{\partial x_\beta} + \frac{\partial \bar{u}_\beta}{\partial x_\alpha}\right). \tag{8}$$

The strain rate tensor can be locally computed from the non-equilibrium distribution functions $f^\mathrm{neq} = \mathsf{M}^{-1}(\mathsf{M}f - m^\mathrm{eq})$ as

$$S_{\alpha\beta} = -\frac{3}{2\tau_\mathrm{total} c^2} \Pi_{\alpha\beta}^\mathrm{neq} \tag{9}$$

with

$$\Pi_{\alpha\beta}^\mathrm{neq} = \sum_i \left(e_{i\alpha}\, e_{i\beta}\, f_i^\mathrm{neq}\right). \tag{10}$$

With this and

$$\tau_\mathrm{total} = \frac{3}{c^2}\nu_\mathrm{total} + \frac{1}{2}\Delta t = \frac{3}{c^2}(\nu_0 + \nu_\mathrm{T}) + \frac{1}{2}\Delta t \tag{11}$$

a quadratic equation for the locally and temporally variable τ_total can be derived with the solution

$$\tau_\mathrm{total} = \frac{3}{c^2}\nu_0 + \frac{1}{2}\Delta t + \frac{\sqrt{\tau_0^2 + \frac{18 C_S^2 \Delta t^2 Q}{c^2}} - \tau_0}{2} \tag{12}$$

where $Q = \sum_{\alpha\beta} 2\Pi^{\mathrm{neq}}_{\alpha\beta}\Pi^{\mathrm{neq}}_{\alpha\beta}$. Alternatively, the strain rate tensor can easily be computed from the corresponding moments p_{xy}, p_{yz}, p_{xz}.

3 LBM-SGS on Irregular Grids

Due to the emergence of boundary layers in turbulent flows, high velocity gradients occur near solid walls. To capture these features of the flow, the resolution required near walls is much higher than the resolution required in the rest of the simulation domain. Thus, local refinement of the computational grid is a mandatory part of an efficient method for solving turbulent flow problems. In this work, a block-structured type of grid is used. In each block the resolution is fixed. At the boundary between two blocks the spatial resolution can change by a factor of two.

The hyperbolic scaling $\Delta t_l c = \Delta x_l$ with a level independent microscopic velocity c is employed. The Mach and the Reynolds number are thus equal on all grids. This naturally leads to a nested time stepping scheme. One time step on a coarse grid corresponds to two time steps on the next finer grid level. An approach where the Mach number is scaled down on finer grid levels to ensure convergence to the *incompressible* Navier-Stokes equations can be found in Ref. [15]. That approach requires four time steps on the fine grid during one time step on the coarse mesh.

A typical grid interface is shown in Figure 1. As the propagation step does not provide all distributions on one grid level at the interface, missing distributions on one grid level have to be computed from the adjacent grid level. An overlapping interface is hence required.

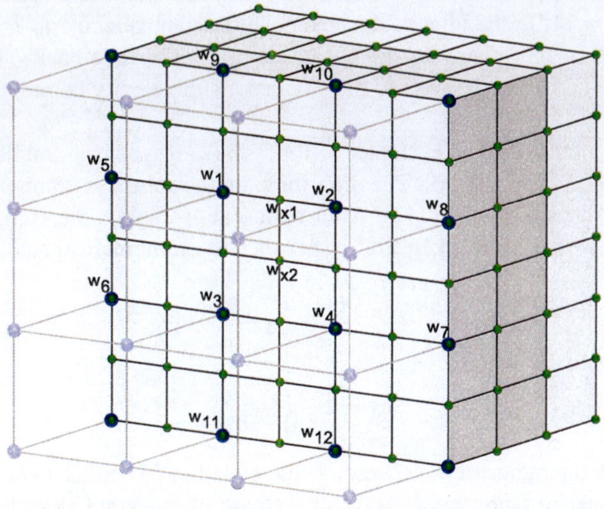

Fig. 1 Schematic view of the interface between two lattice Boltzmann grids with different resolution

The continuity of the stress tensor requires rescaling of the non-equilibrium part $\mathbf{m}^{neq}$ [6, 4]. Rescaling is applied after the streaming step by

$$m_{i,l-1}^{neq} = \frac{s_{l,i,i}}{s_{l-1,i,i}} \frac{\Delta t_{l-1}}{\Delta t_l} m_{i,l}^{neq}. \tag{13}$$

With equations 3 and 4 we get

$$m_{i,l-1}^{neq} = 2\beta_{i,l,l-1} m_{i,l}^{neq} \tag{14}$$

with $\beta_{i,l,l-1} = 1$ for $i \in \{1,2,4,6,8,10,12,16,17,18\}$ and

$$\beta_{i,l,l-1} = \frac{\Delta t_l (6\nu + \Delta t_{l-1} c^2)}{\Delta t_{l-1} (6\nu + \Delta t_l c^2)} \tag{15}$$

for $i \in \{9,11,13,14,15\}$.

Temporal and spatial interpolations is required because of the different mesh spacings Δx_l and time steps Δt_l. A cubic interpolation in space is used for the 'hanging' nodes. Details of the algorithm can be found in [4, 19]. The turbulent viscosity ν_T is proportional to the square of the grid spacing and will thus vanish for sufficiently fine grids, i. e. when the flow is sufficiently resolved.

4 Test Case and Results

The presented test case for the method is a sphere in a square channel (Fig. 2 and 3). The blockage ratio, i.e. the ratio of channel width h and sphere diameter d is $\lambda = h/d = 11$. In the inlet of the channel a constant velocity u_{in} is prescribed. On the surface of the sphere no-slip (i.e. second-order bounce-back [1]) boundary conditions are used. On the wall of the channel a slip velocity boundary condition is applied.

The test case is evaluated for the Reynolds number range $\text{Re} = u_{\text{in}} d/\nu \in [1.000, 1.0000]$. The drag force on the sphere is computed with the momentum-exchange method [12]. After temporal averaging, the computed drags are compared with a curve taken from experimental and numerical studies [2]:

$$\log_{10}(\text{c}_\text{d}) = -2.4571 + 2.5558\log_{10}(\text{Re}) - 0.9295\log_{10}(\text{Re})^2 + 0.1049\log_{10}(\text{Re})^3 \tag{16}$$

with

$$\text{c}_\text{d} = 2\frac{F_d}{\pi \rho u_{\text{in}}^2 r^2} \tag{17}$$

and $r = d/2$ is the radius of the sphere. While equation 16 applies to channels with very large blockage ratios, i.e. $h \gg d$, for the case of moderate blockage ratios [2] gives an empirical correction factor

$$k_c = \frac{1}{1 - 1.6\lambda^{1.6}} \tag{18}$$

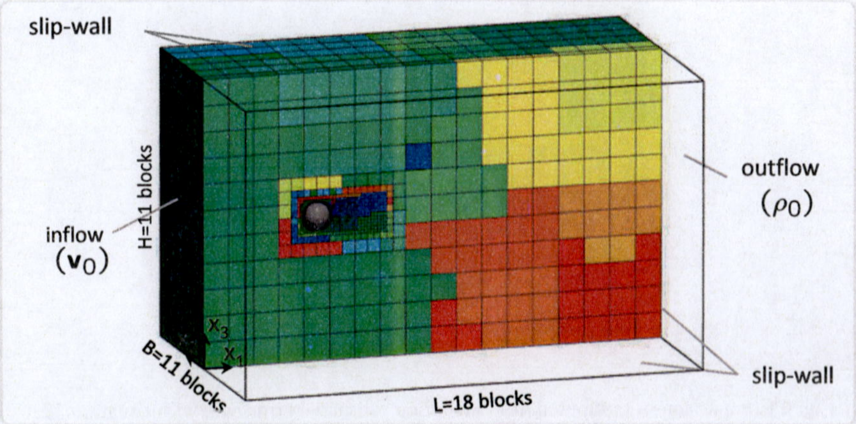

Fig. 2 Setup of the test case: each square in the lower right picture represents a 9x9x9 block of grid points. The total number of grid points is $\approx 2E08$. The block colour represents the corresponding subdomain index.

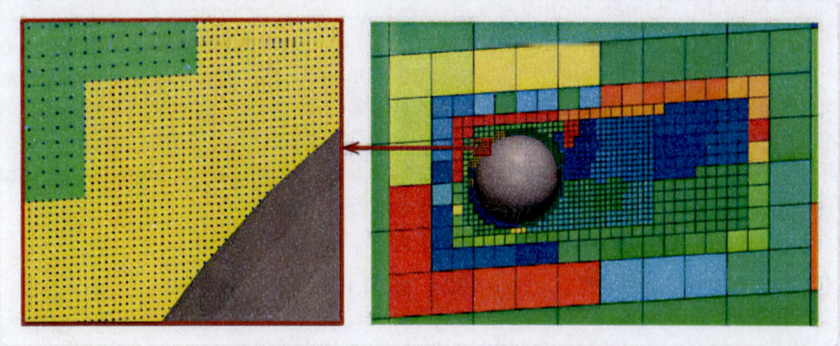

Fig. 3 Block grid next to sphere (right) and individual grid nodes (left)

The results of the comparison can be seen in Fig. 4. Because of the large computational cost of the study, most of the data points where still somewhat noisy. Hence, error bars indicating the remaining uncertainty are used in the Figure. Nonetheless, it can be stated, that the simulation can reproduce the correct trend of the empirical curve and the results for each data point are off by $\leq 5\%$.

The thickness of the boundary layer can be estimated as $1.3d/\sqrt{\text{Re}}$ in the region of the stagnation point upstream of the sphere and as $1.0d/\sqrt{\text{Re}}$ near the center plane of the sphere perpendicular to the main flow direction [2]. We choose the resolution around the sphere such that the sphere diameter is resolved by 256 grid points and therefore the boundary layer is resolved with ≈ 3 points at $\text{Re} = 10.000$ and ≈ 8 points at $\text{Re} = 1.000$.

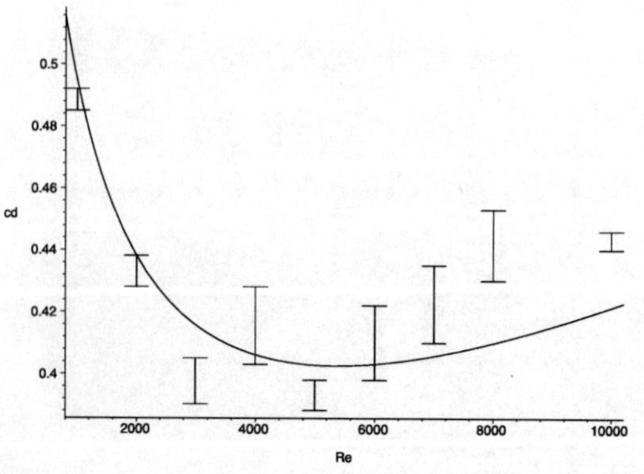

Fig. 4 Comparison of computed drag including estimated error bars with equation 16.

With the given resolution one time step on the coarsest grid level requires $4 \cdot 10^8$ lattice updates. The LB research prototype software package VIRTUALFLUIDS described in detail in [5] was used for the distributed simulations. It has been executed on a 64-bit-AMD-Opteron cluster with 47 dual processor boards (2×1.4 GHz), each connected via a full duplex Myrinet 2000 network interface. On a single core processor, the number of lattice updates per second is $\approx 9 \times 10^5$. 70 processors were used for the simulations in this article and a parallel efficiency of 80% was obtained. The limiting factor for the efficiency were due to specific properties of the given network topology and an imperfect load balancing for the domain decomposition done by METIS [9] for non uniform grids.

4.1 Detailed Look at Case of $\mathrm{Re} = 10.000$

For the case of $\mathrm{Re} = 10.000$ and $\mathrm{Ma} = 0.04/c_s \approx 0.023$ (with $c_s = c \times 1/\sqrt{3}$ being the speed of sound) the flow is still slightly under-resolved at the finest grid level. The turbulent viscosity that would vanish in the limit of infinite resolution is found to be $\nu_{\mathrm{turb}} \approx 0.7\nu$, i.e. given the large velocity gradient in the boundary layer the contribution of the turbulent viscosity to the total viscosity is still significant. Outside of the boundary layer the turbulent viscosity quickly vanishes to a negligible amount (see Fig. 5) acting only as filter eliminating undesired high frequencies in the flow field. In order to satisfy the boundary layer theory the turbulent viscosity near the wall should vanish. To avoid excessive grid-refinement near boundaries wall models such as Van-Driest-damping [18] or more advanced wall models should be employed.

Fig. 6 shows a snapshot of an instantaneous velocity field close to the sphere and Fig. 7 shows the mean pressure distribution.

The distribution of the mean pressure coefficient $c_p = \frac{p - p_\infty}{0.5 \rho u_\infty^2}$ (with p_∞ the free stream pressure and u_∞ the free stream velocity) around the sphere is shown in fig. 8. Its distribution is almost independent of the Reynolds number in the

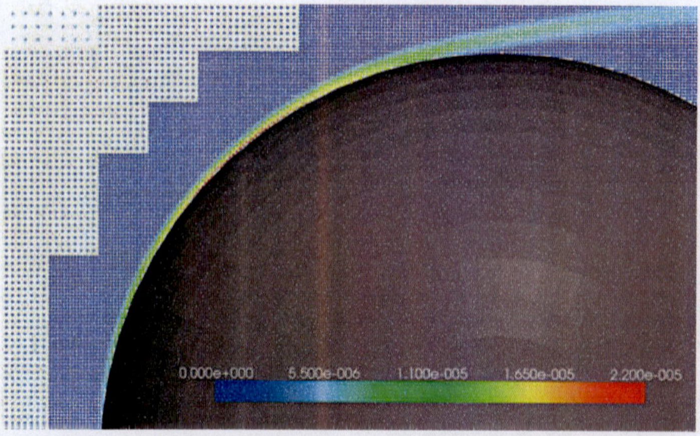

Fig. 5 The colour in the plot indicates the amount of turbulent viscosity produced by the Smagorinsky model. In the boundary layer the contribution of turbulent viscosity to the total viscosity is significant.

Fig. 6 Snapshot of the instantaneous velocity magnitude on a plane cutting through the center of the sphere in the main direction of the flow and perpendicular to the walls.

transitional regime. The computed values agree very well with results from the literature [3].

Fig. 9 shows the contours of one component of the Reynolds stress tensor $\overline{v'_{x_1} v'_{x_2}}$, where the prime denotes temporal fluctuations of the velocities and the overline denotes temporal average, x_1 is the streamwise direction and x_2 the other coordinate direction which is parallel to the shown cutting plane.

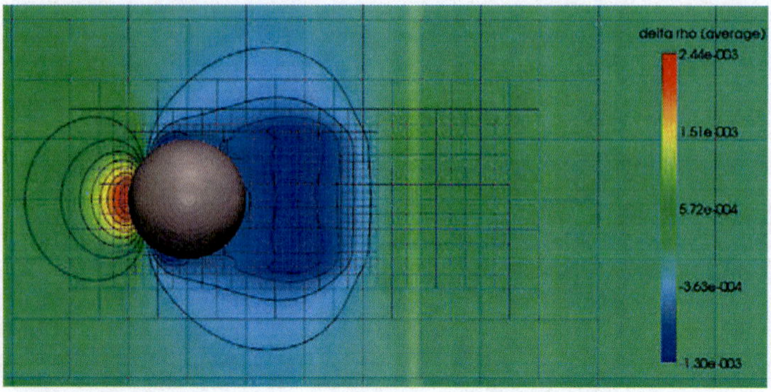

Fig. 7 Pressure time average on a plane cutting through the center of the sphere in the main direction of the flow and perpendicular to the walls.

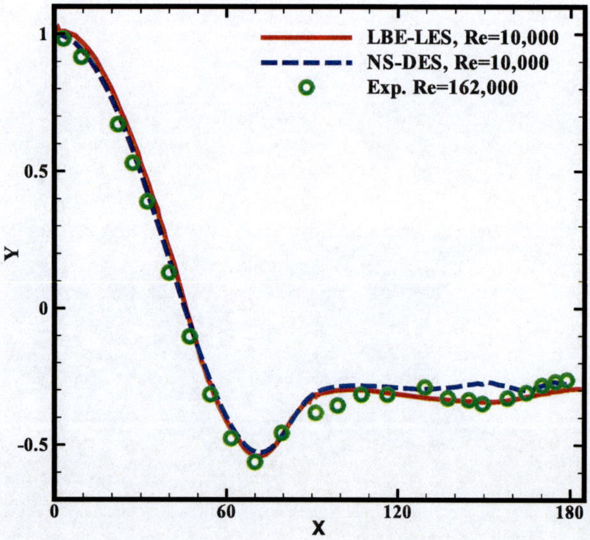

Fig. 8 Comparison of the simulated mean pressure coefficient distribution on the surface of the sphere with literature values.

5 Conclusion

A lattice Boltzmann model and parallel implementation for the computation of transitional flow has been presented. It features LES/Smagorinsky SGS turbulence modelling and non-uniform computational grids, which are necessary for a satisfactory resolution of the boundary layer while keeping the overall computational effort low. The method lends itself to high parallel efficiency.

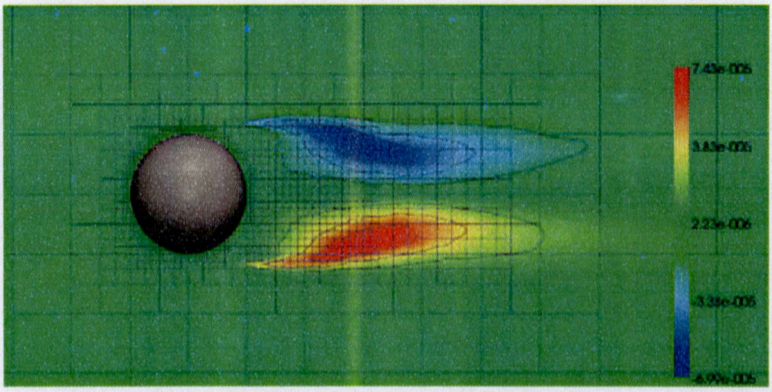

Fig. 9 Reynolds stress $\overline{v'_{x_1} v'_{x_2}}$

For the test case of flow around a sphere in a channel for a range of Reynolds numbers between 1.000 and 10.000 our simulations capture the essential features of these flows. For a further reduction of the computational time the development of a wall model compatible with the LB framework is desirable. In combination with a near wall compatible SGS model as described in [13] we expect an improved performance. These investigations are presently conducted and will be subject to future publication.

Acknowledgements. Financial support from the Federal Ministry of Education and Research (BMBF) under the SKALB project grand 01IH08003E is gratefully acknowledged.

References

1. Bouzidi, M., Firdaouss, M., Lallemand, P.: Momentum transfer of a lattice-Boltzmann fluid with boundaries. Phys. Fluids 13, 3452–3459 (2002)
2. Clift, R., Grace, J.R., Weber, M.E.: Bubbles, Drops and Particles. Academic Press, New York (1978)
3. Constantinescu, G., Squires, K.: LES and DES investigations of turbulent flow over a sphere at Re=10000. Flow Turbul. Combust. 70(1-4), 267–298 (2003)
4. Crouse, B., Rank, E., Krafczyk, M., Tölke, J.: A LB-based approach for adaptive flow simulations. Int. J. Mod. Phys. B 17(1+2), 109–112 (2003)
5. Freudiger, S.: Entwicklung eines parallelen, adaptiven, komponentenbasierten Strömungskerns für hierarchische Gitter auf Basis des Lattice-Boltzmann-Verfahrens, Ph.D. thesis, iRMB, TU Braunschweig (2009)
6. Filippova, O., Hänel, D.: Grid refinement for lattice-BGK models. J. Comp. Phys. 147, 219–228 (1998)
7. Geller, S., Krafczyk, M., Tölke, J., Turek, S., Hron, J.: Benchmark computations based on lattice-Boltzmann, finite element and finite volume methods for laminar flows. Comput. Fluids 35, 888–897 (2006)
8. D'Humières, D., Ginzburg, I., Krafczyk, M., Lallemand, P., Luo, L.: Multiple-relaxation-time lattice Boltzmann models in three dimensions. Phil. Trans. R. Soc. Lond. A 360, 437–451 (2002)

9. Karypis, G., Kumar, V.: METIS - A Software Package for Partitioning Unstructured Graphs, Partitioning Meshes, and Computing Fill-Reducing Orderings of Sparse Matrices - Version 4.0. (1998),
`http://glaros.dtc.umn.edu/gkhome/views/metis`
(last access January 21, 2009)
10. Krafczyk, M., Tölke, J., Luo, L.-S.: Large-eddy simulations with a multiple-relaxation-time LBE model. Int. J. Mod. Phys. B 17(1/2), 33–39 (2003)
11. Lallemand, P., Luo, L.-S.: Theory of the lattice Boltzmann method: dispersion, dissipation, isotropy, Galilean invariance, and stability. Phys. Review E 61, 6546–6562 (2000)
12. Mei, R., Yu, D., Shyy, W., Luo, L.-S.: Force evaluation in the lattice Boltzmann method involving curved geometry. Phys. Review E 65(4), 41203 (2002)
13. Nicoud, F., Ducros, F.: Subgrid-scale stress modelling based on the square of the velocity gradient tensor. Flow Turbul. Combust. 62(3), 183–200 (1999)
14. Qian, Y., d'Humières, D., Lallemand, P.: Lattice BGK models for Navier-Stokes equation. Europhys. Lett. 17, 479–484 (1992)
15. Rheinländer, M.: A consistent grid coupling method for Lattice-Boltzmann schemes. J. Statist. Phys. 121, 49–74 (2005)
16. Sagaut, P.: Large Eddy Simulation for incompressible flows: an introduction. Springer, Heidelberg (2001)
17. Teixeira, C.M.: Incorporating turbulence models into the lattice-Boltzmann method. Int. J. Mod. Phys. C 9(8), 1159–1175 (1998)
18. Van Driest, E.R.: On turbulent flow near a wall. J. Aero. Sci. 23, 1007–1011 (1956)
19. Yu, D., Mei, R., Shyy, W.: A multi-block lattice Boltzmann method for viscous fluid flows. International J. Numer. Meth. Fluids 39(2), 99–120 (2002)

Appendix 1: Orthogonal Basis Vectors $\{\Phi_i, i = 0, \ldots, b-1\}$

The 19 orthogonal basis vectors required for the transformation of the D3Q19 momentum distribution function to its corresponding moments are listed below:

$$\Phi_{0,\alpha} = 1, \quad \Phi_{1,\alpha} = \mathbf{e}_\alpha^2 - c^2, \quad \Phi_{2,\alpha} = 3(\mathbf{e}_\alpha^2)^2 - 6\mathbf{e}_\alpha^2 c^2 + c^4, \tag{19}$$

$$\Phi_{3,\alpha} = e_{\alpha x}, \quad \Phi_{5,\alpha} = e_{\alpha y}, \quad \Phi_{7,\alpha} = e_{\alpha z}, \tag{20}$$

$$\Phi_{4,\alpha} = (3\mathbf{e}_\alpha^2 - 5c^2)e_{\alpha x}, \quad \Phi_{6,\alpha} = (3\mathbf{e}_\alpha^2 - 5c^2)e_{\alpha y},$$

$$\Phi_{8,\alpha} = (3\mathbf{e}_\alpha^2 - 5c^2)e_{\alpha z}, \tag{21}$$

$$\Phi_{9,\alpha} = 3e_{\alpha x}^2 - \mathbf{e}_\alpha^2, \quad \Phi_{11,\alpha} = e_{\alpha y}^2 - e_{\alpha z}^2, \tag{22}$$

$$\Phi_{13,\alpha} = e_{\alpha x}e_{\alpha y}, \quad \Phi_{14,\alpha} = e_{\alpha y}e_{\alpha z}, \quad \Phi_{15,\alpha} = e_{\alpha x}e_{\alpha z}, \tag{23}$$

$$\Phi_{10,\alpha} = (2\mathbf{e}_\alpha^2 - 3c^2)(3e_{\alpha x}^2 - \mathbf{e}_\alpha^2), \quad \Phi_{12,\alpha} = (2\mathbf{e}_\alpha^2 - 3c^2)(e_{\alpha y}^2 - e_{\alpha z}^2), \tag{24}$$

$$\Phi_{16,\alpha} = (e_{\alpha y}^2 - e_{\alpha z}^2)e_{\alpha x}, \quad \Phi_{17,\alpha} = (e_{\alpha z}^2 - e_{\alpha x}^2)e_{\alpha y},$$

$$\Phi_{18,\alpha} = (e_{\alpha x}^2 - e_{\alpha y}^2)e_{\alpha z} \tag{25}$$

Appendix 2: Transformation Matrix M

For convenience the basis vectors are arranged in matrix M which reads:

$$
\begin{bmatrix}
1\cdot & (1 & 1 & 1 & 1 & 1 & 1 & 1 & 1 & 1 & 1 & 1 & 1 & 1 & 1 & 1 & 1 & 1 & 1 & 1) \\
c^2\cdot & (-1 & 0 & 0 & 0 & 0 & 0 & 0 & 1 & 1 & 1 & 1 & 1 & 1 & 1 & 1 & 1 & 1 & 1 & 1) \\
c^4\cdot & (1 & -2 & -2 & -2 & -2 & -2 & -2 & 1 & 1 & 1 & 1 & 1 & 1 & 1 & 1 & 1 & 1 & 1 & 1) \\
c\cdot & (0 & 1 & -1 & 0 & 0 & 0 & 0 & 1 & -1 & 1 & -1 & 1 & -1 & 1 & -1 & 0 & 0 & 0 & 0) \\
c^3\cdot & (0 & -2 & 2 & 0 & 0 & 0 & 0 & 1 & -1 & 1 & -1 & 1 & -1 & 1 & -1 & 0 & 0 & 0 & 0) \\
c\cdot & (0 & 0 & 0 & 1 & -1 & 0 & 0 & 1 & -1 & -1 & 1 & 0 & 0 & 0 & 0 & 1 & -1 & 1 & -1) \\
c^3\cdot & (0 & 0 & 0 & -2 & 2 & 0 & 0 & 1 & -1 & -1 & 1 & 0 & 0 & 0 & 0 & 1 & -1 & 1 & -1) \\
c\cdot & (0 & 0 & 0 & 0 & 0 & 1 & -1 & 0 & 0 & 0 & 0 & 1 & -1 & -1 & 1 & 1 & -1 & -1 & 1) \\
c^3\cdot & (0 & 0 & 0 & 0 & 0 & -2 & 2 & 0 & 0 & 0 & 0 & 1 & -1 & -1 & 1 & 1 & -1 & -1 & 1) \\
c^2\cdot & (0 & 2 & 2 & -1 & -1 & -1 & -1 & 1 & 1 & 1 & 1 & 1 & 1 & 1 & 1 & -2 & -2 & -2 & -2) \\
c^4\cdot & (0 & -2 & -2 & 1 & 1 & 1 & 1 & 1 & 1 & 1 & 1 & 1 & 1 & 1 & 1 & -2 & -2 & -2 & -2) \\
c^2\cdot & (0 & 0 & 0 & 1 & 1 & -1 & -1 & 1 & 1 & 1 & 1 & -1 & -1 & -1 & -1 & 0 & 0 & 0 & 0) \\
c^4\cdot & (0 & 0 & 0 & -1 & -1 & 1 & 1 & 1 & 1 & 1 & 1 & -1 & -1 & -1 & -1 & 0 & 0 & 0 & 0) \\
c^2\cdot & (0 & 0 & 0 & 0 & 0 & 0 & 0 & 1 & 1 & -1 & -1 & 0 & 0 & 0 & 0 & 0 & 0 & 0 & 0) \\
c^2\cdot & (0 & 0 & 0 & 0 & 0 & 0 & 0 & 0 & 0 & 0 & 0 & 0 & 0 & 0 & 0 & 1 & 1 & -1 & -1) \\
c^2\cdot & (0 & 0 & 0 & 0 & 0 & 0 & 0 & 0 & 0 & 0 & 0 & 1 & 1 & -1 & -1 & 0 & 0 & 0 & 0) \\
c^3\cdot & (0 & 0 & 0 & 0 & 0 & 0 & 0 & 1 & -1 & 1 & -1 & -1 & 1 & -1 & 1 & 0 & 0 & 0 & 0) \\
c^3\cdot & (0 & 0 & 0 & 0 & 0 & 0 & 0 & -1 & 1 & 1 & -1 & 0 & 0 & 0 & 0 & 1 & -1 & 1 & -1) \\
c^3\cdot & (0 & 0 & 0 & 0 & 0 & 0 & 0 & 0 & 0 & 0 & 0 & 1 & -1 & -1 & 1 & -1 & 1 & 1 & -1)
\end{bmatrix}
$$

The transformation of the nodal momentum distribution function to its corresponding moments is obtained by $m = \mathsf{M}f$. The multiple-relaxation-time collision operator is applied in moment space according to equation 2.

Combined Approach to Numerical Simulation of Spatial Nonlinear Waves in Shallow Water with Various Bottom Topography

Dmitry G. Arkhipov, Georgy A. Khabakhpashev, and Nurziya S. Safarova

Abstract. This paper deals with the novel approach to the evolution description of nonlinear three-dimensional moderately long disturbances of the liquid free surface. The suggested model consists of one basic second-order differential equation for spatial perturbations and two auxiliary linear differential equations for a determination of the fluid horizontal velocity vector averaged over the layer depth. This vector is contained in the main equation only in the term of the second order of smallness. The suggested model is suitable for finite-amplitude waves running with any angles. This approach is in essence easier than the known systems of equations, where all equations contain both linear and nonlinear items. Some problems of interactions and collisions of waves were solved numerically. A number of test and demonstrational solutions were found in the pools with different topographies. Expectedly it was observed that not only the change of the wave velocities but also the intensification of disturbances propagating towards the lower liquid depth and vice versa their weakening along with the waves motion to the deeper area. It is seen, that the additional peaks and troughs took place over the bottom irregularities.

1 Introduction

Gravity waves in fluids are among the classical objects of the nonlinear phenomena mechanics. In particularly, Korteweg and de Vries derived their famous equation

Dmitry G. Arkhipov · Georgy A. Khabakhpashev · Nurziya S. Safarova
Kutateladze Institute of Thermophysics SB RAS, Lavrentiev Ave. 1,
Novosibirsk, 630090, Russia
e-mail: theory@itp.nsc.ru

Dmitry G. Arkhipov
Novosibirsk Branch of the Nuclear Safety Institute RAS, Lavrentiev Ave. 1,
Novosibirsk, 630090, Russia

Georgy A. Khabakhpashev
Novosibirsk State University, Pirogova Str. 2, Novosibirsk, 630090, Russia

E. Krause et al. (Eds.): Computational Sci., & High Performance Computing IV, NNFM 115, pp. 297–312.
springerlink.com

precisely for disturbances of the free surface of a shallow water layer (see, for example, [1]). Kadomtsev – Petviashvili equation [2] was surely the next major step in the development of evolutionary equations for disturbances of small but finite amplitude. The model assumes that the horizontal scale of the wave processes under consideration is appreciably smaller in the longitudinal rather than in the transverse coordinate.

In recent decades much attention of researchers has been attracted to essentially three-dimensional finite-amplitude disturbances of the water surface (for instance, books [3, 4]). However, almost all simplified models (e. g., articles [5–7]) are applicable only to nonlinear waves propagating chiefly in one direction (therein lies the fundamental restriction of the above-mentioned equations resulting from the fact that the equations of motion involve convective terms). Only in these cases the fluid velocity can be eliminated from these terms, and the problem may be reduced to one equation for the perturbation of the free surface. For this reason, finite-amplitude waves travelling simultaneously in different directions (at arbitrary angles to each other) can be described only by the systems of equations incorporating both the disturbance of the free boundary and the fluid velocity. In the systems proposed before (for example, [8–10]), even the linear terms of all equations involve terms depending on the fluid velocity.

The aim of this paper is to propose a system of equations, which would be more convenient for analysis. The first version of such system was proposed in the article [11], where the results of only two simple calculations for a liquid with the horizontal bottom were given.

2 Problem Formulation and Initial Equations

Since the propagation rate of the gravity disturbance is much lower than the speed of sound in the water, we will use the incompressible liquid approximation. Moreover, we assume that, firstly, the stationary components of the fluid flow equal zero, while the originating flow is potential, that is $\partial \mathbf{u}/\partial z = \nabla w$, where $\mathbf{u}$ is the vector of the horizontal component of the fluid velocity, z is the vertical coordinate, the differential operator $\nabla = (\partial/\partial x, \partial/\partial y)$, and w is the vertical component of the fluid velocity; secondly, the wavelength λ is noticeably larger and the disturbance amplitude η_a is considerably smaller than the equilibrium layer depth h ($h/\lambda \sim \varepsilon^{1/2}$ and $\eta/h \sim \varepsilon$, where ε is a small parameter); thirdly, the bottom is gently sloping ($|\nabla h| \sim \varepsilon^{3/2}$); and, fourthly, effects of viscosity and capillarity are neglected. Therefore, convective terms containing w can be dropped from the equations of motion as a negligible order of smallness. Then, the initial equations (Euler's equations and the continuity equation) can be written in the following simple form:

$$\frac{\partial \mathbf{u}}{\partial t} + \nabla\left(\frac{\mathbf{u}^2}{2} + \frac{p}{\rho}\right) = 0\,, \tag{1}$$

$$\frac{\partial w}{\partial t} + \frac{1}{\rho}\frac{\partial p}{\partial z} + g = 0\,, \tag{2}$$

$$\nabla \cdot \mathbf{u} + \frac{\partial w}{\partial z} = 0\,. \tag{3}$$

Here t is the time, ρ is the liquid density, p is the fluid pressure, and g is the acceleration of free fall. We also impose the conventional kinematic and dynamic boundary conditions on the fixed rigid bottom and on the free surface (see Fig. 1):

$$w = w_b = -\mathbf{u}_b \cdot \nabla h \qquad \text{for} \qquad z = -h(x,y)\,;$$

$$w = w_s = \partial \eta / \partial t + \mathbf{u}_s \cdot \nabla \eta\,, \qquad p = p_s = 0 \qquad \text{for} \qquad z = \eta(t,x,y)\,,$$

where, subscripts b and s refer to the values at the bottom and at the water free boundary, respectively.

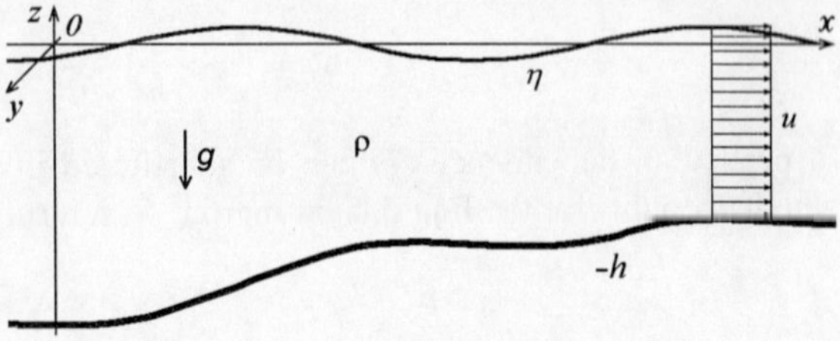

Fig. 1 Schematic of the wave process in the shallow water with a gently sloping bottom

3 Derivation of Model Equations

Let us integrate the continuity equation (3) with respect to coordinate z from $-h$ to η. Using the Leibniz rule for differentiation with respect to a parameter and the kinematic boundary conditions, we find the mass conservation law in the layer

$$\frac{\partial \eta}{\partial t} + \nabla \cdot [\langle \mathbf{u} \rangle (h + \eta)] = 0\,,$$

and in the first approximation (linear waves in a layer of a constant depth) we have

$$\frac{\partial \eta}{\partial t} = -h \nabla \cdot \langle \mathbf{u} \rangle\,. \tag{4}$$

Here angular brackets designate quantities averaged over the layer depth. This formula will be necessary below.

Now we apply scalarly the operator ∇ to the equation of horizontal motion (1):

$$\frac{\partial}{\partial t}(\nabla \cdot \mathbf{u}) + \nabla^2 \left(\frac{\mathbf{u}^2}{2} + \frac{p}{\rho} \right) = 0\,. \tag{5}$$

For the purpose of the pressure vertical profile determination we integrate Eq. (2) with respect to coordinate z from z to $\eta(t,x,y)$ and use the dynamic boundary condition on the free surface. As a result, we obtain

$$\frac{p}{\rho} = g(\eta - z) + \int_z^{\eta} \frac{\partial w}{\partial t}\, dz\,. \tag{6}$$

In the approximation under consideration, the formulas for linear waves in a layer of a constant depth are locally valid. Therefore, the profile of the vertical component of the fluid velocity is linear in the coordinate z and can be written in the form

$$w = \frac{h+z}{h+\eta} w_s - \frac{z-\eta}{h+\eta} w_b\,,$$

and, for the terms of the second-order of smallness, in a more simpler form

$$w = \left(1 + \frac{z}{h}\right) \frac{\partial \eta}{\partial t}\,. \tag{7}$$

In particular, dependence (7) can be substituted in the integral term of Eq. (6), which accounts for the liquid layer inertia. As a result, we find

$$\frac{p}{\rho} = g(\eta - z) - \left(z + \frac{z^2}{2h}\right) \frac{\partial^2 \eta}{\partial t^2}\,. \tag{8}$$

Note that, in the first approximation (very long linear waves), this expression is reduced to hydrostatics: $p = \rho g(\eta - z)$. Integrating the equation of horizontal motion (1) with respect to coordinate z from $-h$ to η, we have in this approximation

$$\frac{\partial \langle \mathbf{u} \rangle}{\partial t} + g\nabla\eta = 0\,, \tag{9}$$

that is the simplest form of the conservation law of the fluid layer momentum.

Now return to Eq. (5) and integrate it over the layer depth, i. e. from $-h$ to η relative to coordinate z. Using the continuity equation (3) and the pressure vertical profile (8), we obtain our fundamental equation in the form

$$\int_{-h}^{\eta} \left(\frac{\partial^2 w}{\partial t \partial z} - \nabla^2 \frac{\mathbf{u}^2}{2} \right) dz - g(h+\eta)\nabla^2\eta - \frac{h^2}{3}\nabla^2 \frac{\partial^2 \eta}{\partial t^2} = 0\,. \tag{10}$$

Due to the Leibniz rule for differentiation with respect to a parameter, the kinematic boundary condition on the free surface of the layer, and the continuity equation (3), the first part of the integral in Eq. (10) can be transformed as follows:

$$\int_{-h}^{\eta} \frac{\partial^2 w}{\partial t \partial z} dz = \left. \frac{\partial w_s}{\partial t} - \frac{\partial \eta}{\partial t} \frac{\partial w}{\partial z} \right|_{z=\eta} = \frac{\partial^2 \eta}{\partial t^2} + \frac{\partial}{\partial t} (\mathbf{u}_s \cdot \nabla \eta) +$$

$$+ \frac{\partial \eta}{\partial t} \nabla \cdot \mathbf{u}_s = \frac{\partial^2 \eta}{\partial t^2} + \frac{\partial \mathbf{u}_s}{\partial t} \cdot \nabla \eta + \nabla \cdot \left(\mathbf{u}_s \frac{\partial \eta}{\partial t} \right) . \tag{11}$$

As for the other part of the integral, we can change the upper limit to zero, take the derivatives out of the integral sign, and use the angular brackets again:

$$\int_{-h}^{0} \nabla^2 \frac{\mathbf{u}^2}{2} dz = \frac{h}{2} \nabla^2 \langle \mathbf{u}^2 \rangle . \tag{12}$$

Under the assumption of long-wave disturbances, the terms of the second order of smallness satisfy the approximate equalities $\langle \mathbf{u}^2 \rangle = \langle \mathbf{u} \rangle^2$ and $\mathbf{u}_s = \langle \mathbf{u} \rangle = \mathbf{u}_b$. Substituting these expressions into Eq. (11), using the formulas (4) and (9), substituting the obtained expression and the Eq. (12) into Eq. (10), we find

$$\frac{\partial^2 \eta}{\partial t^2} - gh\nabla^2 \eta - \frac{g}{2} \nabla^2 (\eta^2) - h\nabla^2 \left(\langle \mathbf{u} \rangle^2 \right) - \frac{h^2}{3} \nabla^2 \frac{\partial^2 \eta}{\partial t^2} = g \nabla h \cdot \nabla \eta . \tag{13}$$

The evolution equation (13) is characterized with the average values of the horizontal components of the fluid velocity entering only in the second-order term. For this reason, the consequence of the flow potentiality and Eq. (4), that is

$$\langle \mathbf{u} \rangle = \nabla \varphi , \qquad \nabla^2 \varphi = -\frac{1}{h} \frac{\partial \eta}{\partial t} , \tag{14}$$

are sufficient for their determination (where φ is the potential of the liquid velocity). We emphasize that the derived model (Eqs. (13) and (14)) is applicable to waves travelling at arbitrary angles to each other. It is radically simpler than the known systems of differential equations (see, for example, [8–10]), in which all equations involve both linear and nonlinear terms.

The integro-differential equation given in the article [12] admits waves that can run counter to each other but only along the x axis. As in the Kadomtsev – Petviashvili model, the author of the paper assumed that the horizontal scale of the considered disturbances is mainly smaller in the longitudinal rather than in the transversal coordinate.

For nonlinear waves travelling chiefly in one direction, the auxiliary equations (14) are not needed, since, in the second-order terms of Eq. (13), $\langle \mathbf{u} \rangle^2$ can be replaced by $g\eta^2/h$. In this case, we arrive at the evolution equation derived in the article [7] for three-dimensional disturbances of the liquid free surface. If, in addition, $\nabla h = 0$, then we obtain the two-dimensional Boussinesq equation [6]. If, on top of it, $\partial^2/\partial y^2 \sim \varepsilon \partial^2/\partial x^2$, then we arrive at the Kadomtsev – Petviashvili equation [2]. As for the slope of the pool bottom, its turns out that its encounting it by a single term included entering in Eq. (13) is sufficient to obtain the Green's law for the plane linear monochromatic wave run-up: the amplitude is inversely proportional to the fourth root of the depth. Finally, if $\eta(t,x,y) = \eta(t,x)$ we derive the modified Boussinesq equation [13]. In particular, Eqs. (13) and (14) has the following solitary solution:

$$\eta = \eta_{sx} = \eta_{ax}/\cosh^2[(x - U_x t)/L_x] \,, \qquad (15)$$

where $U_x = \sqrt{gh(1+\eta_{ax}/3h)/(1-2\eta_{ax}/3h)}$, and $L_x = 2h\sqrt{(1/3 + h/\eta_{ax})/3}$.

4 Discussion of the Numerical Results

For the solution of the main nonlinear equation with partial derivatives (13) the modification of the implicit three-layer finite difference scheme is constructed. The scheme is similar to the one described in the article [14]. Calculations according to this scheme are carried out with Seidel's iteration algorithm in the horizontal plane. The scheme has the second order of approximation in all variables (the time and the horizontal coordinates). Calculations is carried out on three layers at time, and on each layer by five-point template of the cross type. The constructed difference scheme was tested on the analytical solution of the system (13) and (14) with $\nabla h = 0$ in the form of plane solitary disturbance (15). The results of calculations have shown, that expectedly, such a wave was propagating with a constant velocity and was keeping its form. The solution is steady and converges with the second order in all variables. The soliton (15) is extremely sensible for the scheme errors since of its shape is formed by the balance between nonlinear and dispersive terms of the second order of smallness.

For planar problems on each step of time we use also another iteration algorithm: firstly the calculations are made with the velocity vector values determined on the previous step of time, and then the velocity vector is determined with the simple linear auxiliary equations (14). Poisson's equation for determination of the velocity vector is solved by the method of the fast Fourier transformation by both horizontal coordinates at each step of time.

The presetting of a partial time derivative η at the first steps in all points of a pool was carried out with the perturbation displacement according to the characteristic value of the wave propagation velocity. The reflection condition is imposed at the side vertical walls of a pool, i. e. the normal derivatives η equal zero. For that in the performed calculations values of the function $\eta(t,x,y)$ in the boundary sites are equal to its values in the nearest internal sites.

We note that in Figs. 2–9 the dimensionless variables ($\eta^* = \eta/h_0$, $x^* = x/h_0$, $y^* = y/h_0$, $u^* = u/(gh_0)^{1/2}$, and $t^* = t(g/h_0)^{1/2}$) are used.

Firstly we consider interaction of two plane solitons (15) running in the same direction (Fig. 2). The interaction time is sufficiently large because the differences between the disturbances velocities are small. Nevertheless, as it is seen in Fig. 2, solitary waves keep their shapes after interaction. The larger perturbation moved forward in compared with the position calculated by the formulas for one soliton. On the contrary, the smaller disturbance moved a little backward from the corresponding soliton. The scenarios of interaction qualitatively agrees with analytical two-solitons solution of the Korteweg – de Vries equation (see [15]). If the wave amplitudes differ three times we have the overtaking interaction (Figs. 2*a*–2*c*), and if the wave amplitudes differ two times there is the exchange interaction (Figs. 2*d*–2*f*). In the last case two peaks take place during the whole time of calculation.

Now we proceed with a head-on collision of two plane solitons (Fig. 3). We observe a weak nonlinear increase of the disturbances at their interaction followed by the return to the previous wave amplitudes. From the mathematical standpoint the modified Boussinesq equation [13] allows studying a collision of two waves running towards each other. But this equation was derived at the assumption that

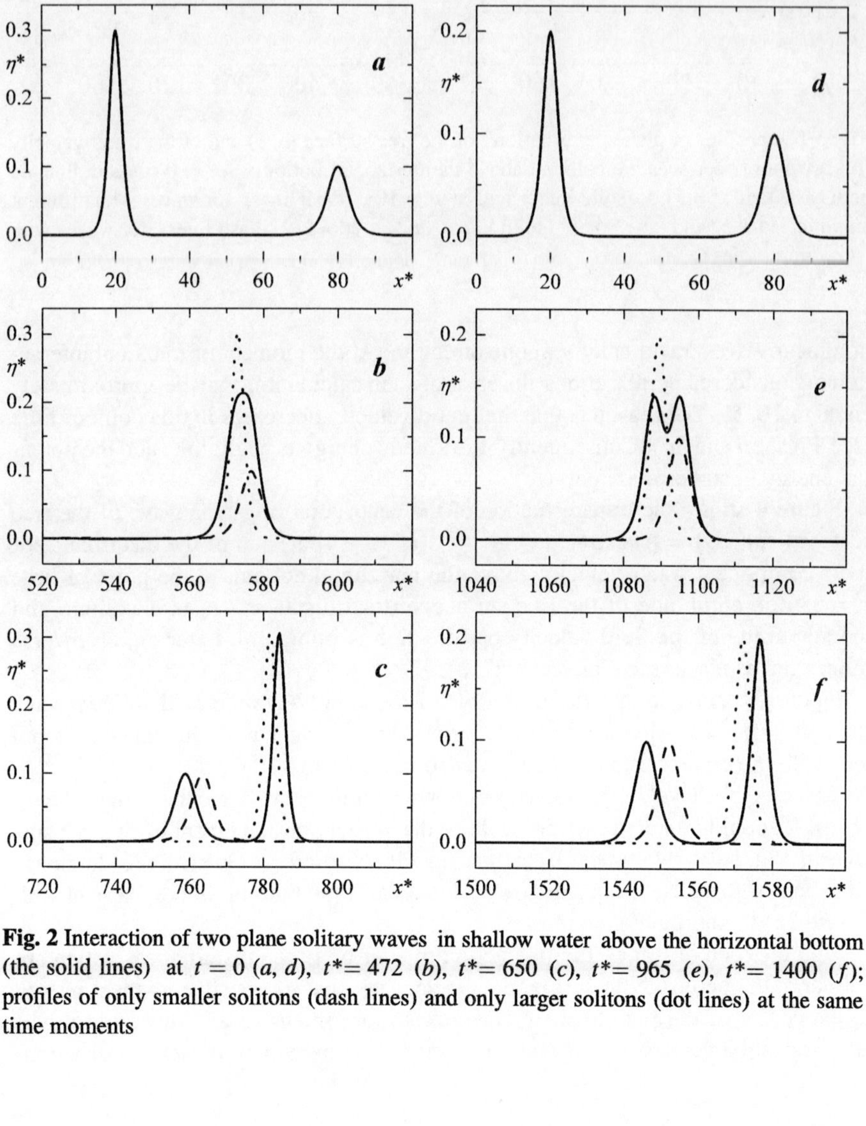

Fig. 2 Interaction of two plane solitary waves in shallow water above the horizontal bottom (the solid lines) at $t = 0$ (*a*, *d*), $t^*= 472$ (*b*), $t^*= 650$ (*c*), $t^*= 965$ (*e*), $t^*= 1400$ (*f*); profiles of only smaller solitons (dash lines) and only larger solitons (dot lines) at the same time moments

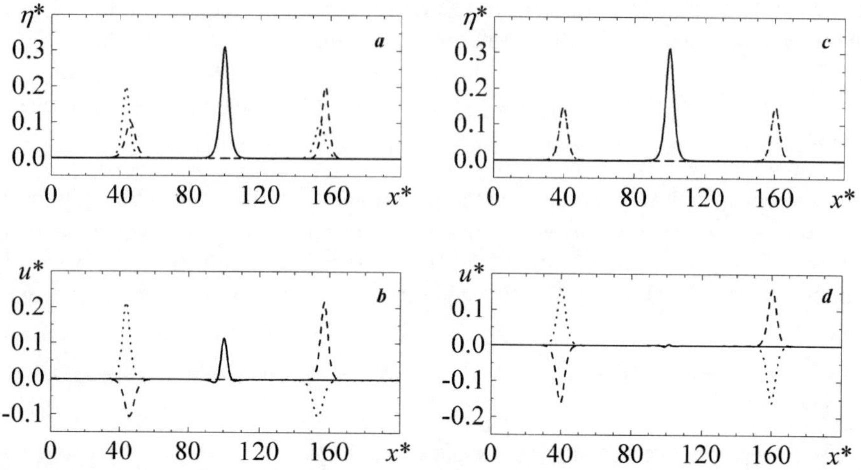

Fig. 3 The profiles of plane perturbations of the free surface (*a*, *c*) and of the water velocity (*b*, *d*), when two waves are collided above the horizontal bottom: at $t = 0$ (the dot lines in both cases), at $t^* = 51.5$ (solid lines) and at $t^* = 103$ (dash lines) for waves with different amplitudes (*a*, *b*) and at $t^* = 56$ (solid lines) and at $t^* = 112$ (dash lines) for waves with equal amplitudes (*c*, *d*)

nonlinear waves travel chiefly in one direction. At the moment of maximal interaction of considered perturbations the error of such calculation may be approximately equal to 10 %. The reason is that the liquid velocity decreases in this point of time (see Figs. 3*b* and 3*d*). Consequently, the kinetic energy is negligible, and the potential energy becomes maximal.

Figure 4 shows the transformation of the centripetal ring-type wave of the free surface ($\eta_0(x,y) = \eta_a \exp(-[(x^{*2} + y^{*2})^{1/2}/10 - 4]^2)$) and of the disturbance of the x-component water velocity above the horizontal bottom. In the point of time $t^* = 33$ the amplitude of the free surface vertical displacement is maximum and the amplitude of the fluid velocity perturbation is minimum. Later on, an inverse process takes place (see Figs. 5*a* and 5*b*).

Figure 5 demonstrates the evolution of the axially-symmetrical solitary wave $\eta_0(x,y) = \eta_a \exp(-[(x^*/20)^2 + (y^*/20)^2])$ on the water with the constant depth and with the crater-bottom: $h(x,y) = h_0[0.6 + 0.4\exp(-[(x^*/50)^2 + (y^*/50)^2])]$. As expected, the level of the liquid goes down behind the wave front and then slowly arises. It should be noted that the scale of the vertical displacement of the free surface in Fig. 5*a* is two times larger than in Figs. 5*b* and 5*c*. One can see the intensification of the wave propagated over the crater-type bottom in comparison with the case of the horizontal bottom.

Figure 6 demonstrates the interaction of four bell-type disturbances which are similar to the perturbation in Fig. 5*a*. One can see that if these disturbances moved to the center of the pool their interaction is more intensive and diverging waves are noticeably weaker. At the moment of maximal interaction of such perturbations

(Fig. 6*e*) the error of the calculation, carried out on the bases of two-dimensional Boussinesq equation [7], equals approximately 12.5 %.

The transformations of the initial disturbance of the cross-type form ($\eta_0(x,y) = 2\eta_{sx} + 2\eta_{sy}$) in the pools with various bottom topographies are interesting as well (Figs. 7 and 8). It is seen, that four quasi-plane waves propagate in four different directions (toward and counter the *x*-axis, as well as the *y*-axis). At the crosses of these fronts, one can see four peaks of the free surface perturbation. If the

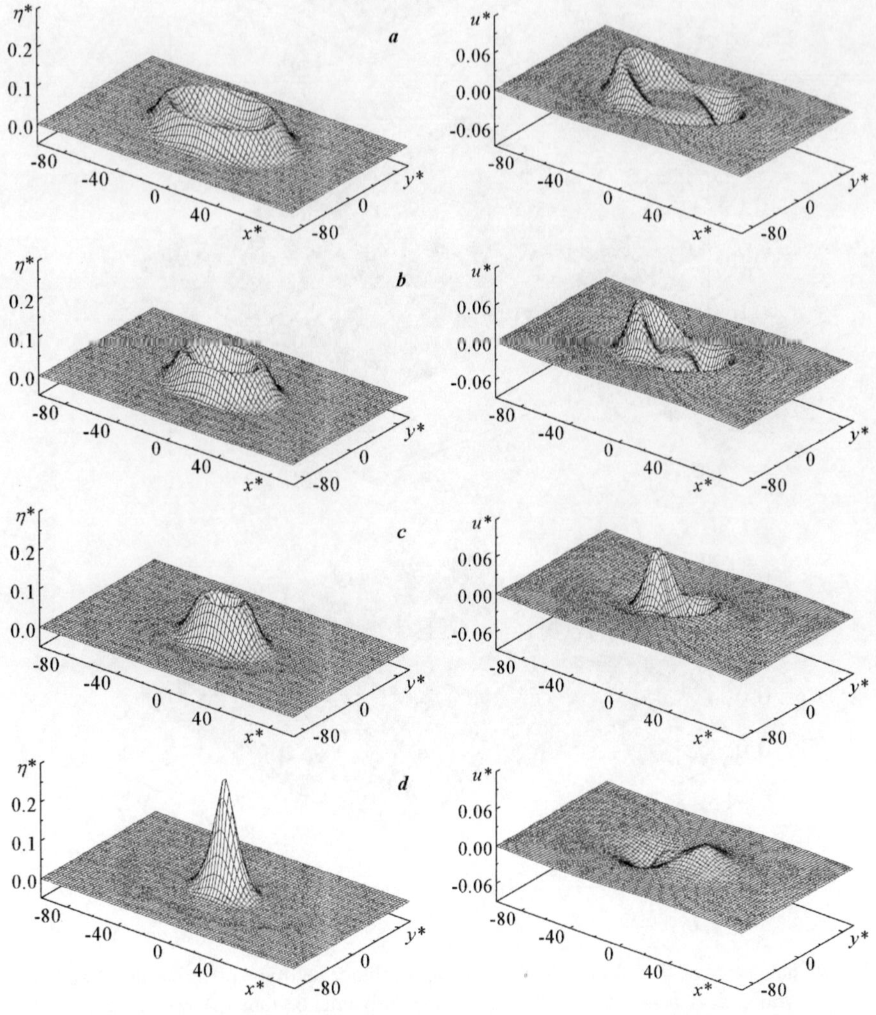

Fig. 4 The shapes of solitary axially-symmetrical perturbations of the water free surface and of the liquid velocity, when the initial ring-type disturbance was characterized by the centripetal motion in the pool with horizontal bottom at $t = 0$ (*a*), $t^* = 11$ (*b*), $t^* = 22$ (*c*), and $t^* = 33$ (*d*)

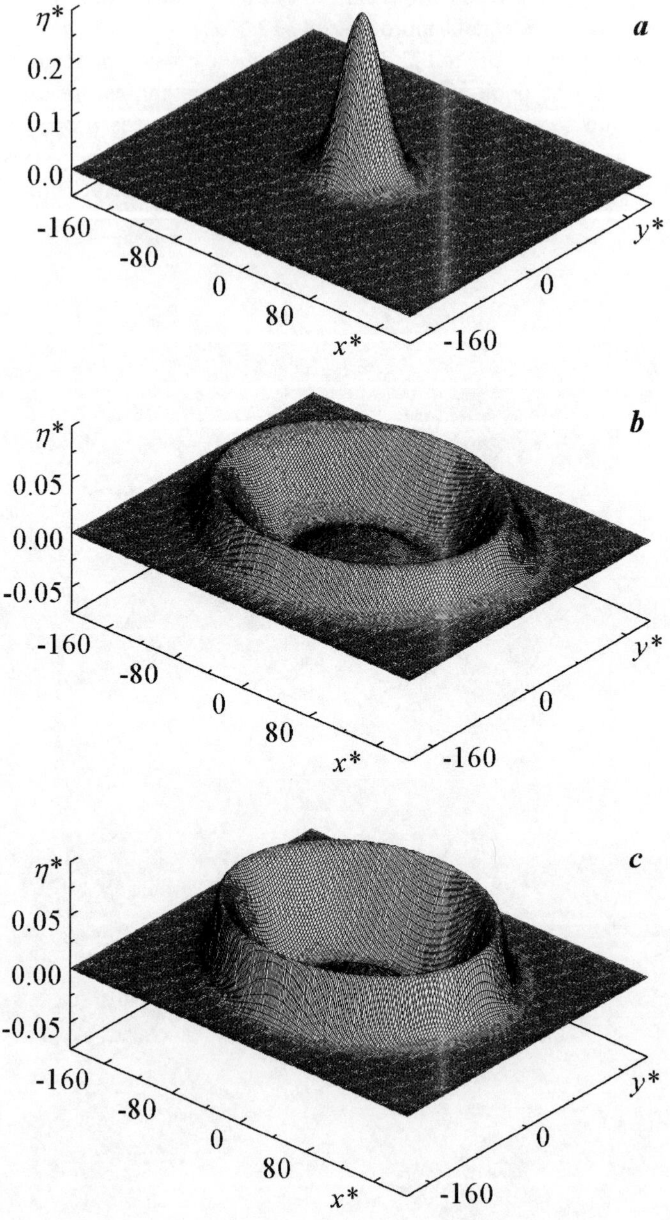

Fig. 5 The shapes of the initial three-dimensional solitary bell-type disturbance (*a*) and the axially-symmetrical wave at $t^* = 120$ above the horizontal bottom (*b*) and the crater-type bottom (*c*)

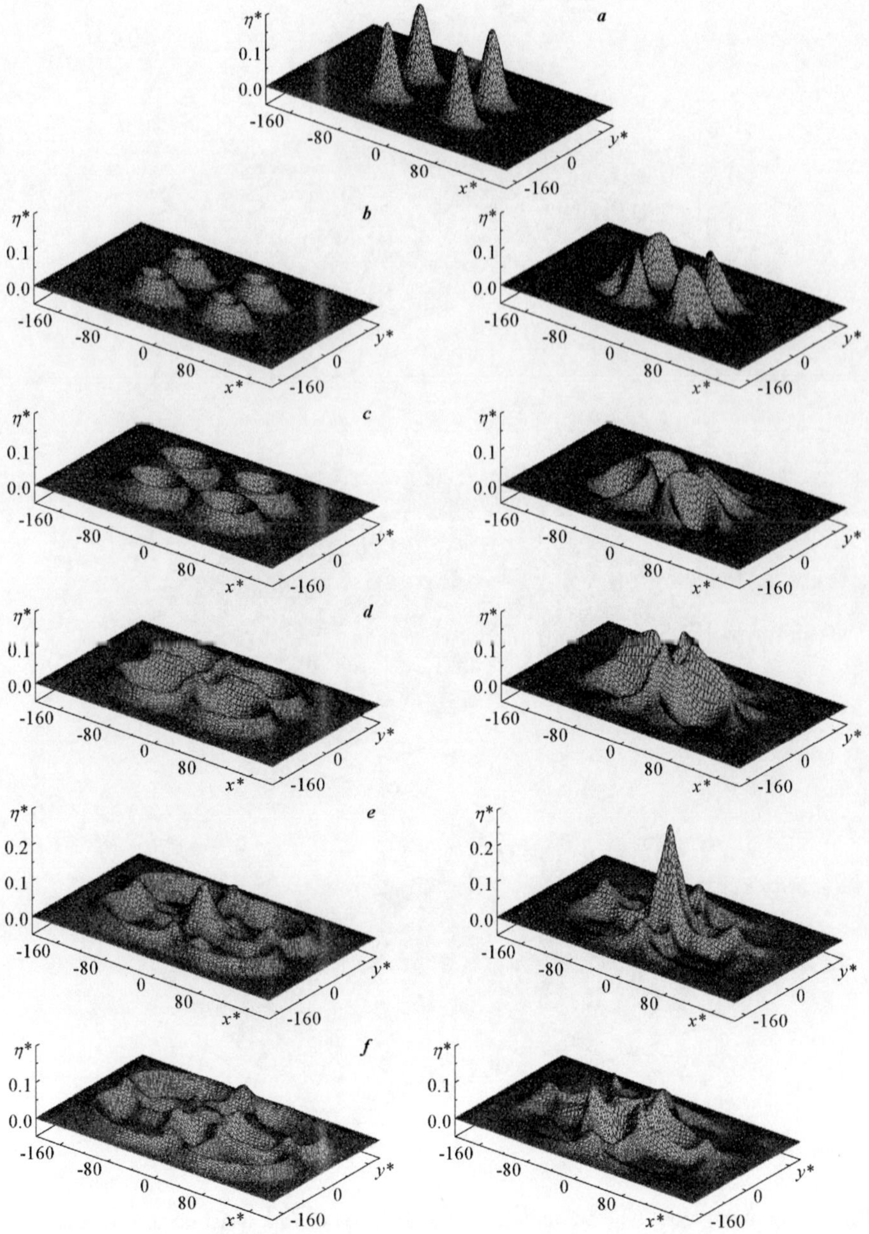

Fig. 6 The shapes of perturbations of the shallow water free surface, when four initial three-dimensional solitary bell-type disturbances (*a*) were at rest (the left illustrations) or were charac- terized by the centripetal motion (the right illustrations) in the pool with the horizontal bottom at five points of time: $t^*= 17$ (*b*), $t^*= 34$ (*c*), $t^*= 51$ (*d*), $t^*= 68$ (*e*), and $t^*= 85$ (*f*)

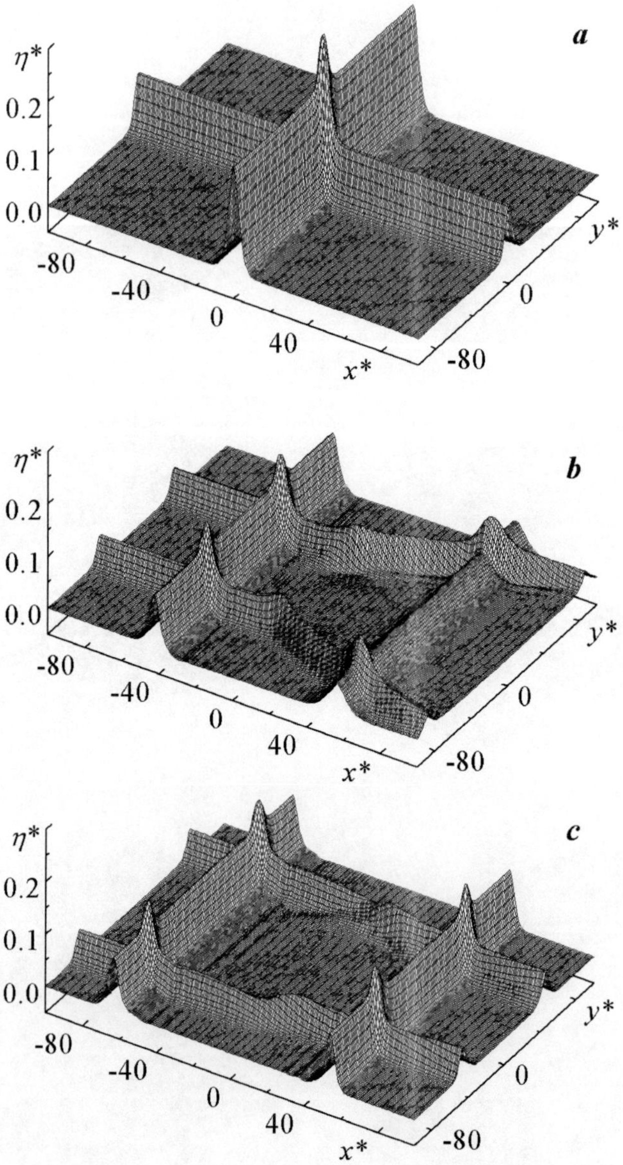

Fig. 7 The initial cross-type perturbation (*a*) and shapes of four quasi-plane waves at $t^*=$ 40 (*b*) and $t^*= 60$ (*c*) on the shallow water in the pools with the partially sloping bottom: $h(x,y) = h_0$ for $x < 0$ in both cases, $h(x,y) = h_0(1+x^*/25)$ (*b*) and $h(x,y) = h_0(1-x^*/200)$ (*c*) for $x > 0$

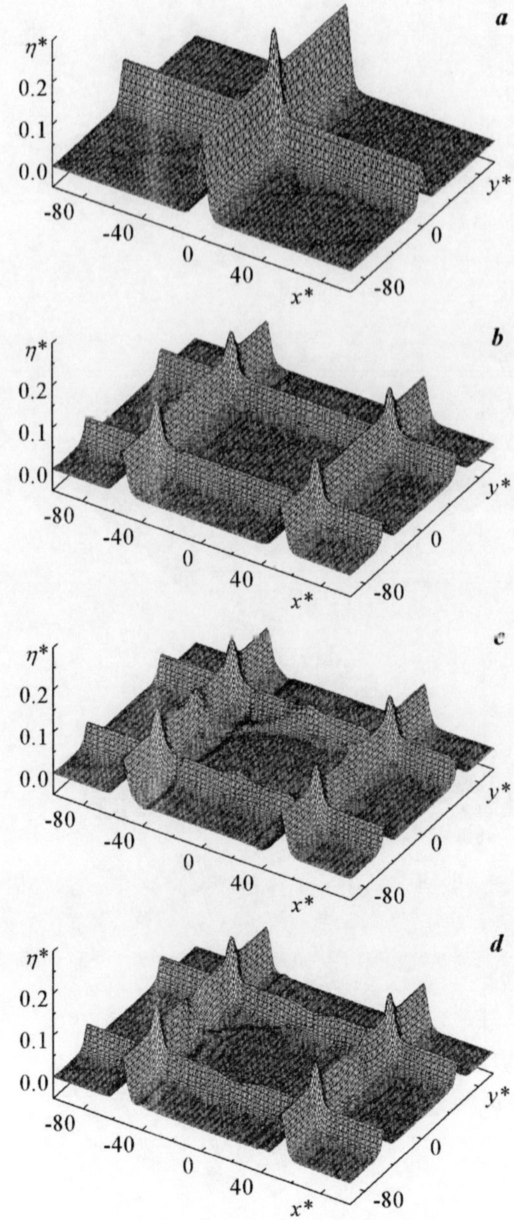

Fig. 8 The initial cross-type perturbation η_0 (a) and the shapes of four quasi-plane waves at $t^*=50$ on water with various bottoms: $h=h_0$ (b), $h=h_0-\eta_0(x,y)$ (c), $h=h_0+\eta_0(x,y)$ (d)

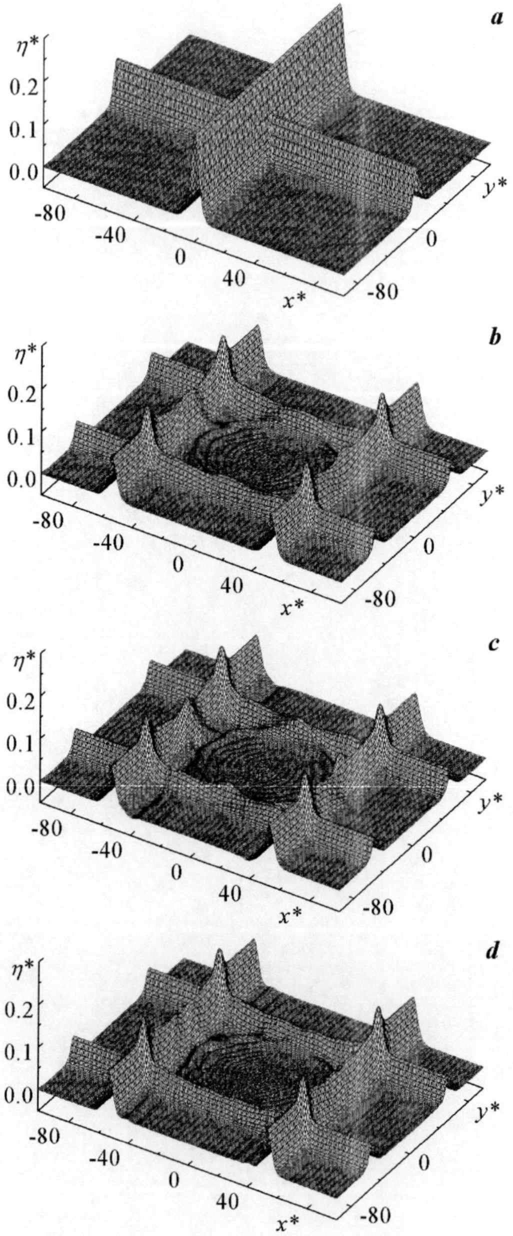

Fig. 9 The initial peakless cross-type perturbation η_0 (*a*) and shapes of four quasi-plane waves at $t^*= 50$ on water with various bottoms: $h = h_0$ (*b*), $h = h_0 - \eta_0(x,y)$ (*c*), $h = h_0 + \eta_0(x,y)$ (*d*)

bottom has the flat slope (Fig. 7) the intensification of waves moving towards the lower water depth and, otherwise, their weakening at the waves move to the deeper area were observed. Figure 8 shows the evolution of the disturbance in the pools with the same cross-type topographies (the ridges and the trenches). One can see the additional peaks and troughs (above the cross at the bottom) and oscillations (behind the fronts of four quasi-plane waves).

Figure 9 shows the evolution of the initial peakless cross-type perturbation (the superposition of two double solitons in the x-axes, as well as the y-axes): $\eta_0(x,y) = 2\,\eta_{sx}$ if $\eta_{sx} > \eta_{sy}$ in the point (x,y), otherwise $\eta_0(x,y) = 2\,\eta_{sy}$. Here one can see the intensification of the additional troughs above the cross at the bottom and oscillations in the center of the pool.

5 Conclusion

The new combined approach to the description of three-dimensional perturbations evolution on the water free surface is suggested. The model consists of one basic nonlinear equation and two linear elementary auxiliary equations for the determination of the horizontal liquid velocity vector. This vector is contained in the main equation only in the term of the second order of smallness. The method is suitable for moderately long waves of small but finite amplitude simultaneously running with any angles. This approach is in essence easier than the known systems of equations in which all equations contain both linear and nonlinear terms.

A validity of this approach to the solution of a number of planar problems of the nonlinear disturbance transformation is shown with the use of numerical experiments. It is important to note, that in cases of uneven bottom (Figs. 4 and 7–9) the topography affects stronger by the change of coefficients of the main linear term and of the dispersive term, than by the term containing gradient of the liquid depth, which is unessential.

Acknowledgements. This research was supported by INTAS – SB RAS (Grant No. 06-9236) and by the Russian Foundation for Basic Research (Project No. 07-01-00574).

References

1. Miles, J.W.: J. Fluid Mech. 106, 131–147 (1981)
2. Kadomtsev, B.B., Petviashvili, V.I.: Sov. Phys. Dokl. 15, 539–542 (1970)
3. Marchuk, A.G., Chubarov, L.B., Shokin, Y.I.: Numerical Simulation of Tsunami Waves, LA-TR-85, Los-Alamos (1985)
4. Khakimzyanov, G.S., Shokin, Y.I., Barakhnin, V.B., Shokina, N.Y.: Numerical Simulation of Currents with Surface Waves. Publ. House SB RAS, Novosibirsk (2001) (in Russian)
5. Kim, K.Y., Reid, R.O., Whitaker, R.E.: J. Comput. Phys. 76, 327–348 (1988)
6. Pelinovsky, D.E.: Stepanyants Yu.A.: JETP 79, 105–112 (1994)
7. Khabakhpashev, G.A.: Comput. Technol. 2, 94–102 (1997) (in Russian)
8. Peregrine, D.H.: J. Fluid Mech. 27, 815–827 (1967)

9. Karpman, V.I.: Nonlinear Waves in Dispersive Media. Pergamon Press, N.Y (1975)
10. Green, A.E., Naghdi, P.M.: J. Fluid Mech. 78, 237–246 (1976)
11. Arkhipov, D.G., Khabakhpashev, G.A.: Doklady Physics 51, 418–422 (2006)
12. Johnson, R.S.: J. Fluid Mech. 323, pp. 65–78 (1996)
13. Khabakhpashev, G.A.: Fluid Dyn. 22, pp. 430–437 (1987)
14. Litvinenko, A.A., Khabakhpashev, G.A.: Comput. Technologies 4, 95–105 (1999) (in Russian)
15. Ostrovsky, L.A., Potapov, A.I.: Modulated Waves: Theory and Applications. Johns Hopkins Univ. Press, Baltimore (1999)

Algorithm for Solving Non-stationary Three-Dimensional Navier-Stokes Equations with Large Reynolds Numbers on Multiprocessor Systems

Nargozy T. Danaev, Dauren B. Zhakebaev, and Abugamil U. Abdibekov

Abstract. In this article was constructed and implemented a new algorithm for solving the three-dimensional Navier - Stokes equation in common with question of inseparability in the Cartesian coordinate system. We describe the technique of decomposition, which used for paralleling problems of numerical simulation of turbulent flows. And we provide the dependence of obtained speed and the scalability coefficient from the number of processors and the size of the grid. We given measurements of productivity on the example of the problem of numerical simulation of turbulent flow on the basis of solutions of non-stationary Navier - Stokes equation in common with the question of inseparability in the Cartesian coordinate system (x_1, x_2, x_3).

1 Introduction

Turbulent flow is a phenomenon, which observed in many streams of liquids and gases, consisting of the formation of numerous eddies of various sizes In consequently of appearance of eddies hydrodynamic and thermodynamic characteristics (speed, temperature, pressure, density) have chaotic fluctuations. For the turbulent motion of liquids with sufficiently large values of the Reynolds number is characterized by extreme infrequency, random changes of speed over time at each point of the stream. Currently, the full quantitative theory of developed turbulence does not exist.

Development and improvement of methods of prediction and description of turbulence by creating new or improving old model equations for the turbulent

Nargozy T. Danaev · Dauren B. Zhakebaev · Abugamil U. Abdibekov
Al-Farabi Kazakh National University, Al-Farabi Av. 71, Almaty, 050038, Kazakhstan
e-mail: Nargozy.Danaev@kaznu.kz, daurjaz@mail.ru,
abugamil@mail.ru

E. Krause et al. (Eds.): Computational Sci., & High Performance Computing IV, NNFM 115, pp. 313–326.
springerlink.com

viscosity, the scale of turbulence, the development of methods of direct numerical simulation of turbulence are the current problems of modern computational hydrodynamics. The most famous methods of numerical simulation of turbulent flows are the direct numerical simulation (DNS), a resolution averaged Navier - Stokes (RANS), and the method of large eddy simulations (LES).

The method of simulation of large eddies is a compromise option between direct numerical simulation and decision averaged Navier - Stokes equations. The method of simulation of large eddies is based on two assumptions. The first assumption consists of the possibility of separation of the speed field on the movement of large and small eddies, and the movement of large eddies can be calculated separately, that linked with the insufficient isotropy and universality of small scale of turbulent motion. The second assumption –in the possibility of approximation of nonlinear interactions between large and small eddies only for the larger eddies using the subgrid models. Thus, large-scale structure is resolved clearly, while small-scale are modelled on the basis of different models and types of filter functions [1].

2 Basic Equations

Numerical modelling of the problem is implemented on the basis of solution of the non-stationary Navier-Stokes equation in common with equation of inseparability in the Cartesian coordinate system(x_1, x_2, x_3):

$$\frac{\partial u_i}{\partial t} + \frac{\partial}{\partial x_j}(u_i u_j) = -\frac{1}{\rho}\frac{\partial p}{\partial x_i} + \nu \frac{\partial^2 u_i}{\partial x_j \partial x_j}, \quad i = 1, 2, 3; \tag{1}$$

$$\frac{\partial u_i}{\partial x_i} = 0, \tag{2}$$

where is used the summation rule for repeated indices. u_1, u_2, u_3 – components of the speed, p – pressure, t – time, i, $j = 1, 2, 3$.

The main idea of the method of large eddies is in the mathematical division of large and small eddies, and this procedure can be carried out through spatial averaging, i.e. it is defined the field of large-scale pulsations through the filter. For any variable $f(x)$ its large-scale part $\bar{f}(x)$ is derived with using the filter function $\bar{G}(x)$:

$$\bar{f}(x) = \int_V \bar{G}(|x - x'|) f(x') dx'. \tag{3}$$

And small-scale part $f'(x)$ is defined like

$$f'(x) = f(x) - \bar{f}(x). \tag{4}$$

There are different approaches to describe the filter:

Filter type "box"

$$G(|\mathbf{r}-\mathbf{r}'|) = \begin{cases} 1/\Delta, & |x-x'| \le \Delta/2, \\ 0, & |x-x'| > \Delta/2. \end{cases}$$

Fourier filter

$$G(|\mathbf{r}-\mathbf{r}'|) = \prod_{i=1}^{3} \frac{\sin(x_i - x_i')/\Delta}{(x_i - x_i')/\Delta}.$$

Gaussian filter

$$G(|\mathbf{r}-\mathbf{r}'|) = (6/\pi\Delta^2)^{1/2} \exp\left(-6(x-x')^2/\Delta^2\right).$$

After applying a filter to equations (1) and (2) we obtain the basic equations for large-scale movements:

$$\frac{\partial \bar{u}_i}{\partial t} + \frac{\partial}{\partial x_j}(\bar{u}_i \bar{u}_j) = -\frac{1}{\rho}\frac{\partial \bar{p}}{\partial x_i} + \nu \frac{\partial^2 \bar{u}_i}{\partial x_j \partial x_j} - \frac{\partial \tau_{ij}}{\partial x_j}, \tag{5}$$

$$\frac{\partial \bar{u}_i}{\partial x_i} = 0, \tag{6}$$

$$\tau_{ij} = \overline{u_i u_j} - \bar{u}_i \bar{u}_j, \tag{7}$$

where $\tau_{i,j}$ –subgrid tensor, responsible for small-scale structures, which necessary modelled.

The idea of the large eddies' method is exactly to provide traffic of subgrid-scale of turbulence. This effect of the influence of subgrid-scale of turbulence is expressed by subgrid member.

To modelling subgrid tensor is used Smagorinsky model, which appears in the form of

$$\tau_{ij} - \frac{\delta_{ij}}{3}\tau_{kk} = -2\nu_T \bar{S}_{ij}, \tag{8}$$

where $\nu_T = C_S \Delta^2 \left(2\bar{S}_{ij}\bar{S}_{ij}\right)^{1/2}$ – turbulence viscosity; C_S – coefficient, depending on the nature of the flow; $\Delta = (\Delta_i \Delta_j \Delta_k)^{1/3}$ – width of grid filter; $\bar{S}_{ij} = \frac{1}{2}\left(\frac{\partial \bar{u}_i}{\partial x_j} + \frac{\partial \bar{u}_j}{\partial x_i}\right)$ – value of strain velocity tensor [1].

3 Algorithm Solutions and Parallelization Algorithm

To solve the problem of turbulent flow is used scheme of splitting for physical parameters [3] and paralleling with decomposition algorithm, which is shown in Fig. 1.

$$\frac{\mathrm{u}^* - \mathrm{u}^n}{\Delta t} = -\left(\mathrm{u}^n \nabla\right)\mathrm{u}^* + \nu \Delta \mathrm{u}^* - \nabla \tau,$$

$$\Delta p^{n+1} = \frac{\nabla \mathrm{u}^*}{\Delta t},$$

$$\frac{\mathrm{u}^{n+1} - \mathrm{u}^*}{\Delta t} = -\nabla p^{n+1}.$$

We suggest the following physical interpretation of the present scheme of splitting. Select one processor root, which will take place the collection of data.

In the first stage, it is assumed that the transfer of traffic's number is only by convection and diffusion. Intermediate velocity field is found by the method of fractional steps, using the run method. This so the intermediate values of velocities are considered on three processors in parallel.

Consider the method of fractional steps for the horizontal component of velocity u_1 at the point of grid $(i+1/2, j, k)$.

In the first stage the velocity u_1 is searched in the direction of the coordinates x_1:

$$\frac{\bar{u}^{n+\frac{1}{3}}_{1i+\frac{1}{2},j,k} - \bar{u}^{n}_{1i+\frac{1}{2},j,k}}{\tau} = \left(\frac{1}{2}\Lambda_1 \bar{u}^{n+\frac{1}{3}}_{1i+\frac{1}{2},j,k} + \frac{1}{2}\Lambda_1 \bar{u}^{n}_{1i+\frac{1}{2},j,k}\right) + \Lambda_2 \bar{u}^{n}_{1i+\frac{1}{2},j,k} + \Lambda_3 \bar{u}^{n}_{1i+\frac{1}{2},j,k} \tag{9}$$

In the second stage the velocity u_1 is searched in the direction of the coordinates x_2:

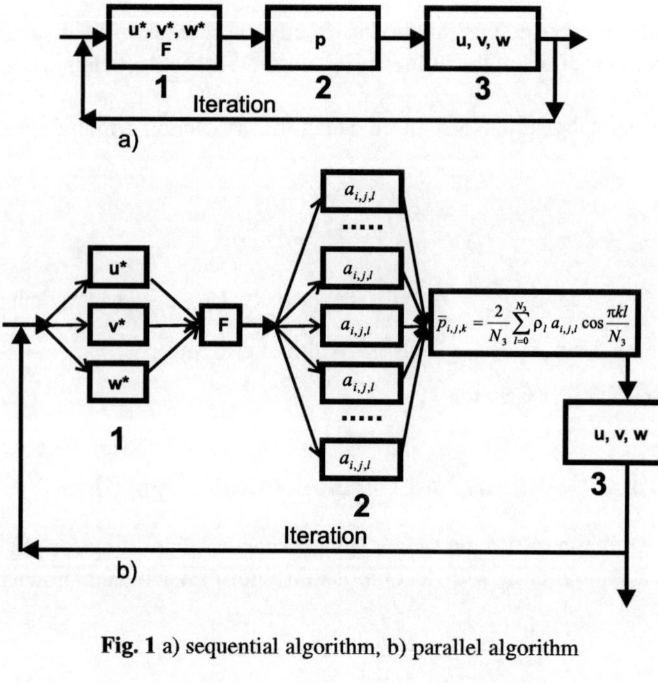

Fig. 1 a) sequential algorithm, b) parallel algorithm

$$\frac{\bar{u}^{n+\frac{2}{3}}_{1\,i+\frac{1}{2},j,k}-\bar{u}^{n+\frac{1}{3}}_{1\,i+\frac{1}{2},j,k}}{\tau}=\left(\frac{1}{2}\Lambda_2\bar{u}^{n+\frac{2}{3}}_{1\,i+\frac{1}{2},j,k}+\frac{1}{2}\Lambda_2\bar{u}^{n}_{1\,i+\frac{1}{2},j,k}\right)-\Lambda_2\bar{u}^{n}_{1\,i+\frac{1}{2},j,k} \tag{10}$$

In the third stage the velocity u_1 is searched in the direction of the coordinates x_3:

$$\frac{\bar{u}^{n+1}_{1\,i+\frac{1}{2},j,k}-\bar{u}^{n+\frac{2}{3}}_{1\,i+\frac{1}{2},j,k}}{\tau}=\left(\frac{1}{2}\Lambda_3\bar{u}^{n+1}_{1\,i+\frac{1}{2},j,k}+\frac{1}{2}\Lambda_3\bar{u}^{n}_{1\,i+\frac{1}{2},j,k}\right)-\Lambda_3\bar{u}^{n}_{1\,i+\frac{1}{2},j,k} \tag{11}$$

where

$$\Lambda_1\bar{u}_{1\,i+\frac{1}{2},j,k}=-\frac{\partial}{\partial x_1}\left(\bar{u}_1\bar{u}_1\right)\Big|_{i+\frac{1}{2},j,k}+\frac{1}{Re}\frac{\partial}{\partial x_1}\left(\frac{\partial\bar{u}_1}{\partial x_1}\right)\Big|_{i+\frac{1}{2},j,k}+\frac{\partial}{\partial x_1}\left(-\tau_{11}\right)\Big|_{i+\frac{1}{2},j,k},$$

$$\Lambda_2\bar{u}_{1\,i+\frac{1}{2},j,k}=-\frac{\partial}{\partial x_2}\left(\bar{u}_2\bar{u}_1\right)\Big|_{i+\frac{1}{2},j,k}+\frac{1}{Re}\frac{\partial}{\partial x_2}\left(\frac{\partial\bar{u}_1}{\partial x_2}\right)\Big|_{i+\frac{1}{2},j,k}+\frac{\partial}{\partial x_2}\left(-\tau_{12}\right)\Big|_{i+\frac{1}{2},j,k},$$

$$\Lambda_3\bar{u}_{1\,i+\frac{1}{2},j,k}=-\frac{\partial}{\partial x_3}\left(\bar{u}_3\bar{u}_1\right)\Big|_{i+\frac{1}{2},j,k}+\frac{1}{Re}\frac{\partial}{\partial x_3}\left(\frac{\partial\bar{u}_1}{\partial x_3}\right)\Big|_{i+\frac{1}{2},j,k}+\frac{\partial}{\partial x_3}\left(-\tau_{13}\right)\Big|_{i+\frac{1}{2},j,k}. \tag{12}$$

For (9) – (11) we use the second scheme with differences of against flow [4] and compact scheme [5, 6] and we get the scheme of high order accuracy (13) – (15):

$$\frac{\alpha_1\bar{u}^{n+\frac{1}{3}}_{1\,i+\frac{3}{2},j,k}+\beta_1\bar{u}^{n+\frac{1}{3}}_{1\,i+\frac{1}{2},j,k}+\gamma_1\bar{u}^{n+\frac{1}{3}}_{1\,i-\frac{1}{2},j,k}}{\tau}-\frac{\alpha_1\bar{u}^{n}_{1\,i+\frac{3}{2},j,k}+\beta_1\bar{u}^{n}_{1\,i+\frac{1}{2},j,k}+\gamma_1\bar{u}^{n}_{1\,i-\frac{1}{2},j,k}}{\tau}=$$
$$=\left(\tfrac{1}{2}\Lambda_1\bar{u}^{n+\frac{1}{3}}_{1\,i+\frac{1}{2},j,k}+\tfrac{1}{2}\Lambda_1\bar{u}^{n}_{1\,i+\frac{1}{2},j,k}\right)+\Lambda_2\bar{u}^{n}_{1\,i+\frac{1}{2},j,k}+\Lambda_3\bar{u}^{n}_{1\,i+\frac{1}{2},j,k} \tag{13}$$

$$\frac{\alpha_2\bar{u}^{n+\frac{2}{3}}_{1\,i+\frac{1}{2},j+1,k}+\beta_2\bar{u}^{n+\frac{2}{3}}_{1\,i+\frac{1}{2},j,k}+\gamma_2\bar{u}^{n+\frac{2}{3}}_{1\,i+\frac{1}{2},j-1,k}}{\tau}-\frac{\alpha_2\bar{u}^{n+\frac{1}{3}}_{1\,i+\frac{1}{2},j+1,k}+\beta_2\bar{u}^{n+\frac{1}{3}}_{1\,i+\frac{1}{2},j,k}+\gamma_2\bar{u}^{n+\frac{1}{3}}_{1\,i+\frac{1}{2},j-1,k}}{\tau}=$$
$$=\left(\tfrac{1}{2}\Lambda_2\bar{u}^{n+\frac{2}{3}}_{1\,i+\frac{1}{2},j,k}+\tfrac{1}{2}\Lambda_2\bar{u}^{n}_{1\,i+\frac{1}{2},j,k}\right)-\Lambda_2\bar{u}^{n}_{1\,i+\frac{1}{2},j,k} \tag{14}$$

$$\frac{\alpha_3 \bar{u}^{n+1}_{1\,i+\frac{1}{2},j,k+1} + \beta_3 \bar{u}^{n+1}_{1\,i+\frac{1}{2},j,k} + \gamma_3 \bar{u}^{n+1}_{1\,i+\frac{1}{2},j,k-1}}{\tau} - \frac{\alpha_3 \bar{u}^{n+\frac{2}{3}}_{1\,i+\frac{1}{2},j,k+1} + \beta_3 \bar{u}^{n+\frac{2}{3}}_{1\,i+\frac{1}{2},j,k} + \gamma_3 \bar{u}^{n+\frac{2}{3}}_{1\,i+\frac{1}{2},j,k-1}}{\tau} = \tag{15}$$
$$= \left(\tfrac{1}{2}\Lambda_3 \bar{u}^{n+1}_{1\,i+\frac{1}{2},j,k} + \tfrac{1}{2}\Lambda_3 \bar{u}^{n}_{1\,i+\frac{1}{2},j,k}\right) - \Lambda_3 \bar{u}^{n}_{1\,i+\frac{1}{2},j,k}$$

For example, to get scheme (13):

1) The second scheme with differences of against flow:

$$\frac{\partial}{\partial x}(uu) = \frac{A_x U_A - B_x U_B}{\Delta x},$$

$$A_x = \frac{u_{i+1}+u_i}{2}, \quad B_x = \frac{u_i+u_{i-1}}{2},$$

$$\frac{\partial}{\partial x}(uu) = \begin{cases} \frac{A_x u_i - B_x u_{i-1}}{\Delta x}, & for\, A_x > 0,\; B_x > 0, \\ \frac{A_x u_{i+1} - B_x u_i}{\Delta x}, & for\, A_x < 0,\; B_x < 0. \end{cases} \tag{16}$$

2) The compact scheme for the case $A_x > 0$, $B_x > 0$:

Compact schemes are using the equation, which links the importance of the results in several neighbouring points with values of data in several neighbouring pixels. This allows increasing order of approximation. An example of the compact scheme for the differentiation:

$$\Lambda_1 \bar{u}_{1\,i+\frac{1}{2},j,k} = -\frac{\partial}{\partial x_1}(\bar{u}_1\bar{u}_1)\Big|_{i+\frac{1}{2},j,k} + \frac{1}{Re}\frac{\partial}{\partial x_1}\left(\frac{\partial \bar{u}_1}{\partial x_1}\right)\Big|_{i+\frac{1}{2},j,k} + \tag{17}$$
$$+ \frac{\partial}{\partial x_1}(-\tau_{11})\Big|_{i+\frac{1}{2},j,k}$$

The traditional representation of the derivative approximation (17) has the form

$$A_x = \frac{u_{i+1}+u_i}{2}, \quad B_x = \frac{u_i+u_{i-1}}{2},$$

$$f_{i,j,k} = \Lambda_1 u_{i,j,k} = -\frac{A_x u_{i,j,k} - B_x u_{i-1,j,k}}{\Delta x} + \frac{1}{Re}\frac{u_{i+1,j,k} - 2u_{i,j,k} + u_{i-1,j,k}}{\Delta x^2} + \tag{18}$$
$$+2(\nu_t)_{i+\frac{1}{2},j,k}\frac{u_{i+1,j,k} - u_{i,j,k}}{\Delta x^2} - 2(\nu_t)_{i-\frac{1}{2},j,k}\frac{u_{i,j,k} - u_{i-1,j,k}}{\Delta x^2}.$$

Unlike (18), compact approximation of the ratio (18) can be written as follows:

$$\alpha \cdot f_{i+1,j,k} + \beta \cdot f_{i,j,k} + \gamma \cdot f_{i-1,j,k} = \Lambda_1 u_{i,j,k} \tag{19}$$

Introduce the symbols:

$$\alpha = \alpha_1 + \alpha_2 + \alpha_3 + \alpha_4,\ \beta = \beta_1 + \beta_2 + \beta_3 + \beta_4,\ \gamma = \gamma_1 + \gamma_2 + \gamma_3 + \gamma_4 \tag{20}$$

and get

$$\begin{aligned}
&(\alpha_1 \cdot f_{i+1,j,k} + \beta_1 \cdot f_{i,j,k} + \gamma_1 \cdot f_{i-1,j,k}) + \\
&+ (\alpha_2 \cdot f_{i+1,j,k} + \beta_2 \cdot f_{i,j,k} + \gamma_2 \cdot f_{i-1,j,k}) + \\
&+ (\alpha_3 \cdot f_{i+1,j,k} + \beta_3 \cdot f_{i,j,k} + \gamma_3 \cdot f_{i-1,j,k}) + \\
&+ (\alpha_4 \cdot f_{i+1,j,k} + \beta_4 \cdot f_{i,j,k} + \gamma_4 \cdot f_{i-1,j,k}) = \Lambda_1 u_{i,j,k}
\end{aligned} \tag{21}$$

Choose the uncertain coefficients in (21) so that the ratio (22) was carried out:

$$\begin{cases}
\left|(\alpha_1 \cdot f_{i+1,j,k} + \beta_1 \cdot f_{i,j,k} + \gamma_1 \cdot f_{i-1,j,k}) - \left(-\frac{A_x u_{i,j,k} - B_x u_{i-1,j,k}}{\Delta x}\right)\right| = O\left(h^3\right) \\
\left|(\alpha_2 \cdot f_{i+1,j,k} + \beta_2 \cdot f_{i,j,k} + \gamma_2 \cdot f_{i-1,j,k}) - \left(\frac{1}{Re}\frac{u_{i+1,j,k} - 2u_{i,j,k} + u_{i-1,j,k}}{\Delta x^2}\right)\right| = O\left(h^4\right) \\
\left|(\alpha_3 \cdot f_{i+1,j,k} + \beta_3 \cdot f_{i,j,k} + \gamma_3 \cdot f_{i-1,j,k}) - \left(2(\nu_t)_{i+\frac{1}{2},j,k}\frac{u_{i+1,j,k} - u_{i,j,k}}{\Delta x^2}\right)\right| = O\left(h^4\right) \\
\left|(\alpha_4 \cdot f_{i+1,j,k} + \beta_4 \cdot f_{i,j,k} + \gamma_4 \cdot f_{i-1,j,k}) - \left(-2(\nu_t)_{i-\frac{1}{2},j,k}\frac{u_{i,j,k} - u_{i-1,j,k}}{\Delta x^2}\right)\right| = O\left(h^4\right)
\end{cases} \tag{22}$$

For this, we substitute in (21) expansion in a Taylor series of functions f(x,y,z) and u(x,y,z) in point $x = x_i$:

$$\begin{aligned}
&\alpha_1 \cdot \left(f_i + \Delta x f_x + \tfrac{\Delta x^2}{2} f_{xx} + \tfrac{\Delta x^3}{6} f_{xxx} + \tfrac{\Delta x^4}{24} f_{xxxx}\right) + \beta_1 \cdot f_i + \\
&+ \gamma_1 \cdot \left(f_i - \Delta x f_x + \tfrac{\Delta x^2}{2} f_{xx} - \tfrac{\Delta x^3}{6} f_{xxx} + \tfrac{\Delta x^4}{24} f_{xxxx}\right) = \\
&= -\tfrac{A_x}{\Delta x} u_{i,j,k} + \tfrac{B_x}{\Delta x}\left(u_i - \Delta x u_x + \tfrac{\Delta x^2}{2} u_{xx} - \tfrac{\Delta x^3}{6} u_{xxx} + \tfrac{\Delta x^4}{24} u_{xxxx}\right) + O\left(h^4\right)
\end{aligned} \tag{23}$$

From (23) we get the system:

$$\begin{cases}
(\alpha_1 + \beta_1 + \gamma_1) = -B_x \\
(\alpha_1 - \gamma_1) = \dfrac{B_x}{2} \\
(\alpha_1 + \gamma_1) = -\dfrac{B_x}{3}
\end{cases} \tag{24}$$

Solving the system (24), we define:

$$\alpha_1 = \frac{B_x}{12},\ \beta_1 = -\frac{2B_x}{3},\ \gamma_1 = -\frac{5B_x}{12}. \tag{25}$$

$$\begin{aligned}
&\alpha_2 \cdot \left(f_i + \Delta x f_x + \tfrac{\Delta x^2}{2} f_{xx} + \tfrac{\Delta x^3}{6} f_{xxx} + \tfrac{\Delta x^4}{24} f_{xxxx}\right) + \beta_2 \cdot f_i + \\
&+\gamma_2 \cdot \left(f_i - \Delta x f_x + \tfrac{\Delta x^2}{2} f_{xx} - \tfrac{\Delta x^3}{6} f_{xxx} + \tfrac{\Delta x^4}{24} f_{xxxx}\right) = \\
&= \tfrac{1}{Re}\tfrac{1}{\Delta x^2}\left(u_i + \Delta x u_x + \tfrac{\Delta x^2}{2} u_{xx} + \tfrac{\Delta x^3}{6} u_{xxx} + \tfrac{\Delta x^4}{24} u_{xxxx}\right) - \\
&- \tfrac{1}{Re}\tfrac{2}{\Delta x^2} u_{i,j,k} + \\
&+ \tfrac{1}{Re}\tfrac{1}{\Delta x^2}\left(u_i - \Delta x u_x + \tfrac{\Delta x^2}{2} u_{xx} - \tfrac{\Delta x^3}{6} u_{xxx} + \tfrac{\Delta x^4}{24} u_{xxxx}\right) + O\left(h^4\right)
\end{aligned} \tag{26}$$

From (26) we get the system:

$$\begin{cases} (\alpha_2 + \beta_2 + \gamma_2) = \dfrac{1}{Re} \\ (\alpha_2 - \gamma_2) = 0 \\ (\alpha_2 + \gamma_2) = \dfrac{1}{6}\dfrac{1}{Re} \end{cases} \tag{27}$$

Hence we find

$$\alpha_2 = \frac{1}{12}\frac{1}{Re},\ \beta_2 = \frac{5}{6}\frac{1}{Re},\ \gamma_2 = \frac{1}{12}\frac{1}{Re}. \tag{28}$$

$$\alpha_3 = \frac{1}{4}(v_t)_{i+\frac{1}{2},j,k},\ \beta_3 = \frac{5}{6}(v_t)_{i+\frac{1}{2},j,k},\ \gamma_3 = -\frac{1}{12}(v_t)_{i+\frac{1}{2},j,k}. \tag{29}$$

$$\alpha_4 = -\frac{1}{12}(v_t)_{i-\frac{1}{2},j,k},\ \beta_4 = \frac{5}{6}(v_t)_{i-\frac{1}{2},j,k},\ \gamma_4 = \frac{1}{4}(v_t)_{i-\frac{1}{2},j,k}. \tag{30}$$

Substituting the founding values (25), (28), (29), (30) in (20), we get:

$$\alpha = \tfrac{B_x}{12} + \tfrac{1}{12}\tfrac{1}{Re} + \tfrac{1}{4}(v_t)_{i+\frac{1}{2},j,k} - \tfrac{1}{12}(v_t)_{i-\frac{1}{2},j,k},$$

$$\beta = -\tfrac{2B_x}{3} + \tfrac{5}{6}\tfrac{1}{Re} + \tfrac{5}{6}(v_t)_{i-\frac{1}{2},j,k} + \tfrac{5}{6}(v_t)_{i-\frac{1}{2},j,k},$$

$$\gamma = -\tfrac{5B_x}{12} + \tfrac{1}{12}\tfrac{1}{Re} - \tfrac{1}{12}(v_t)_{i+\frac{1}{2},j,k} + \tfrac{1}{4}(v_t)_{i-\frac{1}{2},j,k},$$

$$\begin{aligned}
&\alpha \cdot f_{i+1,j,k} + \beta \cdot f_{i,j,k} + \gamma \cdot f_{i-1,j,k} = -\tfrac{(2A_x - B_x)u_{i,j,k} - B_x u_{i-1,j,k}}{\Delta x} + \\
&+ \tfrac{1}{Re}\tfrac{u_{i+1,j,k} - 2u_{i,j,k} + u_{i-1,j,k}}{\Delta x^2} + 4(v_t)_{i+\frac{1}{2},j,k}\tfrac{u_{i+1,j,k} - u_{i,j,k}}{\Delta x^2} - 4(v_t)_{i-\frac{1}{2},j,k}\tfrac{u_{i,j,k} - u_{i-1,j,k}}{\Delta x^2}.
\end{aligned} \tag{31}$$

Introduce the difference operator M by the formula:

$$Mf_{i,j,k} = \alpha \cdot f_{i+1,j,k} + \beta \cdot f_{i,j,k} + \gamma \cdot f_{i-1,j,k}$$

Then (31) can re-write as :

$$Mf_{i,j,k} = -\frac{(2A_x - B_x)u_{i,j,k} - B_x u_{i-1,j,k}}{\Delta x} + \frac{1}{Re}\frac{u_{i+1,j,k} - 2u_{i,j,k} + u_{i-1,j,k}}{\Delta x^2} + \\ +4\left(\nu_t\right)_{i+\frac{1}{2},j,k}\frac{u_{i+1,j,k} - u_{i,j,k}}{\Delta x^2} - 4\left(\nu_t\right)_{i-\frac{1}{2},j,k}\frac{u_{i,j,k} - u_{i-1,j,k}}{\Delta x^2} \tag{32}$$

From (32) we obtain the following compact approximation for convective terms of the third, and for the diffusion member of the fourth order accuracy:

$$f_{i,j,k} = M^{-1}\begin{bmatrix} -\frac{(2A_x - B_x)u_{i,j,k} - B_x u_{i-1,j,k}}{\Delta x} + \frac{1}{Re}\frac{u_{i+1,j,k} - 2u_{i,j,k} + u_{i-1,j,k}}{\Delta x^2} + \\ +4\left(\nu_t\right)_{i+\frac{1}{2},j,k}\frac{u_{i+1,j,k} - u_{i,j,k}}{\Delta x^2} - 4\left(\nu_t\right)_{i-\frac{1}{2},j,k}\frac{u_{i,j,k} - u_{i-1,j,k}}{\Delta x^2} \end{bmatrix} \tag{33}$$

Now, on the basis of (33) we build the simplest a two- layer compact difference scheme (34) of order approximation $O\left(\tau^2\right) + O\left(h^3\right)$:

$$\frac{\alpha\bar{u}^{n+\frac{1}{3}}_{1\,i+\frac{3}{2},j,k} + \beta\bar{u}^{n+\frac{1}{3}}_{1\,i+\frac{1}{2},j,k} + \gamma\bar{u}^{n+\frac{1}{3}}_{1\,i-\frac{1}{2},j,k}}{\tau} - \frac{\alpha\bar{u}^{n}_{1\,i+\frac{3}{2},j,k} + \beta\bar{u}^{n}_{1\,i+\frac{1}{2},j,k} + \gamma\bar{u}^{n}_{1\,i-\frac{1}{2},j,k}}{\tau} = \\ = \left(\tfrac{1}{2}\Lambda_1\bar{u}^{n+\frac{1}{3}}_{1\,i+\frac{1}{2},j,k} + \tfrac{1}{2}\Lambda_1\bar{u}^{n}_{1\,i+\frac{1}{2},j,k}\right) + \Lambda_2\bar{u}^{n}_{1\,i+\frac{1}{2},j,k} + \Lambda_3\bar{u}^{n}_{1\,i+\frac{1}{2},j,k} \tag{34}$$

Difference solution $\bar{u}^{n+\frac{1}{3}}_{1\,i+\frac{1}{2},j,k}$, $\bar{u}^{n+\frac{2}{3}}_{1\,i+\frac{1}{2},j,k}$ and $\bar{u}^{n+1}_{1\,i+\frac{1}{2},j,k}$ efficiently computed by using the three-point run. Other components of velocity are searched on the same scheme.

After the account information from the three processors is collected on the main processor *root*.

At the second stage, on found an intermediate velocity field is searched the field of pressure. Poisson's equation for the pressure field is solved using the Fourier method in common with the matrix run, which is used to determine the Fourier coefficients, is presented in the following final-difference form:

$$\frac{\bar{p}_{i+1,j,k} - 2\bar{p}_{i,j,k} + \bar{p}_{i-1,j,k}}{\Delta x_1^2} + \frac{\bar{p}_{i,j+1,k} - 2\bar{p}_{i,j,k} + \bar{p}_{i,j-1,k}}{\Delta x_2^2} + \frac{\bar{p}_{i,j,k+1} - 2\bar{p}_{i,j,k} + \bar{p}_{i,j,k-1}}{\Delta x_3^2} = -F_{i,j,k} \tag{35}$$

Pressure field is represented as Fourier rows [7]:

$$\bar{p}_{i,j,k} = \frac{2}{N_3}\sum_{l=0}^{N_3}\rho_l\, a_{i,j,l}\cos\frac{\pi kl}{N_3},\; F_{i,j,k} = \frac{2}{N_3}\sum_{l=0}^{N_3}\rho_l\, b_{i,j,l}\cos\frac{\pi kl}{N_3}; \tag{36}$$

where

$$a_{i,j,l} = \sum_{k=0}^{N_3} \rho_k \bar{p}_{i,j,k} \cos\frac{\pi kl}{N_3}, \; b_{i,j,l} = \sum_{k=0}^{N_3} \rho_k F_{i,j,k} \cos\frac{\pi kl}{N_3}; \tag{37}$$

$$\rho_i = \begin{pmatrix} 1, & 1 \le i \le N_3 - 1 \\ 0.5, & i = 0, N_3. \end{pmatrix}$$

After substitution (36) in equation (35) and after a few arithmetic transformation equation (35) leads to the following type of:

$$-\frac{1}{\Delta x_2^2} a_{i,j-1,l} + \left[\begin{matrix} \left(\frac{2}{\Delta x_1^2} + \frac{2}{\Delta x_2^2} - \frac{1}{\Delta x_3^2}\left(2\cos\frac{\pi l}{N_3} - 2\right)\right) a_{i,j,l} - \\ -\frac{a_{i+1,j,l} + a_{i-1,j,l}}{\Delta x_1^2} \end{matrix} \right] - \frac{1}{\Delta x_2^2}\frac{a_{i,j+1,l}}{\Delta x_2^2} = -b_{i,j,l} \tag{38}$$

The vector form of this equation (38) is recorded for each fixed value $l = k,\ k = 0, N_3$ in following type:

$$-A_j \mathbf{a}_{j-1} + B_j \mathbf{a}_j - C_j \mathbf{a}_{j+1} = \mathbf{F}_j \tag{39}$$

Where matrixA_j, B_j, C_j and vectors $\mathbf{F}_j$, $\mathbf{a}_j$ take the following form:

$$A_j = \begin{vmatrix} \frac{1}{\Delta x_2^2} & & 0 \\ & \ddots & \\ 0 & & \frac{1}{\Delta x_2^2} \end{vmatrix}, \quad C_j = \begin{vmatrix} \frac{1}{\Delta x_2^2} & & 0 \\ & \ddots & \\ 0 & & \frac{1}{\Delta x_2^2} \end{vmatrix},$$

$$B_j = \begin{vmatrix} \left(\frac{2}{\Delta x_1^2} + \frac{2}{\Delta x_2^2} - \frac{1}{\Delta x_3^2}\left(2\cos\frac{\pi l}{N_3} - 2\right)\right) & -\frac{2}{\Delta x_1^2} & 0 \\ -\frac{1}{\Delta x_1^2} & \ddots & -\frac{1}{\Delta x_1^2} \\ 0 & -\frac{2}{\Delta x_1^2} & \left(\frac{2}{\Delta x_1^2} + \frac{2}{\Delta x_2^2} - \frac{1}{\Delta x_3^2}\left(2\cos\frac{\pi l}{N_3} - 2\right)\right) \end{vmatrix},$$

$$\mathbf{a}_j = \begin{vmatrix} a_{0,j} \\ \vdots \\ \vdots \\ a_{N_1,j} \end{vmatrix}, \quad \mathbf{F}_j = -\begin{vmatrix} b_{0,j} \\ \vdots \\ \vdots \\ b_{N_1,j} \end{vmatrix}.$$

The equation (39) is solved using matrix run with parallelization at each $l = k,\ k = 0, N_3$.

$l = \left[\frac{(N_3+1)}{Nproc} \cdot rank\right], \left[\left(\frac{(N_3+1)}{Nproc} - 1\right) \cdot (rank + 1)\right]$ – determine if a value l is solved equation (39) on each processor, where *rank*– number of processor, Np – the total number of processor.

Realization algorithm for solving matrix run equation (39) has the form:

$$\alpha_{j+1} = (C_j - A_j\alpha_j)^{-1} B_j, \quad j = 1,2,\ldots N_2 - 1, \quad \alpha_1 = C_0^{-1} B_0,$$

$$\beta_{j+1} = (C_j - A_j\alpha_j)^{-1} (\mathbf{F}_j + A_j\beta_j), \quad j = 1,2,\ldots N_2, \quad \beta_1 = C_0^{-1}\mathbf{F}_0,$$

$$\mathbf{a}_j = \alpha_{j+1}\mathbf{a}_{j+1} + \beta_{j+1}, \quad j = N_2 - 1,\ldots 0, \quad \mathbf{a}_{N_2} = \beta_{N_2+1}.$$

On each processor, after finding the coefficients $a_{i,j,l}$, we gather at the head processor *root* as shown in Fig. 1. Then at the head to the processor value is the pressure field is searched from the formulas (36).

To calculate the sums (36) is applied the method of fast Fourier transform, by which data's sum can calculate $O(N_3 \ln N_3)$the amount of action that let reduce computing time.

In the third stage, it is assumed that the transfer is carried out only by the pressure gradient.

4 Computing Experiment and Analysis of Results

Speedup obtained using a parallel algorithm for *Np* processors, compared with the serial version of the implementation of computation is determined by size:

$$S_p(N) = T_1(N)/T_p(N)$$

(the value $N = N_3$ is used for parameterization of the complexity of computing solvable problems)

Efficiency of processors using parallel algorithms in solving the problem determined by the ratio:

$$E_p(N) = T_1(N)/(pT_p(N)) = S_p(N)/p$$

(the value of efficiency determines the average share of run-time parallel algorithm in during which processors actually used to solve the problem).

Cost of computation: $C_p = pT_p$ (the rate of efficiency of parallel algorithm)

To evaluate the speedups, the cost of computing and efficiency were considered by several parallel versions, implementing a three-dimensional simulation of field sizes: *a*) $N_1 \times N_2 \times N_3 = 125000$, N=50; *b*) $N_1 \times N_2 \times N_3 = 1000000$, N=100; *c*) $N_1 \times N_2 \times N_3 = 8000000$, N=200.

Calculation points in an URSA cluster, the technical characteristics of the equipment are listed in the Table 1. When we parallelize the first stage and compare it with sequence algorithm we determine that parallel algorithm compute it three times faster. Main problem of this work is solving system of equations is parallelization of Poisson's equation. Because of main computational time is spend to solve this three dimensional Poisson's equation. It also shows the results paralleling Fourier

Table 1 Characteristics of using equipment

URSA	
Number of nodes	14 nodes
A node	Intel(R) Xeon(R) CPU E5335 2.00GHz
Memory	4 GB Memory
Network	Gigabit Broadcom NetXtreme II
System	Scientific Linux 4.6, Oscar 5.0
MPI	MPICH2

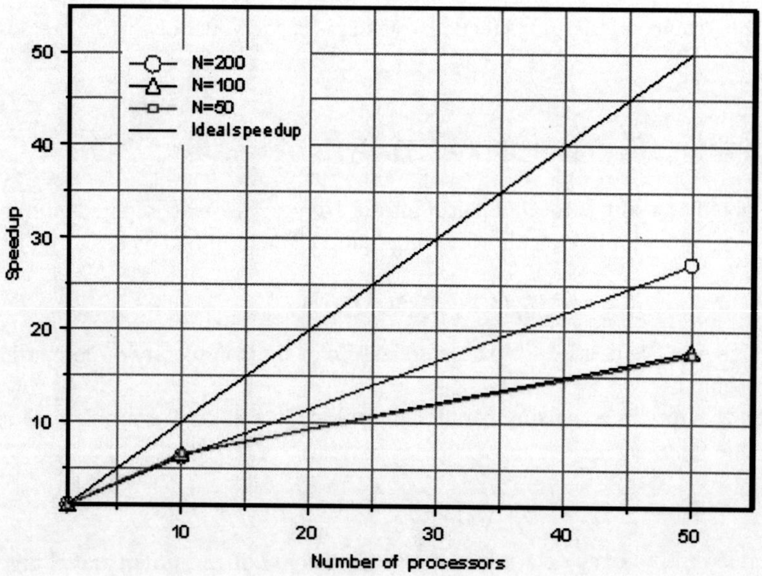

Fig. 2 The dependence of speedup from number of used processors

method for solving Poisson's equation for calculating the pressure field in Table 2, see also Fig. 2. The set task was solved with the distribution calculated on a rectangular grid on computing subdomain for each processor. The necessary exchange of the calculated values between processors is implemented using the library MPI [8, 9].

Table 2 shows the actual speedup value, the cost of computation, the efficiency and time of calculations, which received during the computational experiment in

Table 2 Computing experiment of paralleling the Poisson's equation, $T_p(N)$– time of computation, $C_p(N)$ – cost of computation, $E_p(N)$ – efficiency and $S_p(N)$ – speedup depending on the N_p– number of used processors in various cases

a) $N = 50$				
N_p	$T_p(N)$	$S_p(N)$	$E_p(N)$	$C_p(N)$
1	2,27	1	1	2,27
10	0,346	6,560	0,656	3,46
50	0,13	17,461	0,349	6,5
b) $N = 100$				
N_p	$T_p(N)$	$S_p(N)$	$E_p(N)$	$C_p(N)$
1	70	1	1	70
10	10,79	6,487	0,648	107,9
20	5,88	11,904	0,595	117,6
50	3,93	17,811	0,356	196,5
100	2,35	29,787	0,297	235
c) $N = 200$				
N_p	$T_p(N)$	$S_p(N)$	$E_p(N)$	$C_p(N)$
1	2603,516	1	1	2603,516
10	418,712	6,217	0,621	4187,129
20	232,156	11,214	0,560	4643,132
50	95,668	27,213	0,544	4783,44
100	48,31	53,891	0,538	4831

different cases. As can be seen from consideration of Table 2, the speedup depends linearly on the number of processors, i.e. program is a scalable, see also Fig. 2.

5 Conclusion

For solutions of the Navier - Stokes equation was constructed numerical algorithm on the basis of compact difference schemes of third order accuracy, admitting to solve the problem of turbulent flows at high Reynolds numbers.

In this article on the example of the problem of numerical simulation on the computer cluster of complex three-dimensional turbulent flows are considered the technological aspects of the development of scalable parallel calculations using the library for MPI. As part of this article is reviewed the main technological stages in the development of complex computer programs for massively parallel systems: the choice of model programs and scheme of paralleling, identification scheme calculations and programming problems, compiling, debugging and testing, profiling and tracing programs, holding computational experiment, analysis of results.

Obtained calculated values of speedups, allowing to assess the scalability of the algorithm and its program realization. These results show that the algorithm has significant amount of potential of parallelism, and good, in terms of structure paralleling that offers hope for a speedup close to the line depending on the number of used processors.

Provided values of speedup, efficiency and velocity of calculations, obtained in the numerical experiments, which agree well with theoretical estimates.

References

1. Ferziger, J.H.: Large eddy simulation of turbulent flows. AIAA J. 15(9), 1261–1267 (1977)
2. Sagaut, P.: Large Eddy Simulation for Incompressible Flows. Springer, Heidelberg (2002)
3. Yanenko, N.N.: The Method of Fractional Steps for the Multidimensional Problems of Mathematical Physics. Nauka, Novosibirsk (1967) (in Russian)
4. Roache, P.: Computational Fluid Dynamics. Mir, Moscow (1980) (in Russian)
5. Tolstykh, A.I.: Compact Difference Schemes and Their Applications to Fluid Dynamics Problems. Nauka, Moscow (1990) (in Russian)
6. Vorozhtsov, E.V.: Exercises for the Theory of Difference schemes. Publishing House of NSTU, Novosibirsk (2000) (in Russian)
7. Samarskii, A.A., Nikolaev, E.S.: The methods for Solving Grid Equations. Nauka, Moscow (1978) (in Russian)
8. Antonov, A.S.: Parallel programming using MPI. Moscow State University, Moscow (2004) (in Russian)
9. Shpakovsky, G.I., Serikova, N.V.: Programming for multiprocessor systems in the standard MPI. Belarus State University, Minsk (2002) (in Russian)

Numerical Modeling of Non-homogeneous Turbulence on Cluster Computing System

Ualikhan S. Abdibekov, Dauren B. Zhakebaev, and Bakytzhan T. Zhumagulov

Abstract. This work considers the numerical simulation of inhomogeneous turbulence in an incompressible fluid with zero mean for the entire volume of the liquid velocity. Initial fluid motion was taken isotropic and nonisotropic. The calculations were performed in a Cartesian coordinate system at high Reynolds numbers. To improve the accuracy of the solution used compact scheme. Obtain the statistical characteristics of turbulence.

1 Introduction

At present time the progress of application of numerical methods for solving practical problems the study of turbulent flows, where the required registration of numerous parameters associated with numerical methods of solving the Reynolds-averaged Navier-Stokes equations.

It is known that the numerical simulation of inhomogeneous turbulence require enormous computational costs. With the advent of high-performance systems has become possible to use more accurate methods for simulation of turbulent processes.

This paper considers the numerical simulation of inhomogeneous turbulence by the method of large eddies. Interest in solving this problem is caused by the desire to see how construction of numerical method and algorithm parallelization, and draws its model in large eddy method. Correctness verification would be to obtain turbulent characteristics, which should have similar evolutionary behaviour with the same characteristics adopted in semi-empirical theory.

Ualikhan S. Abdibekov · Dauren B. Zhakebaev
Al-Farabi Kazakh National University,
Al-Farabi Av. 71, Almaty, 050038, Kazakhstan
e-mail: Uali1@mail.ru, daurjaz@mail.ru

Bakytzhan T. Zhumagulov
Kazakh National University,
Al-Farabi Av. 71, Almaty, 050038, Kazakhstan
e-mail: kvt@kaznu.kz

E. Krause et al. (Eds.): Computational Sci., & High Performance Computing IV, NNFM 115, pp. 327–338.
springerlink.com

2 Basic Equations, Problem

To describe the large-scale turbulent flow we take the averaged Navier-Stokes [1]:

$$\frac{\partial \bar{u}_i}{\partial t} + \frac{\partial}{\partial x_j}(\bar{u}_i \bar{u}_j) = -\frac{1}{\rho}\frac{\partial \bar{p}}{\partial x_i} + \nu \frac{\partial^2 \bar{u}_i}{\partial x_j \partial x_j} - \frac{\partial \tau_{ij}}{\partial x_j}, \tag{1}$$

$$\frac{\partial \bar{u}_i}{\partial x_i} = 0, \tag{2}$$

$$\tau_{ij} = \overline{u_i u_j} - \bar{u}_i \bar{u}_j, \tag{3}$$

where$\tau_{i,j}$ – the subgrid tensor, responsible for small-scale structure.

To simulate the subgrid tensor Smagorinsky model is used, which appears in the form

$$\tau_{ij} - \frac{\delta_{ij}}{3}\tau_{kk} = -2\nu_T \bar{S}_{ij}, \tag{4}$$

where $\nu_T = C_S \Delta^2 \left(2\bar{S}_{ij}\bar{S}_{ij}\right)^{1/2}$ – turbulence viscosity; C_S – coefficient, depending on the nature of the flow; $\Delta = (\Delta_i \Delta_j \Delta_k)^{1/3}$ – width of grid filter; $\bar{S}_{ij} = \frac{1}{2}\left(\frac{\partial \bar{u}_i}{\partial x_j} + \frac{\partial \bar{u}_j}{\partial x_i}\right)$ – value of strain velocity tensor.

The problem should be allocated to the final region of space, the question of formulating the boundary conditions for the selected computational region is not trivial. Can approximately assume that the fluid flow is periodic with periods along the axes of x, y and z, equal to the estimated size of the selected area on these axes. The error of this assumption would seem, a small, if the size of calculated area significantly exceeds large eddies turbulent.

Let $N = N_x = N_y = N_z$ - the total number of segments in length Δx, which is broken axle$x$ within the calculated area. Similarly do the same procedure on the other two axes.

All calculations are carried out for cells i, j, k, where $i = 1,2,3,...,N_x$; $j = 1,2,3,...,N_y$; $k = 1,2,3,...,N_z$. From the periodicity conditions, accept for the velocity components u_1:

$$\left.\begin{array}{l}(u_1)_{-1,j,k} = (u_1)_{N_X-4,j,k},\\ (u_1)_{0,j,k} = (u_1)_{N_X-3,j,k},\ (u_1)_{1,j,k} = (u_1)_{N_X-2,j,k},\\ (u_1)_{N_X-1,j,k} = (u_1)_{2,j,k},\ (u_1)_{N_X,j,k} = (u_1)_{3,j,k}\end{array}\right\}\text{for any j,k}$$

$$\left.\begin{array}{l}(u_1)_{i,0,k} = (u_1)_{i,N_Y-3,k}, (u_1)_{i,1,k} = (u_1)_{i,N_Y-2,k},\\ (u_1)_{i,N_Y-1,k} = (u_1)_{i,2,k}, (u_1)_{i,N_Y,k} = (u_1)_{i,3,k}\end{array}\right\}\text{for any i,k}$$

$$\left.\begin{array}{l}(u_1)_{\mathrm{i,j,0}} = (u_1)_{\mathrm{i,j,}N_Z-3},\ (u_1)_{\mathrm{i,j,1}} = (u_1)_{\mathrm{i,j,}N_Z-2},\\ (u_1)_{\mathrm{i,j,}N_Z-1} = (u_1)_{\mathrm{i,j,2}},\ (u_1)_{\mathrm{i,j,}N_Z} = (u_1)_{\mathrm{i,j,3}}\end{array}\right\}\text{for any i,j}$$

For the components of velocity u_2 and u_3 boundary conditions are placed similarly.

Initial velocity field must satisfy the continuity equation, in addition, must meet the requirements that the average over the whole volume and the average rate of angular momentum is zero, the average in terms of the values $\langle u_1^2\rangle$, $\langle u_2^2\rangle$, $\langle u_3^2\rangle$ should be given, and the average values $\langle u_i u_j\rangle$ must be zero if $i \neq j$; must be given some "scale" of motion ("the size of vortices") on all axes.

In meeting these requirements within the initial velocity field can be set arbitrarily. It is convenient to specify the initial field is periodic (the analogy with the action of the lattice), bearing in mind that the further development of the turbulent flow velocity should get a random character. However, it is possible that with a reasonable computation time velocity field has not yet become completely random and it will be a periodic component. Therefore, it is possible to specify the initial field of the periodic [2].

Setting the initial field in the form of a periodic function. Can take, for example, following the initial distribution of velocities $u_x = u_1$, $u_y = u_2$, $u_z = u_3$:

$$\begin{aligned} u_1 &= A_1 \sin k_1 x_1 \cdot \sin k_2 x_2 \cdot \sin k_3 x_3 + M_1 \cos k_1 x_1 \cdot \cos k_2 x_2 \cdot \cos k_3 x_3 \\ &a_1 \sin \bar{k}_1 x_1 \cdot \sin \bar{k}_2 x_2 \cdot \sin \bar{k}_3 x_3 + m_1 \cos \bar{k}_1 x_1 \cdot \cos \bar{k}_2 x_2 \cdot \cos \bar{k}_3 x_3, \end{aligned} \tag{5}$$

$$\begin{aligned} u_2 &= D_2 \sin k_1 x_1 \cdot \sin k_2 x_2 \cdot \cos k_3 x_3 + H_2 \cos k_1 x_1 \cdot \cos k_2 x_2 \cdot \sin k_3 x_3 \\ &d_2 \sin \bar{k}_1 x_1 \cdot \sin \bar{k}_2 x_2 \cdot \cos \bar{k}_3 x_3 + h_2 \cos \bar{k}_1 x_1 \cdot \cos \bar{k}_2 x_2 \cdot \sin \bar{k}_3 x_3, \end{aligned} \tag{6}$$

$$u_3 = B_3 \sin k_1 x_1 \cdot \cos k_2 x_2 \cdot \sin k_3 x_3 + b_3 \sin \bar{k}_1 x_1 \cdot \cos \bar{k}_2 x_2 \cdot \sin \bar{k}_3 x_3. \tag{7}$$

Quantities k_i and $\bar{k}_i$ should be chosen so that the size of the field is calculated along each axis fit an integer number of periods. For example, if the size of the field is calculated along the axisx_i is equalL_i, you can choose $k_i = 2\pi n_i/L_i$, $\bar{k}_i = 2\pi \bar{n}_i/L_i$, where n_i and $\bar{n}_i$- integers, $n_i = m_i/[h_i(N_i-3)]$, $L_i = 1$, $i = 1, 2, 3$, m_i – the number of periods. We will take $\bar{k}_i > k_i$ (so that members with$\bar{k}_i$ at (5)- (7) give the "small-scale disturbance"). The valuen_i, seems to be no less than 4 or 5, and $\bar{n}_i$ 3-4 times more.

Distributions (5) - (7) satisfy the continuity equation, if:

$$\begin{aligned} k_1 A_1 &= k_2 H_2, \quad k_1 M_1 = k_2 D_2 + k_3 B_3, \\ \bar{k}_1 a_1 &= \bar{k}_2 h, \quad \bar{k}_1 m_1 = \bar{k}_2 d_2 + \bar{k}_3 b_3; \end{aligned} \tag{8}$$

easily verify that

$$\begin{aligned} &A_1^2 + M_1^2 + a_1^2 + m_1^2 = 8\langle u_1^2\rangle, \\ &D_2^2 + H_2^2 + d_2^2 + h_2^2 = 8\langle u_2^2\rangle, \\ &B_3^2 + b_3^2 = 8\langle u_3^2\rangle, \\ &\langle u_1 u_2\rangle = \langle u_1 u_3\rangle = \langle u_2 u_3\rangle = 0. \end{aligned} \tag{9}$$

It is convenient to allocate a separate small-scale part of the initial perturbations. For this we write at $t = 0$:

$$\langle u_i^2 \rangle = \langle u_i^2 \rangle_l + \langle u_i^2 \rangle_s \tag{10}$$

In view of (9)

$$\begin{aligned} A_1^2 + M_1^2 &= 8\langle u_1^2 \rangle_l, \; a_1^2 + m_1^2 = 8\langle u_1^2 \rangle_s, \\ D_2^2 + H_2^2 &= 8\langle u_2^2 \rangle_l, \; d_2^2 + h_2^2 = 8\langle u_2^2 \rangle_s \\ B_3^2 &= 8\langle u_3^2 \rangle_l, \; b_3^2 = 8\langle u_3^2 \rangle_s. \end{aligned} \tag{11}$$

From (8) - (11) obtained

$$B_3 = \sqrt{8\langle u_3^2 \rangle_l}, \tag{12}$$

$$D_2^2 = [4/(k_2 k_3 B_3)] \left(k_1^2 \langle u_1^2 \rangle_l - k_2^2 \langle u_2^2 \rangle_l - k_3^2 \langle u_3^2 \rangle_l \right), \tag{13}$$

$$H_2 = \sqrt{8\langle u_2^2 \rangle_l - D_2^2}, \tag{14}$$

$$A_1 = \left({}^{k_2}/_{k_1} \right) H_2, \quad M_1 = \left({}^{k_2}/_{k_1} \right) D_2 + \left({}^{k_3}/_{k_1} \right) B_3 \tag{15}$$

and similar formulas for small-scale perturbations

$$b_3 = \sqrt{8\langle u_3^2 \rangle s}, \tag{16}$$

$$d_2^2 = \left[4/(\bar{k}_2 \bar{k}_3 B_3) \right] \left(\bar{k}_1^2 \langle u_1^2 \rangle_s - \bar{k}_2^2 \langle u_2^2 \rangle s - \bar{k}_3^2 \langle u_3^2 \rangle s \right), \tag{17}$$

$$h_2 = \sqrt{8\langle u_2^2 \rangle s - d_2^2}, \tag{18}$$

$$a_1 = \left({}^{\bar{k}_2}/_{\bar{k}_1} \right) h_2, \quad m_1 = \left({}^{\bar{k}_2}/_{\bar{k}_1} \right) d_2 + \left({}^{\bar{k}_3}/_{\bar{k}_1} \right) b_3 \tag{19}$$

More small-scale perturbations are assumed to be isotropic, that will be taken

$$\langle u_1^2 \rangle s = \langle u_2^2 \rangle s = \langle u_3^2 \rangle s = \mathrm{W}^2 \tag{20}$$

Then, from (16) - (19) obtained

$$b_3 = W\sqrt{8}, d_2 = -W\sqrt{2}, \quad h_2 = W\sqrt{6}, \quad a_1 = W\sqrt{6}, \quad m_1 = W\sqrt{2} \tag{21}$$

We also introduce the notation

$$\frac{1}{3} \left(\langle u_1^2 \rangle_l + \langle u_2^2 \rangle_l + \langle u_3^2 \rangle_l \right)_{\mathrm{t=0}} = u_0^2 \tag{22}$$

3 Numerical Method

Numerical solution of (1) - (4) is realized on the carried grid using the scheme against the flow of the second type [3] and compact approximation [4, 5]. To solve the problem given above, the proposed model of turbulence we use the scheme of splitting by physical parameters. In the first stage, it is assumed that the transfer of momentum is only by convection and diffusion. Intermediate velocity field is the method of fractional steps, using the sweep method [6]. At the second stage, found an intermediate velocity field is the field pressure. Poisson equation for the pressure field is solved by Fourier series [7] in conjunction with the matrix sweep method, which is used to determine the Fourier coefficients. The third phase is supposed that the transfer occurs only through the pressure gradient.

4 Defining Characteristics of Turbulence

To find the turbulent characteristics in the physical space, we introduce the first procedure of averaging over the volume of different scales. Averaged values will be involved in finding themselves already turbulent characteristics. This procedure for calculating the turbulent characteristics similar to performed work [2]. Averaged value calculated for the entire region, which in this work is rectangular, is calculated by the following formula:

$$\langle u_i \rangle = \frac{1}{N_1 N_2 N_3} \sum_{n=1}^{N_1} \sum_{m=1}^{N_2} \sum_{q=1}^{N_3} (u_i)_{n,m,q}.$$

Different correlation coefficients velocities are defined as follows:

$$R_{i1}(r) = \frac{1}{2(N_1 - r) N_2 N_3 \langle \overline{u}_i^2 \rangle} \left\{ \begin{array}{l} \sum\limits_{n=1}^{N_1 - r} \sum\limits_{m=1}^{N_2} \sum\limits_{q=1}^{N_3} \left[\overline{u}_i(n,m,q)\, \overline{u}_i(n+r,m,q) \right] \\ + \sum\limits_{n=1+r}^{N_1} \sum\limits_{m=1}^{N_2} \sum\limits_{q=1}^{N_3} \left[\overline{u}_i(n,m,q)\, \overline{u}_i(n-r,m,q) \right] \end{array} \right\}, \tag{23}$$

$$R_{i2}(r) = \frac{1}{2N_1 (N_2 - r) N_3 \langle \overline{u}_i^2 \rangle} \left\{ \begin{array}{l} \sum\limits_{n=1}^{N_1} \sum\limits_{m=1}^{N_2 - r} \sum\limits_{q=1}^{N_3} \left[\overline{u}_i(n,m,q)\, \overline{u}_i(n,m+r,q) \right] \\ + \sum\limits_{n=1}^{N_1} \sum\limits_{m=1+r}^{N_2} \sum\limits_{q=1}^{N_3} \left[\overline{u}_i(n,m,q)\, \overline{u}_i(n,m-r,q) \right] \end{array} \right\}, \tag{24}$$

$$R_{i3}(r) = \frac{1}{2N_1 N_2 (N_3 - r) \langle \overline{u}_i^2 \rangle} \left\{ \begin{array}{l} \sum\limits_{n=1}^{N_1} \sum\limits_{m=1}^{N_2} \sum\limits_{q=1}^{N_3 - r} \left[\overline{u}_i(n,m,q)\, \overline{u}_i(n,m,q+r) \right] \\ + \sum\limits_{n=1}^{N_1} \sum\limits_{m=1}^{N_2} \sum\limits_{q=1+r}^{N_3} \left[\overline{u}_i(n,m,q)\, \overline{u}_i(n,m,q-r) \right] \end{array} \right\}. \tag{25}$$

Calculations of (23) – (25) must be fulfilled for $r = 1, 2, ..N_i/2$. Formula (23) – (25) are represented by discrete analogues of continuous representations of the correlation coefficients [2].

The scale of turbulence can be written in finite-difference form:

$$l_x = \Delta x_1 \left\{ \frac{\langle \overline{u}_2^2 \rangle}{\langle \overline{u}_2^2 \rangle + \langle \overline{u}_3^2 \rangle} \sum_{r=1}^{N_1/2} \left[R_{21}|_{r-1} + R_{21}|_r \right] + \frac{\langle \overline{u}_3^2 \rangle}{\langle \overline{u}_2^2 \rangle + \langle \overline{u}_3^2 \rangle} \sum_{r=1}^{N_1/2} \left[R_{31}|_{r-1} + R_{31}|_r \right] \right\},$$

$$l_y = \Delta x_2 \left\{ \frac{\langle \overline{u}_1^2 \rangle}{\langle \overline{u}_1^2 \rangle + \langle \overline{u}_3^2 \rangle} \sum_{r=1}^{N_2/2} \left[R_{12}|_{r-1} + R_{12}|_r \right] + \frac{\langle \overline{u}_3^2 \rangle}{\langle \overline{u}_1^2 \rangle + \langle \overline{u}_3^2 \rangle} \sum_{r=1}^{N_2/2} \left[R_{32}|_{r-1} + R_{32}|_r \right] \right\},$$

$$l_z = \Delta x_3 \left\{ \frac{\langle \overline{u}_1^2 \rangle}{\langle \overline{u}_1^2 \rangle + \langle \overline{u}_2^2 \rangle} \sum_{r=1}^{N_3/2} \left[R_{13}|_{r-1} + R_{13}|_r \right] + \frac{\langle \overline{u}_2^2 \rangle}{\langle \overline{u}_1^2 \rangle + \langle \overline{u}_2^2 \rangle} \sum_{r=1}^{N_3/2} \left[R_{23}|_{r-1} + R_{23}|_r \right] \right\}.$$

Where subscripts r and $r-1$ indicate the serial number of the element of the correlation coefficients. You can designate a

$$l = (l_x + l_y + l_z)/3,$$

average integral scale of turbulence and the characteristic average velocity of turbulence.

$$u = \left(\sqrt{\left(\langle \overline{u}_1^2 \rangle + \langle \overline{u}_2^2 \rangle + \langle \overline{u}_3^2 \rangle \right)/3} \right).$$

Energy dissipation is calculated using the formula:

$$\varepsilon = \frac{1}{2\Delta t} \left\{ \left(\langle \overline{u}_i^2 \rangle \right)^t - \left(\langle \overline{u}_i^2 \rangle \right)^{t-\Delta t} \right\}.$$

In this case the dimensionless energy dissipation

$$\overline{\varepsilon} = \varepsilon \frac{l}{u^3},$$

should behave as a constant.

Kinetic energy is the following formula

$$E(t) = \frac{\overline{u}_i^2(t)}{2},$$

however, the dependence over time can be defined as a function

$$s(t) = \frac{E(0)}{E(t)},$$

to be growing a linear function of time-dependent, since over time in the case of decay of turbulence energy decreases.

5 Algorithms to Realize Initial and Boundary Conditions

Small-scale perturbations (20)-(22) are assumed to be isotropic for all tasks.

To illustrate, consider a few examples of the initial conditions:

1) For the simulation of isotropic turbulence, the following parameters to specify the initial conditions are accepted:

$$k_1 = k_2 = k_3 = k, \quad \langle u_1^2 \rangle_l = \langle u_2^2 \rangle_l = \langle u_3^2 \rangle_l = u_0^2.$$

From (12) - (15) obtained $B_3 = u_0\sqrt{8}$, $D_2 = -u_0\sqrt{2}$, $H_2 = u_0\sqrt{6}$, $A_1 = u_0\sqrt{6}$, $M_1 = \left(u_0\sqrt{8} - u_0\sqrt{2}\right)$.

Assume: $\bar{k} = 3k$, $W = 0.1u_0$.

2) To simulate non-isotropic turbulence with the same scale on all axes, we take these parameters:

$$k_1 = k_2 = k_3 = k, \quad \langle u_2^2 \rangle_l = \langle u_3^2 \rangle_l = u_0^2/2, \langle u_1^2 \rangle_l = 2u_0^2.$$

This $B_3 = 2u_0$, $D_2 = 2u_0$, $H_2 = 0$, $A_1 = 0$, $M_1 = 4u_0$.

Assume: $\bar{k} = 3k$, $W = 0.1u_0$.

3) Examining a different speed on all axes, with the scale, varying in proportion to the velocity, we take these parameters:

$$k_1 = k/2,\ k_2 = 2k,\ k_3 = k, \quad \langle u_2^2 \rangle_l = u_0^2/7, \langle u_3^2 \rangle_l = 4u_0^2/7, \langle u_1^2 \rangle_l = 16u_0^2/7.$$

Turns: $B_3 = \sqrt{32/7}\,u_0$, $D_2 = -\sqrt{2/7}\,u_0$, $H_2 = \sqrt{6/7}\,u_0$, $A_1 = 4\sqrt{6/7}\,u_0$, $M_1 = \sqrt{32/7}\,u_0$.

You can take: $\bar{k} = 3k_2 = 6k_3$, $W = 0.1u_0$.

6 Simulation Results

Dimensionless size of the field calculation $L_i = 1, \forall i = 1,2,3$. Smagorinsky coefficient C_S set equal to 0.21, since the lower the numerical values of the score was unstable. Number *Re* is equal to 15000. Dimensions of computational grid are $100 \times 100 \times 100$. The boundary conditions are periodic in all directions. The initial field is given by (5) - (7). Averaging in space by the Gaussian filter.

Fig. 1 shows the initial velocity field and turbulent kinetic energy for isotropic turbulence, as in figure Fig. 4 – non-isotropic turbulence. Figures Fig. 2, Fig. 5, Fig. 7 show the statistical turbulence characteristics. Figures Fig. 3, Fig. 6, Fig. 8 shows the values $s(t)$ and scale turbulence for time. With decreasing $s(t)$ energy decreases proportionally.

Fig. 9 shows the correlation function R_{11}, and R_{22}, R_{33} - approximately equal R_{11} to the isotropic turbulence with the same scale in the initial time. It is seen that with increasing r, they flock to zero. Thus, we can conclude that the initial field was close to the turbulent state.

The most important characteristic is calculated dimensionless dissipation. Refer to Fig. 10, given its behaviour over a long period of the dimensionless time - T = 120. It is seen that its behaviour is more or less permanent, enclosed in a range between one and two, roughly around the value 1.7 - 1.8. In the beginning it is a big and a bit chaotic as a result of making the equilibrium position for the current physical process.

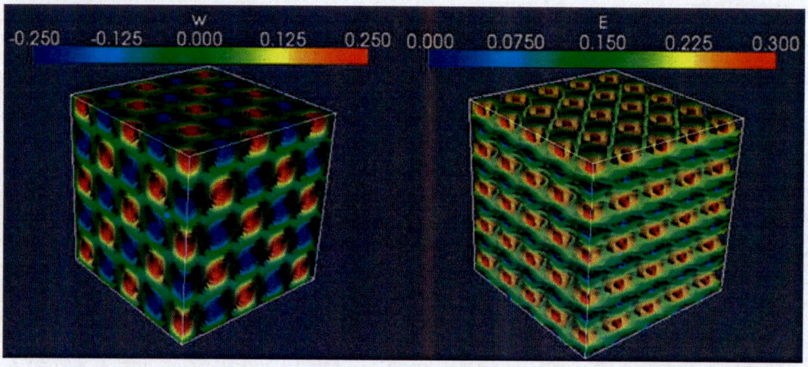

Fig. 1 Components of velocity and kinetic energy in the initial moment, for isotropic turbulence with the same scale in all axes

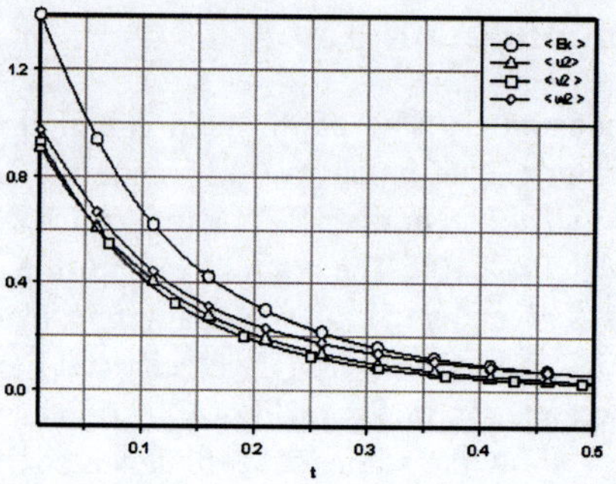

Fig. 2 Graph of changing the values $\langle u2\rangle = \frac{\langle u_1^2\rangle}{u_0^2}$, $\langle v2\rangle = \frac{\langle u_2^2\rangle}{u_0^2}$, $\langle w2\rangle = \frac{\langle u_3^2\rangle}{u_0^2}$, $\langle Ek\rangle = \frac{\langle u_1^2\rangle+\langle u_2^2\rangle+\langle u_3^2\rangle}{2u_0^2}$ for isotropic turbulence

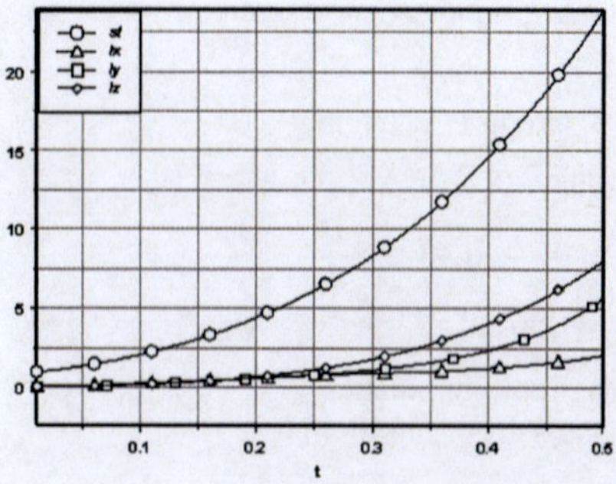

Fig. 3 Graph of changing the scale of turbulence and the functions $s(t) = \frac{E(0)}{E(t)}$ for isotropic turbulence

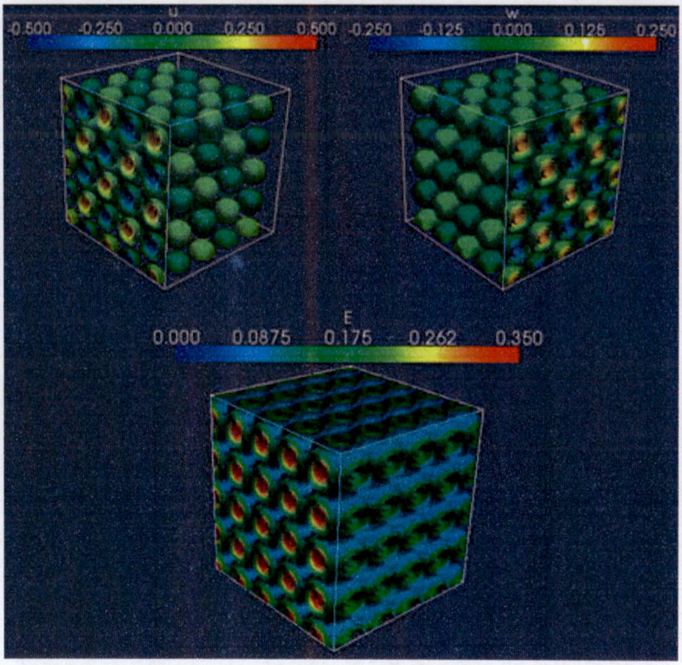

Fig. 4 Components of velocity u, w and kinetic energy in the initial time for non-isotropic turbulence with the same scale in all axes

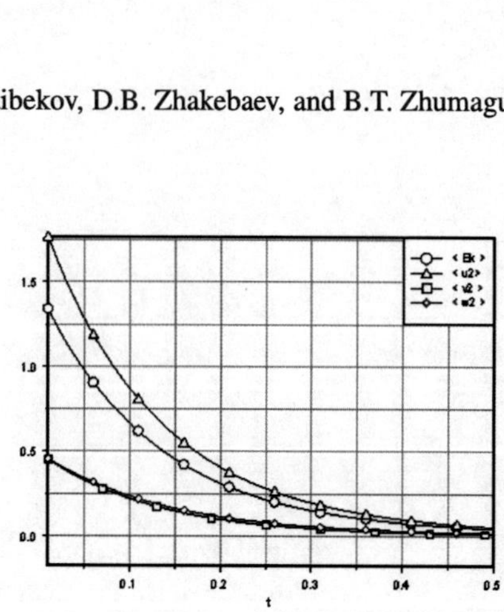

Fig. 5 Graph of changing the values $\langle u2\rangle = \frac{\langle u_1^2\rangle}{u_0^2}$, $\langle v2\rangle = \frac{\langle u_2^2\rangle}{u_0^2}$, $\langle w2\rangle = \frac{\langle u_3^2\rangle}{u_0^2}$, $\langle Ek\rangle = \frac{\langle u_1^2\rangle+\langle u_2^2\rangle+\langle u_3^2\rangle}{2u_0^2}$ for non-isotropic turbulence with the same scale in all axes

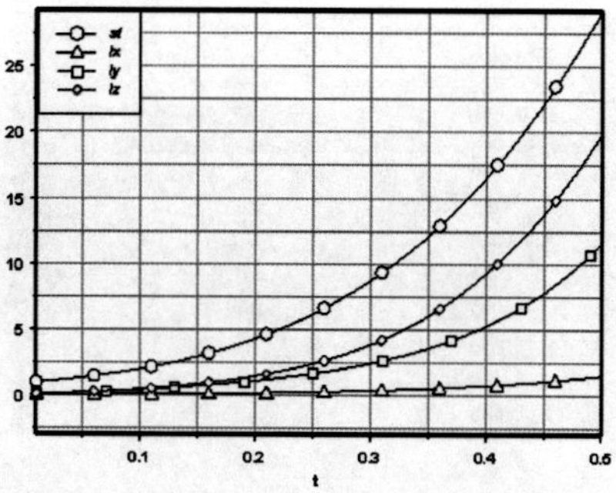

Fig. 6 Graph of changing the scale of turbulence and the functions $s(t) = \frac{E(0)}{E(t)}$ for non-isotropic turbulence with the same scale in all axes

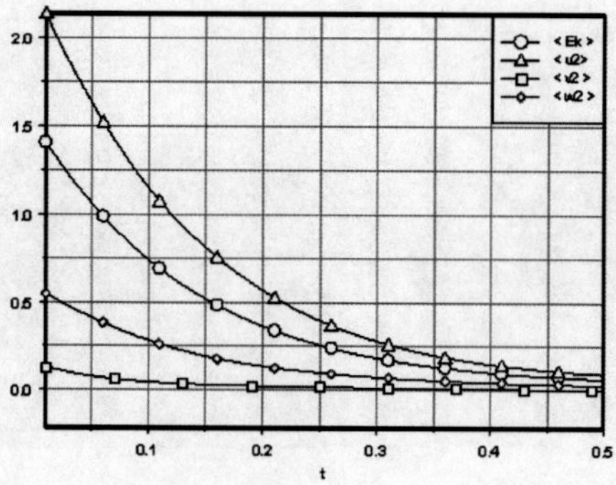

Fig. 7 Graph of changing the values $\langle u2\rangle = \frac{\langle u_1^2\rangle}{u_0^2}$, $\langle v2\rangle = \frac{\langle u_2^2\rangle}{u_0^2}$, $\langle w2\rangle = \frac{\langle u_3^2\rangle}{u_0^2}$, $\langle Ek\rangle = \frac{\langle u_1^2\rangle+\langle u_2^2\rangle+\langle u_3^2\rangle}{2u_0^2}$ for non-isotropic turbulence, with different speeds in all axes, with the scale, varying in proportion to velocity

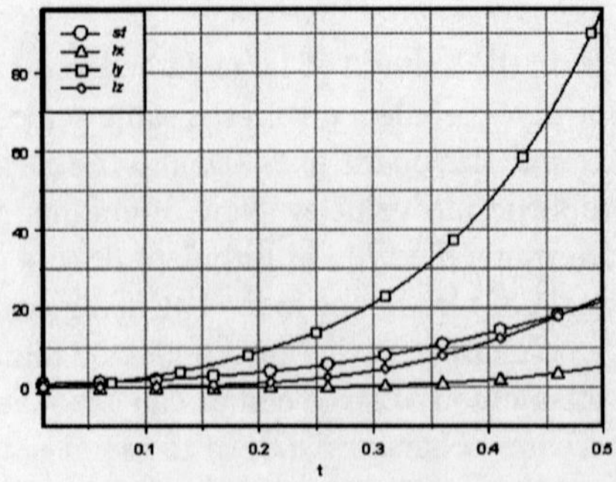

Fig. 8 Graph of changing the scale of turbulence and the functions $s(t) = \frac{E(0)}{E(t)}$ for non-isotropic turbulence, with different speeds in all axes, with the scale, varying in proportion to velocity

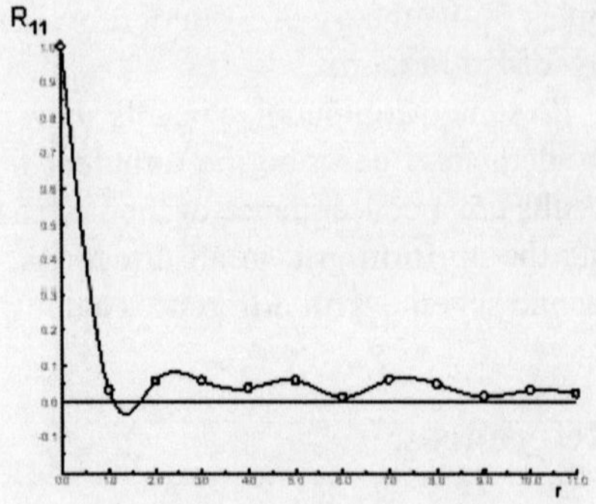

Fig. 9 The curve of the correlation coefficient R_{11} for isotropic turbulence with the same scale at the initial time

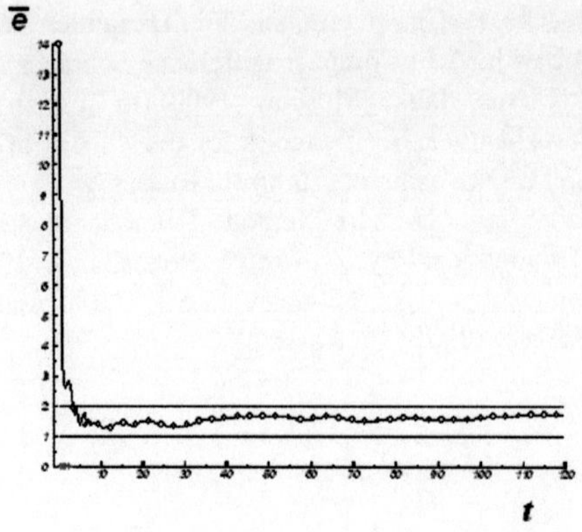

Fig. 10 The dimensionless dissipation $\overline{\varepsilon}(t)$

7 Conclusion

Based on the method of large eddies, produced a large-scale numerical simulations of isotropic turbulence with the same scale in all axes, nonisotropic turbulence with different scales on the axes and at different speeds in all axes, with the scale, varying in proportion to velocity. New offers model can determine the kind of turbulence. To determine the type of turbulent flow is sufficient to know the averaged velocity and scale of turbulence in the entire region is calculated.

These characteristics and conducted numerical experiments allow us to conclude that the method of large eddies can adequately reflect the evolutionary behaviour of the turbulent characteristics of the semi-empirical theory without invoking any empirical parameters. To calculate the turbulent characteristics used the method proposed by V.M. Ievlev's [2]. Revealed that the Smagorinsky model in the averaged Navier-Stokes equation takes into account the investment for its construction principles of the theory of Kolmogorov, namely, "2 / 3 law of Kolmogorov" and "-5 / 3 law of Kolmogorov.

Presented comparative results show that the construction of numerical algorithms to adequately describe the turbulent flow. Some discrepancies with the numerical results are a consequence of the fact that in this study, we used Smagorinsky model and the uniform grid in all directions, to obtain high-order scheme was used: the second scheme with differences against flow and compact circuit.

References

1. Sagaut, P.: Large Eddy Simulation for Incompressible Flows. Springer, Heidelberg (2002)
2. Ievlev, V.M.: Numerical Simulation of Turbulent Flows. Nauka, Moscow (1990) (in Russian)
3. Roache, P.: Computational Fluid Dynamics. Mir, Moscow (1980) (in Russian)
4. Tolstykh, A.I.: Compact Difference Schemes and Their Applications to Fluid Dynamics Problems. Nauka, Moscow (1990) (in Russian)
5. Vorozhtsov, E.V.: Exercises for the Theory of Difference schemes. Publishing House of NSTU, Novosibirsk (2000) (in Russian)
6. Yanenko, N.N.: The Method of Fractional Steps for the Multidimensional Problems of Mathematical Physics. Nauka, Novosibirsk (1967) (in Russian)
7. Samarskii, A.A., Nikolaev, E.S.: The methods for Solving Grid Equations. Nauka, Moscow (1978) (in Russian)

Dynamic Finite Element Analysis of Cancellous Bone Micro Structure

Ralf Schneider, Ulrich Hindenlang, and Peter Copf

Abstract. This study presents dynamic finite element analysis applied to a cancellous bone specimen. The modelled structure of the specimen is extracted from a micro computer tomography dataset with a resolution high enough to cover the cancellous bone micro structure in full detail. An impulse is applied to the bone specimen and its path through the structure is analysed in terms of kinetic and potential energy. To show the concept of the chosen approach at first the results gained from the simulation of a virtual single trabecula are presented. Further on the behaviour of the cancellous bone specimen under dynamic excitation is presented.

1 Introduction

The static simulation of bones and bone-implant-systems is subject to many studies presented in the open literature during the last years. Even though many useful results can be derived from these kind of simulations a big portion of the load situations, a bone has to respond to in everyday life, is dynamical. In addition it was shown by [1] that the process of remodelling depends on the amplitudes of dynamic loads applied to bone cells.

Ralf Schneider
High Performance Computing Center Stuttgart,
Nobelstraße 19, Stuttgart, 70569, Germany
e-mail: schneider@hlrs.de

Ulrich Hindenlang
LASSO Ingenieurgesellschaft mbH, Leinfelderstraße 60,
Leinfelden-Echterdingen, 70771, Germany
e-mail: lasso@lasso.de

Peter Copf
Copf-Bionic GmbH, Schloss Monrepos / Gebäude 1,
Ludwigsburg, 71634, Germany
e-mail: p.copf@copf-bionic.de

E. Krause et al. (Eds.): Computational Sci., & High Performance Computing IV, NNFM 115, pp. 339–347.
springerlink.com

So if one want's to derive a deeper understanding of the behaviour of bone-implant-systems, the so called post-operative primary-stability which is mainly characterised by the clamping of the implant in the bone as well as its osseointegration meaning the ingrowth of the implant to the bone, he or she has to do dynamical simulations.

One of the problems one encounters when planing to do dynamic simulations of bone structures is the application of proper material data to the simulation models. Since cancellous bone is a micro structured material it has to be clarified whether the damping effects of these micro structure have to be taken into account for the evaluation of the considered problem or not and if they are not negligible it has to be carefully evaluated on which scale what kind of effects takes place.

Another problem with the determination of material constants is the completely inhomogeneous density distribution inside living bones, especially in the distal (the joint) regions. In the case of elastic bone material constants it is well known and experimentally proofed [2, 3, 4] that these constants show a strong dependence upon the bone density. So it is evident to expect the same behaviour from the damping coefficients of bone material.

In the context of dynamic bone-implant-compound simulations the aim of this study is to do a first step towards the determination of structural damping behaviour of cancellous bone and the analysis which effects have to be taken into account when switching from the material micro structure to the continuum formulation of bone material. Therefore the propagation of compression waves through the micro structure of cancellous bone material is analysed in terms of kinetic and potential energy. The results are compared to the propagation of the same waves through a continuum material without micro structure.

2 Material and Methods

In this section the subject matter to this study, a cancellous bone specimen taken from the distal region (knee joint region) of a primed human femur along with the modelling techniques and the assumptions made to enable the simulations, are described in detail. In addition the model of a single abstracted trabecula is presented. This model was used to evaluate the wave propagation in a homogeneous material without micro structure but with the same dimensions as the bone specimen.

2.1 Bone Specimen - Geometry Reconstruction

The model that was analysed during this study was derived from a micro computer tomography data set which was taken from a bone specimen of a primed human femur.

The specimen had a physical size of $10mm \, x \, 10mm \, x \, 10mm$ and was scanned with $0.014mm$ isotropic spacial resolution.

From the usable data volume of $504 \, x \, 504 \, x \, 504$ voxels a sub-volume of $200 \, x \, 200 \, x \, 200$ voxels was extracted which lead to a physical volume size of $2.8mm \, x \, 2.8mm \, x \, 2.8mm$ (see fig. 1).

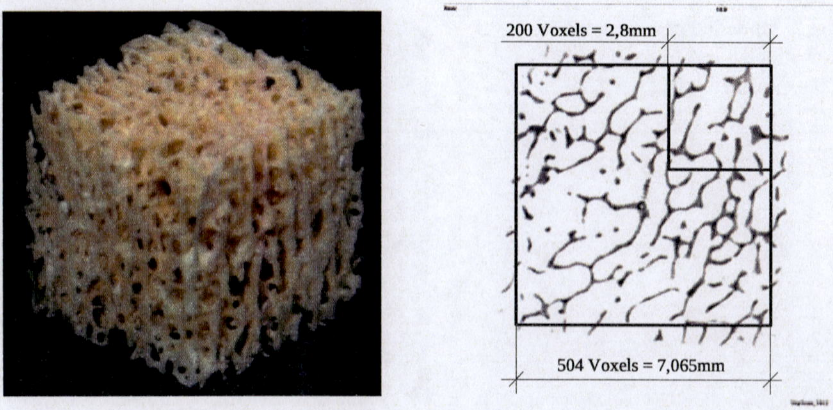

Fig. 1 Bone specimen and micro CT image with marked data area

From the voxel field of the computer tomographic density an iso-surface was calculated by means of a marching cubes algorithm. Since this triangulation method leads to very bad shaped triangles and it thus is not reasonable to derive a volume mesh from it, the surface mesh was in a second step reconstructed to proper triangle quality with the CAE pre-processing tool ANSA [1].

The FE model resulting from this procedure finally consisted of 1339273 linear tetrahedra elements with 860100 degrees of freedom.

2.2 *Model Configurations*

The geometries of both models are shown in fig. 2. It can be seen, that the abstracted trabecula has the same length as one edge of the bone specimen. Since the model dimensions in this study are very small compared to the ones which can usually be found in technical FE-Models it was necessary to determine the needed simulation time and post-processing intervals with the trabecula model. Also this model was used to analyse the numerical behaviour of the commercial FE-solver ABAQUS 6.7-5 [2].

Boundary Conditions

The boundary conditions for the trabecula model where defined as follows

- Lower end: Displacements of all nodes fixed in all directions
- Upper end: Force impulse $F(t)$ of $1 \cdot 10^6 sec$ duration

[1] ANSA is a trademark of BETA CAE Systems S.A., Kato Scholari, Thessaloniki, GR-57500 Epanomi, Greece.

[2] ABAQUS is a trademark of SIMULIA, Rising Sun Mills, 166 Valley Street Providence, RI 02909-2499.

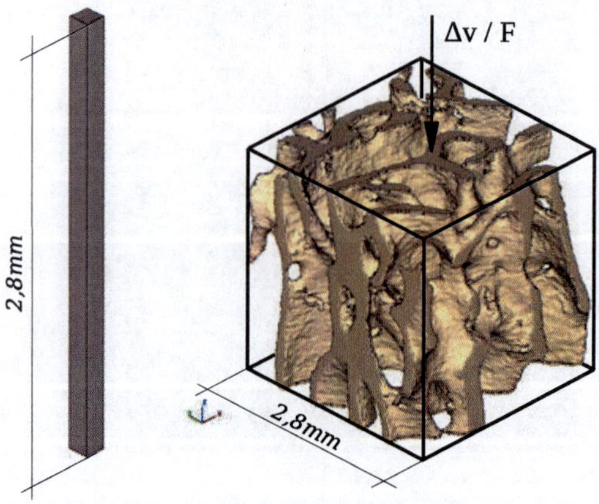

Fig. 2 Modelled geometries of abstracted trabecula and bone specimen

The shape of $F(t)$ can be seen in fig. 3. It should be noted that the maximum amplitude $F(0.5 \cdot 10^{-6}) = 1$ is adjusted to each node by an amplitude multiplication factor which was derived from a static simulation with prescribed displacements on the upper end.

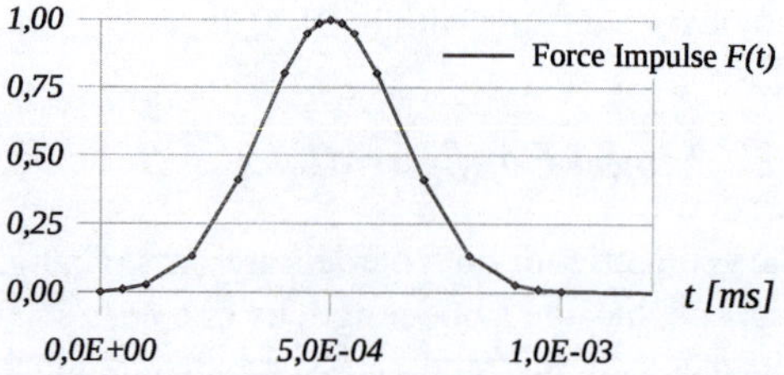

Fig. 3 Normed force impulse applied to each node in the respective model's upper facet

The boundary conditions applied to the bone specimen model where different from the ones given for the trabecula model since in case of the bone specimen it has to be taken into account, that the specimen is surrounded by its neighbouring volume elements. Due to this fact in situ boundary conditions as suggested in [5] where applied to the nodes lying in the side and bottom facets of the cube enclosing the bone specimen model. In these facets all nodal displacements in the facet's normal directions where suppressed while all nodal displacements in the facet's tangential directions where kept unconstrained.

To all nodes lying in the upper facet of the model the force amplitude $F(t)$ was applied also. To keep the cross section of the upper facet planar during the load application, the amplitude multiplication factors of $F(t)$ for each node where determined from a static simulation with the same boundary conditions as for the dynamic load case but with a prescribed displacement of 0.0028*mm* applied to all nodes in the upper facet. As the result of the static simulation the reaction forces of the nodes in the upper facet were recorded. The z-components of the nodal reaction forces then gave the desired amplitude multiplication factors.

Material

Since the aim of this study was the determination whether there is a macroscopic damping effect induced by the cancellous bone's micro structure, the material used to model the bony structures had an ideal linear elastic formulation without any internal material damping effects. The Elastic modulus and the Poisson number are taken from [6], being

- $E = 5600\,MPa$
- $\nu = 0.3$

This means that all macroscopic damping observed in the results of the dynamic simulations can only be accounted to the micro structure of the analysed bone specimen.

On the other hand for the trabecula model it can be expected that the applied force impulse has to be transferred through the model without any energy loss. That ABAQUS can numerically provide this precondition will be shown later in the subsequent section.

3 Results

In this section the results of the dynamic simulations done on the single trabecula and the bone specimen are presented and compared against each other. All results are shown as relative values normed to the maximum amplitudes of the respective load case. All time scales are shown in *ms*.

3.1 Dynamic Analysis of Single Trabecula

The first evaluation done on this model was to analyse whether there is any numerical damping shown by the chosen ABAQUS procedure. That is the energy amount introduced by the applied force impulse hast to remain constant throughout the evaluated analysis time of $2 \cdot 10^{-}6$ *sec*.

In fig. 4 it can be seen that the sum of integral strain- and integral kinetic energy indeed remains constant completely without conversion from one energy form to the other. This indicates, that the shock wave remains compact in its initial form throughout the complete propagation.

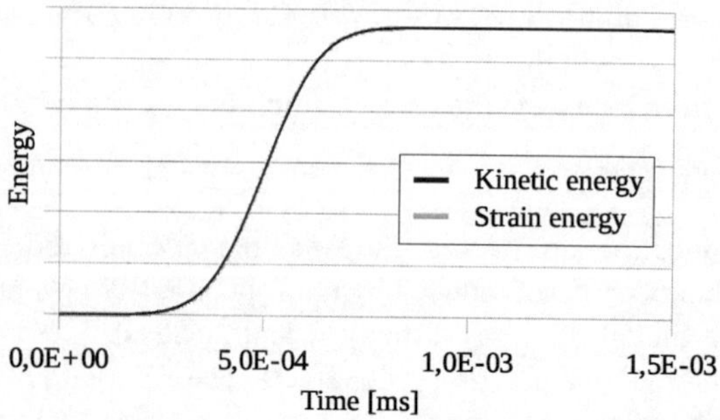

Fig. 4 Trabecula model: Integral energy results

This can also be seen from fig. 5 where the strain energy in three selected cross sections (see also fig. 5) evaluated throughout the analysed simulation time is shown. When the shock wave passes the cross sections for the first time the maximum strain energy as well as the initial gradients are the same for all three evaluated cross sections.

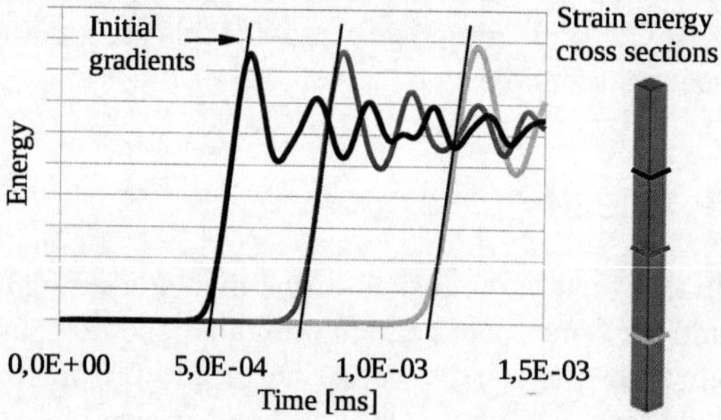

Fig. 5 Strain energy evaluations of the trabecula model in three selected cross sections

In summary it could be derived from this study that ABAQUS is capable of computing a dynamic simulation on the given scale and with the chosen frequency without numerical problems also it should be noted that the analysis procedure was executed in its double precision form since the solver issued a warning message when using the analysis procedure in its standard form. The double precision procedure was necessary because of the excessively small time step size induced by the very small element edge lengths in the model.

3.2 *Static Simulation*

As described in section 2, the first simulation done with the bone specimen model was a static compression load case with prescribed displacements on all cube facets. The main objective of this simulation was the determination of the amplitude multiplication factors needed in the dynamic simulation but the model shows the influence of the micro-structure to the local strain results as well.

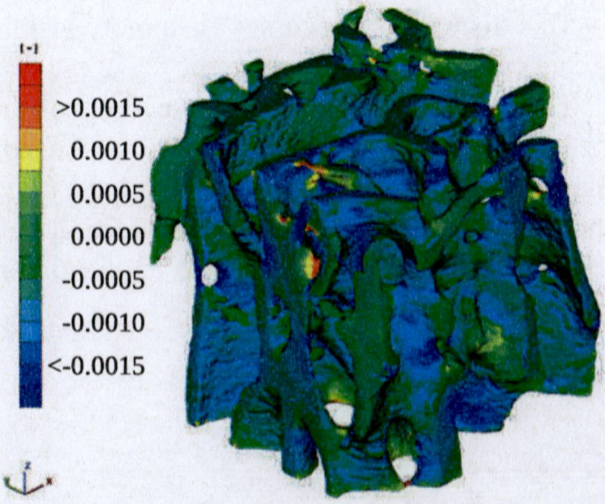

Fig. 6 Static result of the bone specimen model. Shown are the signed major principal stresses

As can be seen in fig. 6 the application of 0.1% macroscopic strain leads to local microscopic strains which are more than 50% higher.

This results leads to the conclusion, that in non-linear simulations e.g. for failure estimation one has to carefully consider this effect before the application of continuum material models which use failure criteria only derived from the averaged calculated values.

3.3 Dynamic Analysis of Cancellous Bone Sample

The dynamic analysis of the bone specimen model was done in the same way as for the trabecula model. First the integral strain and kinetic energy in the complete model were evaluated. In fig. 7 it can be seen, that again the sum of strain- and kinetic energy remains constant throughout the analysed simulation time.

In this model however shortly after the application of the maximum amplitude of the force impulse an alternating conversion of the two energy forms into one another can be observed which is a first indication that the shock wave is disturbed in its propagation.

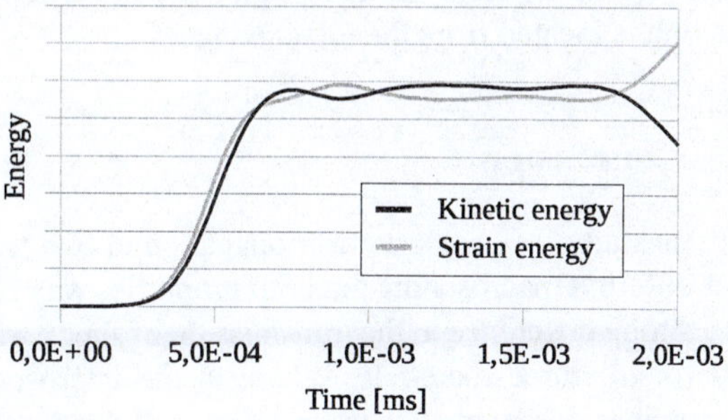

Fig. 7 Bone specimen model: Integral energy results

This disturbance becomes even more visible when again the strain energy in selected cross sections is analysed.

The evaluation in four cross sections of the bone specimen model is shown in fig. 8. The three lower cross sections are positioned similar to the ones used in the evaluation of the trabecula model at 1/3-, 1/2- and 2/3-height of the complete model. Additionally a fourth cross section is located close to the top cube facet to capture the force impulse after its application nearly undisturbed.

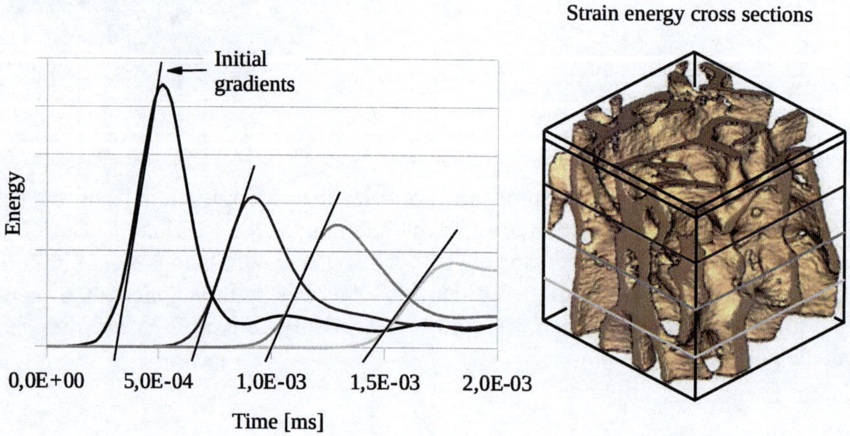

Fig. 8 Strain energy evaluations of the bone specimen model in four selected cross sections

What can clearly be seen is the decay of the maximum strain energy amplitude the further the respective cross section is located from the top cube facet. Since the sum of the integral energies remains constant throughout the analysed simulation time the observed effect has to be induced by a kind of shock wave diffusion that protracts the applied force impulse the further it propagates through the cancellous bone micro structure.

The effect can also be derived from the comparison of the initial strain energy gradients in the cross sections which are less steep the further the respective cross section is located from the top cube facet.

4 Conclusion

In this study the shock wave propagation in two materials with equal microscopic but different macroscopic material properties was analysed. It was shown that there is a strong difference in the transmission of shock waves between homogeneous and micro structured materials. Where in the microscopically homogeneous material the applied force impulse propagates nearly undisturbed a wave front decay can be observed in the material with micro structure.

Since the microscopic material formulation was ideal and undamped this effect can not be accounted to energy absorption.

The decay of the wave front can therefore only be accounted to wave splitting and reflection at bifurcations in the trabecular micro structure, the superposition of the reflected wave parts as well as runtime shifts due to the wave reflections and splits.

In conclusion it can be said that the more cancellous bone substance has to be passed by a force impulse the more the decay and diffusion of the wave front will be so reducing the energy per volume acting on the bone structures.

Due to the observations made in this study there is a strong evidence for the micro structure of cancellous bone not only being a technique of nature to reduce the weight of the bone structure but also fulfilling a kind of shock absorbing function to protect the bone structures and joints behind from vibrations.

References

1. Kaspar, D., Seidl, W., Ignatius, A., Neidlinger-Wilke, C., Claes, L.: In-vitro-Zellverhalten humaner Osteoblasten nach physiologischen dynamischen Dehnungen. Orthopädie 29, 85–90 (2000)
2. Ashman, R.B., Rho, J.Y.: Elastic modulus of trabecular bone material. J. Biomech. 21(3), 177–181 (1988)
3. Wirtz, D.C., Schiffers, N., Pandorf, T., et al.: Critical evaluation of known bone material properties to realize anisotropic FE-simulation of the proximal femur. J. Biomech. 33, 1325–1330 (2000)
4. Ciarelli, M.J., Goldstein, S.A., Kuhn, J.L., et al.: Evaluation of Orthogonal Mechanical Properties and Density of Human Trabecular Bone From the Major Metaphyseal Regions with Materials Testing and Computed Tomography. J. Orthopaedic Research 9, 674–682 (1991)
5. Van Rietbergen, B., et al.: Direct mechanics assessment of elastic symmetries and properties of trabecular bone architecture. J. Biomech. 29(12), 1653–1657 (1996)
6. Van Rietbergen, B., et al.: A new method to determine trabecular bone elastic properties and loading using micromechanical finite-element models. J. Biomech. 28(1), 69–81 (1995)

Three-Dimensional Numerical Modeling of Stratified Flows in Littoral Zone of Israel Using Shallow Water Approximation

Michael Sladkevich, Anna Levin, and Eliezer Kit

Abstract. The paper describes a three-dimensional hydrodynamic model of stratified fluid flow CAMERI3D/HD-ST developed in the Coastal and Marine Engineering Research Institute (CAMERI). The model, which incorporates the shallow water approximation, has been successfully applied for solution of various problems encountered within the coastal waters along the Israeli shore, such as release of cooling water and brine disposal. The empirical coefficients employed in the turbulent model have been selected based on comparison with the measurements of heated water spreading in the field and laboratory experiments. The selected coefficients are model constants and should not be modified in specific applications. Then, the model was thoroughly verified versus field measurements: instrumental temperature measurements in many locations along the Mediterranean coast of Israel including detailed temperature measurements in the Ashkelon region, airborne and satellite imaging of temperature plume propagation.

1 Introduction

The littoral zone along the Israeli shore is subject of permanent discharge of either warm water from cooling system of electrical power stations or brine from desalination plants. Release of cooling water and/or brine disposal result in generation of stratified flows within the near shore region, up to 3 km offshore. An intrusion of heated seawater, which is about 10^oC above the ambient, into the cooling system

Michael Sladkevich · Anna Levin
Coastal and Marine Engineering Research Institute, Technion City,
Haifa, 32000, Israel
e-mail: `michslad@tx.technion.ac.il`, `lanna@tx.technion.ac.il`

Eliezer Kit
Department of Fluid Mechanics and Heat Transfer, Faculty of Engineering,
Tel Aviv University, Tel Aviv, 69978, Israel
e-mail: `kit@eng.tau.ac.il`

E. Krause et al. (Eds.): Computational Sci., & High Performance Computing IV, NNFM 115, pp. 349–361.
springerlink.com

inlets leads to remarkable decrease in power station efficiency and, thus, is highly undesirable. In order to overcome the problem of water shortage that the Israeli community faced during the last years of severe drafts, the number and capacity of the desalination facilities is expected to be significantly increased in the coming years. Brine concentration disposed by salinity plants exceeds approximately twice the salinity of the receiving seawater. Moreover, the desalination plants carry other wastes related to chemicals used for biofouling control, scale control, foam reduction, corrosion inhibition, etc. These substances can have a harmful impact on the marine environment, in particular for large effluent flow rates. All marine organisms that inhabit the coastal zone have a range of tolerance to salinity and temperature, which is related to their ability to regulate their essential functions. In this respect it is very important to ensure a high dilution of the effluents so as to minimize harmful chemical and thermal effects on the marine habitats. Obviously, the discharged warm water behaves as a turbulent buoyant jet that tends to propagate within the upper water layers exposed to wind, while high concentrated salt brine has negative buoyancy with the tendency to progress towards the bottom of the receiving water. In some places the concentrate of the seawater desalination plants is combined with heated seawater leaving the power station's cooling system. In these cases buoyancy of the stratified jet depends on the ratio between the saline and warm water discharges.

The present work describes numerical modelling of stratified flows based on three-dimensional shallow water approximation. The model assumes hydrostatic pressure profile. Vertical eddy viscosity is evaluated according to the mixing length theory, which accounts for stratification both by temperature and salinity. A carefully made model calibration is given in [8]. The required empirical coefficients had been found by comparison of model results with laboratory experiments. Then, the model together with the selected coefficients, which have been kept constant, was verified versus numerous field measurements in the various locations along the Mediterranean coast of Israel. Later on, this model has been successfully applied for flow modelling within the Israeli coastal zone.

2 Model Description

The three-dimensional (3D) flow model in shallow stratified water was developed in CAMERI and is based on mass conservation equations and differential equations for nearly horizontal flows:

$$\frac{\partial u}{\partial x}+\frac{\partial v}{\partial y}+\frac{\partial w}{\partial z}=0 \tag{1}$$

$$\rho(\frac{\partial u}{\partial t}+u\frac{\partial u}{\partial x}+v\frac{\partial u}{\partial y}+w\frac{\partial u}{\partial z})+\rho g\frac{\partial z_s}{\partial x}-\frac{\partial}{\partial z}(\rho v_V\frac{\partial u}{\partial z})+g\int_z^{z_s}\frac{\partial \rho}{\partial x}dz-Cv=0 \tag{2}$$

$$\rho(\frac{\partial v}{\partial t}+u\frac{\partial v}{\partial x}+v\frac{\partial v}{\partial y}+w\frac{\partial v}{\partial z})+\rho g\frac{\partial z_s}{\partial y}-\frac{\partial}{\partial z}(\rho v_V\frac{\partial v}{\partial z})+g\int_z^{z_s}\frac{\partial \rho}{\partial y}dz+Cu=0 \tag{3}$$

where t is the time; x and y are the longitudinal and transverse coordinates, respectively; z is the axis normal to the earth surface; u, v and w are the velocity components in the x-, y- and z-directions, correspondingly; g is the gravitational acceleration; ρ is the water density, Z_s is the free water surface elevation above the datum; ν_V is the vertical eddy viscosity; and C is the Coriolis parameter.

The fourth equation, which is necessary to resolve the four unknown variables (in the case of homogeneous fluid) such as three velocity components u, v and w, and water elevation above the datum, Z_s, can be derived from the kinematic boundary condition on free surface

$$\frac{\partial Z_s}{\partial t} + u_s \frac{\partial Z_s}{\partial x} + v_s \frac{\partial Z_s}{\partial y} = w_s \tag{4}$$

Integration of the continuity equation 1 across the water depth by applying the Newton-Leibniz integration rule, and utilizing appropriate kinematic boundary conditions at the free surface represented by eq. 4 and at the bottom (e.g. [2])

$$u_b \frac{\partial Z_b}{\partial x} + v_b \frac{\partial Z_b}{\partial y} = w_b \tag{5}$$

yields

$$\frac{\partial Z_s}{\partial t} + \frac{\partial \int_{Z_b}^{Z_s} u dz}{\partial x} + \frac{\partial \int_{Z_b}^{Z_s} v dz}{\partial y} = 0 \tag{6}$$

where subscripts "b" and "s" correspond to magnitude of the variable taken at the bottom and surface, respectively.

The velocity components were computed using the equations for horizontal components of vorticity in stratified flow:

$$\frac{d\Omega_x}{dt} - \Omega_x \frac{\partial v}{\partial y} + \Omega_y \frac{\partial u}{\partial y} = \frac{1}{\rho_0} \frac{\partial^2}{\partial z^2} (\nu_V \Omega_x) + \frac{g}{\rho_0} \frac{\partial \rho}{\partial x} + C\Omega_y \tag{7}$$

$$\frac{d\Omega_y}{dt} + \Omega_y \frac{\partial u}{\partial x} - \Omega_x \frac{\partial v}{\partial x} = \frac{1}{\rho_0} \frac{\partial^2}{\partial z^2} (\nu_V \Omega_y) + \frac{g}{\rho_0} \frac{\partial \rho}{\partial y} - C\Omega_x \tag{8}$$

here

$$\Omega_x = \frac{\partial u}{\partial z} \quad \text{and} \quad \Omega_y = \frac{\partial v}{\partial z} \tag{9}$$

and ρ_0 denotes the density of the ambient fluid.

The expressions for the vorticity components given by 9 originate from the very small value of the vertical velocity component and essentially stronger gradients in the vertical direction compared to those in the horizontal plane.

Equations 7 and 8 are subject to various types of boundary conditions. Shear stresses at the free surface are given by

$$\left(\rho \nu_V \frac{\partial u}{\partial z}\right)_s = \tau_{xs} \quad \text{and} \quad \left(\rho \nu_V \frac{\partial v}{\partial z}\right)_s = \tau_{ys} \tag{10}$$

where shear stresses at the bottom can be presented as

$$\left(\rho \nu_V \frac{\partial u}{\partial z}\right)_b = \tau_{xb} \qquad and \qquad \left(\rho \nu_V \frac{\partial v}{\partial z}\right)_b = \tau_{yb} \tag{11}$$

In these equations τ_{xs} and τ_{ys} designate the components of the surface wind stress; τ_{xb} and τ_{yb} are the components of the bed stress.

The shallow water 3D model given by eqs.1-3 and eq.6 is based on the following assumptions, typical to a shallow aquatic system:

* The vertical acceleration is much smaller than the acceleration due to gravity; therefore the pressure is distributed as a hydrostatic pressure;

* The horizontal turbulent momentum exchange and related Reynolds stresses are much smaller than turbulent momentum exchange in the vertical direction;

* The Bousinesq assumption is applicable to calculate horizontal turbulent shear stresses; these stresses are expressed by the vertical velocity gradient and the eddy viscosity;

* The effect of the horizontal turbulent diffusion is much smaller than that of the advection and convection.

The equations are solved by a finite difference method on an unevenly spaced grid. The governing equations used for finite difference discretization are transformed from z-coordinates to σ-coordinates in a usual way (e.g. [1]). This transformation of $\sigma = (Z_s - z)/(Z_s - Z_b)$ converts the variable depth domain to a constant depth domain, with $0 \leq \sigma \leq 1$.

The vertical eddy viscosity ν_V is given according to the mixing length theory by

$$\nu_V = \nu + \nu_t = \nu + l^2 \frac{\partial |\vec{u}|}{\partial z} \tag{12}$$

where l is the mixing length; ν is the kinematic molecular viscosity; and ν_t is the kinematic turbulent viscosity.

In case of a vanishing shear stress on free surface, the mixing length in a homogeneous fluid, l_h, is given in [7] by

$$\frac{l_h}{h} = 0.14 - 0.08 z_1^2 - 0.06 z_1^4 \qquad z_1 < 1 \tag{13}$$

where $z_1 = z/h$.

When the shear stress is imparted to the free surface, better results can be obtained if z_1 is presented as follows

$$z_1 = \frac{\frac{z}{h} - z_0}{1 - z_0} \quad if \quad \frac{z}{h} \geq z_0 \qquad and \qquad z_1 = \frac{\frac{z}{h} - z_0}{z_0} \quad if \quad \frac{z}{h} < z_0 \tag{14}$$

here

$$z_0 = \frac{|\tau_b|}{|\tau_b| + |\tau_s|} \tag{15}$$

The form expressing the effects of variations of the density on the mixing length in the case of stable stratification ($-\partial\rho/\partial z > 0$) is

$$l_s = l_h / \left(1 + Ri_h^{n/2}\right)^{1/n} \tag{16}$$

where l_s is the mixing lengths in stratified fluid and Ri_h is the Richardson number for homogeneous fluid.

In the case of a flow with a density gradient, the general form of the Richardson number, Ri, can be written as

$$Ri = -a\frac{g\partial\rho/\partial z l_s^2}{|\tau|} \tag{17}$$

where a and n are free parameters to be found via the model calibration.

In a highly turbulent flow, where $\nu_t >> \nu$, the expression 17 transforms to a conventional gradient Richardson number

$$Ri = -a\frac{g\partial\rho/\partial z}{\rho\left(\partial|\vec{u}|/\partial z\right)^2} \tag{18}$$

The basic expressions for transport of heat as well as for a conservative contaminant are given approximately as

$$\frac{\partial s}{\partial t} + u\frac{\partial s}{\partial x} + v\frac{\partial s}{\partial y} + w\frac{\partial s}{\partial z} = \frac{\partial}{\partial x}\left(k_H\frac{\partial s}{\partial x}\right) + \frac{\partial}{\partial y}\left(k_H\frac{\partial s}{\partial y}\right) + \frac{\partial}{\partial z}\left(k_V\frac{\partial s}{\partial z}\right) \tag{19}$$

where s is a dummy variable representing the temperature or the contaminant concentration; and k_H and k_V are the horizontal and vertical components of the eddy diffusivity, respectively.

It can be argued (e.g. [6]) that the gradients in the horizontal directions are much smaller than those in the vertical direction, and therefore the corresponding terms in the RHS of eq.19 can be neglected. Moreover, in the following it is assumed that

$$k_V = 0.9\nu_V \tag{20}$$

If temperature and/or concentration differences in the domain are significant then the fluid density can be represented as

$$\rho = \rho_0\left(1 + \alpha_\rho s)\right) \tag{21}$$

where α_ρ is a coefficient.

The boundary conditions at the free surface and the bed for the substances are as follows:

$$k_V\frac{\partial s_s}{\partial z} = f(s_s) \qquad and \qquad k_V\frac{\partial s_b}{\partial z} = 0 \tag{22}$$

where $f(s)$ is the transfer function of heat to the atmosphere.

More detailed description of the model is given in [8].

3 Model Calibration

The main objective of the calibration was to determine the parameters n and a of eq.16 and eq.18. The temperature and velocity distributions in a two-layer flow have been computed for conditions described in [9]. The model parameters were chosen so that to ensure the best match between the numerical and experimental data. Good agreement between the computed and measured velocities and temperature profiles was obtained in a wide range of Froude, Fr, and Reynolds, Re, which indicates that the proposed mathematical model provides an adequate description of stratified flows. For example, the vertical distributions of flow velocities and temperatures are plotted in Fig.1 for n=1, a = 20, Fr=1.6 and Re=7,500. It appears that in a relatively

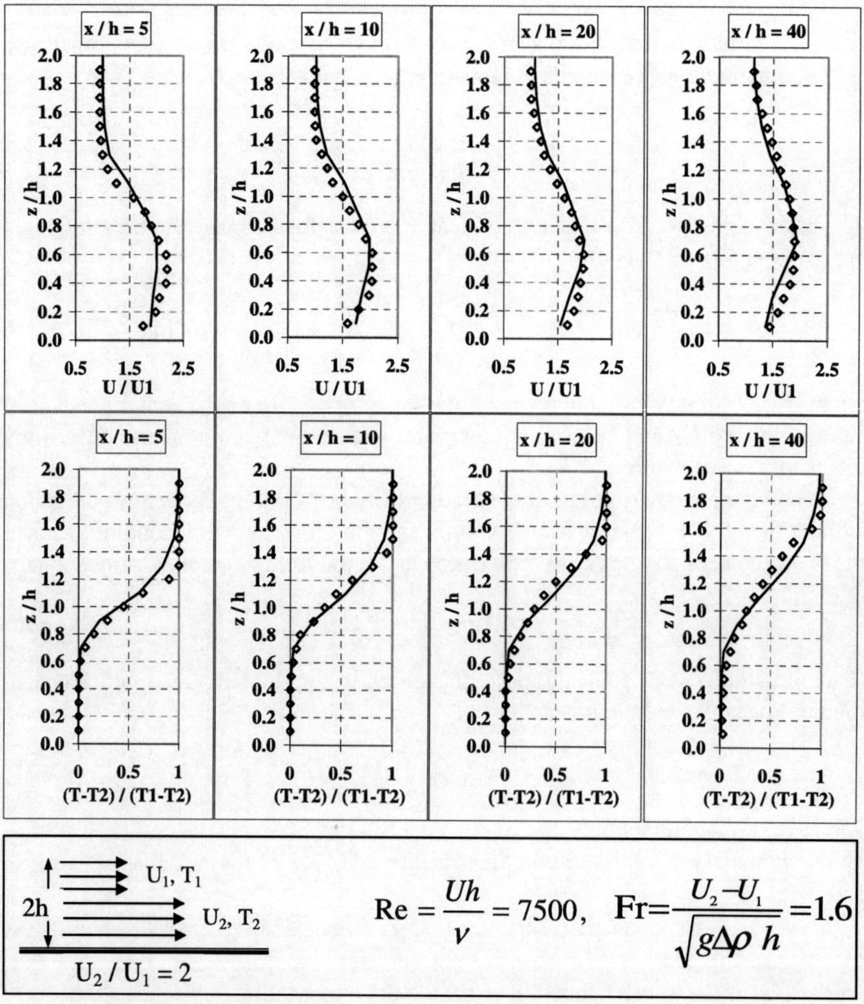

Fig. 1 Model calibration versus laboratory experimental data taken from [9]

wide range of parameters, $0.5<n<2$ and $12<a<30$, the results of numerical computations do not vary significantly showing that the model is robust and not sensitive to precise determination of parameters n and a. Hence, the selected coefficients ($n=1$, $a=20$) are model constants and should not be modified in specific applications.

4 Model Verification

The CAMERI3D/HD-ST model has been successfully implemented in a number of projects and has undergone extensive verification, e.g. comparison with laboratory, field measurements and infrared satellite photographs obtained around several power stations located on the Israeli shore. These stations use the sea water for cooling and release the heated by $\sim 10^{o}C$ water back to the sea. An intrusion of heated seawater into the cooling system causes remarkable decrease in power station efficiency and, thus, is highly undesirable. In the model the temperature drop between the outfall and the intake is specified so that the temperature rise at the intake leads to the temperature increase at the outfall.

The following example considers cooling water recirculation around the Eshkol power station (Israel). The intakes are located within the settling basin, while the outfalls release the heated water at both sides of the basin. An adequate comparison of a model with sea measurements requires conducting continuous simulations for a period of few days prior to the time of measurement in order to generate reliable climate conditions in terms of wind- and wave-induced currents.

The computed velocity fields and temperature patterns (excess temperature above the temperature of the ambient water) at the water surface and at 3m water depth at 9:30 and 12:30 o'clock, 05 March, 1998, are plotted in Figs.2 and 3. It should be kept in mind that the process of measurements took about three hours (from 9:30 to 12:30), i.e. the probes at different points had been taken at different unknown for us time. Obviously, during this even short period the temperature field had been changing. Therefore, both sets of the computed pictures that correspond to 9:30 as well as to 12:30 are provided with the same measured values. It turned out that the temperature measured at 3m water depth at some distance northward the outfall is larger than that in the outfall surroundings. Since there is no an outlet of hot water at this location, this result looks anomalous. However, the mathematical modelling, started two days beforehand for the real weather conditions, provides the similar results. It is believed that the obtained temperature field represents the result of the complicated three-dimensional flow through the region of irregular bathymetry, i.e. the water depths become larger northward with the alongshore distance from the outfall.

Another example demonstrates the brine surface disposal from the onshore desalination facility. The desalination plant releases the brine of $\sim 21{,}000 m^3/hr$, which is approximately twice less than the inflow rate. The salinity of the effluent is 73.5 g/L that is significantly larger than the receiving water body 39.4 g/L. In Fig.4 the vertical distributions of the salinity and turbulent viscosity obtained in our computations are presented for three monitoring points: Point #1, Point #2, and Point

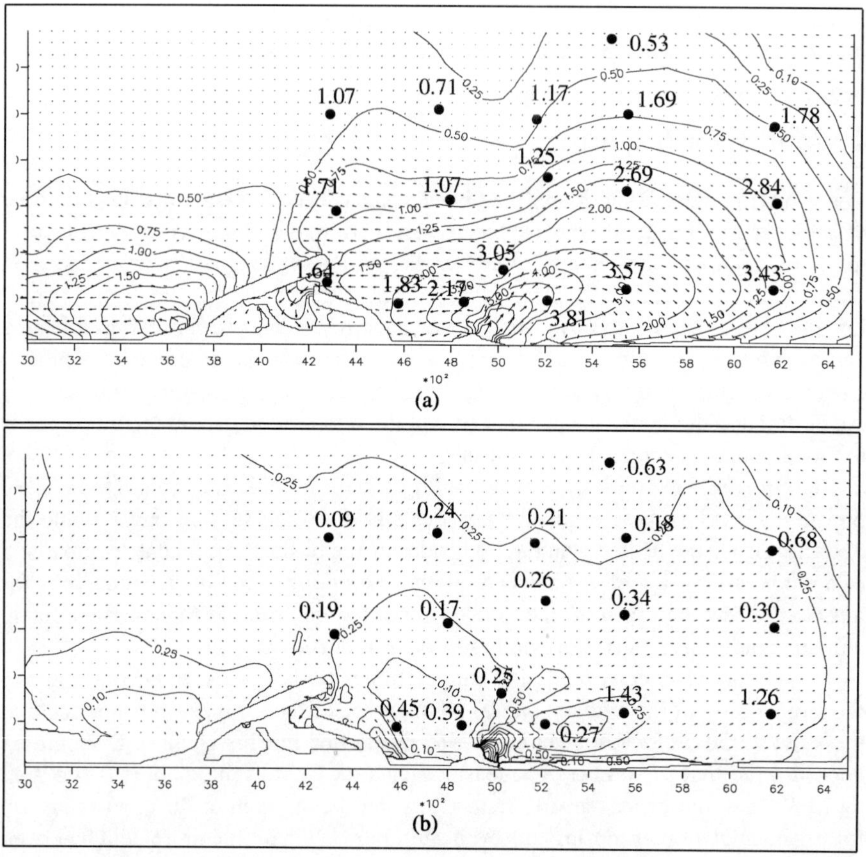

Fig. 2 Temperature patterns and flow fields computed for 05.03.1998 at 9:30: at the water surface (a); and at 3m water depth (b). The measurements had been conducted by the Israel Electricity Corporation

#3 located 80m, 140m and 200m from the outfall, respectively. Some small excess salinity is obtained in the nearest to the outlet Point #1 at 0.5m below the water surface, but for Point #2 the exceeding salinity is practically zero. It appears that the plume of the salinity larger than the ambient tends to propagate in the near the seafloor layer, whereas the water surface salinity is equal to the ambient. In the upper layers the viscosity is almost equal to the molecular viscosity ($10^{-6}m^2/s$), though the near-bottom viscosity is greater that the ambient by few orders of magnitude. It is seen that stratification suppresses the turbulence and at the density interfaces the viscosity decreases and actually reaches the values of molecular viscosity.

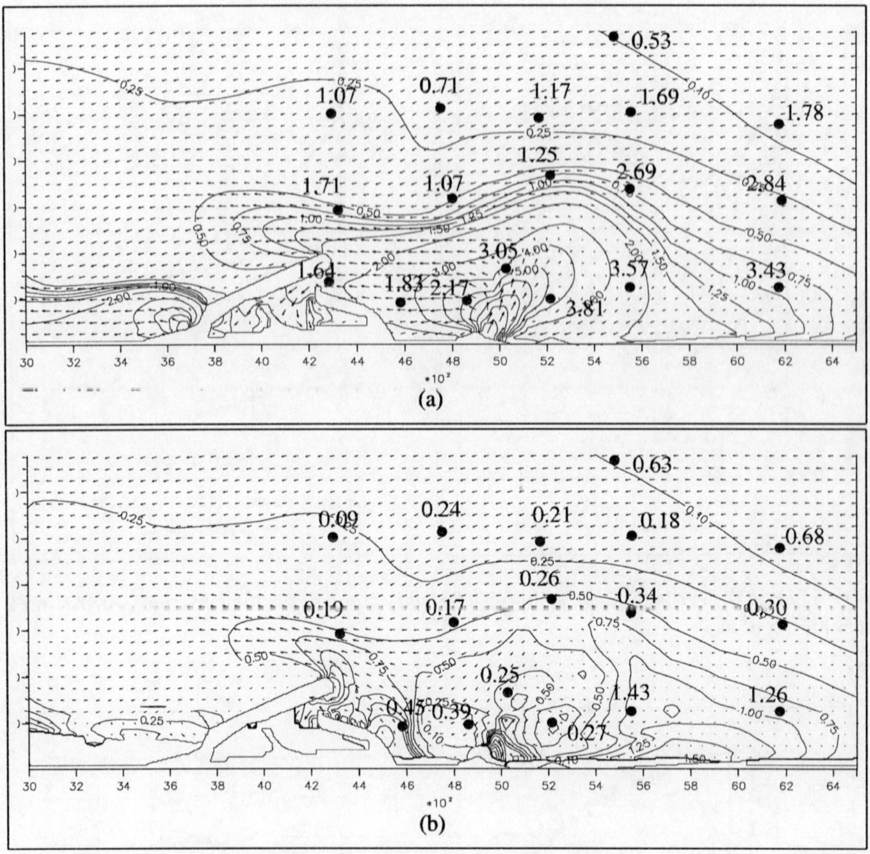

Fig. 3 Temperature patterns and flow fields computed for 05.03.1998 at 12:30: at the water surface (a); and at 3m water depth (b). The measurements had been conducted by the Israel Electricity Corporation

The model sensitivity to the value of the empirical coefficient *a* has been analysed for brine spreading to the sea water from the onshore desalination facility. As expected, the shape of the plume is somewhat effected by the value of the coefficient *a*, however, within the areas limited by the excess salinity below 10% such variations appear to be very small. The salinity distributions near the bottom along a cross-section for *a*=10, 20 and 40 given in Fig.5 show that the plume salinity is almost the same, in particular, within the distances less then ~200m, where the salinity excess is larger than 10%.

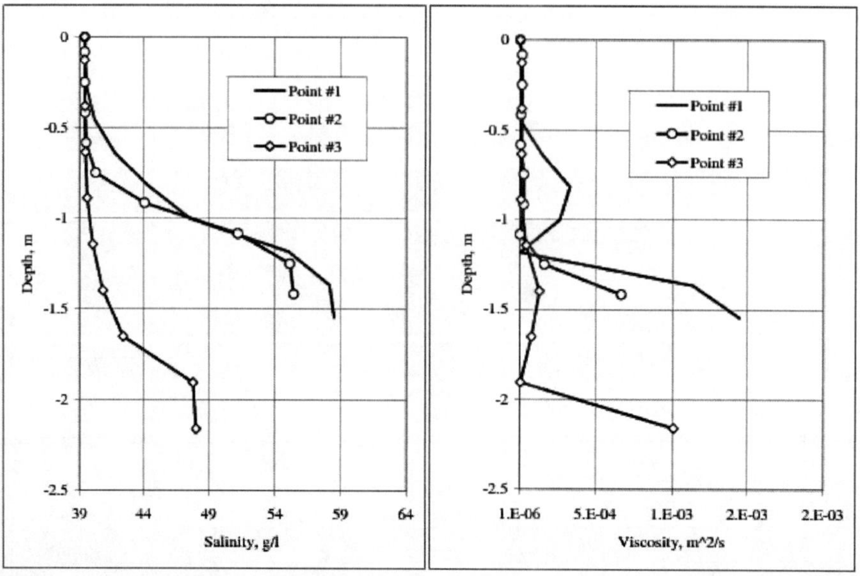

Fig. 4 Vertical distributions of salinity (left) and viscosity coefficient (right) at the three monitoring points

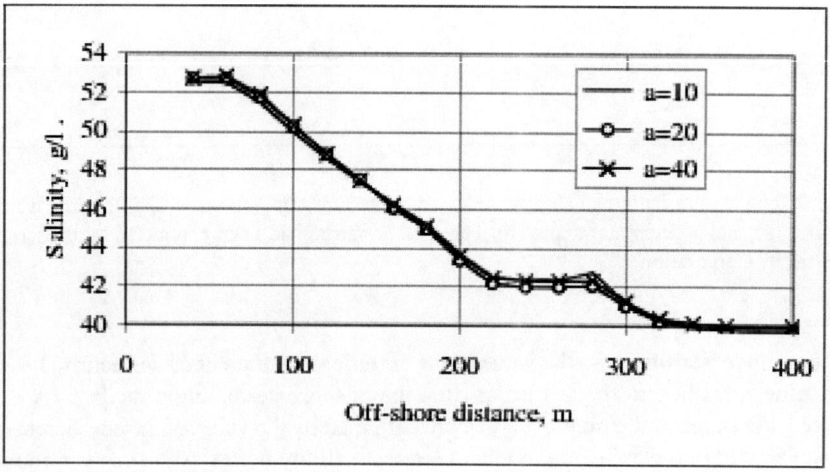

Fig. 5 An example of changes in water salinity near the seafloor for different values of empirical coefficient *a* in eq. 6

5 Model Limitation

The discussed model is a far-field model, which as all models of such type has some limits in the near-field computations.

The hydrodynamics of an effluent continuously discharging into a receiving water body can be conceptualized as a mixing process occurring in two separate regions. In the first region, namely near-field, the initial jet characteristics of momentum flux, buoyancy flux, and outfall geometry influence the jet trajectory and degree of mixing. As the turbulent plume/jet moves further away from the source, the source characteristics become less important and the region referred to as the far-field is attained. Moreover, far-field mixing begins after the effluent flow has interacted with the water surface, bottom, pycnocline, or terminal layer. The far-field region characterized by one or two mixing processes and usually is simulated with the following two conditions: (1) if it contains sufficient buoyancy (either positive or negative) there will be a density current region followed by a passive diffusion region, and (2) if it is non-buoyant or weakly buoyant there is no density current region, only a passive diffusion region ([4] and [5]).

Shallow water hydrodynamic model like CAMERI3D/HD-ST does not account for some processes, which are important in the near-field, for instance, buoyancy. Furthermore, the model horizontal scales of interest should be much larger than the depth of the fluid, thus, the model could not describe properly some initial features of the effluent jet/plume, for example, related to the geometrical characteristics of the orifice. Nevertheless, outside several cells from the outfall, where the fluid flux

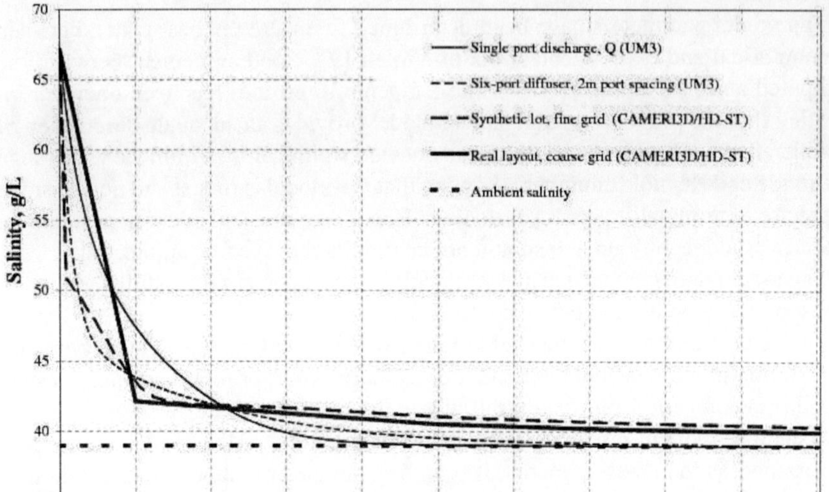

Fig. 6 Plume salinity decrease as a function of horizontal distance from the source, CAMERI3D/HD-ST model vs. UM3-model

characteristics within a single cell become more or less homogeneous, the model is able to provide adequate results as follows from Fig.6.

Different alternatives have been considered including single-port discharge (through a single orifice), diffuser (six-port) discharge (in case of simulations where UM3 model of Visual Plume package is employed [3]) for different port elevations and grid resolutions (in case of the CAMERI3D/HD-ST simulations). The results of the far-field model (CAMERI3D/HD-ST) are presented for a domain of constant water depth 15m once for a synthetic bathymetry with fine grid resolution in horizontal plane (1.64m*1.64m) and for the real layout of the projected desalination facility with coarse grid resolution (20m*20m). The results correspond to 4m port's elevation and the total discharge Q~24,000m^3/hr. Comparing the results outside 30m distance from the outfall (Fig.6), one can conclude that the far-field model (CAMERI3D/HD-ST) provides similar or deficient dilution compared to the near-field model (UM3), i.e. provides a solution with some spare factor.

6 Conclusions

The three-dimensional flow model in shallow stratified water developed in CAMERI and based on mass conservation equations and differential equations for nearly horizontal flows has been described. The model assumes hydrostatic pressure profile. Vertical eddy viscosity is evaluated according to the mixing length theory, which accounts for stratification both by temperature and salinity. The required empirical coefficients have been found by comparison of model results with laboratory experiments.

The model parameters have been determined to ensure the best match between the numerical and experimental data of Violett [9]. Good agreement between the computed and measured velocities and temperature profiles has been obtained indicating that the proposed mathematical model provides an adequate description of stratified flows. The numerical results do not vary significantly within the wide range of Froude and Reynolds numbers showing that the model is robust and not sensitive to precise determination of its parameters. Hence, the selected coefficient might be chosen as model constant and should not be modified in specific applications.

The model underwent extensive verification and has been successfully applied in flow modelling within the Israeli coastal zone.

The discussed model is a far-field model, which, as all models of such type, has some limits in the near-field computations. Therefore, it provides adequate results outside several cells from the outfall where the fluid flux features within a single cell becomes more or less homogeneous. Although the CAMERI3D/HD-ST model is classified as the far-field model, it provides similar or underestimate dilution in the vicinity of the outfall as shown through comparison to the near-field model UM3 (from Visual Plumes package developed by the United States Environmental protection Agency [3]).

The results of model calibration and validation demonstrate that the 3D model developed in the current research using shallow water approximation and mixing length turbulent model accounting for the stratification is capable to provide reasonable simulation of the flow patterns, temperature and salt distributions.

Acknowledgements. This study has been supported by the Israel Electric Corporation (IEC).

References

1. Blumberg, A.F., Mellor, G.L.: A description of a three-dimensional coastal ocean circulation model. In: Heaps, N. (ed.) Three-Dimensional Coastal Ocean Models, pp. 1–16. American Geophysical Union (1987)
2. Falconer, R.A.: An introduction to nearly-horizontal flows. In: Abbot, M.B., Price, W.A. (eds.) Coastal, Estuarial and Harbour Engineers, Reference Book, pp. 27–36. E & FN Spon, London (1994)
3. Frick, W.E., Roberts, P.J.W., Davis, L.R., Keyes, J., Baumgartner, D.J., George, K.P.: Dilution Models for Effluent Discharges (Visual Plumes). In: Ecosystems Research Division, NERL, ORD, U.S. Environmental Protection Agency, 4th edn. (2003)
4. Jirka, G.H.: Integral model for turbulent buoyant jets in unbounded stratified flows. Part 1: Single round jet. Environ. Fluid Mech. 4, 1–56 (2004)
5. Jirka, G.H.: Integral model for turbulent buoyant jets in unbounded stratified flows. Part 2: Plane jet dynamics resulting from multiport diffuser jets. Environ. Fluid Mech. 6, 43–100 (2006)
6. Koutitas, C.G.: Mathematical Models in Coastal Engineering. Pentech Press, London (1988)
7. Schlichting, H.: Boundary Layer Theory, 6th edn. McGraw-Hill, New York (1978)
8. Sladkevich, M., Militeev, A.N., Rubin, H., Kit, E.: Two and three dimensional simulation of transport phenomena in a shallow aquatic environment. J. Hydraulic Eng. 126(2), 123–136 (2000)
9. Violett, P.L.: Turbulent mixing in a two-layer stratified shear flow. In: Proceedings of Second International Symposium on Stratified Flows. Trondheim, Norway (1980)

A Cartesian Cut-Cell Solver for Compressible Flows

Daniel Hartmann, Matthias Meinke, and Wolfgang Schröder

Abstract. A Cartesian cut-cell solver is presented to simulate two- and three-dimensional viscous, compressible flows on arbitrarily refined graded meshes. The finite-volume method uses cut cells at the boundaries rendering the method strictly conservative and is flexible in terms of shape and size of embedded boundaries. A linear least-squares method is used to reconstruct the cell center gradients in irregular regions of the mesh such that the surface flux can be formulated. The accuracy of the method is demonstrated for the three-dimensional laminar flow past a sphere.

1 Introduction

In recent years, numerical methods based on non-boundary-conforming grids to solve the Euler and Navier-Stokes equations have gained in popularity since they offer the possibility to automatically generate the grid and thus make the simulation of flows with complex embedded boundaries very efficiently accessible. One of the challenges of these methods is the representation of the embedded boundaries which do not coincide with grid lines. A decomposition of the existing methods for non-boundary conforming grids into two categories is given by Mittal and Iaccarino [13].

In numerical methods of the first category, the continuous governing equations are modified by adding forcing terms to account for embedded boundaries, such as the immersed boundary method by Peskin [14]. The primary disadvantage of these methods is that the solution on the immersed boundary is smeared out to the width of the local cell size, i.e., a sharp fluid-to-boundary interface cannot be generated.

Numerical methods of the second category rely on a discrete representation of the embedded boundaries without modifying the continuous governing equations. In this second category, finite-difference based methods [12, 3] and finite-volume

Daniel Hartmann · Matthias Meinke · Wolfgang Schröder
Institute of Aerodynamics, RWTH Aachen University,
Wüllnerstr. 5a, 52062 Aachen, Germany
e-mail: `office@aia.rwth-aachen.de`

E. Krause et al. (Eds.): Computational Sci., & High Performance Computing IV, NNFM 115, pp. 363–376.
springerlink.com

based methods also referred to as Cartesian grid or cut-cell methods [15, 4] can be further discriminated. The inherent advantages of finite-volume based methods over the finite-difference counterparts are strict conservation of mass, momentum, and energy even at the embedded boundaries and their susceptibility to local mesh refinement [1, 2, 4]. While in general immersed boundary methods for compressible viscous flows are rather rare, three-dimensional implementations of cut-cell based Cartesian grid methods for the compressible Navier-Stokes equations have not been presented.

The objective of this article is to present a cut-cell Cartesian grid method on locally refined hierarchical meshes for general three-dimensional problems of compressible viscous flow. The major problem towards three-dimensional implementations of cut-cell based Cartesian grid methods is the complexity involved in formulating a finite-volume discretization on cells which are arbitrarily intersected by the embedded boundaries of the computational domain. The least-squares method is used to compute the cell center derivatives such that the proposed method is uniformly formulated and simple to implement. Small cells, which are inherent in cut-cell Cartesian grid methods and lead to numerical instabilities, are removed by a cell-merging/cell-linking procedure based on a simple algorithm [4]. The application of boundary conditions is simplified by introducing ghost cells which can be freely positioned in space such that the proposed method is flexible in terms of shape and size of embedded boundaries.

The organization of this paper is as follows. In Section 2 the governing equations are given, before the cut-cell Cartesian grid method is presented in Section 3. Finally, results are discussed in Section 4 and in Section 5 the findings of this paper are summarized.

2 Governing Equations

Let the density be denoted by ρ, the Cartesian components of the velocity vector $\mathbf{v}$ in the x_j directions of $\mathbf{x}$ be given by v_j ($j = \{1,2,3\}$), and the total specific energy and the pressure be represented by $E = e + \frac{\mathbf{v}^2}{2}$ and p, respectively. The quantity e denotes the specific internal energy. We consider the Navier-Stokes equations for compressible flows in their non-dimensional form, which can be written as

$$\frac{\partial \mathbf{Q}}{\partial t} + \nabla \overline{\mathbf{H}} = \mathbf{0}, \tag{1}$$

with $\mathbf{Q} = [\rho, \rho\mathbf{v}, \rho E]^T$ being the vector of the conservative variables and $\overline{\mathbf{H}} = [\mathbf{F}_1, \mathbf{F}_2, \mathbf{F}_3]^T$ being the flux vector containing an inviscid part $\overline{\mathbf{H}}^i$ and a viscous part $\overline{\mathbf{H}}^v$

$$\overline{\mathbf{H}} = \overline{\mathbf{H}}^i - \overline{\mathbf{H}}^v = \begin{pmatrix} \rho\mathbf{v} \\ \rho\mathbf{v}\mathbf{v} + p \\ \mathbf{v}(\rho E + p) \end{pmatrix} + \frac{1}{Re} \begin{pmatrix} 0 \\ \overline{\tau} \\ \overline{\tau}\mathbf{v} + \mathbf{q} \end{pmatrix}, \tag{2}$$

where $\nabla = \left[\frac{\partial}{\partial x_1}, \frac{\partial}{\partial x_2}, \frac{\partial}{\partial x_3}\right]^T$ denotes the vector of spatial derivatives in the Cartesian coordinate system and the superscript T identifies the vector transpose. Furthermore, let $\mathbf{P}$ define the vector of primitive variables $\mathbf{P} = [\rho, \mathbf{v}, p]^T$. The Reynolds number is given by $Re = \frac{\rho_\infty v_\infty l}{\mu_\infty}$, where l is a characteristic length. The dynamic viscosity μ is computed using Sutherland's law $\mu = \mu_\infty \left(\frac{T}{T_\infty}\right) \frac{T_\infty + S}{T + S}$, where T is the static temperature and the constant S is set $S = 111K$, which is valid at moderate temperatures. We assume a Newtonian fluid and zero bulk viscosity such that the second-rank stress tensor $\overline{\tau}$ can be formulated

$$\overline{\tau} = -2\mu\overline{\mathbf{S}} + \frac{2}{3}\mu(\nabla \cdot \mathbf{v})\overline{\mathbf{I}}, \tag{3}$$

where $\overline{\mathbf{I}}$ is the unit tensor and $\overline{\mathbf{S}}$ is the rate-of-strain tensor

$$\overline{\mathbf{S}} = \frac{1}{2}\left(\nabla\mathbf{v} + (\nabla\mathbf{v})^T\right). \tag{4}$$

The vector of heat conduction $\mathbf{q}$ is expressed by Fourier's law

$$\mathbf{q} = -\frac{k}{Pr(\gamma - 1)}\nabla T, \tag{5}$$

where γ is the ratio of specific heats. The Prandtl number $Pr = \frac{\mu_\infty c_p}{k_\infty}$ contains the specific heat at constant pressure c_p. Finally, the thermal conductivity is evaluated from $k(T) = \mu(T)$, which holds for a constant Prandtl number.

Using the Gauss divergence theorem, the integral form of the Navier-Stokes equations is obtained as

$$\int_V \frac{\partial \mathbf{Q}}{\partial t} dV + \int_A \overline{\mathbf{H}}\, \mathbf{n} dA = \mathbf{0}, \tag{6}$$

where $\mathbf{n}$ is the outward unit normal vector on the surface dA.

3 Numerical Method

The Navier-Stokes equations (6) are integrated in time using a 5-stage second-order accurate Runge-Kutta scheme optimized for stability,

$$\begin{cases} \mathbf{Q}^{(0)} = \mathbf{Q}^w, \\ \mathbf{Q}^{(k)} = \mathbf{Q}^{(0)} - \alpha_k \Delta t L(\mathbf{Q}^{(k-1)}), \\ \mathbf{Q}^{w+1} = \mathbf{Q}^{(N)}, \end{cases} \tag{7}$$

where $N = 5$ and the coefficients $\alpha = \left(\frac{1}{4}, \frac{1}{6}, \frac{3}{8}, \frac{1}{2}, 1\right)$ are used. The superscript k denotes the Runge-Kutta stage, while the superscript w represents the time step Δt computed globally by

$$\Delta t = \min_{n \in \mathscr{D}} \left\{ \min_{j \in \{1,2,3\}} \left(\frac{Ch^n}{|v_j^n| + a^n} \right) \right\}, \tag{8}$$

where n is a cell in the computational domain $\mathscr{D}$, $a^n = \sqrt{\gamma \frac{p^n}{\rho^n}}$ is the local speed of sound, h is the local mesh size, and C is the CFL number. Finally, the operator $L(\mathbf{Q})$ denotes the numerical approximation of the term $\nabla\overline{\mathbf{H}}$ in Eq. (1).

3.1 Evaluation of Surface Fluxes

While the flux evaluation is straightforward in structured grid areas, it becomes more difficult on partial surfaces at fine-coarse cell interfaces and near boundaries. In the proposed formulation, the least-squares method is utilized to obtain a discretization scheme which is flexible in terms of the local mesh topology and the presence and shape of embedded boundaries. Boundary conditions are imposed by using ghost cells which can be freely positioned in space and do not have to conform to the Cartesian grid lines [4]. This allows a unified cell-based formulation of the discretization scheme.

First, the surface integral in Eq. (6) is approximated by second-order accuracy to obtain for a cell n

$$\left(\int_A \overline{\mathbf{H}} \mathbf{n} dA \right)^n = \sum_{m \in \partial n} \overline{\mathbf{H}} (\mathbf{x}_s^m) \mathbf{n}^m A^m + \mathscr{O}(h^n)^2, \tag{9}$$

where $\mathbf{n}^m$ denotes the normal vector on the surface m. The reconstruction of the flux vector $\overline{\mathbf{H}}(\mathbf{x}_s^m)$ in the surface centroid x_s^m is discussed in the following.

3.1.1 Inviscid Flux

The inviscid surface flux $\overline{\mathbf{H}}^i(\mathbf{x}_s^m)$ is computed based on the left and right interpolated states L and R in the surface centroid. It is formulated using a modified version of the advection upstream splitting method (AUSM) originally proposed by Liou et al. [9]. In this method, the inviscid flux is split into a convective component and a pressure term involving the Mach number $M_j = v_j/a$, such that the numerical inviscid flux $\mathbf{F}^i_j(\mathbf{x}_s^m)$, $j = \{1,2,3\}$, on the surface m can be computed by

$$\mathbf{F}^i_j(\mathbf{x}_s^m) = \frac{1}{2} \left\{ M_j^m \left[(\mathbf{f}_j^m)^L + (\mathbf{f}_j^m)^R \right] + |M_j^m| \left[(\mathbf{f}_j^m)^L - (\mathbf{f}_j^m)^R \right] \right\} + \mathbf{p}^m, \tag{10}$$

where

$$M_j^m = \frac{1}{2} \left[(M_j^m)^L + (M_j^m)^R \right] \tag{11}$$

and

$$\mathbf{f}_j = \begin{pmatrix} \rho a \\ \rho a \mathbf{v} \\ a(\rho E + p) \end{pmatrix}. \tag{12}$$

The pressure term $\mathbf{p}^m$ is computed as proposed in [11], such that

$$\mathbf{p}^m = \left\{ (p^m)^L \left[\frac{1}{2} + \chi \left(M_j^m\right)^L\right] + (p^m)^R \left[\frac{1}{2} - \chi \left(M_j^m\right)^R\right]\right\} \begin{pmatrix} 0 \\ \mathbf{n}^m \\ 0 \end{pmatrix}, \tag{13}$$

where a dissipative splitting at $\chi = 0.5$ is used on the fine grid cells at fine-coarse mesh interfaces to damp spurious oscillations appearing at grid interfaces and a low-dissipation centered difference scheme at $\chi = 0$ is used otherwise. The left and right interpolated fluxes $\mathbf{f}_j$, the pressure, and the Mach numbers are computed with second-order accuracy using a monotone upstream centered schemes for conservation laws (MUSCL) approach for the primitive variables [8], i.e.,

$$\mathbf{P}(\mathbf{x}_s^m)^{L/R} = \mathbf{P}(\mathbf{x}^{L/R}) + \nabla\mathbf{P}|^{L/R} \left(\mathbf{x}_s^m - \mathbf{x}^{L/R}\right). \tag{14}$$

The cell center gradient $\nabla\mathbf{P}|^n$, $n = \{L, R\}$, is computed using an unweighted linear least-squares reconstruction scheme which can be written

$$\overline{\mathscr{H}^n}\, \nabla\mathbf{P}|^n = \Delta\mathbf{P}^n, \tag{15}$$

where $\overline{\mathscr{H}^n}$ is a matrix of dimension $s \times d$ and $\Delta\mathbf{P}^n$ is a vector of dimension s,

$$\begin{cases} overline\mathscr{H}^n = \left[\Delta\mathbf{x}^{n,\mathscr{S}_1^n}, \ldots, \Delta\mathbf{x}^{n,\mathscr{S}_s^n}\right]^T, \\ \Delta\mathbf{P}^n = \left[\Delta\mathbf{P}^{n,\mathscr{S}_1^n}, \ldots, \Delta\mathbf{P}^{n,\mathscr{S}_s^n}\right]^T, \end{cases}$$

with $\Delta\phi^{a,b} = \phi^b - \phi^a, \phi \in \{\Delta\mathbf{x}, \Delta\mathbf{P}\}$. $\mathscr{S}^n$ denotes the least-squares stencil of s cells which are clustered around the cell n and $\mathscr{S}_k^n$ is the kth element of $\mathscr{S}^n$.

The least-squares system of equations (15) is rearranged to obtain

$$\nabla\mathbf{P}|^n = \overline{\mathbf{C}}^n \Delta\mathbf{P}^n, \tag{16}$$

where the matrix $\overline{\mathbf{C}}^n = \left(\left(\overline{\mathscr{H}^n}\right)^T \overline{\mathscr{H}^n}\right)^{-1} \left(\overline{\mathscr{H}^n}\right)^T$ has dimension $d \times s$ and contains geometric constants, which can be computed and stored in a preprocessing stage. In structured grid areas the least-squares formulation reduces to standard finite-difference operators.

Usually, an additional constraint is imposed on the least-squares system to ensure the reconstructed linear in-cell distributions to conserve the cell averages represented by the cell center values $\mathbf{P}^n$. Formally, this can be written

$$\frac{1}{V^n} \int_{V^n} \mathbf{p}^n(\mathbf{x}) d\mathbf{x} = \mathbf{P}^n, \tag{17}$$

where

$$\mathbf{p}^n(\mathbf{x}) = \mathbf{P}^n + \nabla\mathbf{P}|^n (\mathbf{x} - \mathbf{x}^n) \tag{18}$$

is the reconstructed linear distribution of $\mathbf{P}$ on the cell n. Inserting Eq. (18) into Eq. (17), one obtains

$$\nabla\mathbf{P}|^n \mathbf{X}^n = 0, \tag{19}$$

where

$$\mathbf{X}^n = \frac{1}{V^n} \int_{V^n} (\mathbf{x} - \mathbf{x}^n) d\mathbf{x}. \tag{20}$$

Since the cell center $\mathbf{x}^n$ is located in the volumetric center of the cell, clearly $\mathbf{X}^n = 0$, such that Eq. (17) is always satisfied. Hence, in the proposed method there is no need for an additional explicit constraint to conserve the mean in the least-squares system of equations (15).

3.1.2 Viscous Flux

The viscous flux $\overline{\mathbf{H}}^v$ is computed using the cell-center gradients available from computing $\overline{\mathbf{H}}^i$. In general, the surface gradient is computed as a distance-weighted convex combination of cell center gradients,

$$\nabla\mathbf{P}|^m = \pi_m^L \, \nabla\mathbf{P}|^L + \pi_m^R \, \nabla\mathbf{P}|^R, \tag{21}$$

where

$$\begin{cases} \pi_m^L = \dfrac{|\mathbf{x}^L - \mathbf{x}^m|}{|\mathbf{x}^L - \mathbf{x}^m| + |\mathbf{x}^R - \mathbf{x}^m|}, \\ \pi_m^R = 1 - \pi_m^L. \end{cases} \tag{22}$$

Formally, this is a first-order accurate approximation in irregular mesh areas, i.e., for instance at grid interfaces and at embedded boundaries, where, however, the reconstructed gradients can only be expected to be first-order accurate as well. In structured grid areas $\pi_m^L = \pi_m^R = 1/2$, such that Eq. (21) reduces to the second-order accurate 4-point center difference scheme

$$\nabla\mathbf{P}|^m = \frac{\nabla\mathbf{P}|^L + \nabla\mathbf{P}|^R}{2}. \tag{23}$$

3.2 Representation of Embedded Boundaries

The boundary cells are reshaped in such a manner that embedded bodies are cut out of the computational domain, and their cell centers are moved to their volumetric center denoted by $\hat{\mathbf{x}}$, Fig. 1a. In the final grid, the embedded boundaries are then represented by the piece-wise linear boundary surfaces. By this procedure, cells with a very small cell volume may be generated. These small cells are an inherent problem in cut-cell Cartesian grid methods and lead to numerical instabilities. This drawback can be removed by a cell-merging/cell-linking procedure [4]. Boundary

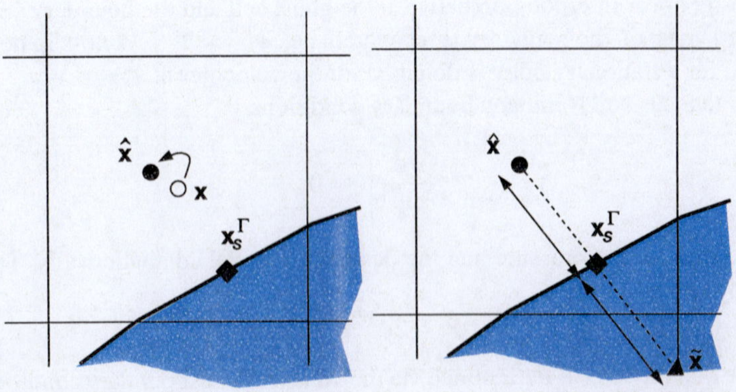

Fig. 1 Representation of curved boundaries using freely positioned ghost cells: (a) location of the volumetric cell center of a cut cell; (b) location of the ghost cell. The diamond identifies the boundary surface centroid, and the triangle marks the location of the ghost cell

conditions are imposed by prescribing the primitive variables on ghost cells one of which is created for each boundary cell. Since the purpose of these ghost cells is to facilitate the application of boundary conditions in the framework of the least-squares reconstruction they are not connected to any other cells by parent-child or neighbour relationships. The only interface they have is the boundary surface Γ, by which they are related to their corresponding boundary cell. The location of the ghost cells is not restricted by the Cartesian grid lines [4] and is computed by

$$\widetilde{\mathbf{x}^n} = \hat{\mathbf{x}}^n + 2\left(\mathbf{x}_s^{\Gamma^n} - \hat{\mathbf{x}}^n\right). \tag{24}$$

That is, the surface centroid is located on the line connecting the boundary cell center and the corresponding ghost cell center, as illustrated in Fig. 1b. In principle, the ghost cell center could also be defined such that it coincides with the centroid of the corresponding boundary surface, which, however, results in a less balanced least-squares stencil for the boundary cell.

3.2.1 Application of Boundary Conditions

With the ghost cells defined above a Dirichlet boundary condition $\phi(\mathbf{x},t) = \Phi(\mathbf{x},t)$ is for a generic quantity ϕ imposed by setting the ghost cell value $\widetilde{\phi^n}$

$$\widetilde{\phi^n} = 2\Phi^{\Gamma^n} - \phi^n, \tag{25}$$

such that by linear interpolation between the ghost cell and the boundary cell n the prescribed value on the boundary is recovered, i.e., $\phi^{\Gamma^n} = \Phi^{\Gamma^n}$. A no-slip boundary condition for a stationary body is obtained for the velocities at $\widetilde{\mathbf{v}^n} = -\mathbf{v}^n$.

Homogeneous von Neumann boundary conditions

$$\left.\frac{\partial \phi}{\partial \mathbf{n}^{\Gamma^n}}\right|^n = 0 \tag{26}$$

are used, e.g., for the pressure and the density, to model adiabatic walls. They are applied by setting

$$\widetilde{\phi^n} = 2\phi^{\Gamma^n} - \phi^n, \tag{27}$$

where the quantity ϕ^{Γ^n} is determined via the von Neumann boundary condition (26) using a truncated Taylor series expansion around the boundary cell,

$$\phi^{\Gamma^n} = \phi^n + \nabla\phi|^n \cdot \left(\mathbf{x}_s^{\Gamma^n} - \mathbf{x}^n\right). \tag{28}$$

What remains to be done is to compute the gradient $\nabla\phi|^n$ required in Eq. (28) taking into account Eq. (26) without using the unknown ghost cell variable $\widetilde{\phi^n}$ in the least-squares gradient reconstruction. The von Neumann boundary condition is accounted for by including into the least-squares reconstruction a second ghost point located at

$$\widetilde{\widetilde{\mathbf{x}^n}} = \hat{\mathbf{x}}^n + 2\left(\left(\mathbf{x}_s^{\Gamma^n} - \hat{\mathbf{x}}^n\right) \cdot \mathbf{n}^{\Gamma^n}\right)\mathbf{n}^{\Gamma^n}, \tag{29}$$

which is obtained by mirroring the boundary cell center at the boundary surface. Eq. (26) then implies $\widetilde{\widetilde{\phi^n}} - \phi^n = 0$, such that the second ghost point only contributes to the least-squares reconstruction for the boundary cell n (Eq. (16)) through the geometric coefficient matrix $\overline{\mathbf{C}}^n$.

The overall procedure can be summarized as follows, provided von Neumann boundary conditions are used for the pressure and the density and Dirichlet boundary conditions are used for the velocities as is the case for an adiabatic no-slip wall.

1. Compute the pressure gradient and the density gradient using on boundary cells the second ghost point at $\widetilde{\widetilde{\mathbf{x}}}$, Eq. (29), instead of the ghost cell.
2. Apply the Dirichlet and von Neumann boundary conditions given by Eq. (25) and Eqs. (27-28), respectively.
3. Compute the velocity gradients and the surface flux.

4 Results

The accuracy of the presented Cartesian grid method is validated for the three-dimensional laminar flow past a sphere. In the simulations, a sponge layer formulation for the pressure and the density as presented in [4] is used to drive the solution variables on the far-field boundaries towards the freestream values as well as to avoid spurious waves in the vicinity of these boundaries.

The solver has been fully parallelized using the Message Passing Interface (MPI) libraries, such that parallel computations on shared and distributed memory systems are possible. The domain decomposition can be automatically performed using METIS or space-filling curves. An example for a METIS decomposition of a lung model grid is illustrated in Fig. 2. All presented three-dimensional simulations have been performed in parallel on up to 16 CPUs.

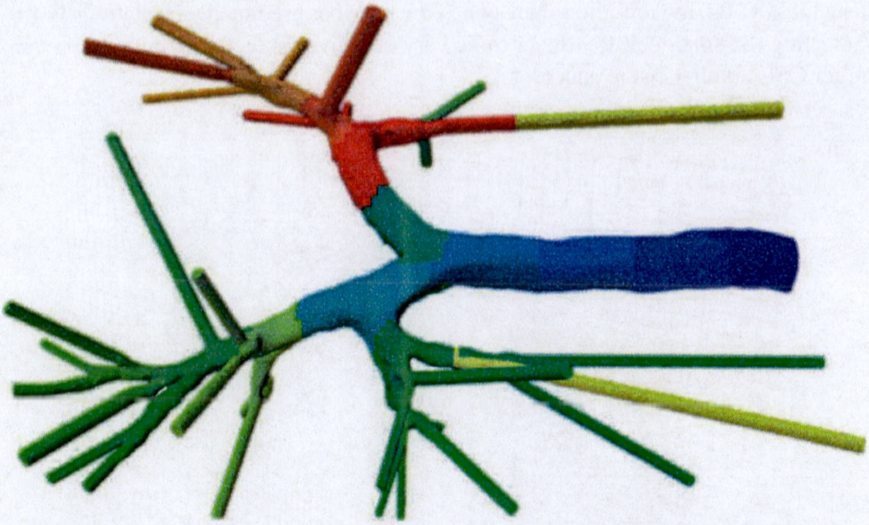

Fig. 2 Domain decomposition for a lung model grid using METIS. The different domains are indicated by different colours

4.1 *Flow Past Sphere*

The numerical simulation of the flow past a sphere is a canonical and challenging test case for Cartesian grid methods since all angles of intersection between the geometry and the computational grid appear. Similar to the circular cylinder, the laminar flow past a sphere exhibits several regimes and the transition between those regimes depends on the Reynolds number of the freestream. Up to $Re_D \approx 210$, where D is the diameter of the sphere, the flow is steady and axisymmetric, whereas the flow at Reynolds numbers between 210 and 270 is steady but non-axisymmetric. The onset of unsteady motion occurs at Reynolds numbers between 270 and 300, where the flow pattern exhibits periodic vortex shedding [6].

We shall investigate the flow at Reynolds numbers in the range of $Re_D = 25$ to $Re_D = 300$ representative of the different laminar flow regimes outlined above. As in the cylinder flow simulations, the Mach number in all sphere flow simulations is $M = 0.1$ such that the results can be compared with solutions obtained for incompressible flow.

The simulations are performed in a three-dimensional computational domain $\Omega : [-15,49] \times [-15,15] \times [-15,15]$ with the center of the sphere located in the origin of the frame of reference. The patch-wise refined grid shown in Fig. 3 with roughly 1.85 million cells at refinement levels 7 through 11 is used. It is refined axisymmetrically about the streamwise central axis in the x direction. As for the cylinder flow simulations, the grid is clustered near the embedded sphere with a mesh spacing of $\Delta_x \approx 0.03$ such that the cell centers of the boundary cells are located at a distance $\Delta < 0.015$ from the sphere. The CFL number used in all simulations is 1.0 indicating the small cell treatment to be very effective. In fact, simulations at even larger CFL numbers were successful.

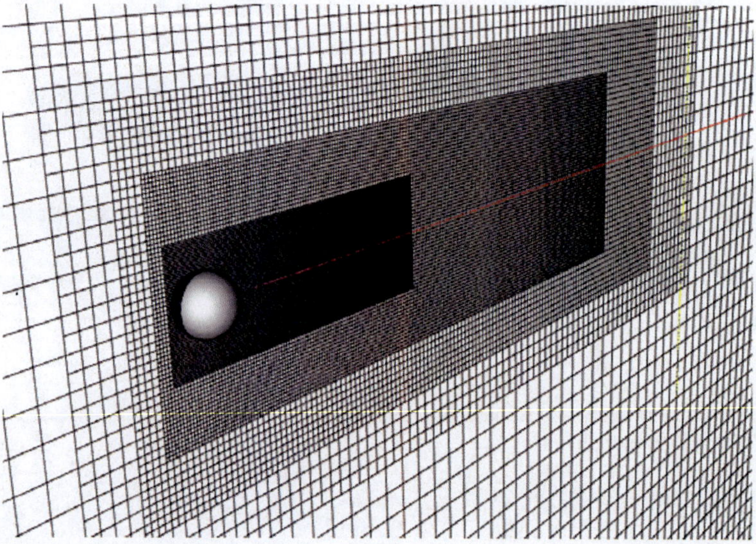

Fig. 3 Topology of the refined Cartesian grid used for the simulations of the laminar flow past a cylinder and past a sphere

The results of the computations summarized and compared with data from the literature in Tables 1 and 2 and in Fig. 4 show excellent agreement of the listed parameters. Note, in these tables the force coefficients are defined by $C_d = F_x / \left(q\pi\frac{D^2}{4}\right)$ and $C_l = F_z / \left(q\pi\frac{D^2}{4}\right)$. The computed instantaneous streamlines corresponding to the different laminar flow regimes are given in Fig. 5. The axisymmetric and steady flow fields at $Re_D = 50$ and $Re_D = 200$ are depicted in Figs. 5a and 5b, where the separating flow can be observed to form an axisymmetric vortex ring which remains attached to the leeward side of the sphere. The length of the recirculation region shown in Figs. 5a and 5b grows with increasing Reynolds number. At $Re_D = 250$, the axisymmetry of the flow field is lost, while it remains steady. In Figs. 5c and 5d, the streamlines in two planes which are perpendicular to each other are compared for the computed flow at this Reynolds number. The streamline pattern in the $x-y$ plane depicted in Fig. 5c exhibits a symmetry with respect to $x=0$, while the flow in the $x-z$ plane shown in Fig. 5d is clearly not symmetric. Note, however, the symmetry of the streamlines in Fig. 5c is artificial since the streamlines which are shown are computed using only the respective in-plane components of the velocity vector, while the component normal to the respective plane is set zero.

The flow at $Re_D = 300$ is unsteady and hairpin vortices are periodically shed from the sphere. To visualize the vortex structure of this flow the Q-criterion [5] is used. The scalar Q is defined

$$Q \equiv \frac{1}{2}(\| \overline{\Omega} \|_2^2 - \| \overline{\mathbf{S}} \|_2^2), \tag{30}$$

where $\overline{\Omega} = \frac{1}{2}(\nabla\mathbf{v} - (\nabla\mathbf{v})^T)$ is the vorticity tensor and $\overline{\mathbf{S}}$ is the rate-of-strain tensor as defined in Eq. (4). A vortex can be identified as the region surrounded by an iso-surface of $Q > 0$, i.e., a region where the Euclidean norm of the square of the vorticity tensor dominates that of the rate-of-strain tensor. The periodically shed vortex loops in the flow at $Re_D = 300$ shown in Fig. 6 visualize an instantaneous solution. Finally, the projected vorticity contours in Fig. 6 evidence the planar symmetry of the vortex shedding.

Table 1 Three-dimensional simulation of steady laminar flow past a sphere. The drag coefficient C_d and the non-dimensional length of the recirculation region L_r/D for the flow at $M = 0.1$ and Reynolds numbers $Re_D = 25$, $Re_D = 50$, $Re_D = 100$, $Re_D = 150$, and $Re_D = 200$ are compared against data from the literature

Contribution	$Re_D = 25$		$Re_D = 50$		$Re_D = 100$		$Re_D = 150$		$Re_D = 200$	
	C_d	L_r/D	C_d	L_r/D	C_d	L_r/D	C_d	L_r/D	C_d	L_r/D
Johnson and Patel [6]	2.36	0.08	1.57	0.41	1.08	0.88	0.90	1.20	0.78	1.46
Marella et al. [10]	-	-	1.56	0.39	1.06	0.88	0.85	1.19	-	-
Present	2.351	0.072	1.568	0.410	1.083	0.880	0.882	1.210	0.764	1.456

Table 2 Three-dimensional simulation of laminar flow past a sphere at $M = 0.1$. The drag coefficient C_d and the lateral force coefficient C_l are listed for the flow at $Re_D = 250$, while the mean drag coefficient $\overline{C_d}$, the mean lateral force coefficient $\overline{C_l}$, and the Strouhal number St are given for the flow at $Re_D = 300$

Contribution	$Re_D = 250$		$Re_D = 300$		
	C_d	C_l	$\overline{C_d}$	$\overline{C_l}$	St
Johnson and Patel [6]	-	0.06	0.656	-0.069	0.137
Kim et al. [7]	0.701	0.059	0.657	-0.067	0.134
Present	0.698	0.065	0.657	-0.069	0.135

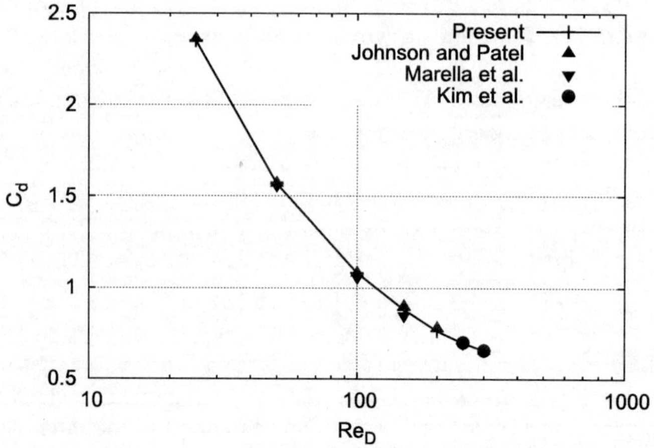

Fig. 4 Drag coefficient C_d over the Reynolds number Re_D for the laminar flow past a sphere. The results are compared with data from Johnson and Patel [6], Marella et al. [10], and Kim et al. [7]

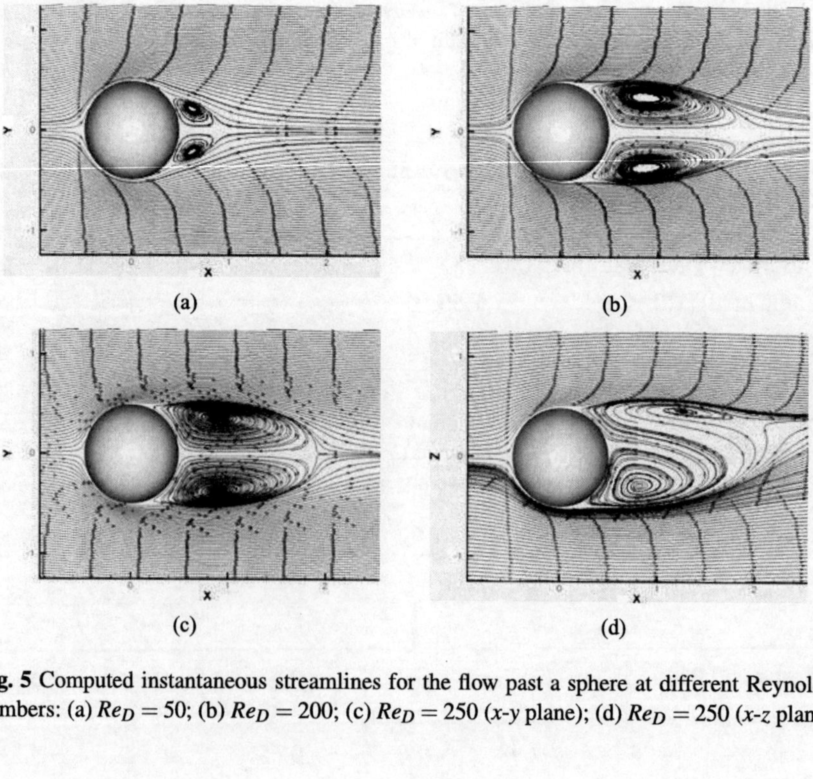

Fig. 5 Computed instantaneous streamlines for the flow past a sphere at different Reynolds numbers: (a) $Re_D = 50$; (b) $Re_D = 200$; (c) $Re_D = 250$ (*x*-*y* plane); (d) $Re_D = 250$ (*x*-*z* plane)

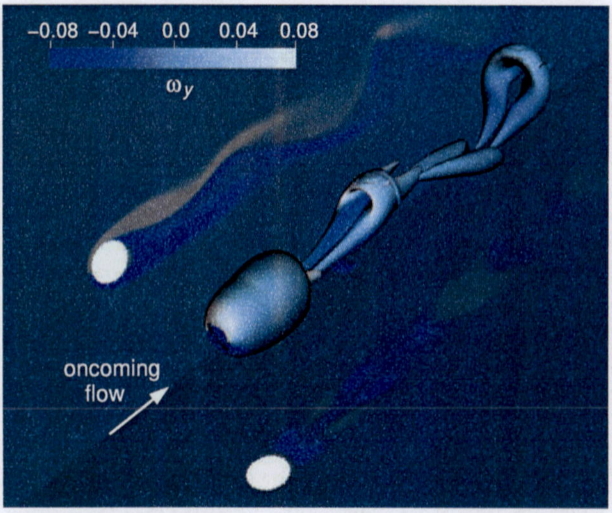

Fig. 6 Visualization of the vortex structure of the flow past a sphere at $Re_D = 300$. Contours of the spanwise vorticity ω_y computed in the center planes with respect to the sphere are projected as well as they are mapped onto an iso-surface of Q as defined in Eq. (30)

5 Conclusions

A cut-cell based Cartesian grid method for two- and three-dimensional compressible viscous flows using local mesh refinement is presented. Using a finite-volume discretization throughout the entire computational domain, the method exhibits strict discrete convergence of mass, momentum, and energy. The small-cell problem inherent in Cartesian cut-cell methods is solved using a versatile cell-merging/cell-linking procedure. The application of boundary conditions is simplified by using ghost cells, which can be freely positioned in space.

As a canonical three-dimensional test case, the laminar flows past a sphere is documented to validate the numerical method. The effective treatment of the small cells enabled the use of rather large CFL numbers in these simulations. The results presented for both steady and unsteady flows show a convincing agreement with established data from the literature.

Acknowledgements. This research was performed as part of the collaborative research center SFB 686, which is funded by the German Research Association (Deutsche Forschungsgemeinschaft (DFG)), and the Cluster of Excellence "Tailor-Made Fuels from Biomass", which is funded by the German Excellence Initiative. The support is gratefully acknowledged.

References

1. Berger, M., Oliger, J.: Adaptive mesh refinement for hyperbolic partial differential equations. J. Comput. Phys. 53, 484–512 (1984)
2. De Zeeuw, D., Powell, K.: An adaptively refined Cartesian mesh solver for the Euler equations. J. Comput. Phys. 104, 56–68 (1993)
3. Ghias, R., Mittal, R., Dong, H.: A sharp interface immersed boundary method for compressible viscous flows. J. Comput. Phys. 225, 528–553 (2007)
4. Hartmann, D., Meinke, M., Schröder, W.: An adaptive multilevel multigrid formulation for Cartesian hierarchical grid methods. Comput. Fluids 37, 1103–1125 (2008)
5. Hunt, J., Wray, A., Moin, P.: Eddies, stream, and convergence zones in turbulent flows. Center for turbulence research report CTR-S88 (1988)
6. Johnson, T., Patel, V.: Flow past a sphere up to a Reynolds number of 300. J. Fluid Mech. 378, 19–70 (1999)
7. Kim, J., Kim, D., Choi, H.: An immersed-boundary finite-volume method for simulations of flow in complex geometries. J. Comput. Phys. 171, 132–150 (2001)
8. van Leer, B.: Towards the ultimate conservative difference scheme. V. A second-order sequel to Godunov's method. J. Comput. Phys. 32, 101–136 (1979)
9. Liou, M.S., Steffen Jr., C.J.: A new flux splitting scheme. J. Comput. Phys. 107, 23–39 (1993)
10. Marella, S., Krishnan, S., Liu, H., Udaykumar, H.: Sharp interface Cartesian grid method I: an easily implemented technique for 3D moving boundary computations. J. Comput. Phys. 210, 1–31 (2005)
11. Meinke, M., Schröder, W., Krause, E., Rister, T.: A comparison of second- and sixth-order methods for large-eddy simulation. Comput. Fluids 31, 695–718 (2002)
12. Mittal, R., Dong, H., Bozkurttas, M., Najjar, F., Vargas, A., von Loebbecke, A.: A versatile sharp interface immersed boundary method for incompressible flows with complex boundaries. J. Comput. Phys. 227, 4825–4852 (2008)
13. Mittal, R., Iaccarino, G.: Immersed boundary methods. Ann. Rev. Fluid Mech. 37, 239–261 (2005)
14. Peskin, C.: Flow patterns around heart valves: a numerical method. J. Comput. Phys. 10, 252–271 (1972)
15. Ye, T., Mittal, R., Udaykumar, H., Shyy, W.: An accurate Cartesian grid method for viscous incompressible flows with complex immersed boundaries. J. Comput. Phys. 156, 209–240 (1999)

Author Index

Notes on Numerical Fluid Mechanics and Multidisciplinary Design

Available Volumes

Volume 115: Egon Krause, Yurii Shokin, Michael Resch, Dietmar Kröner, Nina Shokina (eds.): Computational Science and High Performance Computing IV. ISBN 978-3-642-17769-9

Volume 114: Piotr Doerffer, Charles Hirsch, Jean-Paul Dussauge, Holger Babinsky, and George N. Barakos (eds.): Unsteady Effects of Shock Wave Induced Separation. ISBN 978-3-642-03003-1

Volume 112: Andreas Dillmann, Gerd Heller, Michael Klaas, Hans-Peter Kreplin, Wolfgang Nitsche, and Wolfgang Schröder (eds.): New Results in Numerical and Experimental Fluid Mechanics VII – Contributions to the 16th STAB/DGLR Symposium Aachen, Germany 2008. ISBN 978-3-642-14242-0

Volume 111: Shia-Hui Peng, Piotr Doerffer, and Werner Haase (eds.): Progress in Hybrid RANS-LES Modelling – Papers Contributed to the 3rd Symposium on Hybrid RANS-LES Methods, Gdansk, Poland, June 2009. ISBN 978-3-642-14167-6

Volume 110: Michel Deville, Thien-Hiep Lê, and Pierre Sagaut (eds.): Turbulence and Interactions – Proceedings the TI 2009 Conference. ISBN 978-3-642-14138-6

Volume 109: Wolfgang Schröder (ed.): Summary of Flow Modulation and Fluid-Structure Interaction Findings – Results of the Collaborative Research Center SFB 401 at the RWTH Aachen University, Aachen, Germany, 1997-2008. ISBN 978-3-642-04087-0

Volume 108: Rudibert King (ed.): Active Flow Control II – Papers Contributed to the Conference "Active Flow Control II 2010", Berlin, Germany, May 26–28, 2010. ISBN 978-3-642-11734-3

Volume 107: Norbert Kroll, Dieter Schwamborn, Klaus Becker, Herbert Rieger, Frank Thiele (eds.): MEGADESIGN and MegaOpt – German Initiatives for Aerodynamic Simulation and Optimization in Aircraft Design. ISBN 978-3-642-04092-4

Volume 106: Wolfgang Nitsche, Christoph Dobriloff (eds.): Imaging Measurement Methods for Flow Analysis - Results of the DFG Priority Programme 1147 "Imaging Measurement Methods for Flow Analysis" 2003–2009. ISBN 978-3-642-01105-4

Volume 105: Michel Deville, Thien-Hiep Lê, Pierre Sagaut (eds.): Turbulence and Interactions - Keynote Lectures of the TI 2006 Conference. ISBN 978-3-642-00261-8

Volume 104: Christophe Brun, Daniel Juvé, Michael Manhart, Claus-Dieter Munz: Numerical Simulation of Turbulent Flows and Noise Generation - Results of the DFG/CNRS Research Groups FOR 507 and FOR 508. ISBN 978-3-540-89955-6

Volume 103: Werner Haase, Marianna Braza, Alistair Revell (eds.): DESider – A European Effort on Hybrid RANS-LES Modelling - Results of the European-Union Funded Project, 2004–2007. ISBN 978-3-540-92772-3

Volume 102: Rolf Radespiel, Cord-Christian Rossow, Benjamin Winfried Brinkmann (eds.): Hermann Schlichting – 100 Years - Scientific Colloquium Celebrating the Anniversary of His Birthday, Braunschweig, Germany 2007. ISBN 978-3-540-95997-7

Volume 101: Egon Krause, Yurii I. Shokin, Michael Resch, Nina Shokina (eds.): Computational Science and High Performance Computing III - The 3rd Russian-German Advanced Research Workshop, Novosibirsk, Russia, 23–27 July 2007. ISBN 978-3-540-69008-5

Volume 100: Ernst Heinrich Hirschel, Egon Krause (eds.): 100 Volumes of 'Notes on Numerical Fluid Mechanics' - 40 Years of Numerical Fluid Mechanics and Aerodynamics in Retrospect. ISBN 978-3-540-70804-9

Volume 99: Burkhard Schulte-Werning, David Thompson, Pierre-Etienne Gautier, Carl Hanson, Brian Hemsworth, James Nelson, Tatsuo Maeda, Paul de Vos (eds.): Noise and Vibration Mitigation for Rail Transportation Systems - Proceedings of the 9th International Workshop on Railway Noise, Munich, Germany, 4–8 September 2007. ISBN 978-3-540-74892-2

Volume 98: Ali Gülhan (ed.): RESPACE – Key Technologies for Reusable Space Systems - Results of a Virtual Institute Programme of the German Helmholtz-Association, 2003–2007. ISBN 978-3-540-77818-9

Volume 97: Shia-Hui Peng, Werner Haase (eds.): Advances in Hybrid RANS-LES Modelling - Papers contributed to the 2007 Symposium of Hybrid RANS-LES Methods, Corfu, Greece, 17–18 June 2007. ISBN 978-3-540-77813-4

Volume 96: C. Tropea, S. Jakirlic, H.-J. Heinemann, R. Henke, H. Hönlinger (eds.): New Results in Numerical and Experimental Fluid Mechanics VI - Contributions to the 15th STAB/DGLR Symposium Darmstadt, Germany, 2006. ISBN 978-3-540-74458-0

Volume 95: R. King (ed.): Active Flow Control - Papers contributed to the Conference "Active Flow Control 2006", Berlin, Germany, September 27 to 29, 2006. ISBN 978-3-540-71438-5

Volume 94: W. Haase, B. Aupoix, U. Bunge, D. Schwamborn (eds.): FLOMANIA - A European Initiative on Flow Physics Modelling - Results of the European-Union funded project 2002 - 2004. ISBN 978-3-540-28786-5

Volume 93: Yu. Shokin, M. Resch, N. Danaev, M. Orunkhanov, N. Shokina (eds.): Advances in High Performance Computing and Computational Sciences - The Ith Khazakh-German Advanced Research Workshop, Almaty, Kazakhstan, September 25 to October 1, 2005. ISBN 978-3-540-33864-2

Volume 92: H.J. Rath, C. Holze, H.-J. Heinemann, R. Henke, H. Hönlinger (eds.): New Results in Numerical and Experimental Fluid Mechanics V - Contributions to the 14th STAB/DGLR Symposium Bremen, Germany 2004. ISBN 978-3-540-33286-2

Volume 91: E. Krause, Yu. Shokin, M. Resch, N. Shokina (eds.): Computational Science and High Performance Computing II - The 2nd Russian-German Advanced Research Workshop, Stuttgart, Germany, March 14 to 16, 2005. ISBN 978-3-540-31767-8

Volume 87: Ch. Breitsamter, B. Laschka, H.-J. Heinemann, R. Hilbig (eds.): New Results in Numerical and Experimental Fluid Mechanics IV. ISBN 978-3-540-20258-5

Volume 86: S. Wagner, M. Kloker, U. Rist (eds.): Recent Results in Laminar-Turbulent Transition - Selected numerical and experimental contributions from the DFG priority programme 'Transition' in Germany. ISBN 978-3-540-40490-3

Volume 85: N.G. Barton, J. Periaux (eds.): Coupling of Fluids, Structures and Waves in Aeronautics - Proceedings of a French-Australian Workshop in Melbourne, Australia 3-6 December 2001. ISBN 978-3-540-40222-0

Volume 83: L. Davidson, D. Cokljat, J. Fröhlich, M.A. Leschziner, C. Mellen, W. Rodi (eds.): LESFOIL: Large Eddy Simulation of Flow around a High Lift Airfoil - Results of the Project LESFOIL supported by the European Union 1998 - 2001. ISBN 978-3-540-00533-9

Volume 82: E.H. Hirschel (ed.): Numerical Flow Simulation III - CNRS-DFG Collaborative Research Programme, Results 2000-2002. ISBN 978-3-540-44130-4

Volume 81: W. Haase, V. Selmin, B. Winzell (eds.): Progress in Computational Flow Structure Interaction - Results of the Project UNSI, supported by the European Union 1998-2000. ISBN 978-3-540-43902-8

Volume 80: E. Stanewsky, J. Delery, J. Fulker, P. de Matteis (eds.): Drag Reduction by Shock and Boundary Layer Control - Results of the Project EUROSHOCK II, supported by the European Union 1996-1999. ISBN 978-3-540-43317-0